INDEX OF EXTENDED APPLICATIONS

Sunscreens	38
Calories and Dieting	39
Saving Energy	95
Poiseville's Law	148
Mathematics in the Space Program	192
Mathematics in Biology	230
Rescue Mission	280
Radar	329
Planetary Orbits	382
Market Equilibrium	433
Noise Insulation	476
The Dose-Response Relationship	517

INTERMEDIATE ALGEBRA

INTERMEDIATE ALGEBRA

Marcus McWaters
University of South Florida

You-Feng Lin
University of South Florida

D. C. Heath and Company   Lexington, Massachusetts   Toronto

Cover art: *Alloway,* by Susan C. Hamant.

Copyright © 1985 by D. C. Heath and Company.

All rights reserved. No part of this publication may be reproduced or transmitted in any form or by any means, electronic or mechanical, including photocopy, recording, or any information storage or retrieval system, without permission in writing from the publisher.

Published simultaneously in Canada.

Printed in the United States of America.

International Standard Book Number: 0-669-05123-3

Library of Congress Catalog Card Number: 84-80462

*This book is dedicated to
Marcus and Isola McWaters
and to all of our students.*

Preface

Our intention in writing is to provide students with a textbook for intermediate algebra that is accessible to them in a variety of ways. As the following list of text features indicates, we have made every effort to produce a pedagogically sound book filled with an enthusiasm for understanding algebraic concepts and their application to our everyday lives.

TOPIC TREATMENT   Chapter 1 treats only topics that should be familiar to most students from their beginning algebra course. Chapter 1 can be skimmed, assigned as outside reading, or omitted, depending on the background of the students in the class. Interval notation is introduced in Section 2.5 for use in graphing linear inequalities in Section 2.6. There is considerable emphasis on graphing linear and quadratic functions and relations (see Chapters 7 and 8). Chapter 10 provides separate sections for finding logarithm and antilog values using tables (Section 10.4) and using calculators (Section 10.5). Either section may be omitted or both may be covered, as the instructor prefers. The last three chapters are completely independent and may be taught in any order. Any of these may be omitted for a short course.

CHAPTER FLOWCHARTS   Each chapter opens with a diagram that shows the various possibilities for ordering the coverage of sections within the chapter.

APPLICATIONS   A large number of simple applications designed to be accessible to most students is included in every chapter. More substantial applications are available to challenge the better students. A step-by-step procedure for solving word problems is first introduced in Chapter 2. Thereafter every chapter provides word problems to further develop the problem-solving skills learned in Chapter 2.

EXAMPLES   More than 500 completely worked-out examples are provided to illustrate concepts and demonstrate successful problem-solving techniques.

EXERCISES   More than 4300 exercises and 450 word problems keyed to examples are provided. These exercises are divided into A and B groupings, the A exercises are routine problems emphasizing fundamental skills and the B exercises contain both theoretical problems and problems requiring more difficult mechanics.

STANDARD ASSIGNMENT   A list of exercises designated as a standard assignment appears in the beginning of each exercise set. The exercises in this list are always from the A group and the answers to these exercises appear in the answer section of the text. The standard assignment can be used by graduate students or new instructors as trial assignments until the needs of their particular classes can be determined. They also serve as a guide for students who miss class or are unable to get the assignment given by their instructor.

CALCULATOR EXERCISES   Exercises intended for calculator solutions are included throughout the book. They are optional exercises and are identified by the symbol ▤.

WARNINGS   A warning symbol clearly marks the many Warnings intended to call attention to errors and misconceptions common among students.

CHAPTER SUMMARY   Key words and phrases are identified and keyed to the section in which they are defined. Key concepts and rules are summarized to provide an overview of the chapter.

REVIEW EXERCISES   Numerous exercises, keyed to the appropriate section of the text, provide additional practice on all topics covered in the chapter.

PRACTICE TESTS   Sample exams, comparable in length to that of actual classroom tests, are provided at the end of each chapter.

EXTENDED APPLICATIONS   Each chapter ends with an application that uses topics covered in the chapter. Interested students can be assigned these applications for presentation in class and/or to work the accompanying exercises for extra credit.

SECOND COLOR   A second color is used to indicate key steps in the examples and for the explanatory side comments accompanying the examples. Key formulas, definitions, and procedures are also highlighted using color. A vertical colored bar runs along each example, making the example stand apart from the rest of the text's material.

SUPPLEMENTS   An *Instructor's Guide with Tests* contains five chapter tests for each chapter. Two of the tests for each chapter are in multiple choice format. A *Solutions to Even-Numbered Exercises*, featuring complete solutions to all even-numbered exercises, is available. A *Student Study Guide*, containing problems in a semiprogrammed format, with solutions, provides additional guidance for students requiring extra help. A *Student Solutions Guide* contains solutions to all the odd-numbered exercises in the text. Students can use this as an additional source of examples.

We are grateful to the people who reviewed all or part of the manuscript for their many insightful suggestions: Ann Anderson, Broward Community

College; Arthur Dull, Diablo Valley College; Barbara Juister, Elgin Community College; Theodore Laetsch, University of Arizona; Pamela E. Matthews, Chemeketa Community College; Mary McCammon, Pennsylvania State University; Raymond McGivney, University of Hartford; Bruce Partner, Ball State University; Mark Phillips, Cypress College; B. Louise Whisler, San Bernardino Valley College.

The authors wish to thank the fine staff at D. C. Heath for their assistance and cooperation in preparing the text. Particular thanks are due to Mary Lu Walsh, editor; Peggy J. Flanagan, production editor; Mary LeQuesne, development editor; and Libby Van de Kerkhove, designer.

Special thanks are owed to Carolyn Russell for typing the entire manuscript, and to Helen Medley for her assistance in preparing the supplements.

Marcus McWaters
You-Feng Lin

To the Student

This is your book. Its purpose is to help you learn algebra. To use the book to your best advantage, read with pencil in hand and keep a good supply of paper handy. Every new concept or technique is illustrated by completely worked-out examples. As you work through the examples, pay attention to the comments printed in color on the right-hand side of the examples. Read ahead and be prepared to ask your instructor any questions you may have that are not cleared up after hearing the class lecture.

Always work the homework problems assigned by your instructor. Work alone in order to develop confidence in your ability to understand the topics covered in class. If you have trouble working an exercise, refer to the examples cited in the instructions for that exercise. Review these examples and try again. If you are still unsuccessful, see your instructor for help. If you miss class and cannot find out what your homework assignment is, work the Standard Assignment given for the section or sections you miss.

The Warnings, which appear throughout the book, will point out common errors that students make. Be aware of these common mistakes and make note of the proper technique given in the Warnings.

The review material at the end of each chapter is an important part of the book. Read the Key Words and Phrases in the Chapter Summary and be sure you know what they mean. If you are not clear on the meaning of any listed word or phrase, refer to the section in which that word or phrase is defined (the numbers in brackets give the section number). Review the Key Concepts and Rules and then work through the Review Exercises. These exercises are keyed to an appropriate section so that you may refer to this section if you have any difficulty solving a problem.

Once you have finished the Review Exercises you are ready to take the Practice Test. The answers to these tests are in the answer section at the back of the text. This exam should tip you off to any existing weakness in your skills. Review once again the sections you are having difficulty with. You should now be ready for your class test. Good luck!

Marcus McWaters
You-Feng Lin

Contents

1 The Real Numbers — 1

    1.1 Sets   2
    1.2 Equality and Order   7
    1.3 Properties of Real Numbers   11
    1.4 Addition and Subtraction   19
    1.5 Multiplication and Division   26
    Chapter 1 Summary   33
    Review Exercises   35
    Practice Test   37
    Extended Applications   38

2 First-Degree Equations and Inequalities — 40

    2.1 Linear Equations in One Variable   41
    2.2 Formulas   48
    2.3 Applications   53
    2.4 Equations Involving Absolute Value   64
    2.5 Intervals   69
    2.6 Solving Linear Inequalities   72
    2.7 Compound Sentences of Inequality   79
    2.8 Inequalities Involving Absolute Value   85
    Chapter 2 Summary   89
    Review Exercises   90
    Practice Test   94
    Extended Applications   95

3 Exponents and Polynomials — 97

    3.1 Integral Exponents   98
    3.2 Rules for $(xy)^n$, $\left(\dfrac{x}{y}\right)^n$, and $(x^m)^n$ 105

3.3 Scientific Notation 109
3.4 Polynomials 114
3.5 Product of Polynomials 122
3.6 Common Factors; Grouping 128
3.7 Factoring Techniques 132
3.8 Diagnosing a Factoring Problem 140
Chapter 3 Summary 144
Review Exercises 145
Practice Test 147
Extended Applications 148

4 Algebraic Fractions — 149

4.1 From Rational Numbers to Rational Expressions 150
4.2 Sums and Differences 156
4.3 Products and Quotients 164
4.4 Improper Fractions and the Division Algorithm 170
4.5 Complex Fractions 176
4.6 Synthetic Division 182
Chapter 4 Summary 188
Review Exercises 189
Practice Test 191
Extended Applications 192

5 Exponents, Roots, and Radicals — 195

5.1 Roots and Radicals 196
5.2 Rational Exponents 201
5.3 Operations with Radicals 205
5.4 Rationalizing and Simplifying Radicals 210
5.5 Radical Equations 214
5.6 Complex Numbers 218
5.7 Operations with Complex Numbers 222
Chapter 5 Summary 226
Review Exercises 227
Practice Test 229
Extended Applications 230

6 Second-Degree Equations and Inequalities — 232

    6.1 Solution by Factoring   233
    6.2 Solution of Equations of the Form $x^2 = b$; Completing the Square   240
    6.3 The Quadratic Formula   246
    6.4 Equations Involving Radicals   254
    6.5 Equations Leading to Quadratic Equations   257
    6.6 Applications   261
    6.7 Quadratic Inequalities   269
    Chapter 6 Summary   276
    Review Exercises   277
    Practice Test   279
    Extended Applications   280

7 Linear Equations, Inequalities, Functions — 282

    7.1 The Cartesian Coordinate System   283
    7.2 The Slope of a Line   291
    7.3 Equations of a Line   298
    7.4 Linear Inequalities and Their Graphs   306
    7.5 Functions and Relations   311
    7.6 Direct and Inverse Variation   319
    Chapter 7 Summary   326
    Review Exercises   327
    Practice Test   329
    Extended Applications   329

8 Functions and Conic Sections — 332

    8.1 Graphs of Quadratic Functions   333
    8.2 Graphing Parabolas   340
    8.3 The Distance Formula and Circles   348
    8.4 The Ellipse and the Hyperbola   354
    8.5 The Graphs of Quadratic Inequalities   362
    8.6 Applications   366
    8.7 Inverse Functions   372
    Chapter 8 Summary   378
    Review Exercises   379
    Practice Test   381
    Extended Applications   382

9 Systems of Equations 384

9.1 Linear Systems in Two Variables 385
9.2 Linear Systems in Three Variables 393
9.3 Applications of Linear Systems 398
9.4 Determinants 406
9.5 Cramer's Rule 410
9.6 Solving Linear Systems by Matrix Methods 415
9.7 Nonlinear Systems of Equations 420
9.8 Systems of Inequalities 426
Chapter 9 Summary 429
Review Exercises 430
Practice Test 432
Extended Applications 433

10 Exponential and Logarithmic Functions 436

10.1 Exponential Functions 437
10.2 Logarithm Functions 441
10.3 Properties of Logarithms 446
10.4 Computations with Logarithms 452
10.5 Logarithmic Computation Using Calculators 457
10.6 Natural Logarithms 460
10.7 Further Applications 466
Chapter 10 Summary 472
Review Exercises 473
Practice Test 475
Extended Applications 476

11 Sequences, Series, Permutations, and Combinations 478

11.1 Sequences and Series 479
11.2 Arithmetic Sequences and Series 484
11.3 Geometric Sequences and Series 490
11.4 The Binomial Theorem 497
11.5 Permutations 502
11.6 Combinations 507
Chapter 11 Summary 512
Review Exercises 513
Practice Test 516
Extended Applications 517

Appendixes — A1

    Appendix A   Tables   A1
    Appendix B   Linear Interpolation   A8

Answers to Selected Exercises — A11

Index — A63

INTERMEDIATE ALGEBRA

1

The Real Numbers

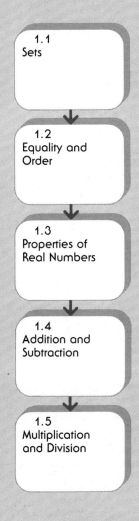

1.1 Sets

The familiar word "set" is used in both of the following sentences: "A complete set of baseball cards is worth a fortune," and "Mom wants a set of fine china." A **set** is a collection of objects. The objects in a set are called the **elements** or **members** of the set. In the study of algebra we are primarily interested in sets of numbers. The numbers used in counting, 1, 2, 3, 4, 5, 6, 7, 8, ... are the elements of the set we call the **natural numbers**.

In order to specify or define a set we either

1. List the elements of the set, or
2. Describe the elements of the set.

When listing the elements of a set, it is customary to enclose the listed elements in braces, { }. We say that two sets A and B are **equal**, or $A = B$, if they contain exactly the same members.

EXAMPLE 1
(a) The set described by the phrase "the first five natural numbers" can be listed as $\{1, 2, 3, 4, 5\}$.
(b) The set described by the phrase "the natural numbers less than 6 and larger than 3" can be listed as $\{4, 5\}$.

To indicate that an object, a, is an element of the set A we use the symbol ϵ. Specifically,

$a \in A$ means "a is an element of A" so $5 \in \{3, 5, 7\}$ means "5 is an element of $\{3, 5, 7\}$."

The set of people in your algebra class who are over eight feet tall is a peculiar set; it has no members. A set with no members is called the **empty**, or **null, set** and is represented by a special symbol.

$\emptyset$ denotes "the empty set"

In algebra we often use a symbol, which is usually a letter, to denote an unspecified element in a given set. When the given set contains more than one element, such a symbol is called a **variable**. A symbol used to represent a specific element in a set is called a **constant**. Physicists use the letter c as a constant to represent the speed of light ($c = 300,000,000$ m/sec).

Variables are helpful in describing sets through **set-builder notation.** In this notation,

$$\{x \mid x \text{ is a natural number less than six}\}$$

describes the set $\{1, 2, 3, 4, 5\}$. The vertical bar following x is read "such that." The variable x represents an unspecified natural number.

The first and last braces are read together as "the set of all." Thus, $\{x \mid x$ is a natural number less than six$\}$ is read: "The set of all x such that x is a natural number less than six." Generally, $\{x \mid x$ has property $P\}$ designates the set of all x such that x has the given property P.

EXAMPLE 2   List the elements of $\{x \mid$ twice x is 8 and x is a natural number$\}$.

Solution: Since twice x is 8 exactly when $x = 4$, the desired set is $\{4\}$.

If each element of a set A is also an element of a set B, we say that A is a **subset** of B. The symbol $\subset$ is used to denote the subset relation, so that $A \subset B$ is read: "A is a subset of B."

A is a subset of B or $A \subset B$ means every element of A is an element of B.

For example, $\{1, 3\} \subset \{1, 2, 3, 4, 5\}$ since 1 and 3 are both in the set $\{1, 2, 3, 4, 5\}$. Similarly, $\{$Russia, Germany, France$\} \subset \{x \mid x$ is a country currently using the metric system$\}$ since Russia, Germany, and France all use the metric system. The slash, /, together with the symbols $\subset$, $=$, and $\in$ is read "not . . .," and thus

$A \not\subset B$ indicates that A is *not* a subset of B,

$A \neq B$ indicates that A is *not* equal to B, and

$x \notin A$ indicates that x is *not* an element of A.

For instance, $\{1, 3, 6\} \not\subset \{1, 2, 3, 4, 5\}$ since 6 is not a member of $\{1, 2, 3, 4, 5\}$. Similarly, $\{1, 3, 6\} \neq \{1, 2, 3, 4, 5\}$, and $7 \notin \{1, 2, 3, 4, 5\}$.

Some Important Sets of Numbers

The natural numbers are adequate for representing the profit in dollars that a successful business earns in a given year. However, if we want to represent the loss a business may show in a given year or a "break-even" year, we need to introduce zero and negative numbers. This leads to our first three sets of numbers: natural numbers, whole numbers, and integers.

The Natural Numbers: N, whose elements are 1, 2, 3, 4, 5, 6, . . . (The three dots, . . ., may be read as "and so on.")

The Whole Numbers: W, whose elements are 0, 1, 2, 3, 4, 5, . . .

The Integers: J, whose elements are 0, ± 1, ± 2, ± 3, ± 4, . . . (Here we use ± 1 to indicate the two numbers 1 and -1.)

If two people equally split a candy bar, however, none of these numbers will properly represent the portion of the bar (one-half) that each person gets. This requires us to introduce the rational numbers.

The Rational Numbers: **Q,** whose elements are all numbers that can be written in the form *p/q,* where *p* and *q* are integers and *q* is not zero. Thus, in set notation, **Q** = {(*p/q*) | *p* and *q* are integers and *q* ≠ 0}. Examples of rational numbers are:

$$\frac{1}{2}, \quad -\frac{4}{17}, \quad \frac{25}{3}, \quad \text{and} \quad \frac{7}{1}$$

The area of a square is obtained by multiplying the length of its side by itself. The area of a square with a side measuring 1 inch is 1 · 1 = 1 square inch. It is geometrically believable that there is a square whose area is 2 square inches. (See Figure 1.1.) However, no rational number represents the length of its side. This is due to the fact, which is proven in more advanced courses, that there is no rational number *b* such that *b* · *b* = 2. We denote the positive number *b* such that *b* · *b* = 2 by √2.

Figure 1.1

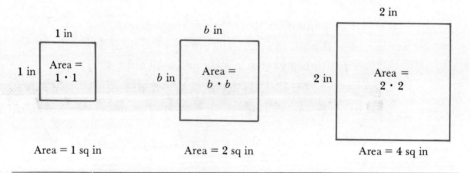

Area = 1 sq in   Area = 2 sq in   Area = 4 sq in

The Irrational Numbers: **H,** whose elements are all the real numbers that cannot be written in the form *p/q,* where *p* and *q* are integers and *q* is not zero. Examples are √2, −3·√5, 1 + √7, and π.

The Real Numbers: **R,** whose elements are all the numbers used to represent lengths (which may be used for measurement) and their negatives. The rational and irrational numbers together form the real numbers. The "length" aspect of the real numbers will be discussed more fully in Section 1.2.

Notice that each natural number is also an integer and that each integer is also a rational number. For instance, 3 = $\frac{3}{1}$ shows that the integer 3 is also the quotient of integers, and hence a rational number.

The subset relationships among these six sets are indicated in Figure 1.2.

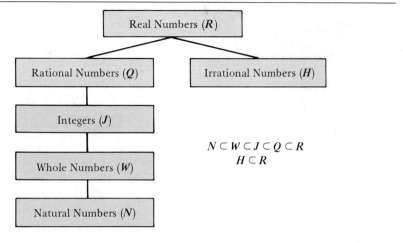

Figure 1.2

$N \subset W \subset J \subset Q \subset R$
$H \subset R$

EXAMPLE 3

List the elements of each of the following sets.

(a) $\{x \mid x \in J$ and x is less than $4\}$
(b) $\{x \mid x \in N$ and x is a multiple of $3\}$

Solution:
(a) Since $x \in J$ and x is less than 4 means that x is an integer and x is less than 4, we can list (a) as $\{\ldots, -4, -3, -2, -1, 0, 1, 2, 3\}$.
(b) Now $x \in N$ and x is a multiple of 3 means that x is a natural number and x is a multiple of 3. We can obtain all such numbers by multiplying each natural number by 3. We can list (b) as $\{3, 6, 9, 12, 15, \ldots\}$.

Exercises 1.1

Standard Assignment: Exercises 3, 5, 7, 9, 11, 13, 17, 19, 21, 27, 29, 33, 43, 45.

A List the elements of each of the following sets. See Examples 1, 2, and 3.

1. $\{x \mid x$ is a natural number $(x \in N)$ and x is less than $10\}$
2. $\{x \mid x \in N$ and x is between 4 and $9\}$
3. $\{x \mid x \in N$ and x is between 4 and $5\}$
4. $\{x \mid x \in N$ and x is odd and less than $17\}$
5. $\{x \mid x \in N$ and x is a multiple of 3 that is less than $28\}$
6. $\{x \mid x \in J$ and twice x is $1\}$
7. $\{x \mid x \in N$ and x is a whole number multiple of $5\}$

8. $\{x \mid x \in N$ and x is an even number less than 33$\}$

9. $\{x \mid x \in Q$ and twice x is 1$\}$

Give a description for each of the following sets.

> EXAMPLE: $\{\ldots, -5, -3, -1, 1, 3, 5, \ldots\}$
> One description is "the set of odd integers."

10. $\{1, 3, 5, 7\}$
11. $\{2, 4, 6, 8\}$
12. $\{2, 4, 6, 8, 10, \ldots\}$
13. $\{-6, -4, -2, 0, 2, 4, 6, \ldots\}$
14. $\{1, 3, 5, 7, 9, \ldots\}$
15. $\{4, 8, 12, 16, 20\}$
16. $\{0\}$
17. $\{1\}$

Let
$$P = \left\{-\frac{22}{17}, -\sqrt{2}, -1, -\frac{1}{3}, 0, \frac{1}{2}, \sqrt{2}, \frac{9}{2}, 5, 6\right\}$$

List the elements of the sets given in Exercises 18–25.

> EXAMPLE: $\{x \mid x \in P$ and $x \in J$ and x is less than 6$\}$
> These are the integer elements of P that are less than 6: $\{-1, 0, 5\}$.

18. $\{x \mid x \in P$ and $x \in N\}$
19. $\{x \mid x \in P$ and $x \in W\}$
20. $\{x \mid x \in P$ and $x \in J\}$
21. $\{x \mid x \in P$ and $x \in Q\}$
22. $\{x \mid x \in P$ and $x \in H\}$
23. $\{x \mid x \in P$ and x is between -2 and 2$\}$
24. $\{x \mid x \in P$ and x is positive$\}$
25. $\{x \mid x \in P$ and x is negative$\}$

Place $\subset$ or $\in$ in the blank to produce a true statement.

26. $\{1\}$ _____ $\{1, 2, 3\}$
27. 1 _____ $\{1, 2, 3\}$
28. $\emptyset$ _____ $\{1, 2, 3\}$
29. 1 _____ $\{1\}$
30. $\{1, 2, 3\}$ _____ W
31. J _____ W

Place $\subset$ or $\not\subset$ in the blank to produce a true statement.

32. $\{2, 4, 6\}$ _____ Q
33. $\{-1, 0\}$ _____ N
34. $\{-1\}$ _____ N
35. $\{-1\}$ _____ J
36. $\{0\}$ _____ W
37. $\{7\frac{1}{2}, 13\}$ _____ Q
38. $\{-1, \sqrt{2}\}$ _____ H
39. Q _____ H
40. H _____ Q
41. N _____ W
42. W _____ J
43. Q _____ J

In Exercises 44–47, use the symbols ∉ and ⊄ to write the given statement in symbolic form.

| EXAMPLE: | The set **R** of real numbers is not a subset of the set **H** of irrational numbers. Symbolically we have: **R** ⊄ **H**.

44. Five is not an element of the empty set.

45. The set **J** of integers is not a subset of the set **W** of whole numbers.

46. The empty set is not an element of the set **J**.

47. The set of whole numbers is not a subset of the set of natural numbers.

B Answer each of the questions in Exercises 48–53 *Yes* or *No* and give an explanation for your answer.

| EXAMPLE: | If $1 \in A$ and $B \subset A$, must we have $1 \in B$? No, taking $A = \{1, 2, 3\}$ and $B = \{2, 3\}$, we see $1 \in A$ and $B \subset A$, but $1 \notin B$.

48. If $A \neq B$ and $B \subset A$, must there be an $x \in A$ with $x \notin B$?

49. If $3 \in A$ and $A \subset B$, must we have $3 \in B$?

50. If $A = \{4\}$, can we have $3 \in A$?

51. If $A \subset B$, can we have $B \subset A$?

52. If $A \subset B$ and $B \subset C$, must we have $A \subset C$?

53. Are $\{1\}$, $\{2\}$, and $\{1, 2\}$ all the subsets of $\{1, 2\}$?

1.2 Equality and Order

Perhaps the most widely recognized symbol of mathematics is the "equals" sign (=). The equals sign is used to assert that two expressions name the same number ($\frac{1}{2} = \frac{2}{4}$), two sets contain the same elements ($A = B$), and many other things. All of its uses, however, are governed by certain fundamental assumptions such as "Anything is equal to itself." We will state and name the basic properties of **equality** next.

Equality Properties

1. $a = a$ — Reflexive Property
2. If $a = b$, then $b = a$. — Symmetric Property
3. If $a = b$ and $b = c$, then $a = c$. — Transitive Property
4. If $a = b$, then we may replace a with b anywhere we choose in a statement without affecting the truth of the statement. — Substitution Property

EXAMPLE 1
(a) *Reflexive Property:* $x + 2 = x + 2$.
(b) *Symmetric Property:* If $y - 1 = 7$, then $7 = y - 1$.
(c) *Transitive Property:* If $4 = x^2$ and $x^2 = 2z + 3$, then $4 = 2z + 3$.
(d) *Substitution Property:* If $x = y + 3$ and $2x + 3y = 1$, then $2(y + 3) + 3y = 1$.

We have all seen numbers associated with marks on a tape measure. An extension of this idea provides a useful geometric representation of the real numbers on a **number line.** It is possible to associate the real numbers with points on a geometric line in such a way that each real number corresponds to exactly one point and each point corresponds to exactly one number. The point is called the **graph** of the corresponding real number and the real number is called the **coordinate** of the point.

To construct a number line, pick any point on the line and label that point 0. Next, pick any point to the right of 0 and label that point 1. Use these two points to determine the unit length that will then be used to scale the rest of the line. By agreement, positive numbers lie to the right of 0 and negative numbers lie to the left of 0. The point with the coordinate $\frac{1}{2}$ is halfway between 0 and 1. The point with the coordinate $\frac{3}{2}$ is reached by drawing a duplicate of the section from 0 to $\frac{1}{2}$ three times end to end, starting at 0 and going to the right. The points corresponding to $-\frac{3}{2}$ and $-\frac{1}{2}$ are the same distance to the left of 0 that $\frac{3}{2}$ and $\frac{1}{2}$ are to the right of 0, respectively. A number line is shown in Figure 1.3.

For any positive integer, q, we may subdivide the unit segment into q equal parts, each of length $1/q$. We mark the length of such a segment p times to the right of 0 (p is a positive integer), mark the resulting point on the line, and assign the number p/q to this point. If we mark the length of this segment p times to the left of 0, we assign the resulting point on the line the real number $-p/q$. (Note $\frac{7}{3}$ and $-\frac{7}{3}$ in Figure 1.3). In this way, the rational numbers, **Q**, are assigned to points on the line.

Note that 1 is to the left of 4 on the line in Figure 1.3. By adding 3 to 1 we obtain 4, or geometrically speaking, by moving 3 units to the right of 1, we arrive at 4. *The number line has been constructed so that a first number is less than a second number precisely when the first number is to the left of the second number.*

Figure 1.3

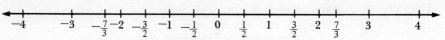

The symbol < is read "is less than," so that 1 < 4 is read "1 is less than 4." In general, for real numbers a and b we define

$a < b$ if and only if $b = a + c$ for some positive number c.

The symbol > is read "is greater than," and the statement $b > a$ has the same meaning as $a < b$. Two other symbols are introduced in the following list.

Symbolic List	Translation
$a < b$	a is less than b
$b > a$	b is greater than a
$a \leq b$	a is less than or equal to b
$b \geq a$	b is greater than or equal to a

The slash, /, together with any of the symbols in the table, or the "equals" sign (=), negates the meaning of the symbol. Thus, $a \neq b$ means "a is not equal to b" and $a \not< b$ means "a is not less than b."

EXAMPLE 2 Notice the reason given for the truth or falsity of the following statements.

(a) $3 \leq 3$ True, since $3 = 3$
(b) $3 \neq 5$ True, since 3 and 5 are not the same number
(c) $-3 > -2$ False, since $-3 < -2$ (-3 is left of -2 on the number line)
(d) $3 \not> 2$ False, since $3 > 2$

WARNING Students often think statements such as $3 \leq 3$ are false since 3 is not less than 3. However, $3 \leq 3$ is true if *either* condition, $3 = 3$ or $3 < 3$ is true. Since $3 = 3$ is true, so is $3 \leq 3$.

We will state next the basic **inequality properties** for arbitrary real numbers a, b, and c. We have already used these properties based on intuition.

Inequality Properties

Trichotomy Property: Exactly one of the following is true: $a < b$, $a = b$, $a > b$.

Transitive Property: If $a < b$ and $b < c$, then $a < c$.

EXAMPLE 3 (a) In Example 2(c), we decided that $-3 > -2$ was false because $-3 < -2$ was true. This uses the Trichotomy Property, which says (substituting $a = -3$ and $b = -2$) that *exactly* one of the statements $-3 < -2$, $-3 = -2$, and $-3 > -2$ is true.

(b) **The Transitive Property** (substituting $a = -1$, $b = 2$, and $c = 3.5$) says that since $-1 < 2$ and $2 < 3.5$, we have $-1 < 3.5$.

Exercises 1.2

Standard Assignment: Exercises 1, 3, 5, 7, 9, 11, 15, 17, 19, 21, 25, 27, 33, 37, 39

A Graph each set in Exercises 1–6 on a number line.

EXAMPLE: $\left\{-\dfrac{3}{2}, 0, \dfrac{2}{3}, 2\right\}$

1. $\{-3, -2, 2, 3\}$
2. $\{-1, 2, 5, 8\}$
3. $\left\{-\dfrac{5}{2}, -2, -\dfrac{3}{2}, -1, -\dfrac{1}{2}\right\}$
4. $\left\{\dfrac{3}{4}, \dfrac{5}{4}, \dfrac{7}{4}, \dfrac{9}{4}\right\}$
5. $\left\{-4, -1, 0, \dfrac{3}{10}, 5\right\}$
6. $\left\{-\dfrac{13}{10}, -\dfrac{3}{5}, \dfrac{24}{9}\right\}$

Use inequality symbols to write the given statements symbolically.

EXAMPLE: 5 is not less than $3x + 1$ $\quad 5 \not< 3x + 1$

7. 3 is greater than -2
8. -3 is less than -2
9. $\dfrac{1}{2}$ is greater than or equal to $\dfrac{1}{2}$
10. 5 is not less than 5
11. 7 is not less than 0
12. x is less than $x + 1$
13. 5 is less than or equal to $2t$
14. $t - 1$ is greater than $p + 2$
15. $-t$ is positive
16. t is negative
17. $2y + 7$ is less than or equal to 14
18. $2t + 3$ is not greater than y
19. z is greater than -6 and less than or equal to -1
20. $2z + 6$ is greater than or equal to -5 and is less than or equal to 0

Fill in the blank with one of the symbols =, <, or >, to produce a true statement.

21. 4 _____ $\dfrac{24}{6}$
22. -3 _____ -2
23. $-\dfrac{1}{2}$ _____ $-\dfrac{3}{2}$
24. 2 _____ -1
25. -4 _____ 0
26. $-\dfrac{5}{2}$ _____ $-2\dfrac{1}{2}$

Do the required arithmetic to determine whether each of the following inequalities is true or false. See Example 2.

27. $(3 \cdot 6) + 9 \leq 27$
28. $(4 \cdot 2) - 3 > 5$
29. $3 - (2 + 7) \neq -5$
30. $2 + 9 \neq 12$
31. $3(6 + 8) \leq 3 \cdot 6 + 3 \cdot 8$
32. $(5 \cdot 25) + 2 = 5(25 + 2)$
33. $-2 \not< 1$
34. $-\frac{1}{2} > -\frac{1}{3}$
35. $17 \geq 17$
36. $17 \not> 17$

Rewrite each of the following statements that uses the symbol $<$ so that it uses the symbol $>$ instead, and rewrite each statement that uses the symbol $>$ so that it uses the symbol $<$.

37. $-6 < -1$
38. $5 > 2$
39. $x < -1$
40. $x > -2$

B Identify the properties that justify each of the following statements. See Example 1.

41. $x^2 + 1 = x^2 + 1$
42. If $2x + 7 = y$, then $y = 2x + 7$
43. If $x^2 = y$ and $x^2 + 1 = 2z - 1$, then $y + 1 = 2z - 1$
44. If $y = 2x - 1$ and $2x + 3y = 2$, then $2x + 3(2x - 1) = 2$
45. If $a^2 = 2a + b - 7$ and $b - a^2 = 3$, then $b - (2a + b - 7) = 3$
46. If $0 = x^2 + 2x + 5$, then $x^2 + 2x + 5 = 0$
47. If $s = 5$ and $2s = 20$, then $2 \cdot 5 = 20$
48. If $5 < 2k + 1$ and $k = 3x$, then $5 < 2(3x) + 1$
49. If $5 < 2k + 1$ and $2k + 1 < x + 7$, then $5 < x + 7$
50. Either $x + 1 > 2$ or $x + 1 = 2$ or $x + 1 < 2$
51. If $-2 + t > 7$ and $7 > s$, then $-2 + t > s$
52. If $s = 5$ and $s + 1 < 27 + t$, then $5 + 1 < 27 + t$
53. If $3x + 7 < 0$ and $x + 5 > 0$, then $3x + 7 < x + 5$

1.3 Properties of Real Numbers

We have discussed some of the properties of the real numbers involving equality and order (inequality). There are legions of useful facts about the real numbers and we will state the most basic properties of the real numbers in the following list. Normally, when a property of real numbers is deduced from the properties of this section together with the equality properties, it is called a **theorem.** The logical argument demonstrating that the theorem is a consequence of the axioms is called a **proof** of the theorem. We will not emphasize proofs in this course, but instead will either give a short informal argument in favor of a theorem or provide illustrations of the meaning of the theorem.

Basic Properties of the Real Numbers

For any real numbers a, b, and c:

1. $a + b$ is a real number — Closure Property for Addition
   ab is a real number — Closure Property for Multiplication
2. $a + b = b + a$ — Commutative Property for Addition
   $ab = ba$ — Commutative Property for Multiplication
3. $(a + b) + c = a + (b + c)$ — Associative Property for Addition
   $(ab)c = a(bc)$ — Associative Property for Multiplication
4. $a(b + c) = ab + ac$ — Distributive Property
5. There exists a unique number 0 such that — Identity Property for Addition

   $$a + 0 = a \text{ and } 0 + a = a$$

6. There exists a unique number 1 such that — Identity Property for Multiplication

   $$a \cdot 1 = a \text{ and } 1 \cdot a = a$$

7. For each real number a, there exists a unique real number $-a$ (called the **negative** of a) such that — Inverse Property for Addition

   $$a + (-a) = 0 \text{ and } (-a) + a = 0$$

8. For each real number a, with $a \neq 0$, there exists a unique real number $1/a$ (called the **reciprocal** of a) with the property — Inverse Property for Multiplication

   $$a(1/a) = 1 \text{ and } (1/a)a = 1$$

EXAMPLE 1 The following list illustrates the use of the basic properties.

(a) $3(-16)$ is a real number — *Closure Property for Multiplication:* $a = 3, b = 16$

(b) $3(2 + 7)$ is a real number — *Closure Property for Addition:* ($a = 2, b = 7$); and *Multiplication:* $a = 3, b = 2 + 7$

(c) $x + 3 = 3 + x$ — *Commutative Property for Addition:* $a = x, b = 3$

(d) $x + (7 + 1) = (x + 7) + 1$ — Associative Property for Addition: $a = x, b = 7, c = 1$

(e) $x(7 + y) = x7 + xy$ — Distributive Property: $a = x, b = 7, c = y$

(f) $x(7 + y) = 7x + yx$ — Distributive Property as in (e); then Commutative Property for Multiplication: $a = x, b = 7$; $a = x, b = y$

(g) $5\left(\dfrac{1}{5}\right) = 1$ and $5 + (-5) = 0$ — Inverse Property for Multiplication and Addition: $a = 5$

The negative of a, or $-a$, is also called the **opposite** of a, or the **additive inverse** of a. On the number line, $-a$ and a are the same distance from zero, but on opposite sides of zero. For nonzero a, the reciprocal $1/a$ is sometimes called the **multiplicative inverse** of a.

EXAMPLE 2 In this example, several numbers are listed with their inverses.

Number	Negative	Reciprocal
0	0	none
3	-3	$\dfrac{1}{3}$
$\dfrac{2}{5}$	$-\dfrac{2}{5}$	$\dfrac{5}{2}$
-2	2	$\dfrac{1}{-2}\left(=-\dfrac{1}{2}\right)$
$-\dfrac{1}{2}$	$\dfrac{1}{2}$	-2

You may have expected to see $-(-2)$ listed as the negative of -2. This would certainly be correct. To see that 2 is also correct, simply recognize that the statement "$-2 + 2 = 0$" says that "2 is the number that, when added to -2, yields zero." The inverse property states that there is only one such number, the negative of -2. Thus we see that $-(-2) = 2$. Similar reasoning yields the following theorem for any real number a.

Double Negative Property

$-(-a) = a$

We also know that $5 \cdot 0 = 0$, $(-3) \cdot 0 = 0$, and $0(0) = 0$. The result, that multiplication of any real number a by zero results in zero, is a useful property (see Exercise 76 at the end of this section).

Zero Factor Property

$a \cdot 0 = 0$

We can now consider an interesting use of the reflexive and substitution properties for equality. By reflexivity, if a and c are real numbers, then $a + c = a + c$. If we are given that $a = b$, then by the substitution property, we may replace a with b on the right-hand side of the equation to obtain $a + c = b + c$. Similarly, from $a = b$ and $ac = ac$, we obtain $ac = bc$. Thus, we have the following property for any three real numbers a, b, and c.

Addition Property of Equality

If $a = b$, then $a + c = b + c$

and

Multiplication Property of Equality

If $a = b$, then $ac = bc$

EXAMPLE 3 Answer the following multiple choice question: $-x$ is

(a) positive
(b) negative
(c) zero
(d) insufficient information to decide

Solution: The correct answer is (d), as the following table should make clear.

x	$-x$	
3	-3	negative
0	0	zero
-3	$-(-3) = 3$	positive

It is extremely important to understand that $-x$ may be negative, zero, or positive before studying absolute value. The definition of the **absolute value** of a real number, x, denoted by $|x|$ is given next.

$$|x| = \begin{cases} x, & \text{if } x \geq 0 \\ -x, & \text{if } x < 0 \end{cases} \quad (1.1)$$

EXAMPLE 4 Find each of the following absolute values.

(a) $|2|$ (b) $|-2|$ (c) $|0|$ (d) $|(-3) + 1|$ (e) $-\left|\dfrac{1}{2}\right|$

Using Equation (1.1) for absolute value requires that we determine whether the expression *inside the absolute value bars* is positive, zero, or negative.

Solution:
(a) Since $2 \geq 0$, the upper line of the equation applies, and thus $|2| = 2$.
(b) Since $-2 < 0$, the lower line of the equation applies, and thus $|-2| = -(-2) = 2$.
(c) Since $0 \geq 0$, the upper line of the equation applies, and thus $|0| = 0$.
(d) Since the expression inside the absolute value bars is $(-3) + 1 = -2$ and $-2 < 0$ the lower line of the equation applies, and thus $|(-3) + 1| = |-2| = -(-2) = 2$.
(e) Since $\dfrac{1}{2} \geq 0$, $\left|\dfrac{1}{2}\right| = \dfrac{1}{2}$, so $-\left|\dfrac{1}{2}\right| = -\dfrac{1}{2}$.

Note that $|2| = 2$ and $|-2| = 2$; this reflects the fact that both 2 and -2 are two units away from 0, as is shown in Figure 1.4.

Figure 1.4

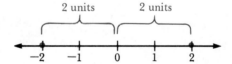

Intuitively, the absolute value of a real number x, $|x|$, gives the distance between 0 and x as represented on a number line. For $x \neq 0$, we know that x and $-x$ are the same distance from 0, but on opposite sides of 0. See Figure 1.5. Thus,

$$|x| = |-x|$$

Figure 1.5

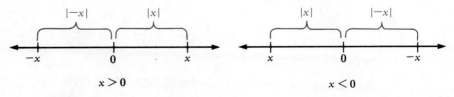

Exercises 1.3

Standard Assignment: Exercises 1, 3, 5, 7, 9, 11, 13, 15, 29, 31, 33, 39, 41, 47, 53, 55, 63, 65, 67

A Specify which real number properties justify the following statements about real numbers. See Example 1.

1. $3(x + 2) = 3x + 3 \cdot 2$
2. $2a + 2y = 2(a + y)$
3. $(-1)(1) = -1$
4. $-(2x) + 2x = 0$
5. $2(3t) = (2 \cdot 3)t$
6. $[a(x + 1)]y = a[(x + 1)y]$
7. $(-2)(x + y) = (x + y)(-2)$
8. $(-2 + x) + y = -2 + (x + y)$
9. $0(3 + 7) = 0$
10. $0\left(\dfrac{1}{2}\right) = 0$
11. $-(-5) = 5$
12. $-(-xy) = xy$
13. $2\left(\dfrac{1}{2}\right) = 1$
14. $1 = \left(\dfrac{1}{x^2 + 1}\right) \cdot (x^2 + 1)$
15. $\left(-\dfrac{1}{2}\right)(2) + \left(\dfrac{1}{2}\right)(2) = 0$
16. $8x$ is a real number
17. $t + [5 + (-5)] = t$
18. $y + 3 = 3 + y$
19. $(x + 5) + 2t = x + (5 + 2t)$
20. $(x + 5) + 2t = 2t + (x + 5)$
21. $3 + 0 = 3$
22. $x(2 + y) + 0 = x(2 + y)$
23. If $m = 3$, then $m + 7 = 3 + 7$
24. If $m = 3$, then $2m = 2 \cdot 3$
25. If $m + 1 = 12$, then $m + 1 + k = 12 + k$
26. If $m + 1 = 12$, then $(m + 1) + (m + 1) = 12 + (m + 1)$
27. $8 \cdot 1 = 8$
28. $2(m + k) \cdot 1 = 2m + 2k$

Give the negative and the reciprocal of each number. See Example 2.

29. -1
30. $\dfrac{3}{7}$
31. 1
32. 4
33. $-\dfrac{2}{9}$
34. 0.25
35. ▦ -0.7524
36. ▦ 3.68
37. ▦ 1.35286
38. ▦ 0.128063

Find a value for x so that the resulting statement becomes an example of the given property.

> EXAMPLE: $x \cdot \sqrt{3} = \sqrt{3}$; Identity property for multiplication. Choose $x = 1$ and then $1 \cdot \sqrt{3} = \sqrt{3}$.

39. $12 + x = 0$; Inverse property for addition
40. $x \cdot 3 = 0$; Zero factor property

41. $2x = 1$; Inverse property for multiplication

42. $3 + 7 = x + 3$; Commutative property for addition

43. $10 \cdot \left(\dfrac{1}{8}\right) = \left(\dfrac{1}{8}\right)x$; Commutative property for multiplication

44. $3(7 + 2) = 3 \cdot 7 + x \cdot 2$; Distributive property

45. $7 \cdot (3y) = (7x)y$; Associative property for multiplication

46. $(5 + x) + y = 5 + (9 + y)$; Associative property for addition

47. $4(1/x) = 1$; Inverse property for multiplication

48. $-6 + x = 3 + (-6)$; Commutative property for addition

49. $3\left(x + \dfrac{1}{3}\right) = 6 + 1$; Distributive property

50. $-3 + 3 = x$; Inverse property for addition

51. $(0.25)4 = x$; Inverse property for multiplication

52. $(7 + 2) + 3 = x + (2 + 3)$; Associative property for addition

Fill in the blanks so that the resulting statement is a true statement obtained by using the given property. Simplify your answer.

EXAMPLE: $2(3x) = $ _____; Associative property for multiplication: $(ab)c = a(bc)$.
$2(3x) = (2 \cdot 3)x = 6x$; where 2 replaces a, 3 replaces b, and x replaces c.

53. $5t + 2t = $ _____; Distributive property

54. $-3 + 0 = $ _____; Identity property for addition

55. $-3 + $ _____ $= 0$; Inverse property for addition

56. $5 + (1 + x) = $ _____; Associative property for addition

57. $3 \cdot ($ _____$) = 1$; Inverse property for multiplication

58. $0(2y + 3y) = $ _____; Zero factor property

59. $-[-(-2)] = $ _____; Double negative property

60. If $x = 2$, then $x + 5 = $ _____; Addition property for equality

61. If $x = 2$, then $2x = $ _____; Multiplication property for equality

62. If $x = 0$, then $\sqrt{2} \cdot x = $ _____; Multiplication property for equality

Rewrite each of the following without using absolute value notation. See Example 4.

63. $|-12|$

64. $\left|-\dfrac{3}{8}\right|$

65. $|(-2) + 2|$

66. $|8 + 1|$

67. $-|-2|$

68. $-\left|\dfrac{2}{7}\right|$

Rewrite the following without using absolute value notation.

B EXAMPLE: $|x - 3|$
From Equation (1.1) we get
$$x - 3, \quad \text{if} \quad x - 3 \geq 0$$
$$-(x - 3), \quad \text{if} \quad x - 3 < 0$$

69. $\left|\dfrac{x}{3}\right|$ 70. $|-t|$ 71. $-|x - 3|$

72. $|2y + 1|$ 73. $|x + 7|$ 74. $|1 - y|$

Give a reason for each step in the following proofs.

EXAMPLE: Prove that $-(-a) = a$.

$(-a) + a = 0$	Inverse property for addition
$(-a) + [-(-a)] = 0$	Inverse property for addition
$-(-a) = a$	Uniqueness of the negative of a number

75. Prove that $(a + b) \cdot c = a \cdot c + b \cdot c$.

$(a + b) \cdot c = c \cdot (a + b)$ _____

$c \cdot (a + b) = c \cdot a + c \cdot b$ _____

$c \cdot a = a \cdot c$ _____

$c \cdot (a + b) = a \cdot c + c \cdot b$ _____

$c \cdot b = b \cdot c$ _____

$c \cdot (a + b) = a \cdot c + b \cdot c$ _____

$(a + b) \cdot c = a \cdot c + b \cdot c$ _____

76. Prove that $a \cdot 0 = 0$.

$0 + 0 = 0$ _____

$a \cdot (0 + 0) = a \cdot 0$ _____

$a \cdot (0 + 0) = a \cdot 0 + a \cdot 0$ _____

$a \cdot 0 = a \cdot 0 + a \cdot 0$ _____

$a \cdot 0 = a \cdot 0 + 0$ _____

$a \cdot 0 + a \cdot 0 = a \cdot 0 + 0$ _____

$a \cdot 0 = 0$ _____

Determine which of the following sets are closed for multiplication and which are closed for addition. A set is *closed for addition* if the sum of two numbers from the set is also a number in the set and a set is *closed for multiplication* if the product of two numbers from the set is also a number from the set.

> EXAMPLE: $\{\ldots, -5, -3, -1, 1, 3, 5, \ldots\}$ is not closed for addition since 1 and 3 are numbers in the set, but $1 + 3 = 4$ is *not* in the set. This set is closed for multiplication since no matter which two numbers are chosen from the set, the product of the numbers is also in the set.

77. $\{0, 1\}$

78. $\{0, -1\}$

79. $\{-1, 0, 1\}$

80. $\{1\}$

81. $\{-1\}$

82. $\{\ldots, -6, -4, -2, 0, 2, 4, 6, \ldots\}$

1.4 Addition and Subtraction

We assume in this section that you know how to add two positive real numbers a and b to obtain a third real number $a + b$. Both a and b are called **addends**, and $a + b$ is called the **sum**. Since $0 + x = x$ and $x + 0 = x$, addition with 0 is quite simple.

To add two negative numbers together, we rely on the following property stated for any pair of real numbers a and b. We will be primarily interested in the cases where a and b are positive.

$$(-a) + (-b) = -(a + b) \qquad (1.2)$$

EXAMPLE 1   Use Equation (1.2) to add the following negative numbers.

(a) $(-5) + (-17) = -(5 + 17) = -22$

(b) $(-3) + (-2) = -(3 + 2) = -5$

(c) $\left(-\dfrac{1}{4}\right) + \left(-\dfrac{3}{4}\right) = -\left(\dfrac{1}{4} + \dfrac{3}{4}\right) = -1$

(d) $(-2) + \left(-\dfrac{1}{2}\right) = -\left(2 + \dfrac{1}{2}\right) = -2\dfrac{1}{2}$

The preceding example suggests the following method for using Equation (1.2) to add negative numbers. We can simply take advantage of the fact that every negative number is the "negative" of a positive number.

Chap. 1 The Real Numbers

> **Addition of Negative Numbers**
>
> If a and b are positive numbers, then
>
> $$(-a) + (-b) = -(a + b)$$

A similar procedure allows us to add three or more negative numbers.

EXAMPLE 2

(a) $(-2) + (-5) + (-1) = -(2 + 5 + 1)$
$= -8$

(b) $(-11) + (-3) + (-7) + (-20) = -(11 + 3 + 7 + 20)$
$= -41$

When adding a positive number and a negative number, we assume that if a and b are both positive and $b \geq a$, you already know how to subtract a from b. For example, $10 - 3 = 7$, $14 - 5 = 9$, and $21 - 15 = 6$.

> **Addition of a Positive and a Negative Number**
>
> To add a positive and a negative number, we
>
> 1. Use their absolute values,
> 2. Subtract the smaller absolute value from the larger, and
> 3. Attach the sign of the number with the larger absolute value to the result.

EXAMPLE 3

(a) $(-3) + 17 = 17 - 3$ ↓ Larger absolute value
                            ↑ Smaller absolute value
              $= +14$ Sign of the number with larger absolute value, $17 = +17$

(b) $(-12) + 5 = -(12 - 5)$ ↓ Larger absolute value
                               ↑ Smaller absolute value
              $= -7$ Sign of the number with larger absolute value, -12

To add three or more numbers

1. Add all the positive numbers together,
2. Add all the negative numbers together, and
3. Add the positive and negative results from (1) and (2). (See Example 4.)

The commutative and associative properties can be used to justify the preceding procedure.

EXAMPLE 4

$$4 + (-7) + 6 + (-3) + (-2) = \overbrace{(4 + 6)}^{\text{positive}} + \overbrace{(-7) + (-3) + (-2)}^{\text{negative}}$$
$$= 10 + (-12)$$
$$= -(12 - 10)$$
$$= -2$$

We will now define *subtraction* for any two real numbers a and b regardless of their relative size.

Subtraction

$$a - b = a + (-b)$$

To subtract b from a we add the negative of b to a. Alternately, we change the sign of the number to be subtracted and add.

EXAMPLE 5

(a) $(-2) - 5 = (-2) + (-5)$ Definition of subtraction
$ = -(2 + 5)$ Addition of negative numbers
$ = -7$

(b) $7 - 12 = 7 + (-12)$ Definition of subtraction
$ = -(12 - 7)$ Addition of a positive and a negative number
$ = -5$

(c) $-10 - (-3) = -10 + (-(-3))$ Definition of subtraction
$ = -10 + 3$ $-(-3) = 3$ from the Double Negative Property
$ = -(10 - 3)$ Addition of a positive and a negative number
$ = -7$

(d) $5 - |-2| = 5 - 2$ $|-2| = 2$
$ = 3$

WARNING Note that $-(3 + 5) \neq -3 + 5$. We could use Equation (1.2) and write $-(3 + 5) = (-3) + (-5)$ or simply work within the parentheses first and obtain $-(3 + 5) = -8$.

There is an agreement among mathematicians that $1 - 3 - 5$ means $(1 - 3) - 5$, which is -7. *In general, when insufficient grouping symbols (such as parentheses, brackets, absolute value bars, and so forth) are present, additions and subtractions are done in order, from left to right.*

EXAMPLE 6

(a) $-2 + (-7) - (-3) = [-2 + (-7)] - (-3)$ Add the first two numbers on the left
$= -9 - (-3)$
$= -9 + 3$
$= -6$

(b) $10 - (-1) - (-2) = [10 - (-1)] - (-2)$ Subtract the first two numbers on the left
$= [10 + 1] - (-2)$ $-(-1) = 1$
$= 11 - (-2)$
$= 11 + 2$ $-(-2) = 2$
$= 13$

(c) $4 + (-3) - (-6) - 13 = [4 + (-3)] - (-6) - 13$ Add the first two numbers on the left
$= 1 - (-6) - 13$
$= [1 - (-6)] - 13$ Subtract the first two numbers on the left
$= [1 + 6] - 13$ $-(-6) = 6$
$= 7 - 13$
$= -6$

(d) $-3 - [-2 + (-9)] - (-3) = -3 - (-11) - (-3)$ Work within grouping symbols first
$= [-3 - (-11)] - (-3)$ Subtract the first two numbers on the left
$= [-3 + 11] - (-3)$ $-(-11) = 11$
$= 8 - (-3)$
$= 8 + 3$ $-(-3) = 3$
$= 11$

A frequently used property involving subtraction is

$$-(a - b) = -a + b \qquad (1.3)$$

This result can be obtained as follows:

$$-(a - b) = -(a + (-b)) \quad \text{Definition of subtraction}$$
$$= (-a) + (-(-b)) \quad \text{by Equation (1.2)}$$
$$= -a + b \quad \text{Since } -(-b) = b$$

We depend upon this rule for numerical calculations such as $-(7 - 10) = -7 + 10 = 3$ and algebraic simplifications such as $-(2 - y) = -2 + y$. We will return to this topic in Chapter 2 and in other parts of the book.

We find an important use of subtraction when determining the distance between two points on a number line. First consider the following example.

EXAMPLE 7 Remember to first evaluate the expression inside the absolute value bars.

(a) $|7 - 1| = |6| = 6$
(b) $|1 - 7| = |-6| = 6$
(c) $|18 - 4| = |14| = 14$
(d) $|4 - 18| = |-14| = 14$

This example illustrates a basic fact about absolute value. For any real numbers a and b,

$$|a - b| = |b - a|$$

Let us now define the distance between two points on the number line with coordinates a and b.

$$\text{Distance between } a \text{ and } b = |a - b|$$

EXAMPLE 8 Find the distance between the points with the coordinates a and b where

(a) $a = -3$ and $b = 2$ Distance $= |-3 - 2| = |-5| = 5$ units

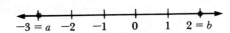

(b) $a = -3$ and $b = -5$ Distance $= |-3 - (-5)| = |5 - 3| = |2| = 2$ units

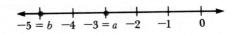

(c) $a = 2$ and $b = 7$ Distance $= |2 - 7| = |-5| = 5$ units

(d) $a = -\dfrac{1}{2}$ and $b = \dfrac{2}{3}$ Distance $= \left|-\dfrac{1}{2} - \dfrac{2}{3}\right| = \left|-\dfrac{3}{6} - \dfrac{4}{6}\right| = \left|-\dfrac{7}{6}\right| = \dfrac{7}{6}$ units

Exercises 1.4

Standard Assignment: Exercises 1, 3, 5, 7, 9, 11, 15, 19, 21, 23, 25, 31, 35, 37, 39, 41, 43

A Add as indicated. See Examples 1–4.

1. $5 + (-2)$
2. $7 + (-5)$
3. $6 + (-8)$
4. $-10 + 4$
5. $-12 + 8$
6. $-5 + (-6)$
7. $-3 + (-9)$
8. $-8 + 3$
9. $\dfrac{1}{2} + \left(-\dfrac{3}{8}\right)$
10. $-\dfrac{1}{2} + \left(-\dfrac{2}{7}\right)$
11. $(-3) + (-7) + (-1)$
12. $(-4) + (-11) + (-2)$
13. $(-10) + (-3) + (-16)$
14. $(-7) + (-5) + (-5)$
15. $(-9) + (-8) + (-4) + (-2)$
16. $(-13) + (-1) + (-6) + (-3)$
17. $(-6) + (-3) + 8$
18. $6 + (-2) + 3$
19. $(-7) + 2 + (-4)$
20. $(-8) + (-14) + 0 + 3 + (-16)$
21. $5 + (-9) + (-2) + (-11) + (-17)$

Subtract as indicated. See Example 5.

22. $22 - 25$
23. $20 - 18$
24. $(-1) - (-7)$
25. $(-7) - (-1)$
26. $3 - 6$
27. $-18 - 4$
28. $-\dfrac{1}{3} - \left(-\dfrac{2}{3}\right)$
29. $-\dfrac{3}{5} - 1$

30. $6 - (2 + 1)$
31. $6 - (2 - 1)$
32. $-2 - 7$
33. $|1 - |-3||$

Perform the indicated operations. See Example 6.

34. $22 - 17 - 20$
35. $-3 + 6 - 9$
36. $7 - 5 + (-3)$
37. $7 + 3 - 5$
38. $-10 - 2 - 3$
39. $-6 - 1 + 2$
40. $14 - (3 - 2) + 16$
41. $(-5 + 9) - 12 - (3 + 6)$
42. $(7 + 2 - 4) - (12 - 3)$
43. $-10 + 18 - (-3 - 4 + 1)$
44. $\frac{1}{2} - \left(\frac{3}{4} - \frac{1}{4}\right)$
45. $\frac{3}{8} - \frac{1}{2} - \left(\frac{3}{4} + \frac{1}{8}\right)$
46. ▦ $24.1723 - 1.6284$
47. ▦ $-123.47103 - 11.07249$
48. ▦ $0.141 - 3.7284 - 16.9854$
49. ▦ $-24.7032 + 32.9 - 2.40371$
50. ▦ $2.4675 - |-21.74326|$
51. ▦ $-108.342 + |12.06 - 0.74321|$

Find the distance between the points with coordinates a and b. See Example 8.

52. $a = -20$ and $b = -6$
53. $a = -4$ and $b = 18$
54. $a = -6$ and $b = 8$
55. $a = 0$ and $b = 14$
56. $a = 5$ and $b = 25$
57. $a = -9$ and $b = 9$
58. $a = \frac{22}{7}$ and $b = -\frac{4}{7}$
59. $a = \frac{16}{5}$ and $b = \frac{3}{5}$

Evaluate the following for $x = 1$, $y = 2$, and $z = -3$. Simplify your answer.

> EXAMPLE: $(x - y/2) - z$
>
> Solution: Replacing x by 1, y by 2, and z by -3, we have
>
> $$(1 - \frac{2}{2}) - (-3) = (1 - 1) - (-3)$$
> $$= 0 - (-3)$$
> $$= -(-3)$$
> $$= 3$$

60. $x - y - z$
61. $x - (y - z)$
62. $\frac{x - y}{2} - z$
63. $z - |y - x|$
64. $|-y - z + x|$
65. $|y + z| - x$
66. $2x - 3y + z - 1$
67. $2(x - 4y + z) - 1$

68. $3(x - |2y + z| - 1)$

69. $x - \dfrac{y + z}{2}$

70. $|x - 3y| - 2x - z$

71. $z - \dfrac{x - y}{2} - x$

B 72. Find a value for x such that $x - 3 \neq 3 - x$.

73. Find a value for x such that $(x - 3) - 2 \neq x - (3 - 2)$.

74. If subtraction were commutative, we could write $x - y = y - x$ for all real values of x and y. Is subtraction commutative?

75. If subtraction were associative, we could write $(x - y) - z = x - (y - z)$ for all real values of x, y, and z. Is subtraction associative?

1.5 Multiplication and Division

Multiplication of two real numbers a and b results in a third number ab or $a \cdot b$, which is called the **product** of a and b. Each of the numbers a and b is called a **factor** of the product. For example, since $3 \cdot 2 = 6$, 3 and 2 are each factors of the product 6. If two numbers are both positive or both negative we say they have **like signs**; if one is positive and the other is negative we say they have **unlike signs**. Thus, -3 and -5 have like signs, whereas -3 and 5 have unlike signs.

We assume you know how to multiply positive numbers. The multiplication of any two real numbers can be accomplished by using the following theorem for any pair of real numbers a and b. We will mostly be concerned with using these rules when a and b are positive.

$$(-1)a = -a$$
$$(-a)b = -(ab)$$
$$a(-b) = -(ab)$$
$$(-a)(-b) = ab \tag{1.4}$$

EXAMPLE 1
(a) $(-5) \cdot 7 = -(5 \cdot 7) = -35$
(b) $(-1) \cdot 13 = -13$
(c) $8 \cdot (-3) = -(8 \cdot 3) = -24$
(d) $\left(-\dfrac{1}{2}\right) \cdot (-2) = \left(\dfrac{1}{2}\right) \cdot 2 = 1$

(e) $(-6) \cdot (-4) = 6 \cdot 4 = 24$

(f) $(-1) \cdot \left(-\dfrac{3}{5}\right) = 1 \cdot \left(\dfrac{3}{5}\right) = \dfrac{3}{5}$

The "sign rule" for multiplication can be stated as: *The product of two numbers with like signs is positive; the product of two numbers with unlike signs is negative.* The rule is illustrated next.

Multiplication

$(+)(+) = (+)$

$(-)(-) = (+)$

$(+)(-) = (-)$

$(-)(+) = (-)$

The rule for a product of more than two numbers is: *If the number of negative factors in a product is even, the product is positive; if the number of negative factors in a product is odd, the product is negative.*

EXAMPLE 2 The product rule follows from the fact that each pair of negative factors results in a positive product.

(a) $(-3)(6)(8)(-2) = (6)(8)(-3)(-2) = 48 \cdot 6 = 288$

(b) $(-4)(3)(-3)(2)(-5) = (-4)(-3)(-5)(3)(2) = (-4)(15)(6) = -360$

In the previous section we defined subtraction in terms of addition. We now define *division* in terms of multiplication. For real numbers a and b with $b \neq 0$

$$\dfrac{a}{b} = q \quad \text{if and only if} \quad a = bq \ (b \neq 0)$$

The number q is called the **quotient,** b the **denominator** or **divisor,** and a the **numerator** or **dividend.** We sometimes write $a \div b$ instead of a/b.

EXAMPLE 3 (a) $\dfrac{8}{2} = 4$ since $8 = 2 \cdot 4$

(b) $\dfrac{-12}{3} = -4$ since $-12 = 3 \cdot (-4)$

(c) $\dfrac{10}{-5} = -2$ since $10 = (-5) \cdot (-2)$

(d) $\dfrac{-3}{-1} = 3$ since $(-3) = (-1) \cdot 3$

Why is there the restriction of $b \neq 0$? Consider this: if $a/0 = q$, then $a = q \cdot 0$. However, if $a \neq 0$, this is impossible since $q \cdot 0 = 0$. On the other hand, if $a = 0$, we get $0 = q \cdot 0$, which is true regardless of what value of q we use. Thus, no single number is selected for a quotient, and no unique quotient is defined. Consequently,

Division by zero is undefined.

Since division is defined in terms of multiplication, it is not surprising that the "sign rule" for multiplication carries over to division. *The quotient of two numbers with like signs is positive; the quotient of two numbers with unlike signs is negative.* The following theorem holds for all real numbers a and b, with $b \neq 0$.

$$\dfrac{a}{-b} = \dfrac{-a}{b} = -\dfrac{a}{b}$$

$$\dfrac{-a}{-b} = \dfrac{a}{b}$$

EXAMPLE 4

(a) $\dfrac{1}{-6} = \dfrac{-1}{6} = -\dfrac{1}{6}$ (b) $\dfrac{-2}{-5} = \dfrac{2}{5}$

(c) $\dfrac{-30}{5} = -\dfrac{30}{5} = -6$ (d) $\dfrac{-21}{-7} = \dfrac{21}{7} = 3$

The sign chart for division is also useful.

Division

$$\dfrac{(+)}{(+)} = (+) \qquad \dfrac{(+)}{(-)} = (-)$$

$$\dfrac{(-)}{(-)} = (+) \qquad \dfrac{(-)}{(+)} = (-)$$

Order of Operations

In the previous section, we discussed the convention for calculating sums and differences when incomplete grouping symbols were given. A similar agreement exists for evaluating expressions such as $2 \cdot 3 + 4$. Do we multiply 2 and 3 and then add 4, or do we add 3 and 4 and multiply the result by 2? The recognized agreements are given next.

Order of Operations

Simplify all expressions enclosed in grouping symbols, starting with the innermost group, using the following rules.
 First, multiplications and divisions are done in order as they are encountered from left to right.
 Second, additions and subtractions are done in order as they are encountered from left to right.
 Finally, when a fraction is encountered, complete all work above and below the bar first, and then simplify if possible.

EXAMPLE 5   Use the rules of operation to simplify the following equations.

(a) $2 \cdot 6 + 4 = 12 + 4$ First multiply from the left
         $= 16$

(b) $2(1 - 5) + 3(-6) + 2 = 2(-4) + 3(-6) + 2$ First work inside grouping symbols
         $= -8 + (-18) + 2$ Multiply from the left $2(-4) = -8$; $3(-6) = -18$
         $= [-8 + (-18)] + 2$ Add from the left
         $= -26 + 2$ $-8 + (-18) = -26$
         $= -24$

(c) $30 \div 6(-2) \cdot 4 \div 2 = [30 \div 6] \cdot (-2)4 \div 2$ First divide from the left
         $= 5 \cdot (-2) \cdot 4 \div 2$ $30 \div 6 = 5$
         $= [5(-2)] \cdot 4 \div 2$ Multiply from the left
         $= (-10) \cdot 4 \div 2$ $5(-2) = -10$
         $= [(-10)4] \div 2$ Multiply from the left
         $= -40 \div 2$ $(-10)4 = -40$
         $= -20$

(d) $\dfrac{(-2)\cdot 4 + (-1)\cdot(-6)}{-8+5\cdot 3} = \dfrac{-8+(-1)\cdot(-6)}{-8+15}$ Multiply from the left in the numerator and in the denominator

$= \dfrac{-8+6}{-8+15}$ Multiply from the left in the numerator

$= \dfrac{-2}{7}$

(e) $\dfrac{7 - 3\left(\dfrac{1+5}{2}\right) + 6}{1 - 3\cdot 2 + 8} = \dfrac{7 - 3\left(\dfrac{6}{2}\right) + 6}{1 - 6 + 8}$ Work inside the parentheses in the numerator; multiply from the left in the denominator

$= \dfrac{(7 - 9) + 6}{(1 - 6) + 8}$ Subtract from the left in the numerator and in the denominator

$= \dfrac{-2 + 6}{-5 + 8}$

$= \dfrac{4}{3}$

(f) $\dfrac{-14 + 6\left(\dfrac{2 - 7}{5}\right) + 3 - 8}{25 - 16 - 9} = \dfrac{-14 + 6(-1) + 3 - 8}{[25 - 16] - 9}$ Work inside the parentheses in the numerator; subtract from the left in the denominator

$= \dfrac{-14 + 6(-1) + 3 - 8}{9 - 9}$ $25 - 16 = 9$

$= \dfrac{-14 + 6(-1) + 3 - 8}{0}$ which is *undefined*!

Exercises 1.5

Standard Assignment: Exercises 3, 9, 11, 13, 17, 21, 23, 25, 29, 31, 33, 35, 43, 47, 55, 59.

A   Find the product or quotient as indicated. See Examples 1–3. In calculator problems, round to the nearest thousandth.

1. $(-3)2$
2. $5(-9)$
3. $(-10)(-8)$
4. $(-1)(-17)$
5. $3(-6)2$
6. $(-8)(-7)6$

7. $4(-9)(-2)$

8. $(-3)(-5)(-8)$

9. $\left(-\frac{1}{2}\right)16$

10. $-3\left(\frac{4}{9}\right)$

11. $(-7)\left(\frac{2}{14}\right)$

12. $\left(-\frac{7}{2}\right)\left(-\frac{3}{5}\right)$

13. $(-2)(-8)(8)(-5)$

14. $(-8)(2)(-4)5$

15. $7(-2)(-3)(-9)$

16. $(-2)(-1)(-7)(-4)(-3)$

17. $\frac{14}{7}$

18. $\frac{-6}{3}$

19. $\frac{6}{-3}$

20. $\frac{-6}{-3}$

21. $\frac{0}{18}$

22. $\frac{0}{-17}$

23. $\frac{13}{0}$

24. $\frac{-23}{0}$

25. $\frac{1}{3} \div \frac{2}{9}$

26. $\frac{-2}{25} \div \frac{12}{5}$

27. $-\left(\frac{4}{7}\right) \div \frac{2}{49}$

28. $-\frac{1}{2} \div \frac{-1}{3}$

29. $-\frac{4}{15} \div \frac{6}{-20}$

30. $\frac{8}{22} \div -\frac{5}{3}$

31. $\frac{2/15}{6/5}$

32. $\frac{-7/36}{14/3}$

33. ▦ $(-1.764)(-32.811)$

34. ▦ $(-3.72) \div (16.248)$

35. ▦ $(13.741) \div (-0.1824)$

36. ▦ $(-3.7609) \div (-4.82)$

Use the order of operation rules to do the following calculations. See Example 5. In calculator problems, round to the nearest thousandth.

37. $2 \cdot 7 + 5$

38. $2(-5) + 7$

39. $(-2) + 5$

40. $(-2)7 - 5$

41. $3[6 - (-2)]$

42. $(-3)[6 + (-2)]$

43. $3 + 7 \cdot 4 + 9$

44. $-3 + 7(-4) + (-9)$

45. $7 - 3(-4)$

46. $-6 - 2(-8)$

47. $-5 + (-3)(-2) - 16$

48. $5 - (-2)(-4) - 7$

49. $(-3)(-6) + 2(-7)$

50. $(-2)(-9) - (-3)(-4)$

51. $(-7 - 1)(3 + 8)$

52. $(-6 - 4)(-3 + 8)$

53. $3(-1) + 2 \cdot 2 - 6$

54. $7 - 6(5 - 2) + 4$

55. $12 - \dfrac{-16}{2} + 13$

56. $12 - \dfrac{16 - 8}{2} - 4$

57. $\dfrac{3(-7 - 2)}{13 - 4} - \dfrac{16 - 2 \cdot 5}{4 - (-2)}$

58. $\dfrac{14}{7 - 1} - \dfrac{8 - 16}{-4}$

59. $\dfrac{8(2 - 7) + 12}{13 - (-13)} + \dfrac{10}{-5}$

60. $\dfrac{14}{25} - \dfrac{2(18 - 6 + 2)}{7 - 1 + 3(-2)}$

61. $\dfrac{(-7)(2) + (-4)(-3) + 3 - 1}{-8 + 2(-32) + 7}$

62. $\dfrac{(-4)(5) + (-8)(-2) + 2(-1)}{(-2)7 + 3 + 4(-3)}$

63. $-3[4 + (1 - 7)] \div [12 \div (-3)]$

64. $14[3 + 3(1 - 2)] \div [32 \div 2(-3)]$

65. $(3 - 5)[5 + 32 \div 8 \div 4]$

66. $(7 - 2)[14 + 22 \div 2 \div (4 - 5 + 1)]$

67. 🖩 $(-6.71)(14.83) + 9.872$

68. 🖩 $(2.84)(-10.327) - 4.891$

69. 🖩 $(-16.433)(-2.81) - 72.328$

70. 🖩 $(-9.764)[(-12.103) - (-7.462)]$

71. 🖩 $\dfrac{-17.31 + 18.204}{2.347 - (7.2)(-3.74)}$

72. 🖩 $\dfrac{-21.382 + 14.794}{13.402 - (-1.06)(8.741)}$

73. 🖩 $\dfrac{21.43(-8.741) - 1.74(-2.74)}{-1.72 + 0.08(-1.609)}$

74. 🖩 $\dfrac{(-4.37)(13.62) - (-16.074)(-37.81)}{(14.393)(-6.27) - (-11.04)}$

Use the order of operation rules to do the following calculations. See Example 5.

75. $\left\{5 - (-2)\left[\dfrac{3 + (-7)}{5(-1)} + \dfrac{12}{6}\right]\right\} - 1$

76. $\left\{2 - 7\left[\dfrac{1 - (-8)}{5 - 3} + \dfrac{8}{2}\right]\right\} + 6$

77. $\dfrac{-3 + \left(\dfrac{6 - 9}{8 + 2 - 7}\right) - 3(-4)}{2\left(\dfrac{-3 - 1}{-5 + 23}\right) - 3\left(\dfrac{8 - 2}{-3 - 1}\right)}$

78. $\dfrac{5 - \left(\dfrac{7 - 4}{-2 + 8 - 7}\right) - (-2)(-6)}{4\left(\dfrac{8 - 2}{9 - 23}\right) - (-2)\left(\dfrac{12 - 8}{4 - 3 + 1}\right)}$

Evaluate each expression for $x = 3$, $y = -5$, and $z = -2$.

EXAMPLE: $\dfrac{2x - 3y + z}{2z - x}$

Solution: Substituting we have

$$\frac{2(3) - 3(-5) + (-2)}{2(-2) - 3} = \frac{[6 - (-15)] - 2}{-4 - 3}$$

$$= \frac{21 - 2}{-7}$$

$$= -\frac{19}{7}$$

79. $2(x + y) - 3y - 2z$

80. $4z - 2(x + y) - 3y$

81. $\dfrac{x - 7y}{z} + xy - 1$

82. $\dfrac{2x - y}{xz} - \dfrac{y}{z} + 3$

83. $\dfrac{2(z - 2x)}{y} - (-x)y + z$

84. $\dfrac{3(x + 2y)}{z} + (-y + 1)z$

85. $\dfrac{4x - 3[z \div 2(x + 1)]}{3 \div 7z - 3}$

86. $\dfrac{2(x - y) - 5(2z \div 3y + z)}{5 - xy \div 3}$

In Exercises 87–90 find an appropriate value for x.

| EXAMPLE: Find a value for x such that $\dfrac{x + 3}{3} \neq x$. Take $x = 0$, then $\dfrac{0 + 3}{3} \neq 0$.

B 87. Find a value for x such that $2(3x) \neq (2 \cdot 3)(2x)$.

88. Find a value for x such that $2(x + 1) \neq 2x + 1$.

89. Find a value for x such that $x \div 3 \neq 3 \div x$.

90. Find a value for x such that $(4 \div x) \div 2 \neq 4 \div (x \div 2)$.

91. If division were commutative, we would have $x \div y = y \div x$ for all nonzero values of x and y. Is division commutative?

92. If division were associative, we would have $(x \div y) \div z = x \div (y \div z)$ for all nonzero values of y and z. Is division associative?

Chapter 1 Summary

Key Words and Phrases

1.1 Set
Element
Member

Natural numbers
Empty set
Variable

Constant
Set-Builder notation
Subset
Whole numbers
Integers
Rational numbers
Irrational numbers
Real numbers
1.2 Equality
Number line
Graph of a number
Coordinate of a point
Inequality
1.3 Theorem

Proof
Negative
Reciprocal
Opposite
Additive inverse
Multiplicative inverse
Absolute value
1.4 Addend
Sum
1.5 Product
Factor
Like Signs
Unlike Signs

Key Concepts and Rules

Equality Properties: If a, b, and c are real numbers, then: (1) $a = a$ (Reflexive Property); (2) If $a = b$, then $b = a$ (Symmetric Property); (3) If $a = b$ and $b = c$, then $a = c$ (Transitive Property); and (4) If $a = b$, then b may replace a anywhere a occurs in a statement without affecting the truth of the statement (Substitution Property).

Inequality Properties: If a, b, and c are real numbers, then: (1) Exactly one of the following is true: $a < b$, $a = b$, $a > b$ (Trichotomy Property); and (2) If $a < b$ and $b < c$, then $a < c$ (Transitive Property).

Double Negative Property: For any real number a, $-(-a) = a$.

Zero Factor Property: For any real number a, $a \cdot 0 = 0$.

Addition Property of Equality: If a, b, and c are real numbers and if $a = b$, then $a + c = b + c$.

Multiplication Property of Equality: If a, b, and c are real numbers and if $a = b$, then $ac = bc$.

Sign Properties: If a and b are real numbers, then (1) $(-a) + (-b) = -(a + b)$; (2) $(-a)(b) = -(ab)$; (3) $a(-b) = -(ab)$; (4) $(-a)(-b) = ab$; and (5) $-a/-b = a/b$.

Review Exercises

If you have difficulty with any of the problems, look in the section indicated by the number in square brackets.

[1.1] List the elements in the given set.

1. $\{x \mid x \in J \text{ and } x \text{ is greater than } -3 \text{ and less than } 3\}$
2. $\{x \mid x \in J, x \text{ is negative, and } x \text{ is greater than } -10\}$
3. $\{x \mid x \in J \text{ and } x = 1\}$

Let
$$P = \left\{-8, -4, -2, -\frac{3}{2}, 0, 2, \frac{5}{2}, 7, 9\right\}$$

List the elements of the following sets.

4. $\{x \mid x \in P \text{ and } x \in N\}$
5. $\{x \mid x \in P \text{ and } x \in J\}$
6. $\{x \mid x \in P \text{ and } x \in Q\}$
7. $\{x \mid x \in P \text{ and } |x| > 3\}$

Place a $\subset$ or $\in$ in the blanks to produce a true statement.

8. $\{-1\}$ ____ $\{-1, 3, 5\}$
9. 2 ____ $\{2, 4, 6, 8\}$

[1.2] Graph each set on a number line.

10. $\left\{-2, -\frac{1}{2}, 1, \frac{5}{3}\right\}$
11. $\left\{\frac{3}{10}, \frac{13}{10}, \frac{23}{10}\right\}$

Use inequality symbols to write the given statement symbolically.

12. x is positive
13. $|x|$ is not less than x
14. $4y$ is less than or equal to $3x + 1$
15. $z - 5$ is greater than -1 and less than 14

Identify the properties that justify each of the following statements.

16. $-5 = -5$
17. If $3x = 7$ and $2y = 7$, then $3x = 2y$
18. If $x > 3t + 1$ and $t = 2y$, then $x > 3(2y) + 1$

[1.3] Specify the real number properties that justify the following statements.

19. $-8 + 3(t + 2) = 3(t + 2) + (-8)$
20. $-(-2x) = 2x$
21. $3(x + 7) = (x + 7)3$
22. If $x = -5$, then $x + 2 = -5 + 2$

23. If $2(x + 1) = 4$, then $x + 1 = 2$

24. $(2x + 1)(5 + y) = (2x + 1)5 + (2x + 1)y$

Give the negative and the reciprocal of each number.

25. $\dfrac{1}{5}$ 26. $-1\dfrac{1}{2}$ 27. $x^2 + 1$

Rewrite each of the following without using absolute value notation.

28. $\left|-\dfrac{3}{10}\right|$ 29. $|7 - (-5)|$

30. $-\left|2 - \dfrac{1}{2}\right|$ 31. $|-3 - 2|$

[1.4] Add or subtract as indicated.

32. $7 + (-3)$ 33. $-14 + (-6)$ 34. $-3 - (-5)$

35. $-8 - 7 + 9$ 36. $(-17 + 12 - 9) - (3 - 18)$

37. $(-4) - (-1) - |6|$ 38. $|-20| - (13 - 7 - 4) + 3$

Find the distance between the points with coordinates a and b.

39. $a = -3$ and $b = 7$ 40. $a = -4\dfrac{1}{2}$ and $b = -1$

41. $a = \dfrac{2}{5}$ and $b = \dfrac{13}{7}$

[1.5] Find the products or quotients as indicated.

42. $(-6)8$ 43. $(-4)(-7)$ 44. $(-2)(-5)(-10)(-4)$

45. $\left(-\dfrac{1}{2}\right)(-4)(3)(2)(-5)$ 46. $\dfrac{18}{-3}$

47. $\dfrac{-42}{-7}$ 48. $-\dfrac{3}{5}$

49. $-\dfrac{2}{3} \div -\dfrac{2}{6}$ 50. $3\dfrac{1}{2} \div (-2)$

Use the order of operation rules to do the following calculations.

51. $5(-8) + (-6)$ 52. $(5 - 3) \cdot [8 - (-2)(-4)]$

53. $14 - 3(7 - 4) + 3$ 54. $13 - 3 \cdot |-5 + 17| + 2$

55. $-(4 - |6| - |-2|) + 3 \cdot 7 - |-3| \cdot |-8|$

56. $2(-3) + 9 \div 3 - 2$

57. $24 \div 8 - |27 - 16 + 3| \div 7 - 1$

58. $\dfrac{-9 + 4(11)}{7}$ 　　　　　　　　59. $\dfrac{-16 - 3(2)}{9}$

60. $\left[\dfrac{12(2 - 5)}{6} - 8 + 3 - 1\right] \div \dfrac{2 + 3(28)}{4 - 16 - 9}$

Evaluate each expression for $x = 0$, $y = -3$, $z = 1$, and $w = 4$.

61. $xy - zw + 3y$ 　　　　　　　　62. $2w + z - x \div y$

63. $\dfrac{2x - 3y}{7 - y}$ 　　　　　　　　64. $2w \div 8z \div x \div 4yz - w$

65. $\dfrac{2z - w + y}{6z + 2}$ 　　　　　　　　66. $\dfrac{25(x - y) + 50z}{12x + 15z}$

Practice Test (60 minutes)

1. List the elements in the following sets.
   (a) $\{x \mid x \in J \text{ and } 2 < x \leq 5\}$
   (b) $\{x \mid x \in N \text{ and } x \text{ is a whole number multiple of } 3\}$
   (c) $\left\{x \mid \dfrac{x}{2} \in J \text{ and } x \cdot x < 10\right\}$

2. Place a $\subset$, $\not\subset$, $\in$, or $\notin$ in the blank to produce a true statement.
   (a) $3 \underline{} \{-1, 0, 1, 2, 3\}$ 　　(b) $\emptyset \underline{} \{1\}$
   (c) $-1 \underline{} \{1, 2, -2\}$ 　　(d) $\{-1, 1\} \underline{} N$

3. Graph each of the following sets on a number line.
   (a) $\{-2, -1, 1, 3\}$ 　　　　　　(b) $\{-3/4, 0, 1/2\}$

4. Identify the properties that justify each of the following statements.
   (a) If $x = 3y$ and $4x + 7y = 10$, then $4(3y) + 7y = 10$.
   (b) If $x - 9 = 48$, then $48 = x - 9$.
   (c) $b = b$
   (d) If $x = 10$ and $10 = 2y - 14$, then $x = 2y - 14$.
   (e) Exactly one of the following statements is true: $10 < 121$, $10 = 121$, or $10 > 121$.

5. Indicate the real number property that justifies the following statements.
   (a) $-5 + (15 + 9) = (-5 + 15) + 9$ 　　　　_____
   (b) $6(x + 3) = 6x + 6 \cdot 3$ 　　　　_____
   (c) $5 \cdot \left(\dfrac{1}{5}\right) = \left(\dfrac{1}{5}\right) \cdot 5 = 1$ 　　　　_____
   (d) $5 \cdot \left(\dfrac{1}{5} \cdot 6\right) = \left[5 \cdot \left(\dfrac{1}{5}\right)\right] \cdot 6$ 　　　　_____
   (e) If $5x = 30$, then $x = 6$ 　　　　_____

6. Find the results of the following operations.
   (a) $(-6 + 2) - 7 - (3 + 4)$
   (b) $-(-(-(-1)))$
   (c) $\dfrac{2}{3} \div \dfrac{4}{3} \div \dfrac{1}{2}$
   (d) $(-3)(-2)(-5)$
   (e) $[7 - (-2)(-3)] \div [5 + (-6) + 1]$
   (f) $7 - (3 - 2)$

7. Remove the absolute value sign from each of the following and simplify if possible.
   (a) $|-5 + 2(3 - 4)|$
   (b) $|-5 \cdot 3 + (-1)(-2)|$

8. Evaluate each expression for $x = 1$, $y = -1$, $z = 0$, and $w = -2$.
   (a) $\dfrac{xy + yz - 4wx}{3y - 2x + w}$
   (b) $\dfrac{13(x - y) - 5(w - z)}{x + y + z + w}$

Extended Applications

Sunscreens

The first noticeable reaction of the skin to exposure to the ultraviolet rays in sunlight is a slight reddening. This reddening indicates that a person has received his or her "minimal erythema dose," or MED, and the amount of exposure time required for this dose should be noted. The FDA suggests that a daily exposure of about double your minimal erythema dose is the fastest way to get a tan.

Sunscreen products selectively absorb radiation so that you can remain in the sunlight for a longer period of time before receiving your minimal erythema dose. The "sunscreen protection factor," or SPF, of a sunscreen product gives an indication of how much longer you can remain in the sun when using the product before receiving your minimal erythema dose.

If, for example, you receive your minimal erythema dose in 20 minutes without protection and apply a sunscreen with an SPF of 6, it will take 6 times 20 minutes, or 2 hours, for your skin to redden slightly. Thus, indicating the time for minimal erythema dose without a sunscreen by MED, we have for a given sunscreen:

$$\text{Recommended Exposure Time} = 2 \times \text{SPF} \times \text{MED}$$

Exercises

1. Find the recommended exposure time for a person with a MED of 30 minutes using a sunscreen with SPF = 8.

2. Find the recommended exposure time for a person with a MED of 22 minutes using a sunscreen with SPF = 6.

3. What SPF is required by a person who wants to sunbathe for 4 hours and has a MED of 40 minutes?

4. What SPF is required by a person who wants to sunbathe for 5 hours and has a MED of 18 minutes?

Calories and Dieting

Your body requires energy to function and your diet supplies your body with that required energy. If your food intake provides more energy than your body needs to function, the excess energy is stored as fat. The unit most commonly used to measure the energy provided in food is the *calorie*.

Individuals vary in the number of calories they require to maintain their body weight. This number can be determined by recording the number of calories consumed in a week where no weight is gained or lost. Once this number is determined, a diet can be formulated for weight gain or weight loss by using the following:

> Add 3500 calories to the maintenance weight number (MWN) for each pound of weight you wish to gain per week, or
>
> Subtract 3500 calories from the MWN for each pound of weight you wish to lose per week.

> EXAMPLE: John discovers that his weight remains unchanged if he consumes 16,000 calories per week. During the next four weeks John wants to lose 8 pounds. What should his calorie intake be per week?
>
> John needs to lose 2 pounds per week so he must cut down by 2 × 3500 (= 7000) calories per week. Thus he can consume 9000 calories per week, or about 1,285 calories per day.

Exercises

1. Mary determined her MWN to be 12,600 calories. She wants to lose 5 pounds during the next four weeks. How many calories per day can she consume?

2. Larry determined his MWN to be 15,400 calories. He wants to lose a pound and a half per week. What should his daily caloric intake be?

3. Mike determined his MWN to be 19,600 calories. He wants to gain 10 pounds during the next three months (about 12 weeks). What should his daily caloric intake be?

2

First-Degree Equations and Inequalities

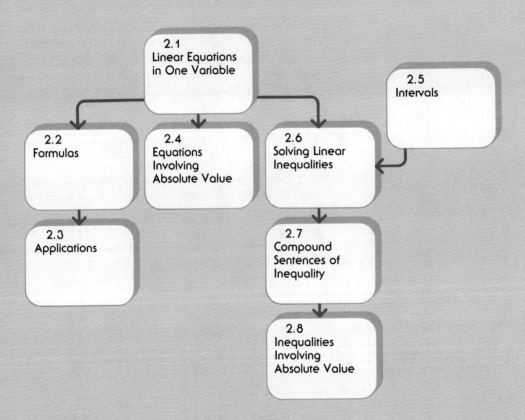

2.1 Linear Equations in One Variable

A sentence that connects two mathematical expressions by means of an equal sign is called an **equation**. If a true statement results when the variable in an equation is replaced by a particular number, that number is called a **solution**, or **root** of the equation. For example, 3 is a solution of the equation $2x - 6 = 0$, because replacing x with 3 results in $2(3) - 6 = 0$, which is a true statement. To solve an equation means to find its **solution set**, the set of all solutions of the equation. As we will soon see, 3 is the only solution of $2x - 6 = 0$, and thus the set $\{3\}$ is the solution set of $2x - 6 = 0$.

 A **linear equation in one variable** such as x is one that can be written in the form $ax + b = c$ where a, b, and c are real numbers, with $a \neq 0$.

EXAMPLE 1 Simplify each side of the following equations so that the resulting equation has the form of a linear equation. Identify the variable and a, b, and c.

(a) $12x + 4x + 9 = 13 + 28$
(b) $3y - 5 + 4y - 2 = 0$
(c) $2(m - 3) + 4 - 4m = -7 + 11$
(d) $8 - (5y - 2) = 2$

Solution: We use the commutative, associative, and distributive properties (see Section 1.3) to simplify the left-hand side of each of the previous equations.

$$\text{Since } 12x + 4x = (12 + 4)x \quad \text{Distributive property}$$
$$= 16x$$

we have

(a) $16x + 9 = 41 \qquad a = 16, \ b = 9, \ c = 41$
    $\ \downarrow \quad\ \ \downarrow \quad\ \ \downarrow$
    $\ ax\ +\ b\ =\ c \qquad$ The variable is x.

Next,

$$3y - 5 + 4y - 2 = 3y + 4y - 5 - 2$$
$$= (3 + 4)y - 7 \qquad \text{Distributive property}$$
$$= 7y + (-7)$$

(b) $7y + (-7) = 0 \qquad a = 7, \ b = -7, \ c = 0$
    $\ \downarrow \quad\quad \downarrow \quad\ \ \downarrow$
    $\ ay\ +\ b\ =\ c \qquad$ The variable is y.

Distributing gives $2(m - 3) = 2m - 6$. Consequently,
$$2(m - 3) + 4 - 4m = -7 + 11$$
becomes
$$2m - 6 + 4 - 4m = -7 + 11$$
$$2m - 4m - 2 = 4 \quad \text{Add and regroup}$$
$$(2 - 4)m + (-2) = 4 \quad \text{Distributive property}$$

(c) $\quad -2m + (-2) = 4 \quad a = -2, \ b = -2, \ c = 4$
$\qquad \downarrow \qquad \downarrow \quad \ \downarrow$
$\qquad am \ + \ b \ = \ c \quad$ The variable is m

We know
$$-(5y - 2) = -5y + 2 \quad \text{By Equation (1.3) in Section 1.4}$$
So
$$8 - (5y - 2) = 8 - 5y + 2$$
$$= -5y + 10$$
Thus

(d) $\quad -5y + 10 = 2 \quad a = -5, \ b = 10, \ c = 2$
$\quad \ \ \downarrow \quad \ \ \downarrow \quad \ \downarrow$
$\quad \ \ ay \ + \ b \ = \ c \quad$ The variable is y

WARNING

Correct Method	Incorrect Method
$2(m - 3) = 2m - 2 \cdot 3$	$2(m - 3) \neq 2m - 3$

The usual method for solving equations is to produce successively simpler equations with identical solution sets. The process stops when an equation is produced whose solution set is obvious. Equations with identical solution sets are called **equivalent equations.** Thus, $x + 3 = 5$ and $x = 2$ are equivalent equations, each having solution set $\{2\}$.

The addition and multiplication properties of equality from Section 1.3 can be used to justify the following useful methods of producing equivalent equations.

An equivalent equation results from a given equation if we add the same expression to both sides of the equations, and multiply both sides of the equation by the same nonzero expression.

These rules are stated symbolically for expression P, Q, and R containing the same variable (and only one variable).

2.1 Linear Equations in One Variable

For expression P, Q, and R, the equation
$$P = Q$$
is equivalent to
$$P + R = Q + R$$
and to
$$P \cdot R = Q \cdot R \quad \text{if } R \neq 0$$

EXAMPLE 2   Solve $4x + 7 = 2x - 1$.

Solution:
$$4x + 7 + (-2x) = 2x - 1 + (-2x) \quad \text{Add } -2x \text{ to both sides}$$
$$2x + 7 = -1$$
$$2x + 7 + (-7) = -1 + (-7) \quad \text{Add } -7 \text{ to both sides}$$
$$2x = -8$$
$$\tfrac{1}{2}(2x) = \tfrac{1}{2}(-8) \quad \text{Multiply by } \tfrac{1}{2}$$
$$x = -4$$

The solution set is $\{-4\}$. We check to make sure that -4 is a solution to the original equation by direct substitution.

Check:
$$4(-4) + 7 = 2(-4) - 1 \quad ?$$
$$-16 + 7 = -8 - 1 \quad ?$$
$$-9 = -9 \quad \checkmark$$

EXAMPLE 3   Solve $6y - (7y - 5) - 3 = 6 - (2y + 3)$.

Solution: Begin by simplifying each side by removing parentheses.

$$6y - 7y + 5 - 3 = 6 - 2y - 3$$
$$-y + 2 = 3 - 2y \quad \text{Commute and subtract}$$
$$2y - y + 2 = 3 - 2y + 2y \quad \text{Add } 2y \text{ to both sides}$$
$$y + 2 = 3$$
$$y + 2 + (-2) = 3 + (-2) \quad \text{Add } -2 \text{ to both sides}$$
$$y = 1$$

The solution set is $\{1\}$. We check by substituting 1 for y in the original equation.

Check:
$$6(1) - [7(1) - 5] - 3 = 6 - [2(1) + 3] \quad ?$$
$$6 - (7 - 5) - 3 = 6 - (2 + 3) \quad ?$$
$$6 - 2 - 3 = 6 - 5 \quad ?$$
$$4 - 3 = 1 \quad ?$$
$$1 = 1 \quad \checkmark$$

WARNING

Correct Method	Incorrect Method
$-(7y - 5) = -7y + 5$	$-(7y - 5) \neq -7y - 5$
$-(2y + 3) = -2y - 3$	$-(2y + 3) \neq -2y + 3$

EXAMPLE 4 Solve $\dfrac{-2k}{3} = k + \dfrac{k}{2}$.

Solution: Eliminate the fractions to make the remaining work simpler.

$$6\left(\frac{-2k}{3}\right) = 6\left(k + \frac{k}{2}\right) \quad \text{Multiply by 6 (the smallest integer multiple of both 3 and 2)}$$

$$-4k = 6k + 3k \quad \text{Simplify}$$

$$-4k + 4k = 6k + 3k + 4k \quad \text{Add } 4k \text{ to both sides}$$

$$0 = 13k$$

$$\frac{1}{13}(0) = \frac{1}{13}(13k) \quad \text{Multiply by } \frac{1}{13}$$

$$0 = k \quad \text{Zero Factor Property}$$

The solution set is $\{0\}$. Check by substituting 0 for k in the original equation.

Check:
$$\frac{-2(0)}{3} = 0 + \frac{0}{2} \quad ?$$
$$0 = 0 + 0 \quad ?$$
$$0 = 0 \quad \checkmark$$

EXAMPLE 5 Solve $5 - \dfrac{6}{m - 3} = \dfrac{-2m}{m - 3}$

Solution: We attempt to eliminate the fractions as in the previous example.

$$(m - 3)\left[5 - \frac{6}{m - 3}\right] = \left[\frac{-2m}{m - 3}\right](m - 3) \quad \text{Multiply by } m - 3$$

$$(m - 3)5 - 6 = -2m \quad \text{Distribute } m - 3 \text{ and simplify}$$

$$5m - 15 - 6 = -2m$$

$$5m + 2m - 21 = -2m + 2m \qquad \text{Add } 2m \text{ to both sides}$$

$$7m - 21 = 0$$

$$7m - 21 + 21 = 0 + 21 \qquad \text{Add 21 to both sides}$$

$$7m = 21$$

$$\frac{1}{7}(7m) = \frac{1}{7}(21) \qquad \text{Multiply by } \frac{1}{7}$$

$$m = 3$$

Check: Checking $m = 3$ in the original equation we have

$$5 - \frac{6}{3-3} = \frac{-2(3)}{3-3} \quad ?$$

$$5 - \frac{6}{0} = \frac{-6}{0}$$

This is a nonsense sentence because division by zero is undefined. The apparent solution, 3, is *not* a solution; thus there is no number m that solves the equation. In this case we write that the solution set is $\emptyset$.

What went wrong in Example 5? When we multiplied by $m - 3$, we presumed that $m - 3 \neq 0$, or $m \neq 3$. Thus, for $m = 3$, we violated the condition that the *multiplier* (in this case, $m - 3$) for an equation should not be 0.

Whenever we multiply both sides of an equation by an expression containing a variable, we must check any apparent solution that results to be sure it is not a root of the multiplier. It is always a good idea to check solutions to detect arithmetic or algebraic errors.

Steps for Solving a Linear Equation in One Variable

1. Eliminate any fractions by multiplication.
2. Add appropriate expressions to both sides so that all addends containing the variable are on one side and the numbers are on the other side.
3. Simplify both sides of the equation.
4. Isolate the variable so that only it appears on one side.
5. Check by substituting the apparent solution into the original equation.

Exercises 2.1

Standard Assignment: Exercises 1, 3, 5, 7, 11, 13, 15, 17, 19, 21, 23, 25, 27, 29, 31, 33, 39, 41, 43, 45, 47, 49, 51, 53, 57, 59

A Simplify each side of the following equations so that the resulting equation has the form of a linear equation. See Example 1.

1. $2x - 4x = -3$
2. $7x - 2x + 1 = 4$
3. $6y + 5y - 3 = 14$
4. $8y - 4y + 2 = -5$
5. $2(p + 1) - 5p = 0$
6. $9p - 3(p + 2) = 16 - 7$
7. $3 - 3[3m - 2(1 - m)] = 4$
8. $3(m - 1) - 2(m - 4) = -11$
9. $z - 2(z + 1) - 3(2 - 3z) = 14 - 8$
10. $-4z + 2[3 - (2 - 6z) + 12] = 21 - 42$

Solve the following equations. See Examples 2 and 3.

11. $3x + 5 = 14$
12. $2x - 17 = 7$
13. $10x = 32$
14. $0.2x = 3$
15. $3 - y = -4$
16. $2 - 7y = 23$
17. $7x + 7 = 2(x + 1)$
18. $3(x + 2) = 4 - x$
19. $3(2 - y) + 5y = 3y$
20. $9y - 3(y - 1) = 6 + y$
21. $k - (4 - k) = 2$
22. $2k + 11 = 2 - (k - 3)$
23. $4y - 3y + 7 - y = 2 - (7 - y)$
24. $3(y - 1) = 6y - 4 + 2y - 4y$
25. $3(z - 2) + 2(3 - z) = 1$
26. $2z - 3 - (3z - 1) = 6$
27. $2x + 3(x - 4) = 7x + 10$
28. $3(2 - 3x) - 4x = 3x - 10$
29. $4[k + 2(3 - k)] = 2k + 1$
30. $3 - [k - 3(k + 2)] = 4$
31. $3(4y - 3) = 4[y - (4y - 3)]$
32. $5 - (6y + 9) + 2y = 2(y + 1)$
33. $2z - 3(2 - z) = (z - 3) + 2z + 1$
34. $5(z - 3) - 6(z - 4) = -5$
35. ▦ $-(3.602x + 4.091) = 1.743$
36. ▦ $8.629 - 3.71(4.643x - 27.05) = 6.17$
37. ▦ $5.222y - (7.416 - 2.774y) + 13.021 - 6.827y = -9.401y$
38. ▦ $3.748y - 2.167(7.224y - 16.841) = 27.438 - 20.94y$

Solve the following equations. See Examples 4 and 5.

39. $\frac{7}{3}x = -49$
40. $-\frac{2}{5}x + 25 = 0$
41. $1 - \frac{2}{5}x = 0$
42. $3 + \frac{1}{5}x = 7$
43. $\frac{y}{2} - \frac{y}{3} = 2$
44. $\frac{y}{5} - 1 = \frac{y}{2}$

45. $\dfrac{2y}{3} + \dfrac{3y}{5} = 19$

46. $\dfrac{5y}{2} - \dfrac{y}{6} = \dfrac{7}{2}$

47. $2z + \dfrac{z}{3} = 1 - \dfrac{z}{4}$

48. $\dfrac{z}{5} - \dfrac{3z}{10} + \dfrac{5z}{2} = \dfrac{3}{10}$

49. $3z + \dfrac{z}{5} - 2 = \dfrac{1}{10} + 2z$

50. $\dfrac{1}{4} + 5z - \dfrac{z}{7} = \dfrac{5z}{14} + \dfrac{z}{2}$

51. $\dfrac{k+2}{3} - \dfrac{k}{2} = 7$

52. $\dfrac{k-1}{3} + \dfrac{2k}{5} = k + 2$

53. $\dfrac{2x+1}{9} - \dfrac{x+4}{6} = 1$

54. $\dfrac{2-3x}{7} + \dfrac{x-1}{3} = \dfrac{3x}{7}$

55. $\dfrac{1-x}{4} + \dfrac{5x+1}{2} = 3 - \dfrac{2(x+1)}{8}$

56. $\dfrac{x+4}{3} + 2x - \dfrac{1}{2} = \dfrac{3x+2}{6}$

57. $\dfrac{3}{x} = 9$

58. $\dfrac{-3}{x} = \dfrac{1}{3}$

59. $\dfrac{5}{x-2} = 1$

60. $\dfrac{-3}{12-x} = 0$

61. $\dfrac{4}{y-1} - \dfrac{1}{2} = 3$

62. $\dfrac{2}{3} + \dfrac{1}{y+3} = \dfrac{1}{2}$

63. $2 - \dfrac{5y}{y-2} = \dfrac{10}{2-y}$

64. $\dfrac{4y}{2-y} - 4 = \dfrac{-8}{y-2}$

65. $\dfrac{1}{y-3} + \dfrac{4}{2y-1} = \dfrac{3}{y-2}$

66. $\dfrac{4}{t-1} = \dfrac{3}{t-4} + \dfrac{6}{4-t}$

B EXAMPLE: Find a value of k so that $3x - 1 = k$ and $7x + 2 = 16$ are equivalent.

Solution: Solve $7x + 2 = 16$ Solve $3x - 1 = k$

$\qquad 7x = 14 \qquad\qquad\qquad 3x = k + 1$

$\qquad x = 2 \qquad\qquad\qquad\quad x = \dfrac{k+1}{3}$

$\{2\}$ is the solution set. $\left\{\dfrac{k+1}{3}\right\}$ is the solution set.

We want identical solution sets, so

$$\dfrac{k+1}{3} = 2$$

$$k + 1 = 6$$

or $k = 5$

67. Find a value of k so that $2x + 4 = k$ and $8x - 1 = 7$ are equivalent.
68. Find a value of k so that $2y - 3 = k$ and $5y + 2 = 15$ are equivalent.
69. Find a value of k so that $5x + k = 2(x - 1)$ has solution set $\{-3\}$.
70. Find a value of k so that $\dfrac{5}{y - 4} = \dfrac{6}{y + k}$ has solution set $\{9\}$.

2.2 Formulas

Solving problems that arise from everyday experiences is one of the most important uses of algebra. The equations that express relationships between variables representing quantities of a practical nature are called **formulas**.

EXAMPLE 1   The formula for converting (approximately) liters, l, to gallons, g, is

$$g = (0.264)l$$

If the gas pump shows 50 liters ($l = 50$), you have pumped

$$0.264(50) = 13.2 \text{ gallons}$$

EXAMPLE 2   The formula for converting the temperature in degrees Celsius (C) to Fahrenheit (F) is

$$F = \frac{9}{5}C + 32 \quad \text{or} \quad F = 1.8(C) + 32$$

If the time-temperature display at the bank shows 23° Celsius ($C = 23$), the temperature is

$$\frac{9}{5}(23) + 32 = 73.4° \text{ Fahrenheit}$$

EXAMPLE 3   Suppose the temperature is 86° Fahrenheit. What is the temperature in degrees Celsius?

Solution: First, substitute 86 for F in the formula $F = \frac{9}{5}C + 32$, then use the methods shown in Section 2.1 to solve for C.

$$86 = \frac{9}{5}C + 32$$

$$86 - 32 = \frac{9}{5}C$$

2.2 Formulas

$$\frac{5}{9}(54) = C$$

$$30 = C$$

Thus, 86°F converts to 30°C.

In the previous examples we specified that $F = 86$ and solved for C. A more general result could have been obtained by assuming that F was a fixed number, or constant, and solving for C. This process is called **solving for a specified symbol (or variable)**.

EXAMPLE 4 Solve $F = \frac{9}{5}C + 32$ for C. We treat F as a number and proceed as before.

Solution:

$$F - 32 = \frac{9}{5}C \qquad \text{Add } -32 \text{ to both sides}$$

$$\frac{5}{9}(F - 32) = C \qquad \text{Multiply by } \frac{5}{9}$$

or

$$C = \frac{5}{9}(F - 32)$$

EXAMPLE 5 The equation for the volume, V, of a rectangular box of height, h, length, l, and width, w, is

$$V = hlw$$

Find the formula for the height of a box of known volume, length and width.

Solution: Treating V, l, and w as (nonzero) numbers

$$\left(\frac{1}{lw}\right)V = hlw\left(\frac{1}{lw}\right) \qquad \text{Multiply by } \frac{1}{lw}$$

$$\frac{V}{lw} = h \qquad \text{Simplify}$$

or

$$h = \frac{V}{lw}$$

EXAMPLE 6 The formula $I = prt$ gives the simple interest, I, in terms of the principal invested, p, the yearly rate of interest, r, and the time in years, t.

(a) Solve for t.

(b) How long must you leave $3000 on deposit to earn $60 if the interest rate is 8%?

Solution:
(a) $\quad I = prt$

$$\left(\frac{1}{pr}\right)I = prt\left(\frac{1}{pr}\right) \qquad \text{Multiply by } \frac{1}{pr}$$

$$\frac{I}{pr} = t$$

(b) Since the rate is 8%, or $\frac{8}{100}$, $r = 0.08$, $p = 3000$, and $I = 60$.

$$\frac{60}{(3000)(0.08)} = t$$

$$\frac{3}{12} = t \text{(years), or 3 months}$$

Solving for a Specified Variable

1. Rearrange the equation, if necessary, so that the specified variable occurs in every addend on one side and does not occur at all on the other side.
2. Simplify the side containing the specified variable to obtain a product of the specified variable and a second factor.
3. Divide both sides by the second factor (or multiply by its reciprocal).

Exercises 2.2

Standard Assignment: Exercises 1, 3, 5, 7, 9, 11, 13, 15, 17, 19, 21, 23, 27, 31

A Solve for the specified variable. See Examples 4–6.

1. $d = rt$ for r
2. $F = ma$ for m
3. $C = 2\pi r$ for π
4. $ax + b = 0$ for x
5. $I = \dfrac{E}{R}$ for R
6. $F = \dfrac{mv}{t}$ for v
7. $g = G\dfrac{m}{R^2}$ for G
8. $F = k\dfrac{q_1 q_2}{r^2}$ for k
9. $\dfrac{1}{f} = \dfrac{1}{a} + \dfrac{1}{b}$ for f
10. $P = 2L + 2W$ for W

11. $S = 2\pi r^2 + 2\pi rh$ for h

12. $A = \frac{1}{2}h(a + b)$ for b

In Exercises 13–21, find the correct geometric formula in the Appendix and solve the formula for the specified variable. (See Examples 4 and 5.) Then find the value of the specified variable. (See Example 3.)

13. The volume of a swimming pool (box shaped) is 2808 cubic feet. If it is 18 feet long and 12 feet deep, find the width.

14. If a box-shaped hole 7 feet long and 3 feet wide holds 168 cubic feet of dirt, how deep must it be?

15. If a circle has a circumference of 114π centimeters, find the radius of the circle.

16. If a rectangle has a perimeter of 28 meters and a width of 5 meters, find the length of the rectangle.

17. A triangle has a perimeter of 47 inches. Two of the sides measure 21 and 13 inches. Find the third side.

18. A can has a surface area of 2π square meters and a radius of one meter. Find the height of the can.

19. A can has a volume of 148π cubic centimeters and a radius of 2 centimeters. Find the height of the can.

20. A trapezoid has an area of 66 square feet and a height of 6 feet. If one base is 3 feet, what is the length of the other base?

21. A trapezoid has an area of 35 square centimeters. If one base is 9 centimeters and the other base is 11 centimeters, find the height of the trapezoid.

22. If P dollars are invested at a simple interest rate, r, the amount, A, that will be available after t years is

$$A = P + Prt$$

If $500 is invested at a rate of 6%, how long will it be before the amount of money available is $920?

23. Using the formula from Exercise 22, determine the amount of money that was invested if $2482 resulted from a ten-year investment at 7%.

24. A rule of thumb for estimating the maximum affordable monthly mortgage note, M, for prospective home owners is

$$M = \frac{m - b}{4}$$

where m is the gross monthly income and b is the total of the monthly bills. What must the gross monthly income be to justify a monthly note of $427 if monthly bills are $302?

25. Among a group of engineers who had received master's degrees, the annual

salary, S, was related to the number of years of work experience before, b, and after, a, receiving the degree by

$$S = 18{,}917 + 1080b + 604a$$

If an engineer had worked 8 years before earning her master's degree and was earning $30,577, how many years had she worked after receiving her degree?

26. Use the formula $I = prt$ from Example 6 to determine what principal was invested to result in $1572 interest at a rate of 12% for 5 years.

27. Use the formula $I = prt$ from Example 6 to determine what interest rate would result in $378 in interest if $1200 was invested for $3\frac{1}{2}$ years.

28. The price, P, (in dollars) for which a manufacturer will sell a radio cassette player is related to the number of players ordered, q, by the formula

$$P = 200 - 0.02q \quad \text{for } 100 \leq q \leq 2000$$

How many players must be ordered in order to pay $170 per player?

29. (*Boyle's Law*) If a volume, V, of a dry gas under pressure, P_1, is subjected to a new pressure, P_2, the new volume, V_2, of the gas is given by

$$V_2 = \frac{V_1 P_1}{P_2}$$

If V_2 is 200 cubic centimeters and V_1 is 600 cubic centimeters and P_1 is 400 millimeters of mercury, find P_2.

30. The voltage gain, V, of a transistor is related to the generator resistance, R_1, the load resistance, R_2, and the current gain, α, by

$$V = \alpha \frac{R_1}{R_2}$$

If V is 950, R_1 and R_2 are 100,000 and 100 ohms, respectively, find α.

31. The quick-ratio, Q, of a business is related to its current assets, A, its current inventory, I, and its current liabilities, L, by

$$Q = \frac{A - I}{L}$$

If $Q = 37{,}000$, $L = \$1500$, and $I = \$3200$, find the current assets.

32. ▦ Use the formula $I = prt$ to determine the interest on $2347 at a rate of 8.65% for 7 months.

33. ▦ Find the amount of principal that must be invested at 12.72% to produce $175.27 interest in 96 days. (Assume a 365-day year.)

34. ▦ Use the formula $A = P + Prt$ from Exercise 22 to find the amount resulting from a principal of $1247.65 invested at a rate of 13.91% for a period of 567 days. (Assume a 365-day year.)

35. ▦ A triangle has a perimeter of 14.372 feet. Two of the side measurements are 6.47 feet and 3.027 feet. Find the length of the third side.

36. ▦ Estimate the volume of a can of height 14.062 centimeters and radius 8.916 centimeters, using the approximation $\pi = 3.142$.

B Solve for the specified variable. See Examples 4–6.

37. $t = a + (m - 1)d$ for d

38. $S = P(1 + rt)$ for r

39. $r = \dfrac{d}{1 - dt}$ for d

40. $r = \dfrac{2mI}{B(n + 1)}$ for I

41. $r = \dfrac{2mI}{B(n + 1)}$ for m

42. $\dfrac{1}{R} = \dfrac{1}{R_1} + \dfrac{1}{R_2}$ for R

43. $\dfrac{1}{R} = \dfrac{1}{R_1} + \dfrac{1}{R_2}$ for R_1

44. $y = mx + b$ for m

45. $ax + by = c$ for y

2.3 Applications

Practical problems are often introduced through verbal descriptions and thus they become "word problems." In this section we will develop a strategy for solving word problems.

We must first learn to translate simple English sentences and phrases into mathematical equations or expressions. The following table indicates how certain phrases are translated into symbols.

English Expressions	Mathematical Symbol
Equals, is equal to, is, was, will be, becomes, gives, leaves, the result is, yields	=
Plus, added to, the sum of, more than, increased by	+
Minus, subtracted from, less, decreased by, the difference is, take away, reduced by, less than	−
Times, the product of, multiplied by	× or ·
Divided by, quotient of, ratio	÷ or /

EXAMPLE 1 The variables x and y are used to represent the unknown quantities in the following English expressions.

English Expressions	Mathematical Expression
5 more than a number	$5 + x$ or $x + 5$
A number increased by 3	$x + 3$ or $3 + x$
The sum of two numbers	$x + y$ or $y + x$
A number decreased by 9	$x - 9$
A number reduced by 14%	$x - 0.14x$
The difference of two numbers	$x - y$ or $y - x$
7 times a number	$7x$
A number multiplied by 7	$7x$
Twice a number	$2x$
The product of two numbers	xy or yx
A number divided by 21	$x/21$
2 divided by a number	$2/x$
The ratio of a number to 2	$x/2$
The quotient of two numbers	x/y or y/x

EXAMPLE 2 Translate the following English sentences into mathematical equations.

English Sentence	Mathematical Equation
3 times the sum of a number and 7 is 30.	$3(x + 7) = 30$
7 times a number decreased by 4 is 77.	$7(x - 4) = 77$
The product of 7 and a number, when decreased by 4, is 77.	$7x - 4 = 77$
The quotient of a number and the sum of 3 and 5 times the number is 16.	$\dfrac{x}{3 + 5x} = 16$
The sum of a number and the number plus 2, when increased by 3, equals 29.	$x + (x + 2) + 3 = 29$

The following outline is designed to help you organize your work into a productive sequence of steps leading to the solution of a word problem. We are only concerned here with problems in which one variable will suffice for a solution.

> **Steps for Solving Word Problems**
>
> 1. Read the problem carefully. Be sure you understand what is given and what must be found.
> 2. Choose a variable to represent an unknown quantity.
> 3. Draw a picture or make a chart that summarizes the known and unknown quantities.
> 4. Express all unknown quantities in terms of the chosen variable.
> 5. Form an equation relating the known and unknown quantities.
> 6. Solve the equation and use the solution to answer all parts of the original problem.
> 7. Check the solution in the *original* problem to see if it is reasonable. Do not check the solution in the equation obtained from the problem.

In the following exercises, several different types of word problems are illustrated. You should gain sufficient insight from these sample problems to be able to solve a wide range of problems of a practical nature.

Exercises 2.3

Standard Assignment: Exercises 1, 3, 5, 7, 9, 11, 13, 15, 17, 19, 21, 23, 25, 27, 29, 31, 33, 35, 37, 39, 41

A Translate the following English expressions and sentences into mathematical expressions and equations. See Examples 1 and 2.

1. 3 less than a number

2. A number divided by $\frac{1}{4}$

3. The sum of a number and twice that number

4. 3 times a number minus one-fourth that number

5. The quotient of a number and the number plus 2

6. A number increased by 10%

7. The sum of a number and 7, decreased by half the number gives 20.

8. The difference between -3 and a number is 17.

9. The product of a number and 7 is 2 more than half the number.

10. The ratio of a number to 4 more than the number will be 15.

11. The quotient of a number and 4 plus that number, when added to 2 is 123.

12. The ratio of a number to 3 less than the number equals 7.

13. A number decreased by 20% results in 465.

14. A number times the sum of 9 and 3 times the number yields 214.

15. The sum of a number and 9, multiplied by 3 times the number yields 214.

16. Seven times the quotient of a number and the number plus 4 is 57.

17. The difference of 6 times a number and 3, when reduced by 3, equals 6 times one less than the number.

In each of the following problems:

(a) Set up an equation that represents the conditions of the problem.

(b) Solve the equation.

(c) Check the solution in the original problem, and not just in an equation derived from the original problem.

> EXAMPLE: Alice is offered $10,000 to settle an insurance claim. A lawyer tells her that he could negotiate a higher settlement figure. He wants one-third of that figure as his fee. What amount must he negotiate for Alice to get the $10,000 already offered?
>
> *Solution:*
>
> Let x = negotiated settlement
>
> then $\frac{1}{3}x$ = lawyer's fee
>
> so $\frac{2}{3}x$ = Alice's portion of the settlement
>
> thus (a) $\frac{2}{3}x = 10,000$
>
> (b) $x = \frac{3}{2}(10,000) = 3(5000) = 15,000$ dollars
>
> (c) *Check:* If Alice gets $15,000 and gives $\frac{1}{3}(15,000) = \$5000$ to her lawyer, she will indeed have the $10,000 initially offered.

A 18. The lawyer in the previous example tells Alice that if the case has to be tried in court, he wants 40% of the settlement awarded by the court. What amount must be awarded for Alice to receive the $10,000 originally offered?

19. When a number is increased by 10%, the result is 330. What is the number?

20. When a number is decreased by 30%, the result is 14. What is the number?

21. A used car salesman buys a car at the wholesale price and adds 20% of this price to obtain a retail price of $2400. What is the wholesale price?

CONSECUTIVE INTEGER PROBLEMS

Consecutive integers are integers that differ by one. A list of consecutive integers starting from the integer x can be expressed by $x, x + 1, x + 2, \ldots$. Consecutive even or odd integers differ by 2 and can be expressed by $x, x + 2, x + 4, \ldots$.

> EXAMPLE: Find three consecutive integers such that when 3 times the largest integer is reduced by 31, the result is the sum of the first and second integers.
>
> Solution: $\quad x =$ first integer
>
> $\qquad x + 1 =$ second integer
>
> $\qquad x + 2 =$ largest integer
>
> (a) $\quad 3(x + 2) - 31 = x + (x + 1)$
>
> (b) $\qquad 3x + 6 - 31 = 2x + 1$
>
> $\qquad\qquad 3x - 25 = 2x + 1$
>
> $\qquad\qquad 3x - 2x = 25 + 1$
>
> $\qquad\qquad\qquad x = 26$
>
> $\qquad\qquad x + 1 = 27$
>
> $\qquad\qquad x + 2 = 28$
>
> (c) Check: $3(28) - 31 = 84 - 31 \qquad$ Three times the largest reduced by 31
>
> $\qquad\qquad\qquad\qquad = 53$
>
> and $\qquad 26 + 27 = 53 \qquad$ Sum of the first and second

A 22. Find three consecutive integers whose sum is 84.

23. Find three consecutive integers if 5 times the smallest integer is 7 more than twice the sum of the other two.

24. Find three consecutive even integers whose sum is 48.

25. Find two consecutive odd integers such that half their sum is 24 less than twice the smaller integer.

INTEREST PROBLEMS

Simple interest problems use the fact that the interest (I) earned in t years is equal to t times the product of the amount invested (P) and the annual rate of interest (r); thus $I = Prt$. It is useful to organize the information from the problem in a table as is shown in the following example.

> EXAMPLE: A man invests a total of $10,000 in blue chip and glamour stocks. At the end of a year the blue chips returned 12%, and the glamour stock returned 8% on the original investments. How much was invested in each type of stock if the total interest earned was $1060?

Solution: Let

$$x = \text{amount invested in blue chip stocks}$$

then

$$10{,}000 - x = \text{amount invested in glamour stocks}$$

Investment	Amount P	Rate r	Time t	Interest $I = Prt$
Blue chip	x	0.12	1	$0.12x$
Glamour	$10{,}000 - x$	0.08	1	$0.08(10{,}000 - x)$

Now (a) $\begin{pmatrix}\text{interest from}\\ \text{blue chip}\end{pmatrix} + \begin{pmatrix}\text{interest from}\\ \text{glamour}\end{pmatrix} = \begin{pmatrix}\text{total}\\ \text{interest}\end{pmatrix}$

(b) $0.12x + 0.08(10{,}000 - x) = 1060$

$12x + 8(10{,}000 - x) = 106{,}000$ Multiply by 100

$12x + 80{,}000 - 8x = 106{,}000$

$4x = 26{,}000$

$x = 6500$ dollars in blue chip stocks

$10{,}000 - x = 3500$ dollars in glamour stocks

(c) *Check:* Does $(0.12)(6500) + (0.08)(3500) = 1060$? Yes

A 26. A real estate agent invested a total of $4200. Part was invested in a high-risk real estate venture and the rest was invested in a savings and loan. At the end of a year, the high-risk venture returned 15% on her investment and the savings and loan returned 8% on her investment. Find the amount invested in each if her total income from investment was $448.

27. Mr. Smith received an inheritance of $7000. He put part of it in a tax shelter paying 9% and part in a bank paying 6%. If his annual interest totals $540, how much was invested at each rate?

28. Ms. Baker invested $4900, part at 6% and the rest at 8%. If the yearly interest on each investment is the same, how much interest does she receive at the end of the year?

29. If $5000 is invested in a bank that pays 5% interest, how much more must be invested in bonds at 8% to earn 6% on the total investment?

GEOMETRY PROBLEMS

Solving problems involving geometric figures often requires the use of formulas that are listed in the Appendix.

EXAMPLE: The length of a rectangular dining room table is three times its width. The perimeter of the table is 16 meters. Find the dimensions of the dining room table.

Solution: Draw a rectangle to help visualize the problem.

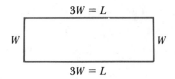

Let
W = width of the table
L = length of the table
P = perimeter of the table

From the Appendix we find $\quad 2L + 2W = P$

That is, $\quad 2L + 2W = 16$

Since $\quad L = 3W$

(a) $\quad 2(3W) + 2W = 16$
$6W + 2W = 16$
$8W = 16$

(b) $\quad W = 2$ meters
$L = 3W = 6$ meters

(c) Check: Does the length equal 3 times the width? Does $6 = 3 \cdot 2$? Yes. Does $2(6) + 2(2) = 16$? Yes.

A 30. The length of a rectangle is 5 meters more than twice its width. The perimeter of the rectangle is 28 meters. Find the dimensions.

31. In an isosceles triangle each base angle measures twice as much as the remaining angle. Find the number of degrees in each angle of the triangle.

32. When the side of each leg of an equilateral triangle is increased by 2 centimeters, the resulting triangle has a perimeter of 27 centimeters. How long is each leg of the original triangle?

33. In a triangle, one angle is twice as large as the smallest angle and the other angle is 20° larger than the smallest angle. Find the measure of each angle.

UNIFORM RATE PROBLEMS

There are three elements in a motion problem: distance (d), rate (r), and time (t). We know from physics that the distance traveled by an object moving at a uniform rate of speed for a specified period of time is distance = rate × time. The formula expressing this relationship is $d = rt$. If a car travels at a uniform rate of 50 miles per hour for 3 hours, it will travel a distance of 50(3) = 150 miles. (The same is true if the *average* rate of speed of the car is 50 mph over 3 hours.)

When working out motion problems, it is a good idea to remember that:

(a) Rate refers to uniform speed or average speed.

(b) Units of distance, rate, and time must be consistent throughout the problem. Thus, if the rate is given in *miles* per *hour*, time must be expressed in *hours*, and distance in *miles*.

EXAMPLE: A policeman on a motorcycle is pursuing a car that is speeding at 70 mph. The policeman is 3 miles behind the car and is traveling 80 mph. How long will it be before the policeman overtakes the car?

Solution: Draw a sketch to help visualize the problem.

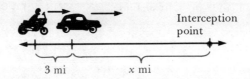

Let x = distance the car travels before being overtaken

$x + 3$ = distance the motorcycle travels before overtaking the car

We can organize our information in the following table.

Object	d	r	$t \left(= \dfrac{d}{r} \right)$
Car	x	70	$\dfrac{x}{70}$
Motorcycle	$x + 3$	80	$\dfrac{(x + 3)}{80}$

Since the time from observation to interception is the same for both the car and the motorcycle, we have

(a) $\dfrac{x}{70} = \dfrac{x + 3}{80}$

(b) $80x = 70(x + 3)$

$80x = 70x + 210$

$10x = 210$

$x = \dfrac{210}{10} = 21$ miles

$$t = \frac{x}{70} = \frac{21}{70} = \frac{3}{10} \text{ hour}$$

$$= \left(\frac{3}{10}\right)(60) = 18 \text{ minutes}$$

Check: If the policeman travels for $\frac{3}{10}$ of an hour at 80 mph, he goes a distance of

$$d = (80)\left(\frac{3}{10}\right) = 24 \text{ miles}.$$

The car traveling for this same $\frac{3}{10}$ of an hour at 70 mph goes a distance of

$$d = (70)\left(\frac{3}{10}\right) = 21 \text{ miles}.$$

Since the policeman travels 3 more miles during this time than the car does because he started from 3 miles behind the car, he does indeed overtake the car at this time.

A 34. Mary jogs 100 meters in the same time that John bicycles 150 meters. If John bicycles 15 meters per minute faster than Mary jogs, find the rate of each.

35. A car leaves New Orleans traveling 50 km/hr. An hour later a second car leaves New Orleans following the first car and traveling 70 km/hr. How long will it take the second car to overtake the first?

36. Two planes leave an airport traveling in opposite directions. One plane travels at 470 km/hr and the other at 430 km/hr. How long will it take them to be 2250 km apart?

37. Joe is $\frac{1}{3}$ of a mile away from home, bicycling at 20 mph, when his brother takes off on his bike to catch up. How fast must Joe's brother go to catch him a mile from home?

MIXTURE PROBLEMS

EXAMPLE: A 6 qt radiator presently contains 75% water and 25% pure antifreeze. How much of this solution should be drained and replaced by pure antifreeze so that the resulting 6 qt solution is 50% pure antifreeze?

Solution: Since we are interested in the quantity of solution drained, let

$$x = \text{number of quarts of solution drained}$$

then also

$$x = \text{number of quarts of antifreeze added}$$

and

$$25\% \text{ of } x = \left(\frac{1}{4}\right)x = \text{number of quarts of antifreeze drained}$$

Now $\begin{pmatrix} \text{antifreeze in} \\ \text{final mixture} \end{pmatrix} = \begin{pmatrix} \text{antifreeze in} \\ \text{original mixture} \end{pmatrix} - \begin{pmatrix} \text{antifreeze} \\ \text{drained} \end{pmatrix} + \begin{pmatrix} \text{antifreeze} \\ \text{added} \end{pmatrix}$

(50% of 6) = (25% of 6) − (25% of x) + x

(a) $\left(\dfrac{1}{2}\right)6 = \left(\dfrac{1}{4}\right)6 - \left(\dfrac{1}{4}\right)x + x$

$12 = 6 - x + 4x$

$6 = 3x$

(b) $2 = x$

(c) *Check:* Does draining two quarts from the radiator and adding two quarts of antifreeze result in a 50% solution? Draining 2 qt leaves 4 qt, containing 1 qt antifreeze and 3 qt water. Adding 2 qt of antifreeze produces 3 qt of antifreeze and 3 qt of water. This is the desired 50% pure antifreeze solution.

A

38. Two gallons of milk contain 2% butterfat. How many quarts of milk should be drained and replaced by pure butterfat to produce a solution of two gallons of milk containing 5% butterfat?

39. A goldsmith has 120 grams of gold alloy containing 75% pure gold. How much pure gold must be added to this alloy to obtain an alloy that is 80% pure gold?

40. A coffee wholesaler wants to blend a coffee containing 35% chicory with a coffee containing 15% chicory to produce a 500 kilogram blend of coffee containing 18% chicory. How much of each type is required?

41. A mixture of nuts contains almonds worth 50¢ a pound, cashews worth $1.00 per pound, and pecans worth 75¢ per pound. If there are three times as many pounds of almonds as cashews in a 100-pound mixture worth 70¢ a pound, how many pounds of each nut does the mixture contain?

In the previous exercises, some hints about how to solve the problems were provided by the examples immediately preceding the exercise. The following exercises contain some problems that are similar to those just given as well as some that will require an extension or modification of the techniques in the examples.

B

42. A retailer's cost for a microwave oven is $480. If he wants to make a profit of 20% of the selling price, at what price should the oven be sold?

43. Find two consecutive integers such that one-third of the smaller is 3 more than one-fifth of the larger.

44. Mary's dad is four times as old as Mary. Two years ago her dad was five times as old as she was at that time. How old is Mary?

45. The Beckly Company manufactures shaving sets for $3 and sells them for $5 each. How many shaving sets must be sold in order for the company to recover an initial investment of $40,000 and earn an additional $30,000 as profit?

46. The coin box on a pay phone contained $17.90 in nickels, dimes, and quarters. It contained the same number of nickels as dimes and contained 136 coins in all. How many of each coin did it hold?

47. Two cars start out together from the same place. They travel in opposite directions, with one of them traveling 7 mph faster than the other. After 3 hours, they are 621 miles apart. How fast is each car traveling?

48. Find three consecutive even integers whose sum is 102.

49. A swimming pool has two drain pipes. One can empty the pool in 3 hours and the other can empty the pool in 7 hours. If both drains are open, how long will it take for the pool to drain?

50. A 3% increase in the population of a city results in a population of 314,900. What was the former population?

51. An open box is to be constructed from a rectangular sheet of tin 3 meters wide by cutting out a one-meter square from each corner and folding up the sides. The volume of the box is to be 2 cubic meters. What is the length of the tin rectangle?

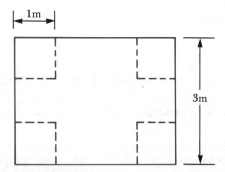

52. A new card sorter works twice as fast as the old one. Both sorters working together complete a job in 8 hours. How long would it have taken the new card sorter to do the job alone?

53. Jane rode her bicycle from her house to a friend's house and averaged 16 kilometers per hour. It became dark before she was ready to return home so her friend drove her home at a rate of 80 kilometers per hour. If Jane's total traveling time was 3 hours, how far away did her friend live?

54. John's grandfather is 57 years older than he is. Five years from now, his grandfather will be 4 times as old as John is at that time. How old is the grandfather?

55. A vending machine coin box contains nickels, dimes, and quarters. It contains 3 times as many nickels and 4 times as many quarters as it does dimes. If the box contains $96.25 in all, how many coins of each kind does it contain?

56. A small theater company with 22 seats sold advance sale tickets for $3.75 and tickets at the door for $4.50. If they collected $84.00 and sold out completely, how many of each type of ticket did they sell?

57. Mr. Lyons invests one sum of money at a certain rate of interest and half again that sum at twice the first rate of interest. This turns out to yield 8% on the total investment. What are the two rates of interest?

58. A farmer can plow his field by himself in 15 days. If his son helps, they can do it in 6 days. How long would it take his son to plow it by himself?

59. In a jar of red and green gumdrops, 12 more than half the gumdrops are red and the number of green gumdrops is 19 more than half the number of red gumdrops. How many of each color are in the jar?

60. A real estate investor bought acreage for $7200. After reselling three-fourths of the acreage at a profit of $30 per acre, she recovered the $7200. How many acres were sold?

2.4 Equations Involving Absolute Value

In Section 1.4 we saw that $|x|$ was either x or $-x$, and that

$$|x| = \begin{cases} x \text{ if } x \geq 0 \\ -x \text{ if } x < 0 \end{cases}$$

We also noted that the **absolute value** has a geometric interpretation,

$|x|$ = distance between x and 0 on the number line.

The numbers that solve the equation $|x| = 3$ are exactly those numbers that are 3 units away from 0 on the number line. There are two such numbers, 3 and -3. (See Figure 2.1.)

Figure 2.1

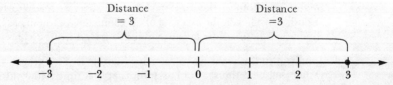

The solution of $|x| = 3$ can also be found by using the fact that $|x| = x$ or $|x| = -x$. Thus $|x| = 3$ yields the two equations $x = 3$ and $-x = 3$, from which we obtain $x = 3$ or $x = -3$.

We can use identical arguments to establish that for any positive number a, the solution set for the equation $|x| = a$ is $\{-a, a\}$. Since 0 is the only number with absolute value 0, we know that $|x| = 0$ has the solution set $\{0\}$. The equation $|x| = -a$, for positive a, has $\emptyset$ as its solution set, since distance, $|x|$, is never negative. We will summarize these observations next.

2.4 Equations Involving Absolute Value

> **For Any Positive Number a**
>
> | Equation | Roots | Solution Set |
> |---|---|---|
> | $\lvert x \rvert = a$ | $x = \pm a$ | $\{-a, a\}$ |
> | $\lvert x \rvert = 0$ | $x = 0$ | $\{0\}$ |
> | $\lvert x \rvert = -a$ | none | $\emptyset$ |

EXAMPLE 1 Solve the following equations.

(a) $\lvert x \rvert = 23$.
(b) $\lvert t \rvert = -4$.

Solution:
(a) There are two numbers that are 23 units from 0: one to the left of 0, $x = -23$, and one to the right of 0, $x = 23$. The solution set is $\{-23, 23\}$.
(b) Distance is never negative, so there is no number whose distance from 0 equals -4. The solution set is $\emptyset$.

EXAMPLE 2 Solve the equation $\lvert y - 3 \rvert = 10$.

Solution: There are two numbers with absolute value 10, namely 10 and -10. If $\lvert y - 3 \rvert$ is 10, then $y - 3$ may be either one of those numbers.

$$y - 3 = 10 \quad \text{or} \quad y - 3 = -10$$
$$y = 13 \quad \text{or} \quad y = -7$$

The solution set is $\{-7, 13\}$. The line graph may be helpful in visualizing this result.

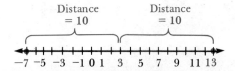

EXAMPLE 3 Solve $\lvert 2k + 3 \rvert = 0$.

Solution: The only number with absolute value 0 is 0 itself. Thus

$$2k + 3 = 0$$
$$2k = -3$$
$$k = -\frac{3}{2}$$

The solution set is $\left\{-\dfrac{3}{2}\right\}$.

It is a good idea to check by substituting the answer into the original equation.

Check:
$$\left|2\left(-\frac{3}{2}\right) + 3\right| = 0 \quad ?$$
$$|-3 + 3| = 0 \quad ?$$
$$|0| = 0 \quad ?$$
$$0 = 0 \quad ✓$$

EXAMPLE 4

Solve $|-3x + 9| = 6$.

Solution: The only two numbers having absolute value 6 are -6 and 6. Thus

$$\begin{aligned}
-3x + 9 &= 6 & \text{or} & & -3x + 9 &= -6 \\
9 - 6 &= 3x & & & 9 + 6 &= 3x \\
3 &= 3x & & & 15 &= 3x \\
1 &= x & & & 5 &= x
\end{aligned}$$

The solution set is $\{1, 5\}$.

Check:
$$\begin{aligned}
x &= 1 & & & x &= 5 \\
|-3(1) + 9| &= 6 \quad ? & & & |-3(5) + 9| &= 6 \quad ? \\
|-3 + 9| &= 6 \quad ? & & & |-15 + 9| &= 6 \quad ? \\
|6| &= 6 \quad ? & & & |-6| &= 6 \quad ? \\
6 &= 6 \quad ✓ & & & 6 &= 6 \quad ✓
\end{aligned}$$

EXAMPLE 5

Solve $2\left|\frac{1}{2}t - 5\right| - 3 = 3$.

Solution: First isolate $\left|\frac{1}{2}t - 5\right|$ on one side of the equation.

$$2\left|\frac{1}{2}t - 5\right| - 3 = 3$$
$$2\left|\frac{1}{2}t - 5\right| = 6$$
$$\left|\frac{1}{2}t - 5\right| = 3$$

The only two numbers having absolute value 3 are -3 and 3. Thus,

$$\frac{1}{2}t - 5 = 3 \quad \text{or} \quad \frac{1}{2}t - 5 = -3$$
$$t - 10 = 6 \quad\quad\quad t - 10 = -6$$
$$t = 16 \quad\quad\quad t = 4$$

The solution set is $\{4, 16\}$.

Check: We leave this for the student.

Suppose x and y are real numbers and $|x| = |y|$. Recall that $|x| = \pm x$ and $|y| = \pm y$. Thus $|x| = |y|$ becomes either $x = y$, $x = -y$, $-x = y$, or $-x = -y$. In any case, $x = y$ or $x = -y$.

$$|x| = |y|$$

if and only if

$$x = y \quad \text{or} \quad x = -y$$

EXAMPLE 6 Solve $|x + 2| = |6 - x|$.

Solution: In order for two numbers to have the same absolute value, they must either be the same number or one must be the negative of the other. This leads to two equations:

$$x + 2 = 6 - x \quad \text{or} \quad x + 2 = -(6 - x)$$
$$2x = 4 \quad\quad\quad x + 2 = -6 + x$$
$$x = 2 \quad\quad\quad 8 = 0 \quad \text{(nonsense!)}$$

The first equation results in the solution $x = 2$, whereas the second equation yields a false statement, producing no solution. The solution set is $\{2\}$.

Check:
$$|2 + 2| = |6 - 2| \quad ?$$
$$|4| = |4| \quad \checkmark$$

EXAMPLE 7 Solve $|3x + 3| = |8 - 2x|$.

Solution: As before, this equation leads to two equations not involving absolute value.

$$3x + 3 = 8 - 2x \quad \text{or} \quad 3x + 3 = -(8 - 2x)$$
$$5x = 5 \quad\quad\quad 3x + 3 = -8 + 2x$$
$$x = 1 \quad\quad\quad x = -11$$

The solution set is $\{-11, 1\}$.

Check:
$$x = -11$$
$$|3(-11) + 3| = |8 - 2(-11)| \quad ?$$
$$|-33 + 3| = |8 + 22| \quad ?$$
$$|-30| = |30| \quad ?$$
$$30 = 30 \quad \checkmark$$

$$x = 1$$
$$|3(1) + 3| = |8 - 2(1)| \quad ?$$
$$|3 + 3| = |8 - 2| \quad ?$$
$$|6| = |6| \quad ?$$
$$6 = 6 \quad \checkmark$$

Exercises 2.4

Standard Assignment: Exercises 1, 3, 5, 7, 8, 9, 17, 19, 21, 23, 25, 27, 29, 31, 33, 35, 39, 41

A Solve the following equations. See Examples 1–7.

1. $|x| = 1$
2. $|y| = 7$
3. $|k| = 0$
4. $|t| = 10$
5. $|y| = -(-6)$
6. $|x| = -(-2)$
7. $|k| = -1$
8. $|t| = -5$
9. $|x| + 1 = 4$
10. $|y| - 3 = 2$
11. $|t| + 9 = 6$
12. $|k| + 3 = 1$
13. $|r| + 2 = 2$
14. $|s| - 6 = -6$
15. $|x - 1| = 0$
16. $|y - 3| = 0$
17. $|t + 4| = 1$
18. $|k + 7| = -5$
19. $|3y - 6| = 15$
20. $|2t + 3| = 7$
21. $|1 - 2s| = 7$
22. $|2r + 5| = 3$
23. $|2x - 1| = 8$
24. $|3y - 11| = 7$
25. $|1 - 5t| = 4$
26. $|3k + 10| = 10$
27. $|x - 4| + 5 = 6$
28. $|y + 6| - 3 = 9$
29. $|3x + 5| - 2 = 2$
30. $|2 - 3s| + 1 = 3$
31. $\left|\dfrac{x}{5} - 1\right| = 2$
32. $\left|\dfrac{1}{7}y - 4\right| = 3$
33. $\left|1 - \dfrac{1}{6}t\right| = \dfrac{1}{3}$
34. $\left|2 - \dfrac{3}{4}r\right| = \dfrac{1}{2}$
35. $\left|\dfrac{3}{7}k - \dfrac{1}{3}\right| = 0$
36. $\left|\dfrac{2 - 5y}{25}\right| = 0$
37. $\left|\dfrac{2x - 1}{9}\right| = -4$
38. $|t - 3| = |t + 15|$
39. $|k| = |k - 4|$
40. $|k + 1| = |2k + 8|$
41. $|2y - 3| = |5 - 2y|$
42. $|1 - 2x| = |4x - 2|$
43. $\left|5 - \dfrac{2}{3}t\right| = \left|\dfrac{1}{3}t + 7\right|$
44. $\left|\dfrac{2s - 3}{3}\right| = \left|\dfrac{1}{3}s - 1\right|$
45. $\left|\dfrac{1}{2}r - \dfrac{2}{3}\right| = \left|\dfrac{3}{4}r + \dfrac{1}{12}\right|$
46. $\left|\dfrac{3}{8}k + 2\right| = \left|\dfrac{1}{4}k + 3\right|$
47. 📱 $|2.031x - 1.414| = 0.446$
48. 📱 $|1.7302y - 12.406| = 2.843$

49. $\left|\dfrac{2.7 - 6.432t}{0.5216}\right| = 0$ 50. $\left|\dfrac{8.411 + 2.7562k}{-3.421}\right| = 2.86054$

B Find all numbers satisfying the given description.

51. The absolute value of the number is 103.

52. The sum of -3 and the absolute value of the number equals 5.

53. The absolute value of -3 and the number is 5.

54. The absolute value of the sum of twice a number and b is 14.

55. The absolute value of a number is the same as the absolute value of the difference of 1 and two-thirds the number.

2.5 Intervals

The result of graphing a point for each number in a set of numbers is called the graph of the set of numbers. Each of the sets that we graph in this section is an **interval** of numbers. The graph of $\{x \mid x > 1\}$ consists of all points on the number line having coordinates greater than 1, or all points "to the right of 1." (See Figure 2.2.)

Figure 2.2

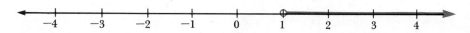

The small open circle at 1 indicates that one is *not* included in the graph. The heavy line extending to the right of 1 shows that all such points are included in the graph. There is an alternate notation for $\{x \mid x > 1\}$, which is $(1, +\infty)$. The symbol $+\infty$ is read "plus infinity" and does not represent a real number. The symbol $+\infty$ in this context indicates that the graph extends indefinitely to the right. The parenthesis, (, next to the 1 indicates that 1 is not included on the graph.

EXAMPLE 1 Graph $\{x \mid x \geq -2\}$.

Solution: We use a heavy line over the part of the number line starting at -2, including -2, and extending to the right of -2. The solid circle at -2 indicates that -2 is on the graph. Alternate notation for

$\{x \mid x \geq -2\}$ is $[-2, +\infty)$. The bracket, [, next to the -2 indicates that -2 is included on the graph.

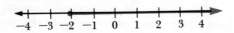

EXAMPLE 2   Graph (a) $\{x \mid x < 3\}$ (b) $\{x \mid x \leq 3\}$

Solution:
(a)

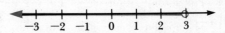

The graph of $\{x \mid x < 3\}$ uses the open circle indicating that 3 is not included on the graph. Alternate notation is $\{x \mid x < 3\} = (-\infty, 3)$. The symbol $-\infty$ is read "minus infinity" and indicates the indefinite extension of the graph to the left; it is not to be interpreted as a number. The parenthesis,), following the 3 indicates that 3 is not included on the graph.

(b)

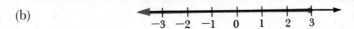

The graph of $\{x \mid x \leq 3\}$ uses the solid circle indicating that 3 is on the graph. Alternate notation is $\{x \mid x \leq 3\} = (-\infty, 3]$. Here, the bracket,], following the 3 indicates that 3 is included on the graph.

Frequently we want to graph the set of numbers that is *between* two given numbers. Now x is a real number between -2 and 3 if and only if $-2 < x$ and $x < 3$. A more compact way to assert these two inequalities is $-2 < x < 3$.

EXAMPLE 3   Graph $\{x \mid -2 < x < 3\}$.

Solution: The graphs of $\{x \mid -2 < x\}$ and $\{x \mid x < 3\}$ are indicated *above* the number line. The points on the number line that are common to both of these graphs constitute the graph of $\{x \mid -2 < x < 3\}$. We use the open circles to indicate that neither -2 nor 3 is included on the graph.

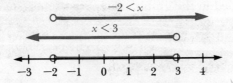

Alternate notation is $\{x \mid -2 < x < 3\} = (-2, 3)$. The parentheses preceding -2 and following 3 indicate that -2 and 3 are not included on the graph.

EXAMPLE 4   Graph $\{x \mid -3 \leq x \leq 1\}$.

Solution: The graphs of $\{x \mid -3 \leq x\}$ and $\{x \mid x \leq 1\}$ are indicated above the number line. The points on the number line that are common to both of these graphs constitute the graph of $\{x \mid -2 \leq x \leq 3\}$. The solid circles show that -3 and 1 are on the graph.

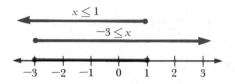

Alternate notation is $\{x \mid -3 \leq x \leq 1\} = [-3, 1]$.

The alternate notations that we have presented are examples of **interval notations**. In this notation, a bracket indicates inclusion and a parenthesis indicates exclusion of the adjacent point. In this sense the brackets correspond to the $\leq$ and $\geq$ signs, while the parentheses correspond to $<$ and $>$.

EXAMPLE 5

The following chart illustrates interval graphs described in both interval and set-builder notation.

Set-builder Notation	Interval Notation	Graph
$\{x \mid -1/2 < x \leq 2\}$	$\left(-\dfrac{1}{2}, 2\right]$	
$\{x \mid x < 5/2\}$	$\left(-\infty, \dfrac{5}{2}\right)$	
$\{x \mid 1 \leq x < 3\}$	$[1, 3)$	

Exercises 2.5

Standard Assignment: All odd-numbered exercises

A Graph each set of the given intervals on a separate number line and give the interval notation for each. See Example 5.

1. $\{x \mid x \geq -1\}$
2. $\{y \mid y > 2\}$
3. $\{y \mid y < 3\}$
4. $\{x \mid x \leq -3/2\}$
5. $\{x \mid -1 < x < 3\}$
6. $\{t \mid 3 \leq t \leq 7\}$
7. $\{x \mid -2 \leq x < 0\}$
8. $\{y \mid -2 < y \leq 0\}$
9. $\{x \mid 4 \leq x\}$
10. $\{y \mid -1 < y\}$
11. $\{z \mid -3 \geq z > -6\}$
12. $\{t \mid 2 > t > -2\}$

Graph each of the given intervals on a separate number line and give the set-builder notation for each. See Example 5.

13. [1, 4]
14. (−2, 2)
15. (−3, 1]
16. [−6, −2)
17. [−3, +∞)
18. (0, +∞)
19. (−∞, 5]
20. (−∞, −1)
21. (14, 28)
22. $\left[\frac{1}{2}, \frac{9}{2}\right]$
23. $\left(-\frac{3}{4}, \frac{9}{4}\right)$
24. $\left(-3, -\frac{1}{2}\right)$

Graph the given interval pairs on the same number line and give the interval notation for each interval.

EXAMPLE: $\{x \mid x < -2\}; \{x \mid 1 \leq x \leq 3\}$
Interval notation: $(-\infty, -2); [1, 3]$

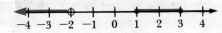

25. $\{x \mid x > 0\}; \{x \mid x < 0\}$
26. $\{x \mid x > 1\}; \{x \mid -3 < x \leq -2\}$
27. $\{y \mid -2 < y < 3\}; \{y \mid y \geq 4\}$
28. $\{x \mid x \leq -2\}; \{x \mid 0 \leq x \leq 3\}$

Graph the given interval pairs on the same number line and give the set-builder notation for each interval.

29. (−∞, 2); [3, +∞)
30. [−4, 0); [0, 2)
31. (0, 1]; (2, +∞)
32. (−∞, 1]; (2, 5]

2.6 Solving Linear Inequalities

In Section 2.1 we learned how to solve linear equations such as $x - 10 = 2$ and $\frac{3}{7}x - 4 = x + 3$. If we replace the equal sign in a linear equation with one of the four inequality signs ($<$, $\leq$, $>$, or $\geq$), the resulting expression is called a **linear inequality**. Thus $x - 10 \leq 2$ and $\frac{3}{7}x - 4 > x + 3$ are linear inequalities in the variable x. Remember that $ax + b = c$, with $a \neq 0$, is the general form of a linear equation. Replacing the equal sign here with one of the four inequality signs generates the general form for linear inequalities.

If the substitution of a particular number for the variable in the inequality results in a true statement, the number is called a **solution** of the inequality. (For example, 10 is a solution of $x - 10 \leq 2$.) To solve an inequality we find all solutions of the inequality, which is called the **solution set** of the inequality. As when solving an equation, we solve a given inequality by producing a sequence of simpler **equivalent inequalities** (inequalities with the same solution set) until we produce an inequality whose solution set is obvious. The inequalities, as opposed to equations, that we will solve will usually have solution sets made up of intervals of numbers.

Properties that are similar to those used in solving equations are used to solve inequalities. Notice that $1 < 5$ and $1 + 2 < 5 + 2$. *In geometric terms, adding 2 to a number corresponds to shifting the number 2 units to the right.* (See Figure 2.3.)

Figure 2.3

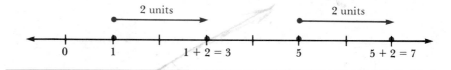

In general, if a is to the left of b ($a < b$), the point "c units to the right of a" is to the left of the point "c units to the right of b" ($a + c < b + c$). This leads to the following property of inequalities, which is stated for the expressions P, Q, and R that contain the same variable or no variable.

Addition Property for Inequalities

$$P < Q$$

and

$$P + R < Q + R$$

are equivalent inequalities.

The preceding inequalities remain equivalent if we replace $<$ by any of the signs $>$, $\leq$, or $\geq$.

EXAMPLE 1

We use the Addition Property for Inequalities to solve the inequality $3x + 5 < 2x + 7$. Notice that we collect the addends containing x on one side of the inequality and place those not involving x on the other side, as would be done when solving equations.

$3x + 5 < 2x + 7$	
$3x + 5 - 5 < 2x + 7 - 5$	Add -5 to both sides
$3x < 2x + 2$	Simplify
$3x - 2x < 2x + 2 - 2x$	Add $-2x$ to both sides
$x < 2$	Simplify

The solution set consists of all x less than 2, or $\{x \mid x < 2\}$; or, using the interval notation $(-\infty, 2)$. The graph of $\{x \mid x < 2\}$, or $(-\infty, 2)$ is shown next.

There is one important difference between the techniques for solving linear equations and those for solving inequalities. This difference involves multiplication by negative numbers. We know that $2 > 1$ and $-2 < -1$. Geometrically, these inequalities state that 2 is to the right of 1 and -2 is to the left of -1. This is a direct consequence of the way we set up a coordinate system on a number line. (See Figure 2.4.)

Figure 2.4

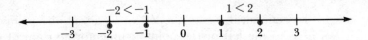

We can view the inequality $-2 < -1$ as being obtained from the inequality $2 > 1$ by multiplying both sides of $2 > 1$ by -1 and reversing the inequality sign.

Suppose we start with $2 < 3$, then multiply each side by -2, and *reverse* the inequality sign. We get $(-2)2 > (-2)3$ or $-4 > -6$ which is correct. (See Figure 2.5.)

Figure 2.5

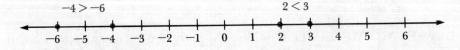

We should note, however, that multiplying both sides of the inequality $2 < 3$ by positive 2 results in a correct statement *without* reversing the inequality sign. Thus

$$2 < 3$$

and

$$2 \cdot 2 < 2 \cdot 3$$

or

$$4 < 6$$

These examples suggest the general rule that we state for expressions P, Q, and R that contain the same variable or no variable. The rule remains valid if we replace $<$ by any of the signs $>$, $\leq$, or $\geq$.

Multiplication Property for Inequalities

I.
$$P < Q$$
and
$$P \cdot R < Q \cdot R \quad \text{for } R > 0$$
are equivalent inequalities.

II.
$$P < Q$$
and
$$P \cdot R > Q \cdot R \quad \text{for } R < 0$$
are equivalent inequalities.

EXAMPLE 2 Solve $-3x < 9$ and graph the solution set.

Solution:

From $\quad -3x < 9$

we have $\quad x > -3 \quad$ Multiply both sides by $-\frac{1}{3}$ and change $<$ to $>$.

The solution set is then $\{x \mid x > -3\}$ or $(-3, \infty)$, and is shown in the graph.

EXAMPLE 3 Solve $4x + 2 \leq x + 8$ and graph the solution set.

Solution: $\quad 4x + 2 - 2 \leq x + 8 - 2 \quad$ Add -2 to both sides

$\qquad\qquad\quad 4x \leq x + 6$

$\qquad\qquad\quad 4x - x \leq x + 6 - x \quad$ Add $-x$ to both sides

$\qquad\qquad\quad 3x \leq 6$

$\qquad\qquad\quad x \leq 2 \qquad\qquad\qquad$ Multiply by $\frac{1}{3}$

The solution set $\{x \mid x \leq 2\}$ or $(-\infty, 2]$, is shown in the graph.

EXAMPLE 4 Solve $-2(2y - 3) + 1 \geq 5y - 2$ and graph the solution set.

Solution: First we must use the distributive law to get

$$-4y + 6 + 1 \geq 5y - 2$$

Simplify each side

$$-4y + 7 \geq 5y - 2$$

Collect addends containing y on one side and place those not containing y on the other side by using the addition property.

$$-4y + 7 - 5y \geq 5y - 2 - 5y \qquad \text{Add } -5y \text{ to both sides}$$
$$-9y + 7 \geq -2$$
$$-9y + 7 - 7 \geq -2 - 7 \qquad \text{Add } -7 \text{ to both sides}$$
$$-9y \geq -9$$

Now use the multiplication property.

$$\left(-\frac{1}{9}\right)(-9y) \leq \left(-\frac{1}{9}\right)(-9) \qquad \text{Multiply by } -\frac{1}{9}, \text{ and change } \geq \text{ to } \leq.$$

$$y \leq 1$$

The graph of the solution set $\{y \mid y \leq 1\}$, or $(-\infty, 1]$ is shown.

EXAMPLE 5 Solve $\frac{5}{2}k - 1 \geq \frac{1}{5}(8k + 1)$ and graph the solution set.

Solution: First we multiply by 10, the smallest common multiple of 5 and 2, to eliminate the fractions.

$$10\left[\frac{5}{2} \cdot k - 1\right] \geq 10\left[\frac{1}{5} \cdot (8k + 1)\right] \qquad \text{Multiply by 10}$$

$$25k - 10 \geq 16k + 2 \qquad \text{Distribute and simplify}$$

$$25k - 10 + 10 \geq 16k + 2 + 10 \qquad \text{Add 10 to both sides}$$

$$25k \geq 16k + 12$$

$$25k - 16k \geq 16k + 12 - 16k \qquad \text{Add } -16k \text{ to both sides}$$

$$9k \geq 12$$

$$k \geq \frac{4}{3} \qquad \text{Multiply by } \frac{1}{9}$$

The solution set $\left\{k \mid k \geq \frac{4}{3}\right\}$, or $\left[\frac{4}{3}, \infty\right)$, is shown in the graph.

EXAMPLE 6

A civic club wants to produce and sell a certain poster as a fund-raising device. The production cost is $1.50 per poster and the selling price will be $4 per poster. There are additional expenses of $1200 independent of production or sales. How many posters must be sold in order for the club to make a profit?

Solution: Let m = number of posters that must be sold. Then (in dollars)

$$\text{Total cost} = 1.5m + 1200$$

$$\text{Total revenue} = 4m$$

We want profit > 0. Since profit = total revenue − total cost, we want

$$\text{total revenue} - \text{total cost} > 0$$

$$4m - (1.5m + 1200) > 0$$

$$2.5m - 1200 > 0$$

$$2.5m > 1200$$

$$m > \frac{1200}{2.5} = 480$$

At least 481 posters must be sold in order for the club to earn a profit.

EXAMPLE 7

(a) Solve $2(x + 1) < 2x - 4$.
(b) Solve $3(2x - 1) < 6x + 7$.

Solution:
(a) $2x + 2 < 2x - 4$
   $2 < -4$ 　　　Add $-2x$ to both sides

The last statement is false. There are no solutions to this inequality; the solution set is $\emptyset$.

(b) $6x - 3 < 6x + 7$
   $-3 < 7$

The last statement is true. Any real number is a solution of this inequality; the solution set is **R**.

Exercises 2.6

Standard Assignment: Exercises 1, 3, 5, 7, 11, 13, 15, 19, 25, 31, 35, 37, 41, 51, 53, 57, 59

A Solve each inequality and graph its solution set. See Examples 1–5.

1. $2x < 14$ 　　　2. $3y \geq 9$ 　　　3. $5k \leq -10$

4. $-4z > 16$
5. $-3x < 7$
6. $-2s \geq -8$
7. $\dfrac{3y + 11}{4} < 2$
8. $\dfrac{5t - 6}{3} > 3$
9. $-\dfrac{2}{3}x + 1 \leq 7\left(1 - \dfrac{2}{21}x\right)$
10. $2 - \dfrac{3}{7}z \leq 1$
11. $-\dfrac{2}{5}k > -1$
12. $-\dfrac{3}{4}s \leq 18$
13. $-2x + 5 < -\dfrac{1}{2}$
14. $3y - \dfrac{1}{2} > 0$
15. $-t \geq -\dfrac{3}{5} + t$
16. $\dfrac{k - 2}{2} - 1 < 4$
17. $3 - \dfrac{z + 6}{7} \geq 2$
18. $2 - \dfrac{5}{3}x \leq 13$
19. $\dfrac{2}{17}x \leq 0$
20. $\dfrac{-9}{25}y > 0$
21. $3 - 0.2t < 20$
22. $-1.2s \leq 2.4$
23. $\dfrac{2}{3}k > \dfrac{7}{5}k$
24. $-\dfrac{7}{2}z \leq \dfrac{5}{8}z$
25. $\dfrac{x - 2x}{3} \leq 7$
26. $\dfrac{3y - 5y}{4} \geq -2$
27. $\dfrac{2t - 7t}{3} < -15$
28. $3k + 4 > 9 + 2k$
29. $z + 3 \leq 7 - z$
30. $2(3x - 1) > 2(5x + 8)$
31. $5 - 2(y - 1) \leq 2(4 + y)$
32. $2(t + 4) < 7t + 2$
33. $4(2x + 1) + 5 \geq 2(3x) + 5$
34. $\dfrac{z}{3} - 4 \geq \dfrac{z}{5}$
35. $s + 1 < \dfrac{3}{2}s - 3$
36. $36(3 - k) \leq 15(k - 3)$
37. $\dfrac{2(t - 4)}{10} > 2 - \dfrac{t}{5}$
38. $\dfrac{3(4y - 3)}{4} \geq y - (4y - 3)$
39. $1 - \dfrac{x}{3} < \dfrac{1}{3}(2x - 3)$
40. $\dfrac{1}{2}(z - 9) - \dfrac{1}{6}(z - 3) \geq -1$
41. $\dfrac{2}{5}(k - 5) - \dfrac{2}{3}(2 - k) \leq 2$
42. $\dfrac{2}{3}(2s - 1) + \dfrac{2}{5}\left(s - \dfrac{1}{2}\right) > 0$
43. ▦ $6.732y \leq -25.874$
44. ▦ $12.421z - 16.327 > -10.9044$
45. ▦ $52.316x - 14.711 < 302.106$
46. ▦ $25.1007 - 32.1264t \geq 10.2007$
47. ▦ $2.361(1.4043 - 5.712k) \leq -14.6055$
48. ▦ $-13.6061y + 8.7723 > 46.8y - 10.10101$
49. ▦ $6.5031(24.306z - 13.4662) < -7.4551z - 18.604$

50. $\dfrac{2.431t}{13.475 - 16.222} < 8.6116 - 4.723t$

Find the number described in each of the following statements.

51. Three times the number when added to 1 exceeds 16.
52. Ten times two-thirds of the number is negative.
53. One-half of the difference of twice the number and three is no more than 5.
54. When four times the number is added to six, the result is positive.
55. One-third of the number minus 4 is less than one-fifth of the number.

Solve the following word problems. See Example 6.

56. Elaine has taken three exams and earned scores of 85, 72, and 77. She must have an average of at least 80 to earn a B in the course. What may she score on the fourth (and last) 100-point exam to obtain a B?
57. John has three times as many nickels and twice as many quarters as dimes. How many dimes could he have if he has at least 48 coins?
58. How much cream that is 30% butterfat must be added to milk that is 3% butterfat in order to have 270 quarts that are at least 4.5% butterfat?
59. The cost of an amplifier to a retailer is $340. At what price can the amplifier be sold if the retailer wants to make a profit of at least 20% of the selling price?
60. A company produces a coin purse at a cost of $3 each and sells the purse for $5 each. If the company had initial fixed costs of $4000, how many purses must be sold to earn a profit in excess of $3000?

2.7 Compound Sentences of Inequality

A sentence that is formed by joining two sentences with a connective word such as *and* or *or* is called a **compound sentence**. In Section 2.5 (see Example 3) we introduced the notation $-2 < x < 3$ as a more compact way of asserting the compound sentence $-2 < x$ and $x < 3$. The solution of this compound sentence is the set of numbers that are common to the solution sets of $-2 < x$ and $x < 3$. The top graph in Figure 2.6 (next page) shows the solution set of $-2 < x$, the middle graph shows the solution set of $x < 3$, and the bottom graph shows the numbers that are common to both graphs.

The set of elements that are common to two sets is called the **intersection** of the two sets. The symbol ∩ is used to indicate the intersection of two sets. We saw earlier that

$$\{x \mid -2 < x < 3\} = \{x \mid -2 < x\} \cap \{x \mid x < 3\}$$

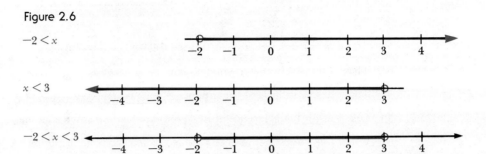

Figure 2.6

In general, for any two sets A and B,

$$A \cap B = \{x \mid x \in A \text{ and } x \in B\}$$

EXAMPLE 1

Let $A = \{1, 2, 3, 4, 5, 6\}$, $B = \{2, 4, 6, 8, 10\}$, and $C = \{1, 2, 3, 7, 8, 9\}$. Find

(a) $A \cap B$
(b) $A \cap C$
(c) $(A \cap B) \cap C$

Solution:
(a) The elements common to both A and B form

$$A \cap B = \{2, 4, 6\}$$

(b) The elements common to both A and C form

$$A \cap C = \{1, 2, 3\}$$

(c) The elements common to both $A \cap B$, and C form

$$(A \cap B) \cap C = \{2\}$$

EXAMPLE 2

Solve $2x + 5 \leq 3$ and $14 - 5x > 4$.

Solution: First we solve $2x + 5 \leq 3$ and $14 - 5x > 4$ separately:

$$2x + 5 \leq 3 \qquad 14 - 5x > 4$$
$$2x \leq -2 \qquad -5x > -10$$
$$x \leq -1 \qquad x < 2$$

Now we graph $\{x \mid x \leq -1\}$ and $\{x \mid x < 2\}$ separately, since they are helpful in determining that the intersection of these two sets is $\{x \mid x \leq -1\}$. The solution set for the compound sentence is $\{x \mid x \leq -1\}$.

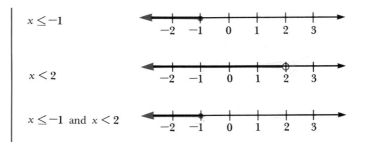

EXAMPLE 3 Solve $\frac{1}{4}(16 - 11x) \geq 4$ and $3(2x - 1) - 4 \geq 7$.

Solution: First we solve each part separately.

$$\frac{1}{4}(16 - 11x) \geq 4 \qquad 3(2x - 1) - 4 \geq 7$$
$$16 - 11x \geq 16 \qquad 6x - 3 - 4 \geq 7$$
$$-11x \geq 0 \qquad 6x \geq 14$$
$$x \leq 0 \qquad x \geq \frac{7}{3}$$

Then we graph $\{x \mid x \leq 0\}$ and $\{x \mid x \geq \frac{7}{3}\}$ separately as an aid to determining the intersection of these two sets.

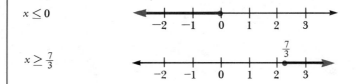

We see that there is no number that is common to these two graphs. Thus $\{x \mid x \leq 0\} \cap \{x \mid x \geq \frac{7}{3}\} = \emptyset$. The empty set, $\emptyset$, is the solution set for the given compound sentence.

If we form a compound sentence by joining two inequalities by the word *or* instead of *and*, we completely change the solution set being sought. The connective *and* requires us to find the set of numbers that satisfy *both* inequalities. The connective *or* requires us to determine the set of numbers that satisfy *either* (or both) of the two inequalities.

EXAMPLE 4 Solve $3x - 5 > 2x - 3$ or $-2x + 7 \geq 9$.

Solution: First we solve each part separately.

$$3x - 5 > 2x - 3 \qquad -2x + 7 \geq 9$$
$$3x > 2x + 2 \qquad -2x \geq 2$$
$$x > 2 \qquad x \leq -1$$

The solution set of the compound inequality is the set of numbers that solve either part, namely $\{x \mid x \leq -1 \text{ or } x > 2\}$. Unfortunately, there is no more compact way to say $x \leq -1$ or $x > 2$. The graphs of the three solution sets are shown.

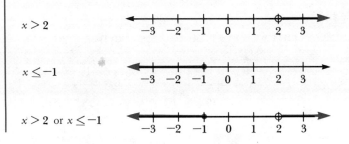

Correct Method	Incorrect Method
$x \leq -1$ or $x > 2$	$2 < x \leq -1$
	There are no such numbers!

The set of elements that belong to either or both of two sets A and B is called the **union** of A and B. The symbol $\cup$ is used to indicate the union of two sets. Thus

$$A \cup B = \{x \mid x \in A \text{ or } x \in B\}.$$

In Example 4 we saw that

$$\{x \mid x > 2\} \cup \{x \mid x \leq -1\} = \{x \mid x \leq -1 \text{ or } x > 2\}.$$

EXAMPLE 5 Let $A = \{1, 2, 3, 4, 5, 6\}$, $B = \{2, 4, 6, 8, 10\}$, and $C = \{1, 2, 3, 7, 8, 9\}$. Find

(a) $A \cup B$
(b) $(A \cup B) \cup C$
(c) $(A \cup B) \cap C$

Solution:
(a) $A \cup B = \{1, 2, 3, 4, 5, 6, 8, 10\}$.
(b) The elements appearing in either $A \cup B$ or C form

$$(A \cup B) \cup C = \{1, 2, 3, 4, 5, 6, 7, 8, 9, 10\}.$$

(c) The elements common to $A \cup B$ and C form
$$(A \cup B) \cap C = \{1, 2, 3, 8\}$$

EXAMPLE 6 Solve $3y - 13 \leq -7$ or $1 - 2y > 1 - y$

Solution: First we solve $3y - 13 \leq -7$ and $1 - 2y > 1 - y$ separately.

$$3y - 13 \leq -7 \qquad 1 - 2y > 1 - y$$
$$3y \leq 6 \qquad -2y > -y$$
$$y \leq 2 \qquad 0 > y$$

The solution of the given compound statement is the union of the separate solution sets: $\{y \mid y \leq 2\} \cup \{y \mid y < 0\} = \{y \mid y \leq 2\}$. The three graphs are shown next.

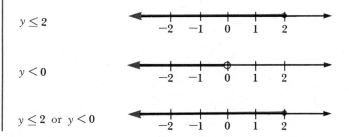

$y \leq 2$

$y < 0$

$y \leq 2$ or $y < 0$

Solving a Compound Sentence

1. Solve each inequality in the compound sentence separately.
2. If the connecting word is *and*, the solution set is the *intersection* of the solution sets of the individual inequalities.
3. If the connecting word is *or*, the solution set is the *union* of the solution sets of the individual inequalities.

Exercises 2.7

Standard Assignment: Exercises 1, 3, 5, 7, 9, 11, 13, 15, 19, 21, 25, 27, 29, 31, 33, 35, 37, 41, 43, 51, 55

A Let $A = \{a, c, e, g\}$, $B = \{b, d, f, h\}$, $C = \{a, d, e, f\}$, and $D = \{a\}$. List the elements in each of the following sets. See Examples 1 and 5.

1. $A \cap B$
2. $A \cap C$
3. $A \cup C$
4. $A \cup B$
5. $A \cap D$
6. $A \cup D$
7. $B \cup C$
8. $B \cap C$

9. $B \cap D$ 10. $B \cup D$ 11. $A \cup A$ 12. $A \cap A$
13. $B \cap \emptyset$ 14. $B \cup \emptyset$ 15. $(A \cap C) \cup B$
16. $(A \cup C) \cap B$ 17. $(A \cap B) \cap D$ 18. $(A \cup B) \cup C$
19. $(A \cap \emptyset) \cup D$ 20. $(A \cap D) \cup D$ 21. $(B \cup B) \cap B$
22. $(C \cup \emptyset) \cap \emptyset$ 23. $(\emptyset \cap \emptyset) \cup D$ 24. $(\emptyset \cup \emptyset) \cap C$

Solve each of the following compound sentences and graph each solution set. See Examples 2, 3, 4, and 6.

25. $x > 1$ and $x < 3$
26. $y \geq -2$ and $y < 1$
27. $z < 1$ or $z \geq 3$
28. $t < 0$ or $t > 2$
29. $k \geq -2$ and $k > 1$
30. $s > -3$ and $s \geq -1$
31. $x + 3 > 5$ or $2 < 1 - x$
32. $t + 4 > 9$ and $t - 3 < 0$
33. $2y + 4 \leq 6$ and $12 - y \leq 11$
34. $2z + 4 \leq 6$ or $12 - z \leq 11$
35. $9k - 2 < 1$ or $8 - 3k < 7$
36. $6s + 7 \geq -2$ and $15s + 17 \leq 2$
37. $3 - 2x \leq 4$ and $4x + 9 < 7$
38. $12y + 5 > 5$ or $6 + 3y < 3$
39. $6t + 5 < -3$ or $-6 - 5t > 4$
40. $2k - 7 < 3k$ and $-3k \geq 4k + 21$
41. $2x - 7 < 6$ and $2x - 7 > -6$
42. $-2s - 2 > 4$ and $3s - 1 > 0$
43. $y + 3 \leq 1 - y$ or $5y + 3 \geq 12 + 2y$
44. $3x + 2 < 2x + 1$ or $-x - 1 > 1 - 2x$
45. $2t < 2(t + 6)$ or $3t - 1 > 5 + 2t$
46. $5k - 12 \leq 12 - k$ and $3k - 7 < 9 - k$
47. $\frac{1}{2}(3x - 2) < \frac{5}{4}$ or $\frac{1}{5}(2x - 1) - 1 > \frac{2}{5}$
48. $y + 1 \geq -\frac{1}{3}(y + 5)$ and $\frac{1}{2}(3y + 7) - 1 < 0$
49. $5 < 4(s + 2) - 5s$ and $\frac{1}{2}s + 3 > 2$
50. $t + 5 < \frac{1}{3}(2t + 5)$ or $\left(t + \frac{3}{2}\right) + 2t > 2(t - 1)$

Graph each of the following sets.

51. $\{x \mid x \leq 0\} \cup \{x \mid x > 2\}$
52. $\{x \mid x < -3\} \cup \{x \mid x \geq -1\}$
53. $\{x \mid x < 5\} \cap \{x \mid x \geq 3\}$
54. $\{x \mid x > 0\} \cap \{x \mid x < 9\}$
55. $\{x \mid x < -2\} \cup \{x \mid x \leq 4\}$
56. $\{x \mid x \geq 3\} \cup \{x \mid x > -1\}$

57. $\{x \mid x \geq 13\} \cap \{x \mid x \leq 9\}$

58. $\{x \mid x \geq \frac{3}{2}\} \cap \{x \mid x < \frac{3}{2}\}$

59. $\{x \mid x \leq 2\} \cup \{x \mid x > 2\}$

60. $\{x \mid x < -1\} \cup \{x \mid x > -1\}$

B Decide whether each of the following is true or false for arbitrary sets A and B.

61. $A \cap A = A$

62. $A \cap B = A \cup B$

63. $A \cup A = A$

64. $A \cap \emptyset = \emptyset$

65. $A \cup \emptyset = A$

66. $A \cup \emptyset = A \cap \emptyset$

67. If $A \subset B$, then $A \cup B = A$

68. If $A \subset B$, then $A \cup B = B$

69. If $A \subset B$, then $A \cap B = A$

70. If $A \subset B$, then $A \cap B = B$

2.8 Inequalities Involving Absolute Value

In Section 2.4 we saw that there are exactly two solutions of the equation $|x| = 2$, namely -2 and 2. If x is neither 2 nor -2, then $|x| < 2$ or $|x| > 2$. (Recall the trichotomy property of Section 1.2.) The geometric content of these statements is shown next.

Statement	Geometric Interpretation	Graph
$\|x\| = 2$	x is 2 units from 0	
$\|x\| < 2$	x is less than 2 units from 0	
$\|x\| > 2$	x is more than 2 units from 0	

As we have seen, the statement $|x| < 2$ is equivalent to the compound statement $-2 < x < 2$, so that $-2 < x$ and $x < 2$. Similarly, the statement $|x| > 2$ is equivalent to the compound statement $x < -2$ or $x > 2$. The exchange of an inequality involving absolute value for an equivalent compound statement that does not involve absolute value allows us to employ the methods of the previous section in solving such inequalities.

EXAMPLE 1 Solve $|y| < 4$ and graph the solution set.

Solution: As we have seen, $|y| < 4$ describes the number y that is less than 4 units from 0. Thus, y is between -4 and 0, between 0 and 4, or $y = 0$, so that $-4 < y < 4$.

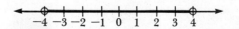

EXAMPLE 2 Solve $|x| \geq 4$ and graph the solution set.

Solution: We seek all x for which $|x| = 4$ or $|x| > 4$. Now $|x| > 4$ precisely when x is more than 4 units away from 0, or $x < -4$ or $x > 4$. Of course $|x| = 4$ when $x = -4$ or $x = 4$. Consequently $|x| \geq 4$ is equivalent to $x \leq -4$ or $x \geq 4$.

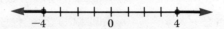

The result that the statements $|x| = 2$, $|x| < 2$, and $|x| > 2$ can be exchanged for equivalent compound statements not involving absolute value is correct if *any* positive real number a is substituted for 2. This result is stated next.

Absolute Value Inequality	Equivalent Statement	Graph
$\|x\| = a$	$x = \pm a$	
$\|x\| < a$	$-a < x < a$	
$\|x\| > a$	$x < -a$ or $x > a$	

These statements remain valid if we replace $<$ by $\leq$, and $>$ by $\geq$, and replace open circles by solid circles in the graphs.

EXAMPLE 3 Solve $\frac{1}{2}|k| \leq 3$ and graph the solution set.

Solution: First we multiply both sides of the inequality by 2 to obtain $|k| \leq 6$.

Now we use the equivalent compound statement, $-6 \leq k \leq 6$, whose graph is the following interval.

EXAMPLE 4 Solve $|2t - 3| < 4$ and graph the solution set.

Solution: We can solve this inequality in two different ways.

Solution One: We know $|x| < 4$ if and only if $-4 < x < 4$. Replacing x by $2t - 3$ we have $|2t - 3| < 4$ if and only if $-4 < 2t - 3 < 4$. Solving separately

$$-4 < 2t - 3 \quad \text{and} \quad 2t - 3 < 4$$
$$-1 < 2t \quad \text{and} \quad 2t < 7$$
$$-\frac{1}{2} < t \quad \text{and} \quad t < \frac{7}{2}$$

2.8 Inequalities Involving Absolute Value   87

The solution set is all t such that $-\frac{1}{2} < t$ and $t < \frac{7}{2}$, or $\left\{ t \mid -\frac{1}{2} < t < \frac{7}{2} \right\}$. This set is shown in the graph.

Solution Two: As in solution one, we arrive at $|2t - 3| < 4$ if and only if $-4 < 2t - 3 < 4$. This time we will work simultaneously on both parts.

$$-4 < 2t - 3 < 4$$

$$-1 < 2t < 7 \qquad \text{Add 3 to left, center, and right}$$

$$-\frac{1}{2} < t < \frac{7}{2} \qquad \text{Divide by 2 on left, center, and right}$$

Of course $\left\{ t \mid -\frac{1}{2} < t < \frac{7}{2} \right\} = \left(-\frac{1}{2}, \frac{7}{2} \right)$ is the solution set. The method of working simultaneously on both parts is usually much more convenient.

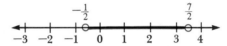

EXAMPLE 5 Solve $|4y - 6| \geq 6$ and graph the solution set.

Solution: We know $|x| \geq 6$ if and only if $x \leq -6$ or $x \geq 6$. Replacing x by $4y - 6$ we have $|4y - 6| \geq 6$ if and only if $4y - 6 \leq -6$ or $4y - 6 \geq 6$. Solving separately

$$4y - 6 \leq -6 \qquad \text{or} \qquad 4y - 6 \geq 6$$
$$4y \leq 0 \qquad \text{or} \qquad 4y \geq 12$$
$$y \leq 0 \qquad \text{or} \qquad y \geq 3$$

The solution set $\{ y \mid y \leq 0 \text{ or } y \geq 3 \}$, which is the union of $\{ y \mid y \leq 0 \}$ and $\{ y \mid y \geq 3 \}$, is shown in the graph.

EXAMPLE 6 Solve $|z - 3| \leq 2$ by interpreting the inequality in terms of distance on the number line.

Solution: We recall from Section 1.4 that $|z - 3|$ gives the distance between z and 3 on the number line. There are two such points, z, whose distance from 3 is 2: the number 1, which is two units to the left of 3, and 5, which is two units to the right of 3. Any point between 1 and 5 will be within 2 units of 3; any other point will be more than two units from 3. Thus, $|z - 3| \leq 2$ exactly when $1 \leq z \leq 5$; this is shown in the graph.

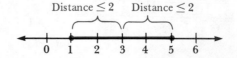

Lastly, note that the solution set of an inequality of the form $|ax + b| \geq c$ where c is negative is the entire set of real numbers since absolute value is never negative. Similarly, the solution set of an inequality of the form $|ax + b| \leq c$ where c is negative is the empty set.

Exercises 2.8

Standard Assignment: Exercises 1, 3, 5, 7, 9, 11, 17, 19, 21, 25, 27, 29, 31, 33, 35, 37, 39, 41

A Solve the following inequalities. See Examples 1, 3, 4, and 6.

1. $|x| < 3$
2. $|y| \leq 5$
3. $|t| < 1$
4. $|k| \leq 0$
5. $|x + 1| \leq 6$
6. $|t + 2| \leq -3$
7. $|k - 1| < -2$
8. $|y - 2| \leq 1$
9. $|2x - 5| < 3$
10. $|3k - 3| \leq 15$
11. $|3y + 4| < 19$
12. $|9 - 7t| < 23$
13. $\left|\dfrac{x}{2} - 1\right| < 3$
14. $\dfrac{1}{2}|y + 3| \leq 5$
15. $\left|\dfrac{k}{4} + \dfrac{1}{2}\right| < \dfrac{3}{4}$

Solve the following inequalities. See Examples 2 and 5.

16. $|z| > 5$
17. $|y| \geq 2$
18. $|t| > 0$
19. $|k| \geq 0$
20. $|x - 7| > 2$
21. $|z + 3| \geq 5$
22. $|2 - k| \geq 4$
23. $|7 - t| > 3$
24. $|3y + 2| \geq 4$
25. $|8x - 7| > -1$
26. $|2z - 1| > 7$
27. $|4x + 1| > 0$
28. $\left|2t - \dfrac{7}{2}\right| - \dfrac{1}{2} > 0$
29. $|2k - 1| - 3 \geq 2$
30. $\left|\dfrac{2}{3}z - 1\right| - 2 > \dfrac{1}{3}$

Solve the following inequalities.

31. $|x - 1| \leq 3$
32. $|t + 2| > 7$
33. $|4 - k| < 9$
34. $|3s + 12| > 0$
35. $|2y - 1| \leq 0$
36. $|2z + 3| + 1 \geq 6$
37. $\left|\dfrac{3}{5}x - 2\right| \leq 4$
38. $\left|\dfrac{t}{2} - \dfrac{1}{3}\right| > \dfrac{2}{3}$
39. $|7 - 2y| < 9$
40. $|3k - 15| \geq 9$
41. $\left|3t - \dfrac{7}{3}\right| < 5$
42. $|1 - 2x| + 3 > 10$
43. $\left|\dfrac{z}{2} - 1\right| - 1 \leq \dfrac{1}{2}$
44. $\left|\dfrac{y}{3} - \dfrac{1}{2}\right| < 0$
45. $\left|\dfrac{x - 8}{2}\right| \leq 4$
46. $\left|\dfrac{2k + 15}{3}\right| > 4$
47. $-3|1 - 4t| < -3$

Solve the following inequalities by using a calculator.

48. ▦ $|3.769x - 4.321| \geq 3.2074$
49. ▦ $|14.6302y - 3.7261| < 12.326$
50. ▦ $|0.8721z - 0.6432| \leq 2.30581$
51. ▦ $|152.071 + 20.342t| > 57.411$
52. ▦ $|3.7162x + 1.0471| < 7.9426$
53. ▦ $|2.074z - 4.3271| > 6.9432$
54. ▦ $|5.4731x + 2.94106| > 11.7049$
55. ▦ $|-3.752 - 8.6141x| \geq 0$

Find all numbers described in each of the following statements.

B   56. The absolute value of a number when increased by 3 is greater than 1.

57. Twice the absolute value of a number is less than 7.

58. The absolute value of the sum of a number and 9 is no more than 11.

59. The absolute value of a number when increased by one-fourth of itself exceeds 35.

60. The absolute value of one-half a number when decreased by 4 is not less than 50.

Chapter 2 Summary

Key Words and Phrases

2.1 Equation
Solution
Root
Solution set
Linear equation in one variable
Equivalent equations

2.2 Formula
Solving for specified symbol (or variable)

2.4 Absolute value

2.5 Interval
Interval notation

2.6 Linear inequality
Equivalent inequalities

2.7 Compound sentence
Intersection
Union

Key Concepts and Rules

For expressions P, Q, and R, the equation
$$P = Q$$

is equivalent to
$$P + R = Q + R$$
and to
$$P \cdot R = Q \cdot R \quad \text{if } R \neq 0$$

For expressions P, Q, and R, the inequality
$$P < Q$$
is equivalent to
$$P + R < Q + R,$$
$$P \cdot R < Q \cdot R \quad \text{if } R > 0,$$
and to
$$P \cdot R > Q \cdot R \quad \text{if } R < 0$$

For positive c

$|ax + b| = c$ is equivalent to $ax + b = c$ or $ax + b = -c$

$|ax + b| < c$ is equivalent to $-c < ax + b < c$

$|ax + b| > c$ is equivalent to $ax + b < -c$ or $ax + b > c$

Review Exercises

If you have difficulty with any of the problems, look in the section indicated by the numbers in square brackets.

[2.1] Solve the following equations.

1. $5x - 12 = 23$
2. $0.3y - 1 = 0.2$
3. $7 - z = 13$
4. $2(1 - 2t) = 23 + 3t$
5. $1 - 2k = k - 3$
6. $6x - 3x + 2 - x = 6$
7. $9 - y - 2 = 3y - 7 - 2y - 6$
8. $5(2z - 1) - 3 = 4z - 3$
9. $6t - 7 = 5(t - 2) + 3$
10. $3(2x + 10) = 7(5 + 3x)$
11. $1 - (5 + 9k) = 6 - (4k - 1)$
12. $10(y - 5) - 6(5 - y) = 0$
13. $7(5z + 2) = 7 - 3(5z - 9)$
14. $\dfrac{7x}{6} = \dfrac{x}{2} + 2$
15. $\dfrac{4(t + 2)}{9} = \dfrac{t - 5}{3}$
16. $\dfrac{y - 2}{4} = \dfrac{y + 1}{7} - \dfrac{3}{4}$
17. $\dfrac{15x + 9}{5} - 2x - 12 = -10$
18. $\dfrac{-7}{z} = \dfrac{1}{4}$

[2.2] Solve for the specified variable.

19. $y = 3x - 7$, for x

20. $W_1 D_1 = W_2 D_2$, for D_2

21. $T = \dfrac{2B}{B - 1}$, for B

22. $S = \dfrac{a}{1 - r}$, for r

Solve for the requested value.

23. A circular lens has a circumference of 22 centimeters. Find its radius.

24. A rectangle has a perimeter of 18 inches and a length of 5 inches. Find the width of the rectangle.

25. A trapezoid has an area of 32 square meters and a height of 8 meters. If one base is 5 meters, what is the length of the other base?

26. What principal must be deposited for 4 years at 7% annual simple interest in order to earn $354.20 in interest?

27. The volume of a box is 4212 centimeters. If it is 27 centimeters long and 12 centimeters wide, how high is the box?

28. The ratio of the current assets of a business to its current liabilities is called the *current ratio* of the business. If the current ratio is 2.7 and the current assets total $256,500, find the current liabilities of the business.

29. A cylindrical shaft has a volume of 8750π cubic centimeters. If the radius is 5 centimeters, how long is the shaft?

[2.3] Express the following statements as mathematical expressions and equations.

30. The sum of a number decreased by 4 and 5

31. The number minus twice its negative equals -4

32. 8 less than one-half a number

33. The difference of a number and its reciprocal

34. A number when increased by 15% equals 450

35. Five times a number equals the quotient of that number and the sum of the number and 4

Solve the following word problems.

36. A car dealer deducted 15% of the selling price of a car he had sold on consignment and gave the balance, which was $2210, to the owner. What was the selling price of the car?

37. The third angle in an isosceles triangle measures 40 degrees more than twice either base angle. Find the number of degrees in each angle of the triangle.

38. A 600-mile trip took 12 hours in all. Half of the time was across flat terrain and half was through hilly terrain. If the average rate over the hilly section was

20 miles per hour slower than the average rate over the flat section, find the two rates and the distance traveled at each rate.

39. Find three consecutive integers such that when twice the smallest is subtracted from 4 times the largest, the result is 25 less than 3 times the middle integer.

40. A total of $30,000 is invested in two stocks. At the end of the year one returned 6% and the second 8% on the original investment. How much was invested in each if the total profit was $2160?

41. The monthly note on a car that was leased for 2 years was $250 less than the monthly note on a car that was leased for a year and a half. The total income from the two leases was $21,300. Find the monthly note on each.

42. Two solutions, one containing $4\frac{1}{2}$% iodine and the other containing 12% iodine are to be mixed to produce 10 liters of a 6% iodine solution. How many liters of each are required?

43. In 20 years Jack will be 5 times as old as he is now. How old is he now?

44. A child's coin bank contains pennies, nickels, and dimes. There are 3 more nickels than pennies and twice as many dimes as there are pennies. If the coins total $8.73, how many of each type does the bank contain?

45. Two cars leave from the same place at the same time, traveling in opposite directions. One travels 5 kilometers per hour faster than the other. After 3 hours they are 495 kilometers apart. How fast is each car traveling?

[2.4] Solve the following equations

46. $|x| = 3$

47. $|-y| = 7$

48. $|3 - z| = \frac{7}{2}$

49. $|2t - 3| = 15$

50. $\left|\frac{8x - 3}{5}\right| = 7$

51. $|3(k + 2) + 4| = 10$

52. $|3(y + 2)| + 4 = 10$

53. $\left|\frac{1}{2} - z\right| = |z|$

Find the number described in the following statements.

54. The absolute value of the sum of a number and 5 is 11.

55. The difference of 3 and the absolute value of a number is 14.

[2.5] Graph the following intervals.

56. $(-\infty, 6]$

57. $[3, 7)$

58. $[-5, 0]$

59. $\{y \mid y \geq -1\}$

60. $\left\{x \mid -\frac{3}{2} < x \leq \frac{3}{2}\right\}$

61. $\{t \mid -2 < t < 2\}$

[2.6] Solve the following inequalities and graph the solution set.

62. $7y \geq 28$
63. $3z - 1 < 14$
64. $2 - t \leq -8$
65. $7x - 2 > 12$
66. $-\left(\dfrac{3}{4}y + 1\right) \leq 17$
67. $2k + 2 > 7(k + 1)$
68. $(5z + 2) + (1 + z) < 0$
69. $3(t - 4) + 4t \leq 2(t - 1)$
70. $\dfrac{z}{2} - 5 \geq \dfrac{4z}{9}$
71. $\dfrac{x}{5} - \dfrac{x}{7} \leq -2$
72. $(s + 2) - \dfrac{2 - s}{2} > 3(s - 2)$
73. $6[y - (4y - 3)] \leq 3(4y - 3)$
74. $\dfrac{7z + 3}{4} > 3 + \dfrac{9z - 8}{8}$
75. $\dfrac{1}{2}t - 6 < \dfrac{1}{4}(3 + 2t)$

Find all numbers described in the following statements.

76. The negative of a number is less than -3.

77. One minus two-thirds of a number is less than minus eight.

78. If the length of each side of a square is increased by 2 inches, the perimeter of the resulting square is at least 24 inches. What numbers could serve as the length of the sides of the original square?

79. A man bought a collection of used TVs from a retail outlet for $75 each. Three were beyond repair, but the rest could be repaired and sold for $125 each. How many TVs could he have originally bought if his revenue from selling them was over $3000?

[2.7] Solve the following compound statements and graph the solution sets.

80. $x \leq 5$ and $x \geq 2$
81. $y > 4$ or $y \leq 7$
82. $1 - t \leq 3$ or $2t + 1 \leq 5$
83. $2z - 3 \leq -4$ and $13 \geq 3(5 - 4z)$
84. $4(2y + 1) \geq 4$ or $3y + 2 < -8$
85. $3x - 7 < 8$ and $14 - 2x < 0$
86. $5k + 3 \leq 7$ or $5(3k - 2) > 3$

Graph the following sets.

87. $\{t \mid t > -3\} \cap \{t \mid t \leq 1\}$
88. $\{z \mid z \geq 2\} \cup \{z \mid z < 2\}$
89. $\{x \mid x \leq 5\} \cap \{x \mid x \leq -1\}$
90. $\{y \mid y < -4\} \cup \{y \mid y \geq 0\}$

[2.8] Solve the following inequalities and graph the solution sets.

91. $|x| < 3$
92. $|y| > 1$
93. $|9 - 3k| > 6$
94. $\left|\dfrac{t - 3}{17}\right| \geq 0$
95. $|3z - 24| < 21$
96. $|3x + 1| - 2 \leq 2$

97. $\left|\dfrac{1-y}{2}\right| \leq 1$ 98. $\left|\dfrac{1}{2}k + 14\right| > 6$ 99. $\left|\dfrac{t}{7} - 24\right| < -5$

100. $\left|\dfrac{x}{2} + 5\right| = \left|\dfrac{7-x}{2}\right|$

Find all numbers described by the following statements.

101. Five times the absolute value of a number is not more than 25.

102. The absolute value of the sum of a number and 5 is less than 13.

103. If one-half the absolute value of a number is reduced by 3, the result is greater than 9.

104. The absolute value of three-fifths of the sum of a number and 4 is not less than 2.

Practice Test (60 minutes)

Solve the following.

1. $3y + 2 = 2y - 5$
2. $4 - 6(z - 2) = 4z$
3. $5(x + 1) - 3x = -2$
4. $\dfrac{t}{3} - \dfrac{t}{3} = 1$
5. Solve $I = \dfrac{E}{R + (r/n)}$ for n

6. A bus travels 1600 kilometers in 21 hours and 20 minutes. Find the rate of the bus.

7. A rectangle has a perimeter of 42 meters and a length of 13 meters. Find the width of the rectangle.

8. A total of $25,000 is invested in two ventures. One pays 7% and the other $8\frac{1}{2}$% at the end of one year. The total return on the $25,000 investment is $1877.50. How much was invested at each rate?

9. Find three consecutive integers whose sum is 90.

10. How many grams of pure copper must be added to 20 grams of an alloy containing 40% copper to produce an alloy containing 60% copper?

Solve the following equations.

11. $|-z| = 3$
12. $|5x + 3| = 8$

Solve the following inequalities and graph the solution sets.

13. $t + 7 \leq 4$
14. $7x + 2(x - 1) \geq x$
15. $2y - 1 < -6$ or $2y - 1 > 6$
16. $-5 < 2x - 3$ and $2x - 3 > 5$
17. $8k - 21 \leq -7$ or $8k - 21 \geq 7$
18. $|t| > 4$
19. $|2z + 28| < 18$
20. $|2x - 5| \geq 17$

Extended Applications

Saving Energy

The effectiveness of a given material as an insulator against heat loss is indicated by a number called its R-value. The larger the R-value, the more effective the material is as an insulator. When assessing the reduction in heating costs resulting from the installation of insulation with R-value R_1 in a structure with existing R-value R_0, the following formula for seasonal savings, in dollars, is often used

$$S = \frac{R_1 CFWA}{R_0(R_0 + R_1)}$$

The variables take into account certain factors determined by the location of the structure and the type and cost of the fuel used to heat the structure. In particular,

C = cost of fuel (per gallon, kwh, or other designated unit) in dollars
F = fuel factor (accounts for amounts of heat produced by the various fuels)
W = climate factor (accounts for weather conditions during the heating season)
A = area to be heated (in square feet)
R_0 = R-value for existing structure
R_1 = R-value of the added insulation
S = reduction in heating cost (resulting from the added insulation)

Exercises

1. Solve $S = \dfrac{R_1 CFWA}{R_0(R_0 + R_1)}$ for R_1.

2. The ceiling of an oil-heated warehouse in northern Maine ($W = 230$) has an R-value of 3. The area of the ceiling is 11,000 sq ft. The fuel factor for oil is 0.009 and the price is \$1.52 per gallon. If the ceiling is covered with 2 inches of cellulose (R-value $= 3.7$ per inch), what reduction in heating costs results?

3. If the situation described in Exercise 2 occurred in north Louisiana ($W = 62$), what reduction in heating costs would result?

4. If the added insulation in Exercise 2 is perlite (R-value $= 2.6$ per inch) instead of cellulose, what reduction in heating costs would result?

5. The owner of an apartment added one inch of insulation to the existing ceiling, which had R-value 2 and an area of 1500 sq ft. The apartment is located in Tennessee ($W = 86$) and is oil-heated with $C = \$1.46$ and $F = 0.009$. The owner projects that over the next heating season the heating cost reduction would be about \$550. If he is correct, what must the R-value of the new insulation be?

3
Exponents and Polynomials

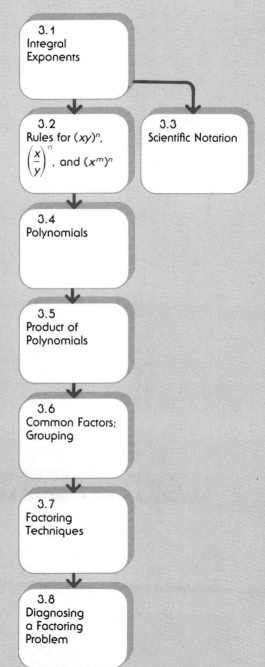

3.1 Integral Exponents

3.2 Rules for $(xy)^n$, $\left(\dfrac{x}{y}\right)^n$, and $(x^m)^n$

3.3 Scientific Notation

3.4 Polynomials

3.5 Product of Polynomials

3.6 Common Factors; Grouping

3.7 Factoring Techniques

3.8 Diagnosing a Factoring Problem

3.1 Integral Exponents

You will recall that if $c = ab$, then a and b are called **factors** of c. Thus, since $15 = 3 \cdot 5$, 3 and 5 are factors of 15. Other factors of 15 are -3, -5, 1, 15, -1, and -15.

We use exponents to denote products of repeated factors. For example, $3^4 = 3 \cdot 3 \cdot 3 \cdot 3$.

The number 4 indicates that 3 appears as a factor 4 times. In this context, the number 4 is called the **exponent,** and 3 is called the **base.** In general, for any real number x and any positive integer n, we write

$$x^n = \underbrace{x \cdot x \cdot x \cdot \cdots \cdot x}_{n \text{ factors of } x}$$

and $\qquad x^1 = x \qquad$ (3.1)

The symbol x^n is read "the nth **power** of x."

EXAMPLE 1
(a) $9^1 = 9$
(b) $5^3 = 5 \cdot 5 \cdot 5 = 125$
(c) $0^2 = 0 \cdot 0 = 0$
(d) $\left(-\dfrac{1}{5}\right)^3 = \left(-\dfrac{1}{5}\right) \cdot \left(-\dfrac{1}{5}\right) \cdot \left(-\dfrac{1}{5}\right) = -\dfrac{1}{125}$
(e) $2x^3 = 2 \cdot x \cdot x \cdot x$

An exponent applies to the base to which it is attached. Thus, using Definition (3.1) $10x^2 = 10 \cdot x \cdot x$ while $(10x)^2 = 10x \cdot 10x$.

EXAMPLE 2
Name the base and the exponent.
(a) $(10x)^2$
(b) $10x^2$
(c) $(-3)^6$
(d) -3^6

Solution:
(a) The base is 10x and the exponent is 2.
(b) The base is x and the exponent is 2.
(c) The base is -3 and the exponent is 6.
(d) Now $-3^6 = -(3^6)$, consequently the base is 3 and the exponent is 6.

WARNING

In fact,
$$(-3)^6 \neq -3^6$$
$$(-3)^6 = (-3) \cdot (-3) \cdot (-3) \cdot (-3) \cdot (-3) \cdot (-3)$$
$$= 729$$

but
$$-3^6 = -(3 \cdot 3 \cdot 3 \cdot 3 \cdot 3 \cdot 3) = -729$$

EXAMPLE 3

Express the following by using exponents.

(a) $x \cdot x \cdot y \cdot y \cdot y$
(b) $5x \cdot x \cdot x \cdot y \cdot y$
(c) $(2x) \cdot (3x) \cdot (4x)$
(d) $x^2 \cdot x^5$

Solution:
(a) Using the associative property for the product of real numbers, we have
$$x \cdot x \cdot y \cdot y \cdot y = (x \cdot x) \cdot (y \cdot y \cdot y)$$
$$= x^2 \cdot y^3$$

(b) Again, by the associative property
$$5x \cdot x \cdot x \cdot y \cdot y = 5 \cdot (x \cdot x \cdot x) \cdot (y \cdot y)$$
$$= 5 \cdot x^3 \cdot y^2$$

Recall that we may write $5x^3y^2$ for $5 \cdot x^3 \cdot y^2$.

(c) By repeated use of the associative and commutative properties for the product of real numbers, we have
$$(2x) \cdot (3x) \cdot (4x) = (2 \cdot 3 \cdot 4)(x \cdot x \cdot x)$$
$$= 24x^3$$

(d) $x^2 \cdot x^5 = (x \cdot x) \cdot (x \cdot x \cdot x \cdot x \cdot x)$
$$= x \cdot x \cdot x \cdot x \cdot x \cdot x \cdot x$$
$$= x^7$$
$$= x^{2+5}$$

Example 3(d) suggests that *to multiply factors with the same base we add the exponents.* That is,

$$x^m \cdot x^n = \underbrace{x \cdot x \cdot x \cdots x}_{m \text{ factors}} \underbrace{x \cdot x \cdot x \cdots x}_{n \text{ factors}}$$

$$= \underbrace{x \cdot x \cdot x \cdots \cdot x}_{m+n \text{ factors}}$$

$$= x^{m+n}$$

Thus, for any real number x and any natural numbers m and n, we have the following rule.

Product Rule

$$x^m x^n = x^{m+n} \tag{3.2}$$

EXAMPLE 4

(a) $x^{25} \cdot x^{26} = x^{25+26} = x^{51}$
(b) $(5z^3)(-6z^7) = 5(-6)z^{3+7} = -30z^{10}$
(c) $(-3y^4)(-2y^5) = (-3)(-2)y^{4+5} = 6y^9$
(d) $2^3 \cdot 2^7 \cdot 2^5 = 2^{3+7+5} = 2^{15}$

Often we want to simplify quotients such as

$$\frac{5^9}{5^9}, \quad \frac{6^9}{6^4} \quad \text{and} \quad \frac{10^4}{10^9}$$

The first quotient is obviously equal to 1; the second is calculated by using Equation (3.2) to write $6^9 = 6^4 \, 6^5$:

$$\frac{6^9}{6^4} = \frac{6^4 \, 6^5}{6^4} = 6^5 = 6^{9-4}$$

so the original exponents, 9 and 4, determine the result. Also

$$\frac{10^4}{10^9} = \frac{10^4}{10^4 \, 10^5} = \frac{1}{10^5} = \frac{1}{10^{9-4}}$$

again, the original exponents, 9 and 4, determine the result. These calculations suggest the general rules

$$\frac{x^m}{x^n} = \begin{cases} 1 & \text{if } m = n \\ x^{m-n} & \text{if } m > n \\ \dfrac{1}{x^{n-m}} & \text{if } m < n \end{cases} \tag{3.3}$$

where m and n are positive integers and $x \neq 0$. These rules are more complicated than is necessary. If the purpose of mathematics is to make our life simpler, we ought to simplify these rules by combining all three cases ($m = n$, $m > n$, and $m < n$) into one. This is done in the following way.

Quotient Rule

$$\frac{x^m}{x^n} = x^{m-n} \tag{3.4}$$

This is correct for all integer exponents m and n ($x \neq 0$). All that is needed now is to make the right-hand side of Equation (3.4) meaningful for $m \leq n$. This requires the introduction of zero and negative exponents. For any real number $x \neq 0$ and positive integer n, we define

$$x^0 = 1$$

and

$$x^{-n} = \frac{1}{x^n} \tag{3.5}$$

As for 0^0, it is undefined.

We now see that the quotient rule in Equation (3.4),

$$\frac{5^9}{5^9} = 5^{9-9} = 5^0 = 1$$

gives the same result as Equation (3.3).

Again, in

$$\frac{10^4}{10^9} = 10^{4-9} = 10^{-5} = \frac{1}{10^5}$$

Equation (3.4) gives the same result as Equation (3.3). Similarly, for any $x \neq 0$, the results of Equation (3.3) and Equation (3.4) are identical.

$$\frac{x^m}{x^m} = x^{m-m} = x^0 = 1$$

and

$$\frac{x^m}{x^n} = \frac{1}{x^{n-m}} = x^{-(n-m)} = x^{m-n}$$

EXAMPLE 5 Eliminate negative exponents, using Equations (3.5), and simplify.
(a) -5^{-2} (b) $(-5)^{-2}$ (c) $(-5)^{-3}$

Solution:
(a) The exponent -2 applies to the base 5.

$$-5^{-2} = -\frac{1}{5^2} = -\frac{1}{25}$$

(b) The exponent -2 applies to the base -5.

$$(-5)^{-2} = \frac{1}{(-5)^2} = \frac{1}{25}$$

(c) $(-5)^{-3} = \dfrac{1}{(-5)^3} = \dfrac{1}{-125}$

EXAMPLE 6 Eliminate negative exponents.
(a) $y^5 \cdot y^{-3}$

(b) $\dfrac{9^{-2}}{9^5}$

(c) $\dfrac{5^{-3}}{5^{-7}}$

(d) $\dfrac{k^{m+1}}{k^{m+1}}$

Solution:
(a) $y^5 \cdot y^{-3} = y^5 \cdot \left(\dfrac{1}{y^3}\right)$ if $y \neq 0$

$$= \frac{y^5}{y^3}$$

$$= y^{5-3}$$

$$= y^2 \quad \text{if } y \neq 0$$

(b) $\dfrac{9^{-2}}{9^5} = 9^{-2-5} = 9^{-7} = \dfrac{1}{9^7}$

(c) $\dfrac{5^{-3}}{5^{-7}} = 5^{-3-(-7)} = 5^{-3+7} = 5^4$

(d) $\dfrac{k^{m+1}}{k^{m-1}} = k^{m+1-(m-1)} = k^2$ if $k \neq 0$

EXAMPLE 7   If $x \neq 0$, and m and n are nonnegative integers, then

(a) $x^m x^{-n} = x^{m-n}$
(b) $x^{-m} x^n = x^{-m+n}$
(c) $x^{-m} x^{-n} = x^{-m-n}$

Solution: Using Equations (3.5), the Quotient Rule, Equation (3.4), and the Product Rule, Equation (3.2), we have

(a) $x^m x^{-n} = x^m \dfrac{1}{x^n} = \dfrac{x^m}{x^n} = x^{m-n}$

(b) $x^{-m} x^n = \dfrac{1}{x^m} \cdot x^n = \dfrac{x^n}{x^m} = x^{n-m} = x^{-m+n}$

(c) $x^{-m} x^{-n} = \dfrac{1}{x^m} \dfrac{1}{x^n} = \dfrac{1}{x^{m+n}} = x^{-(m+n)} = x^{-m-n}$

The Product Rule is now generalized to:

$$x^m x^n = x^{m+n}$$

for any integers m and n, positive, negative, or zero.

Exercises 3.1

Standard Assignment: Exercises 1, 3, 5, 7, 9, 11, 13, 15, 19, 21, 23, 25, 27, 29, 31, 33, 35, 43, 45, 53, 55, 57, 63, 67

A   For Exercises 1–18, name the base and the exponent. See Examples 1 and 2.

1. 9^{10}
2. 10^9
3. 413^2
4. 215^3
5. $(-2)^4$
6. $(-3)^7$
7. -2^5
8. -3^7
9. x
10. y
11. $5x^2$
12. $10y^3$
13. a^{-11}
14. b^{-13}
15. $(a+b)^4$
16. $(x-3)^5$
17. $(3x^2 - 2x + 5)^{-3}$
18. $(\sqrt{7}\,a^3 + \sqrt{2}\,a - \pi)^{-12}$

Evaluate the expressions in Exercises 19–52.

19. 2^5
20. -3^4
21. $(-2)^4$
22. $(-7)^6$
23. $-(-3)^3$
24. $-(-2)^5$
25. $\left(\dfrac{1}{3}\right)^4$
26. $\left(\dfrac{1}{5}\right)^3$
27. 18^0
28. 25^0
29. 0^{18}
30. 0^{101}
31. 0^0
32. 0^{-2}
33. 4^{-2}
34. 5^{-3}

35. $(-2)^{-5}$ 36. $(-3)^{-3}$ 37. $\dfrac{1}{4^{-2}}$ 38. $\dfrac{1}{5^{-3}}$

39. $\dfrac{16}{4^{-2}}$ 40. $\dfrac{10}{5^{-3}}$ 41. $3^{-1} - 4^{-1}$ 42. $4^{-1} - 5^{-1}$

43. $\dfrac{3^{-1}}{4^{-1}}$ 44. $\dfrac{4^{-1}}{5^{-1}}$ 45. $\left(\dfrac{2}{3}\right)^{-3}$ 46. $\left(\dfrac{3}{2}\right)^{-2}$

47. 3.14159×10^4 48. 1.73205×10^5

49. ▦ $(1.414213)^5$

50. ▦ $(9.876543)^4$

51. ▦ $(0.141421)^{-3}$

52. ▦ $(0.132465)^{-6}$

In Exercises 53–78, eliminate negative exponents and simplify. Variable exponents represent natural numbers. See Examples 5 and 6.

53. $5^{10}\, 5^{-3}$ 54. $7^{-4}\, 7^6$

55. $3^{-2}\, 3^5$ 56. $2^7\, 2^{-5}$

57. $(-10)^3 (-10)^{-5}$ 58. $(-5)^{-3} (-5)^2$

59. $100^9 \cdot 100^{-4} \cdot 100^{-5}$ 60. $60^{13}\, 60^{-10}\, 60^{-3}$

61. $\dfrac{9^{11}}{9^7}$ 62. $\dfrac{10^9}{10^2}$

63. $\dfrac{13^{-3}}{13^4}$ 64. $\dfrac{14^{-2}}{14^5}$

65. $\dfrac{11^{-2}}{11^{-4}}$ 66. $\dfrac{12^{-3}}{12^{-5}}$

67. $\dfrac{(-3)^{-7}}{(-3)^{-4}}$ 68. $\dfrac{(-5)^{-8}}{(-5)^{-3}}$

B 69. $\dfrac{x^{k+2}}{x^{k-1}}, \quad (x \neq 0)$ 70. $\dfrac{a^n}{a^{n-3}}, \quad (a \neq 0)$

71. $\dfrac{a^{n-5}}{a^{n-1}}, \quad (a \neq 0)$ 72. $\dfrac{x^{k+1}}{x^{k+5}}, \quad (x \neq 0)$

73. $a^{15}\, a^{-18}\, a^6, \quad (a \neq 0)$ 74. $b^{-5}\, b^{14}\, b^{-6}, \quad (b \neq 0)$

75. $c^k(-2c^{5-2k})(3c^k), \quad (c \neq 0)$ 76. $3d^{3-2k}(5d^{5+2k}), \quad (d \neq 0)$

77. $\dfrac{e^{-2m}\, e^{-3m}}{e^{-6m}}, \quad (e \neq 0)$ 78. $\dfrac{5^{-3m}\, s^{-n}}{s^{-4m-2n}}, \quad (s \neq 0)$

3.2 Rules for $(xy)^n$, $\left(\dfrac{x}{y}\right)^n$, and $(x^m)^n$

We often encounter expressions such as $(xy)^n$, $\left(\dfrac{x}{y}\right)^n$, and $(x^m)^n$. For the first two expressions, we observe that

$$(xy)^n = \underbrace{(xy)(xy)\cdots(xy)}_{n \text{ factors of } xy}$$

$$= \underbrace{x \cdot x \cdots x}_{n \text{ factors of } x} \cdot \underbrace{y \cdot y \cdots y}_{n \text{ factors of } y} = x^n y^n$$

and

$$\left(\frac{x}{y}\right)^n = \underbrace{\frac{x}{y} \cdot \frac{x}{y} \cdots \frac{x}{y}}_{n \text{ factors}} = \frac{x^n}{y^n}, \quad (y \neq 0)$$

In the previous demonstrations, n designates any positive integer, but the results remain valid for all integers, positive, negative, or zero.

$$(xy)^n = x^n y^n$$
$$\left(\frac{x}{y}\right)^n = \frac{x^n}{y^n}, \quad y \neq 0 \qquad (3.6)$$

EXAMPLE 1   Use Equations (3.6) to simplify the following expressions:

(a) $\left(\dfrac{3}{5}\right)^4 = \dfrac{3^4}{5^4} = \dfrac{81}{625}$

(b) $(2a)^6 = 2^6 a^6 = 64 a^6$

(c) $(-3b)^5 = (-3)^5 b^5 = -243 b^5$

(d) $\dfrac{218^7}{109^7} = \left(\dfrac{218}{109}\right)^7 = 2^7 = 128$

(e) $\left(\dfrac{2x}{3}\right)^5 = \dfrac{(2x)^5}{3^5} = \dfrac{2^5 x^5}{243} = \dfrac{32 x^5}{243}$

EXAMPLE 2

(a) $(5^2)^3 = 5^2 \cdot 5^2 \cdot 5^2 = 5^{2+2+2}$
$= 5^6 = 5^{2 \cdot 3}$

(b) $(x^7)^2 = x^7 \cdot x^7 = x^{7+7} = x^{14} = x^{7 \cdot 2}$

Example 2 suggests that for any positive integers m and n, $(x^m)^n = x^{mn}$. In fact, the formula $(x^m)^n = x^{mn}$ remains true for both zero and negative integers m and n provided $x \neq 0$.

Power Rule

$$(x^m)^n = x^{mn} \qquad (3.7)$$

EXAMPLE 3

Use Equation (3.7) to simplify the following expressions.

(a) $(10^5)^6 = 10^{5 \cdot 6} = 10^{30}$

(b) $(2x^3)^5 = 2^5 x^{3 \cdot 5} = 32x^{15}$

(c) $(2^{-2})^3 = 2^{(-2)3} = 2^{-6} = \dfrac{1}{2^6} = \dfrac{1}{64}$

(d) $(3x^{-3})^4 = 3^4 x^{(-3)4} = 81x^{-12} = \dfrac{81}{x^{12}}, \qquad (x \neq 0)$

(e) $(a^{-1} x^{-6})^{-9} = a^{(-1)(-9)} x^{(-6)(-9)} = a^9 x^{54}, \qquad (x \neq 0)$

EXAMPLE 4

Let a and b represent nonzero real numbers. Eliminate negative exponents and simplify the following.

(a) $\dfrac{a^2}{b^{-2}}$

(b) $(3a^{-2}b^2)^3$

(c) $\dfrac{(2ab)^{-2}}{2^{-2}a^{-2}b^{-4}}$

Solution:

(a) $\dfrac{a^2}{b^{-2}} = a^2 \cdot \dfrac{1}{b^{-2}} = a^2(b^{-2})^{-1} = a^2 \cdot b^{(-2) \cdot (-1)} = a^2 b^2$

(b) $(3a^{-2}b^2)^3 = 3^3(a^{-2})^3(b^2)^3 = 27a^{(-2)(3)}b^{2 \cdot 3} = 27b^6 \cdot a^{-6}$
$= 27b^6 \cdot \dfrac{1}{a^6} = \dfrac{27b^6}{a^6}$

(c) $\dfrac{(2ab)^{-2}}{2^{-2}a^{-2}b^{-4}} = \dfrac{2^{-2}a^{-2}b^{-2}}{2^{-2}a^{-2}b^{-4}} = \dfrac{b^{-2}}{b^{-4}} = b^{-2-(-4)} = b^{-2+4} = b^2$

3.2 Rules for $(xy)^n$, $\left(\dfrac{x}{y}\right)^n$, and $(x^m)^n$

EXAMPLE 5   Let x, y, and z be nonzero real numbers. Eliminate negative exponents and simplify the following.

(a) $\dfrac{(2x^5y)^{-2}(-5x^4y^2)^2}{125(xy^4)^3(2x^2y^3)^{-1}}$ (b) $(2y^2z^{-3})^2 \left(\dfrac{yz^2}{x^2}\right)^{-1}$

Solution:

(a) $\dfrac{(2x^5y)^{-2}(-5x^4y^2)^2}{125(xy^4)^3(2x^2y^3)^{-1}} = \dfrac{2^{-2}x^{5(-2)}y^{-2}(-5)^2 x^{4\cdot 2}y^{2\cdot 2}}{125 x^3 y^{4\cdot 3} 2^{-1}x^{2(-1)}y^{3(-1)}}$

$= \dfrac{2^{-2} 25 x^{-10} y^{-2} x^8 y^4}{2^{-1} 125 x^3 y^{12} x^{-2} y^{-3}}$

$= \dfrac{2^{-2} 5^2 x^{-10} x^8 y^{-2} y^4}{2^{-1} 5^3 x^3 x^{-2} y^{12} y^{-3}}$

$= \dfrac{2^{-2-(-1)} 5^{2-3} x^{-10+8} y^{-2+4}}{x^{3+(-2)} y^{12+(-3)}}$

$= \dfrac{2^{-2+1} 5^{-1} x^{-2} y^2}{x^1 y^9}$

$= 2^{-1} 5^{-1} \dfrac{x^{-2}}{x} \cdot \dfrac{y^2}{y^9}$

$= (2\cdot 5)^{-1} x^{-2-1} y^{2-9}$

$= 10^{-1} x^{-3} y^{-7}$

$= (10x^3y^7)^{-1}$

$= \dfrac{1}{10x^3y^7}$

(b) $(2y^2z^{-3})^2 \left(\dfrac{yz^2}{x^2}\right)^{-1} = 2^2(y^2z^{-3})^2 \cdot \dfrac{(yz^2)^{-1}}{(x^2)^{-1}}$

$= 4\, y^{2\cdot 2} z^{(-3)\cdot 2} \cdot \dfrac{y^{-1} z^{2(-1)}}{x^{2(-1)}}$

$= 4y^4 z^{-6} \cdot y^{-1} z^{-2} \cdot \dfrac{1}{x^{-2}}$

$= 4y^4 y^{-1} z^{-6} z^{-2} x^{-(-2)}$

$= 4x^2 y^{4-1} z^{-6-2}$

$= 4x^2 y^3 \dfrac{1}{z^8}$

$= \dfrac{4x^2 y^3}{z^8}$

Now a^{-1} denotes the reciprocal of a, and the reciprocal of x/y is y/x; consequently $(x/y)^{-1} = y/x$. This is a useful observation in problems such as Example 5(b), where it would be helpful to observe that

$$\left(\frac{yz^2}{x^2}\right)^{-1} = \frac{x^2}{yz^2}$$

Exercises 3.2

Standard Assignment: Exercises 1, 3, 5, 7, 9, 11, 13, 15, 17, 19, 21, 23, 25, 27, 29, 31, 33, 35, 41, 45, 47

A In Exercises 1–49, eliminate negative exponents and simplify. All variables may be assumed to be nonzero. Variable exponents represent natural numbers.

1. $(a^3)^{-4}$
2. $\left[\left(\frac{3}{5}\right)^3\right]^4$
3. $(a^{-3})^4$
4. $\left[\left(\frac{3}{5}\right)^{-3}\right]^4$
5. $(2^{-2})^{-3}$
6. $(a^{-2})^{-3}$
7. $(a^{-2}b^{-1})^{-3}$
8. $(5^{-2}7^{-1})^{-3}$
9. $(31)^0 \cdot 4^3 \cdot 8^2 \cdot 16^{-4}$
10. $a^0 \cdot b^3 \cdot (2b)^2(4b)^{-4}$
11. $\left(\frac{a^{-4}}{b^6}\right)^{-5}$
12. $\left(\frac{3^{-4}}{4^6}\right)^{-5}$
13. $x \cdot x^2 \cdot x^3 \cdot x^4 \cdot x^5 \cdot x^6$
14. $10 \cdot 10^2 \cdot 10^3 \cdot 10^4 \cdot 10^5$
15. $(a^2 b^{-3} x)^4$
16. $(a^{-4}b^3 x^7)(a^2 b^4 x^9)$
17. $\left(\frac{a^3 b^4}{x^5}\right)^6$
18. $\left(\frac{a^{-4}b^{-3}}{x^5}\right)^{-7}$
19. $\dfrac{x^{-1}y^{-1}}{x^{-1} - y^{-1}}$
20. $\dfrac{x^{-3} + y^{-3}}{(xy)^{-3}}$
21. $\left(\frac{a^3 x^4}{b^6}\right)^3 \Big/ \left(\frac{a^2 x^3}{b^5}\right)^4$
22. $\left(\frac{a^3 x^2}{b^5}\right)^4 \Big/ \left(\frac{a^4 x^3}{b^2}\right)^3$
23. $(x^{-1} + y^{-1})(x^{-1} - y^{-1})$
24. $(a^{-2} - b^{-2})(a^{-2} + b^{-2})$
25. $\left(\dfrac{(x+y)^{-5}z^7}{x^3(z-2)}\right)^0$
26. $\left(\dfrac{(x-y)^3 z^4}{y(z-x)^2}\right)^0$
27. $\left(\dfrac{15x^{-2}}{4y^{-1}}\right)^{-3}\left(\dfrac{2y^2}{5x^3}\right)^{-3}$
28. $\left(\dfrac{a^{-4}}{2b^4}\right)^{-1}\left(\dfrac{b^2}{2a}\right)^{-3}$
29. $\left(\dfrac{a^2 b}{2c}\right)^{-2}\left(\dfrac{b^{-1}c^{-2}}{a^{-3}}\right)^3\left(\dfrac{ca^2}{b^3}\right)^{-4}$
30. $\left(\dfrac{xy^2}{z}\right)^{-1}\left(\dfrac{y^{-2}z^{-1}}{x^{-1}}\right)^2$

31. $\left(\dfrac{y^{-2}}{y^{-4}}\right)^{-1}$

32. $\left(\dfrac{3^2 a^2 b^0}{27 a^{-1}}\right)^{-3} \left(\dfrac{a^{-2}}{a^{-5}}\right)^2$

33. $\left[\left(\dfrac{t^{-3} x^3 y}{t^{-4} x^{-2} y^2}\right)^2\right]^{-1}$

34. $\dfrac{6}{2^{-1} + 3^{-2}}$

35. $(2^{-1} + 3^{-1})^2$

36. $(y + 3)^{-2}$

37. $\dfrac{(-7k^3 s^5 p)^2 (3^{-1} k^2 s^6)^{-1}}{21(p^{-1} s^{-1} k^2)^3 (s^{-3} k^1 p^{-4})^{-2}}$

38. $a^2 y^{-1} + a^{-1} y$

39. $\left[\dfrac{(-x^{-5} y^6)^3}{(x^8 y^4)^2}\right] (x^{-15} y^5)^{-2}$

40. $\dfrac{4^{-1} + 2^{-4}}{4^{-2}}$

41. $[(x^2 y^4)^{-2} z^{-2}]^3$

42. $\dfrac{(2a^3 b^{-1})^2}{8(2a^3 b^2)^{-1}}$

43. $\left[\dfrac{t^{-1} + q^{-1}}{q^{-1}}\right] (t^{-1} + q^{-1})^{-1}$

44. $\dfrac{x^{3n} z^{n+1}}{x^{2n} z}$

45. $\dfrac{k^{n+1} s^{2n+5}}{k^{1-n} s^{10}}$

46. $\left(\dfrac{4x^{-5} y^{m+1}}{5}\right)^{m-1}$

47. $\dfrac{(a^n b^{n-1})^3}{a^{3n+1} b^{3n}}$

48. $\dfrac{x^{n+2} (y^{n+1})^2}{(x^2)^n y^{2n-1}}$

49. $\left(\dfrac{x^{m-1} y^{-2}}{10}\right)^{m+1}$

[Hint: $(m - 1)(m + 1) = m^2 - 1$]

B In Exercises 50–55 find a value of x that makes the statement true.

50. $\left(\dfrac{2}{3}\right)^x = \dfrac{8}{27}$

51. $\left(\dfrac{1}{2}\right)^x = 32$

52. $(2y)^x = 8y^3$

53. $16^x \cdot z^4 = \left(\dfrac{z}{2}\right)^4$

54. $(3x)^2 = \dfrac{1}{9}$

55. $4^x = 2^{50}$

3.3 Scientific Notation

Scientific calculations or measurements often involve extremely large and extremely small numbers. For example, one gram of hydrogen contains approximately

$$602{,}000{,}000{,}000{,}000{,}000{,}000{,}000 \text{ atoms}$$

and the mass of a hydrogen atom is approximately

$$0.00000000000000000000000167 \text{ gram}.$$

Such numbers are tedious to write and awkward to work with in calculations. A convenient way of recording such large and small numbers in a simplified form uses integer powers of the base 10. For example,

$$602{,}000{,}000{,}000{,}000{,}000{,}000{,}000 = 6.02 \times 10^{23}$$

$$0.00000000000000000000000167 = 1.67 \times 10^{-24}$$

In each case, we have denoted a number as the product of a number whose absolute value is at least 1 but less than 10, and a power of 10. This method of denoting numbers is known as **scientific notation**.

EXAMPLE 1

$$49.5 = 4.95 \times 10^1 \qquad 0.83 = 8.3 \times 10^{-1}$$

$$-206 = -2.06 \times 10^2 \qquad -0.069 = -6.9 \times 10^{-2}$$

$$1990 = 1.99 \times 10^3 \qquad 0.0018 = 1.8 \times 10^{-3}$$

The examples suggest that the exponent of 10 is determined by counting the number of places that the decimal point is moved from its original position. Every integer carries an invisible decimal point behind its unit digit; for example, $8 = 8.0$ and $-206 = -206.0$. If the decimal point is moved to the left n places, the exponent is n; if it is moved to the right m places, the exponent is $-m$. In order to write 0.0018 in scientific notation, the decimal point is moved three places to the right. The exponent of 10 is -3, so that

$$0.0018 = 1.8 \times 10^{-3}$$

3 places

Method of Conversion to Scientific Notation

1. Move the decimal point of the given number n places to the left (or to the right) until the absolute value of the resulting number is at least 1 but less than 10.
2. The resulting number times 10^n (10^{-n} if the decimal point was moved to the right) is the scientific notation of the given number.

EXAMPLE 2

Convert the following to scientific notation.

(a) $321{,}000 = 3.21000 \times 10^5 = 3.21 \times 10^5$

5 places

(b) $-981{,}000{,}000{,}000 = -9.81000000000 \times 10^{11}$

  11 places
  $= -9.81 \times 10^{11}$

(c) $0.0000136 = 000001.36 \times 10^{-5} = 1.36 \times 10^{-5}$

  5 places

(d) $-0.0000000103 = -000000001.03 \times 10^{-8} = 1.03 \times 10^{-8}$

  8 places

A number that is written in scientific notation can be converted to ordinary decimal notation by reversing the procedure. That is

> Move the decimal point n places to the right if the exponent of 10 is n, or n places to the left if the exponent is $-n$.

EXAMPLE 3

(a) $3.14 \times 10^4 = 3.1400 \times 10^4 = 31{,}400$

  4 places

Two zeros were attached behind 3.14 without changing its magnitude. For example,

$$1.2 = 1.20 = 1.200 = 1.2000 = \cdots$$

(b) $-5.26 \times 10^6 = -5.260000 \times 10^6 = -5{,}260{,}000$

  6 places

(c) $1.9 \times 10^{-3} = .0019 = 0.0019$

  3 places

(d) $1.36 \times 10^{-5} = .0000136 = 0.0000136$

  5 places

[Compare this with Example 2(d).]

(1) Can you express the number 0 in scientific notation?
(2) Would it matter, even if you are unable to write 0 in scientific notation?

Very often computations are simplified when scientific notation is used.

EXAMPLE 4 Evaluate $\dfrac{0.027 \times 0.004 \times 65}{0.009 \times 26 \times 0.0001}$.

112   Chap. 3   Exponents and Polynomials

Solution: First convert each number into scientific notation; then use commutative properties, associative properties, and the rules of exponents. We have

$$\frac{0.027 \times 0.004 \times 65}{0.009 \times 26 \times 0.0001} = \frac{2.7 \times 10^{-2} \times 4 \times 10^{-3} \times 6.5 \times 10^{1}}{9 \times 10^{-3} \times 2.6 \times 10^{1} \times 1 \times 10^{-4}}$$

$$= \frac{2.7 \times 4 \times 6.5}{9 \times 2.6 \times 1} \times 10^{-2-3+1-(-3+1-4)}$$

$$= \frac{70.2}{23.4} \times 10^{2}$$

$$= 3 \times 10^{2} \quad \text{or } 300$$

Scientific notation can be used to speed up an estimation of a complicated operation by first rounding off all the numbers involved. Recall that the symbol ≈ means "approximately equal."

EXAMPLE 5   Give a rough estimate of $\dfrac{587{,}329 \times 0.00193}{295.84}$

Solution: $\dfrac{587{,}329 \times 0.00193}{295.84} \approx \dfrac{600{,}000 \times 0.002}{300}$

$$= \frac{6 \times 10^{5} \times 2 \times 10^{-3}}{3 \times 10^{2}}$$

$$= \frac{6 \times 2}{3} \times 10^{5-3-2}$$

$$= 4$$

To enter a number such as 1.35×10^{8} (= 135,000,000) in most calculators with scientific notation capability, you first enter 1.35; next press the key $\boxed{EE}$; then press 8. The calculator will display

$$1.35 \quad 08$$

If you press

$$6 \;\boxed{\cdot}\; 02 \;\boxed{EE}\; 23 \;\boxed{+/-}$$

the display will show the following (for 6.02×10^{-23}):

$$6.02 \quad -23$$

Exercises 3.3

Standard Assignment: Exercises 1, 3, 5, 7, 9, 11, 13, 15, 17, 19, 21, 23, 25, 27, 29, 31, 33, 35, 37, 45, 51

A In Exercises 1–12, write each number in scientific notation. See Examples 1 and 2.

1. 243
2. 1250
3. −305,000
4. −750,000
5. 0.003
6. 0.0034
7. 0.00000378
8. 0.0000049
9. −0.000109
10. −0.0000369
11. 103,000,000,000
12. 981,000,000,000

13. Complete the following table by filling all blank spaces.

Celestial Body	Equatorial Diameter (km)	Scientific Notation
Earth	12,700	
Moon		3.48×10^3 (km)
Sun	1,390,000	
Jupiter		1.34×10^5 (km)
Mercury	4800	

14. The mass of the earth is about 5,980,000,000,000,000,000,000,000 kilograms, and the distance from the earth to the moon is about 380,000,000 meters. Write these two numbers in scientific notation.

In Exercises 15–24, convert each number in scientific notation to ordinary decimal notation. See Example 3.

15. 2.7×10^2
16. 7.2×10^3
17. 1.5×10^{-2}
18. 5.1×10^{-3}
19. -1.23×10^5
20. -3.21×10^6
21. 6.06×10^{-7}
22. 3.05×10^{-8}
23. -8.19×10^{-9}
24. -9.18×10^{-10}

In Exercises 25–44, use scientific notation to evaluate each expression. See Example 4.

25. $\dfrac{0.04 \times 0.002}{0.064}$
26. $\dfrac{0.05 \times 0.004}{0.016}$
27. $\dfrac{5000}{0.025}$
28. $\dfrac{6000}{0.015}$
29. $\dfrac{0.039}{13,000}$
30. $\dfrac{0.096}{320,000}$

31. $\dfrac{0.064 \times 500}{0.04}$

32. $\dfrac{0.016 \times 600}{0.05}$

33. $\dfrac{200 \times 125}{0.002}$

34. $\dfrac{325 \times 255}{0.005}$

35. $\dfrac{0.003 \times 0.012}{0.04 \times 0.09}$

36. $\dfrac{0.006 \times 0.004}{0.012 \times 0.01}$

37. $\dfrac{0.016 \times 9 \times 0.0014}{0.032}$

38. $\dfrac{0.032 \times 6 \times 0.0056}{0.128}$

39. $\dfrac{1.2 \times 0.093 \times 0.0084}{0.0062 \times 0.00021}$

40. $\dfrac{2.5 \times 8.8 \times 0.0013}{0.05 \times 0.022}$

41. $\dfrac{3.5 \times 74{,}000}{0.00032 \times 1254}$

42. $\dfrac{7.21 \times 34{,}000}{0.00016 \times 4521}$

43. $\dfrac{53.63 \times 147{,}000}{0.893 \times 0.0724 \times 3270}$

44. $\dfrac{2369 \times 0.323 \times 0.0013}{0.0147 \times 0.0053}$

In Exercises 45–50, give a rough estimate of each expression. See Example 5.

45. $\dfrac{19{,}998 \times 0.298}{399.9}$

46. $\dfrac{29{,}897 \times 0.389}{598.9}$

47. $\dfrac{2023 \times 1009}{4{,}000{,}121}$

48. $\dfrac{5021 \times 2012}{2{,}500{,}099}$

49. $\dfrac{8987 \times 0.0401}{0.019 \times 2989}$

50. $\dfrac{7950 \times 0.00509}{0.0097 \times 4900}$

51. Astronomers measure the distance between stars by means of a unit called "light-years." One light-year is the distance traveled by light in 1 year (365 days), and the speed of light is about 3×10^5 kilometers per second. Write in scientific notation the approximate number of kilometers in one light-year.

52. The nearest star to the earth, Alpha Centauri, is about 4.3 light-years away. How far apart are the earth and Alpha Centauri? Express your answer in scientific notation and meters.

3.4 Polynomials

The height (in feet) of a golf ball above the driving range t seconds after being driven from the tee (at 70 ft/sec) is $70t - t^2$. After 2 seconds ($t = 2$), the ball is $70(2) - (2)^2 = 136$ feet above the ground.

The expression $70t - t^2$ is an example of a **polynomial**. It is important to learn the algebra of polynomials, as they frequently appear in applications.

A polynomial is an expression that can be obtained from constants and variables using only addition, subtraction, and multiplication.

The polynomial $70t - t^2$ results from subtracting $t^2 \; (= t \cdot t)$ from the product of 70 (a constant) and t.

EXAMPLE 1   Show that each of the following is a polynomial.

(a) $-\pi x^2 + xy$ (b) $\sqrt{3}\, y^2 - \dfrac{5}{8}y + 13$

Solution: In (a) we use: $-\pi$ (a constant), x and y (variables), addition, and multiplication

$$(-\pi) \cdot x \cdot x + x \cdot y = -\pi x^2 + xy$$

In (b) we use: $\sqrt{3}, \tfrac{5}{8}$, and 13 (constants), y (a variable), addition, subtraction, and multiplication

$$(\sqrt{3}) \cdot y \cdot y - \left(\dfrac{5}{8}\right) \cdot y + 13 = \sqrt{3}\, y^2 - \dfrac{5}{8}y + 13$$

It should be noted that division by a nonzero number results in a real number, such as $\tfrac{5}{8}$ in Example 1b, and is then legitimately used in the formation of a polynomial. Similarly, the square root of a non-negative number is a real number and may properly occur in a polynomial.

WARNING

These may be used in polynomials.

$3 \div 7 = \dfrac{3}{7}$ is a real number.

$\sqrt{6}$ is a real number.

These may not be used in polynomials.

$3 \div x = 3/x$ is neither a variable nor a real number.

$\sqrt{x}$ is neither a variable nor a real number.

EXAMPLE 2   Show that the expressions

$$x + \dfrac{1}{x} + 2$$

and

$$x^2 + \sqrt{x} + 1$$

are not polynomials.

Solution: In the formation of $x + 1/x + 2$, division by x must be used to obtain $1/x$; and in $x^2 + \sqrt{x} + 1$, the operation of extracting a root must be applied to the variable x to obtain $\sqrt{x}$. Neither division by a variable nor extraction of roots of variables is permitted. Therefore, these are not polynomials. Most of the polynomials we will be dealing with contain only one letter (variable), and there is a simple form for such polynomials.

A (real) *polynomial* in x is an expression that can be written in the form

$$a_n x^n + a_{n-1} x^{n-1} + \cdots + a_2 x^2 + a_1 x + a_0, \qquad (3.8)$$

where $a_n, a_{n-1}, \ldots, a_0$ are real numbers called the **coefficients**.

If $a_n \neq 0$, then n, the largest exponent of the variable x, is called the **degree** of the polynomial. Any *nonzero* number is considered to be a polynomial of degree 0. By agreement, the number 0 has no degree. Each of $a_n x^n, a_{n-1} x^{n-1}, \ldots, a_0$ is called a **term** of the polynomial. See Definition (3.8).

WARNING Since Definition (3.8) contains only + signs, a polynomial such as $2x^2 - 3x + 1$ must be rewritten as $2x^2 + (-3)x + 1$ in order to recognize -3 as the coefficient of x.

EXAMPLE 3 Indicate the degree, the terms, and the coefficients.

(a) $2x^2 - 4x^3 + x - 13$
(b) $5 - y + 2y^2 - \sqrt{6}\, y^5$

Solutions:
(a) Degree: 3

   Terms: $-4x^3, 2x^2, x\ (x = 1 \cdot x)$, and -13
   Coefficients: $-4, 2, 1$, and -13

(b) Degree: 5

   Terms: $5, -y, 2y^2, 0y^3, 0y^4$, and $-\sqrt{6}\, y^5$
   Coefficients: $5, -1, 2, 0, 0$, and $-\sqrt{6}$

The polynomial in Example 3(b) is written in **ascending order** because it is written in such a way that the powers of the variable increase from left to right. (Here 5 is regarded as $5y^0$.) In contrast, the polynomial in Definition (3.8) is said to be written in **descending order**.

EXAMPLE 4 Express the following in descending order and then in ascending order.

(a) $-x^2 + 14 + 2x^3 - 7x$
(b) $\pi y - \sqrt{13}\, y^5 + 2y^2 - \sqrt{2}\, y^4 - 99$

Solution:
(a) By the commutative property of addition, we have

$$-x^2 + 14 + 2x^3 - 7x$$
$$= 2x^3 - x^2 - 7x + 14 \qquad \text{Descending order}$$
$$= 14 - 7x - x^2 + 2x^3 \qquad \text{Ascending order}$$

(b) The expression may be rewritten as

$$-\sqrt{13}\, y^5 - \sqrt{2}\, y^4 + 2y^2 + \pi y - 99 \qquad \text{Descending order}$$

or

$$-99 + \pi y + 2y^2 - \sqrt{2}\, y^4 - \sqrt{13}\, y^5 \qquad \text{Ascending order}$$

We can easily extend the use of the word **term** to polynomials of more than one variable by agreeing that the parts of the polynomials that are separated by either a + or a − sign, *together with the sign,* will be called the terms of the polynomial.

Special names are given to polynomials consisting of one term, two terms, or three terms; they are called **monomials, binomials,** and **trinomials,** respectively.

EXAMPLE 5
(a) 5, $-2x$, $17y^2$, and $9x^4$ are monomials.
(b) $3x - 7$, $6x^2 + 5$, $2x - 9y^2$, and $x^2y + 1$ are binomials.
(c) $x^3 - 2x + 1$, $3x^2 - xy + y^2$, $x + 2y - 3z$, and $ax^2 + bx + c$ are trinomials.
(d) Each term in the polynomial in Definition (3.8) is itself a monomial.

Addition and Subtraction

Two terms that contain the same variable or variables that are raised to the same power are called **like terms**. Thus, $8x$ and $3x$ are like terms and $5xy^2$ and $-2xy^2$ are like terms. The distributive property makes the addition and subtraction of like terms just as easy as adding and subtracting their coefficients

$$8x + 3x = (8 + 3)x$$
$$= 11x$$
$$8x - 3x = (8 - 3)x$$
$$= 5x$$
$$5xy^2 + (-2xy^2) = (5 - 2)xy^2$$
$$= 3xy^2$$
$$5xy^2 - (-2xy^2) = [5 - (-2)]\, xy^2$$
$$= 7xy^2$$

EXAMPLE 6 Perform the indicated operations and simplify.
(a) $5x^2 + 7x^2 - 11x^2$
(b) $3xy - 15xy + 5xy$
(c) $-5y^2 + 3y^2 + 2y^2$

Solution:
(a) $5x^2 + 7x^2 - 11x^2 = (5 + 7 - 11)x^2$ Distributive Property
$= 1 \cdot x^2$
$= x^2$

(b) $3xy - 15xy + 5xy = (3 - 15 + 5)xy$ Distributive Property
$= (-7)xy$
$= -7xy$

(c) $-5y^2 + 3y^2 + 2y^2 = (-5 + 3 + 2)y^2$ Distributive Property
$= 0y^2$
$= 0$

Addition or subtraction of polynomials involves the collection of like terms; this can be accomplished by the use of the associative, commutative, and distributive properties.

EXAMPLE 7

Perform the indicated operations and simplify.

(a) $(6x^3 - 8x^2 + 7x + 10) + (10x^2 + 11x - 13)$
(b) $(10y^3 + 7y^2 - 4y - 2) - (5y^3 - 2y + 3)$

Solution:
(a) $(6x^3 - 8x^2 + 7x + 10) + (10x^2 + 11x - 13)$
$= 6x^3 + (-8x^2 + 10x^2) + (7x + 11x) + (10 - 13)$
$= 6x^3 + (-8 + 10)x^2 + (7 + 11)x + (-3)$
$= 6x^3 + 2x^2 + 18x - 3$

(b) $(10y^3 + 7y^2 - 4y - 2) - (5y^3 - 2y + 3)$
$= 10y^3 + 7y^2 - 4y - 2 - 5y^3 + 2y - 3$
$= (10y^3 - 5y^3) + 7y^2 + (-4y + 2y) - 2 - 3$
$= 5y^3 + 7y^2 - 2y - 5$

Alternative Solution (Column Method)

The alternative method presented here is to list the polynomials to be added or subtracted in "column form" in such a way that like terms appear in the same columns.

(a) $\begin{array}{r} 6x^3 - 8x^2 + 7x + 10 \\ 10x^2 + 11x - 13 \\ \hline 6x^3 + 2x^2 + 18x - 3 \end{array}$ Add
                                                (Answer)

(b) $10y^3 + 7y^2 - 4y - 2$
$5y^3 - 2y + 3$ Subtract
─────────────────────
$5y^3 + 7y^2 - 2y - 5$ (Answer)

EXAMPLE 8   Find the sum of polynomials $-2x^3 - 3x^2 + 5x$, $4x^3 - 6x + 9$, and $7x^3 + 8x^2 + 10$.

Solution: This problem is more easily handled by the column method.

$$\begin{array}{r} -2x^3 - 3x^2 + 5x \\ 4x^3 - 6x + 9 \\ 7x^3 + 8x^2 + 10 \\ \hline 9x^3 + 5x^2 - x + 19 \end{array}$$ Add   (Answer)

Exercises 3.4

Standard Assignment: Exercises 1, 3, 5, 7, 11, 13, 15, 17, 19, 21, 23, 25, 27, 31, 33, 35, 39, 43, 45, 47, 49, 51, 59, 61

A   In Exercises 1–10, which of the given expressions are polynomials? (See Examples 1 and 2.)

1. (a) $x^2 + 2x - 1$ (b) $x - \dfrac{1}{x}$ (c) 5

2. (a) $\dfrac{1}{y} + y$ (b) $1 - 2y - y^2$ (c) -7

3. (a) z (b) $\sqrt{z^2 + 1}$ (c) 0

4. (a) $\sqrt{1 - x^2}$ (b) $3z - 5z$ (c) $\sqrt{5}$

5. (a) $\dfrac{1}{6}t^2 - \dfrac{1}{2}t$ (b) $\dfrac{t^2}{6} + \dfrac{t}{2} + 1$ (c) t^{-2}

6. (a) $\dfrac{1}{3}n + \dfrac{1}{8}n^2$ (b) $\dfrac{1}{2} - \dfrac{u}{3} - \dfrac{u^2}{8}$ (c) $n^{-1} - 3$

7. (a) $x + y$ (b) $x - \dfrac{1}{y}$ (c) xy

8. (a) $2y - x$ (b) $\dfrac{2}{x} + y$ (c) $x(y + 1)$

9. (a) $\dfrac{x}{y} + \dfrac{y}{x}$ (b) $\dfrac{2x - 5y}{\sqrt{3}}$ (c) $\dfrac{3x - 2y}{\sqrt{z}}$

$\left[\text{Hint for 9(b): } \dfrac{2x}{\sqrt{3}} = \dfrac{2}{\sqrt{3}}x\right]$

10. (a) $\dfrac{x^2 + y^2}{xy}$ (b) $\dfrac{53 - 65x}{\sqrt{2}}$ (c) $\dfrac{5y + 3z}{\sqrt{x}}$

In Exercises 11–22, indicate the degree and coefficient of each monomial.

11. $5x^2$
12. $6y^3$
13. -7
14. 0
15. $\dfrac{1}{2}x^3$
16. $\dfrac{1}{3}y^5$
17. $-\dfrac{1}{8}m^7$
18. $-\dfrac{3}{5}b^9$
19. $\dfrac{t^3}{12}$
20. $\dfrac{u^2}{30}$
21. πv^{20}
22. $\sqrt{21}\, x$

In Exercises 23–30, rewrite each polynomial first in descending order and then in ascending order. (See Example 4.)

23. $x^3 - 5 + x - x^2$
24. $7 - y^3 - 5y + 2y^2$
25. $x^5 - 2x^2 + \sqrt{3}\, x^3 + \dfrac{1}{2}x$
26. $\dfrac{1}{3}y + 16 - \sqrt{2}\, y^4 - y^2 - y^6$
27. $t^{99} - 3t^3 + 5t^{91} - \dfrac{\pi}{6}t^{49}$
28. $u^{50} - \dfrac{\pi}{3}u^{92} - \sqrt{3}\, u + u^{61}$
29. $v^{10} + \sqrt{2} - v^8 - 5v - \dfrac{1}{2}v^4$
30. $z^{20} - 5 + \sqrt{3}\, z^{10} - 6z^{15} - 7z^5$

In Exercises 31–42, rewrite each polynomial in its simplest form by combining like terms; name each result as a monomial, binomial, or trinomial; and indicate the degree of each. (See Examples 3, 5, and 6.)

31. $x^2 - 5 - 2x^2 + 5$
32. $6 - y^2 - 7 + 3y^2 + 1$
33. $2z^2 - 3 - 6z^2 + 5z + 4z^2 + 3$
34. $7t^3 - 5 - 3t - 8t^2 + 3t + t^2$
35. $x^2 + 9 + 5x^2 - 20$
36. $25 + y^2 - 20 - 55y^2$
37. $\dfrac{1}{2}x^3 - \dfrac{2}{3}x - \dfrac{1}{3}x^3 + \dfrac{1}{5} + \dfrac{2}{3}x + 10$
38. $\dfrac{1}{3}y^3 + \dfrac{1}{2}y + \dfrac{2}{3} - \dfrac{5}{2}y - \dfrac{1}{2}y^3 + \dfrac{1}{2} + \dfrac{1}{6}y^3$
39. $2 + 3x - 5x^2 + 7x - 15$
40. $3y^2 - 5y - 10 + 15y - 6y^2$
41. $4u^3 - 16u + 17u^2 + 7u + 15 + 9u$
42. $v - 13v^2 + 25 - v + 12v^2 - v^3 + v^2$

In Exercises 43–58, perform the indicated operations and simplify. (See Examples 7 and 8.)

43. $(x^3 + 2x^2 - 5x + 3) + (-x^3 + 2x - 4)$

44. $(x^3 - 4x + 2) + (x^3 - x^2 + x - 5)$

45. $(2x^3 - x^2 + x - 5) - (x^3 - 4x + 4)$

46. $(-x^3 + 2x - 4) - (x^3 + 2x^2 - 5x + 3)$

47. $(2x + 5y + 3zw^2) + (4x - 6y - 2zw^2)$

48. $(3t^2 - 2tu + u^2 - v) + (-2t^2 + 5tu - u^2 + u + v - 6)$

49. $\begin{array}{l} x^3 + 5x^2 + 6x - 1 \\ \underline{3x^3 - 4x^2 - 8x + 6} \end{array}$ (Addition)

50. $\begin{array}{l} 5y^3 - 6y^2 - 12y - 13 \\ \underline{-3y^3 + 2y^2 + 10y - 22} \end{array}$ (Addition)

51. $\begin{array}{l} 2x^3 + 4x^2 - 7 \\ 3x^3 - 9x + 10 \\ \underline{ - 8x^2 - 11x + 5} \end{array}$ (Addition)

52. $\begin{array}{l} -2y^4 + 3y^2 - 7y \\ - 5y^3 - 3y + 6 \\ \underline{-8y^4 + 6y^3 - 9y^2 - 17} \end{array}$ (Addition)

53. $\begin{array}{l} x^3 - 4x^2 - 8x - 6 \\ \underline{3x^3 + 5x^2 + 6x - 1} \end{array}$ (Subtraction)

54. $\begin{array}{l} -3y^3 - 2y^2 + 15y - 33 \\ \underline{-6y^3 - 9y^2 - 93y + 45} \end{array}$ (Subtraction)

55. $\begin{array}{l} 8x^2 - 2xy - 9y^2 - 13x - 31y + 50 \\ \underline{-7x^2 - 3xy + 8y^2 - 25x - 60y - 19} \end{array}$ (Subtraction)

56. $\begin{array}{l} u^2 + 4uv - v^2 - 14u + 15v + 26 \\ \underline{6u^2 + 4uv - v^2 + 15u + 15v - 5} \end{array}$ (Subtraction)

57. 🖩 $\begin{array}{l} 3.142x^2 - 0.537x + 2.193 \\ -0.375x^2 + 1.972x - 3.247 \\ \underline{4.126x^2 - 6.529x + 0.599} \end{array}$ (Addition)

58. 🖩 $\begin{array}{l} -5.312y^2 + 7.814xy - 2.687x + 8.197y + 0.112 \\ -0.588y^2 - 3.414xy + 1.876x - 9.985y - 3.929 \\ \underline{-3.456y^2 + 0.323xy + 1.132x + 0.135y + 0.012} \end{array}$ (Addition)

EXAMPLE: Let $P(x)$ be a polymonial in x. Then $P(a)$ denotes the **value of the polynomial** $P(x)$ when x is replaced by the number a. If $P(x) = x^2 - 2x - 3$, then $P(4) = 4^2 - 2 \cdot 4 - 3 = 5$, $P(1) = 1^2 - 2 \cdot 1 - 3 = -4$ and $P(0) = 0^2 - 2 \cdot 0 - 3 = -3$.

59. Find $P(2)$ if $P(x) = x^3 - 5x^2 + 3x - 1$.

60. Find $P(3)$ if $P(x) = 2x^3 + 5x^2 - 10x - 5$.
61. Find $P(-1)$ if $P(x) = ax^3 + bx^2 - ax - b$.
62. Find $P(-2)$ if $P(x) = cx^3 + dx^2 - 8x - 2d$.
63. ▦ Find $P(3.14)$ if $P(x) = x^3 - 28x^2 + 1.47x - 7.56$.
64. ▦ Find $P(-2.13)$ if $P(x) = -x^3 - 3.12x^2 + 11x - 59$.

3.5 Product of Polynomials

The multiplication of monomials is just a simple application of the Product Rule in Equation (3.2) and the associative and commutative properties.

EXAMPLE 1

(a) $(3x^2)(7x^3) = 21x^{2+3} = 21x^5$

(b) $(-5z^3)(-2z^5) = (-5)(-2)z^{3+5} = 10z^8$

(c) $(-3x^3y^2)(5x^2y^5) = (-3)5\,(x^3x^2)(y^2y^5)$
$= -15x^{3+2}\,y^{2+5}$
$= -15x^5y^7$

To multiply a monomial and a polymonial, first apply the distributive property and then use the Product Rule of exponents in Equation (3.2) to work the problem as in Example 1.

EXAMPLE 2

$(2x^2)(3x^3 + 4) = (2x^2)(3x^3) + (2x^2)(4)$
$= 2 \cdot 3\, x^{2+3} + 2 \cdot 4\, x^2$
$= 6x^5 + 8x^2$

EXAMPLE 3

$(5y^3 - 7y)(-2y^2) = (5y^3)(-2y^2) + (-7y)(-2y^2)$
$= 5(-2)y^{3+2} + (-7)(-2)y^{1+2}$
$= -10y^5 + 14y^3$

To multiply two binomials, first use the distributive property and then work the problem as in Example 3.

EXAMPLE 4

$(3x + 4)(2x + 5) = (3x + 4)(2x) + (3x + 4) \cdot 5$
$= (3x)(2x) + 4 \cdot (2x) + (3x) \cdot 5 + 4 \cdot 5$
$= 6x^2 + (8x + 15x) + 20$
$= 6x^2 + 23x + 20$

The following method, known as **FOIL**, is a short cut for finding the product of two binomials. Here FOIL stands for the sum of the products of the *First* terms, the *Outer* terms, the *Inner* terms, and the *Last* terms.

EXAMPLE 5

$$(3x + 4)(2x + 5) = 6x^2 + (15x + 8x) + 20$$
$$= 6x^2 + 23x + 20$$

(a) Multiply the "first" terms: $(3x)(2x) = 6x^2$
(b) Add the products of "outer" and "inner" terms:

$$(3x) \cdot 5 + 4 \cdot (2x) = 23x$$

(c) Multiply the "last" terms: $4 \cdot 5 = 20$. Then add the results of (a), (b), and (c):

$$6x^2 + 23x + 20$$

EXAMPLE 6

Using the FOIL method we find

$$(a + b)(a - b) = a^2 - ab + ab - b^2 = a^2 - b^2$$

The identity in Example 6 and its reverse are so frequently used in algebra that the student should always remember them.

The Difference of Two Squares

$$(a + b)(a - b) = a^2 - b^2$$
$$a^2 - b^2 = (a + b)(a - b) \qquad (3.9)$$

EXAMPLE 7

The square of a binomial can be calculated by the FOIL method as follows

$$(a + b)^2 = (a + b)(a + b) = a^2 + ab + ab + b^2$$
$$= a^2 + 2ab + b^2$$

Similarly,
$$(a - b)^2 = a^2 - 2ab + b^2$$

Perfect Squares
$$(a + b)^2 = a^2 + 2ab + b^2$$
$$(a - b)^2 = a^2 - 2ab + b^2 \qquad (3.10)$$

WARNING $(a + b)^2 \ne a^2 + b^2$ and $(a - b)^2 \ne a^2 - b^2$; the correct formulas are $(a + b)^2 = a^2 + 2ab + b^2$ and $(a - b)^2 = a^2 - 2ab + b^2$.

EXAMPLE 8 Find the product $(ax + b)(cx + d)$, where a, b, c, and d are fixed real numbers.

Solution: Following steps (a), (b), and (c) in Example 5, we have

$$(ax + b)(cx + d) = (ac)x^2 + (ad + bc)x + (bd)$$

with acx^2, bd, bcx, adx indicated.

Multiplication of two polynomials can be achieved systematically by the "column method" that is shown in the next example.

EXAMPLE 9 Multiply $4x^3 - 3x + 7$ by $x^2 + 2x - 5$.

Solution:

$$\begin{array}{r} 4x^3 \quad - 3x + 7 \\ x^2 + 2x - 5 \end{array} \quad \text{(Multiplication)}$$

(a) x^2 times $(4x^3 - 3x + 7)$: $\quad 4x^5 \quad - 3x^3 + 7x^2$
(b) $2x$ times $(4x^3 - 3x + 7)$: $\quad 8x^4 \quad - 6x^2 + 14x$
(c) -5 times $(4x^3 - 3x + 7)$: $\quad - 20x^3 \quad + 15x - 35$
add the results of (a), (b), and (c). $\quad 4x^5 + 8x^4 - 23x^3 + x^2 + 29x - 35$

A pair of formulas that are almost as useful as Equation (3.9) is the products $(a + b)(a^2 - ab + b^2)$ and $(a - b)(a^2 + ab + b^2)$. We find these products by the column method.

$$\begin{array}{r} a^2 - ab + b^2 \\ a + b \\ \hline a^3 - a^2b + ab^2 \\ +) \quad a^2b - ab^2 + b^3 \\ \hline a^3 \qquad\qquad + b^3 \end{array} \qquad \begin{array}{r} a^2 + ab + b^2 \\ a - b \\ \hline a^3 + a^2b + ab^2 \\ +) \quad - a^2b - ab^2 - b^3 \\ \hline a^3 \qquad\qquad - b^3 \end{array}$$

We summarize these in

> **Sum and Difference of Cubes**
>
> $(a + b)(a^2 - ab + b^2) = a^3 + b^3$
>
> $(a - b)(a^2 + ab + b^2) = a^3 - b^3$ (3.11)

or reversing the equalities:

> **Sum and Difference of Cubes**
>
> $a^3 + b^3 = (a + b)(a^2 - ab + b^2)$
>
> $a^3 - b^3 = (a - b)(a^2 + ab + b^2)$ (3.11′)

WARNING $a^3 + b^3 \neq (a + b)^3$, and $a^3 - b^3 \neq (a - b)^3$

EXAMPLE 10 Use Equations (3.9) to multiply

(a) $(5x + 6y)(5x - 6y) = (5x)^2 - (6y)^2$
$= 25x^2 - 36y^2$

(b) $(-7p + 9q)(9q + 7p) = (9q + 7p)(9q - 7p)$
$= (9q)^2 - (7p)^2$
$= 81q^2 - 49p^2$

EXAMPLE 11 Use Equations (3.10) to expand

(a) $(3m + 8n)^2 = (3m)^2 + 2(3m)(8n) + (8n)^2$
$= 9m^2 + 48mn + 64n^2$

(b) $(4x - 7y)^2 = (4x)^2 - 2(4x)(7y) + (7y)^2$
$= 16x^2 - 56xy + 49y^2$

(c) $(-5p + 6q)^2 = (-5p)^2 + 2(-5p)(6q) + (6q)^2$
$= 25p^2 - 60pq + 36q^2$

EXAMPLE 12 Use Equations (3.11) to multiply

(a) $(3x + 2y)(9x^2 - 6xy + 4y^2)$
$= (3x + 2y)[(3x)^2 - (3x)(2y) + (2y)^2]$
$= (3x)^3 + (2y)^3$
$= 27x^3 + 8y^3$

(b) $(4m - 5n)(16m^2 + 20mn + 25n^2)$
$= (4m - 5n)[(4m)^2 + (4m)(5n) + (5n)^2]$
$= (4m)^3 - (5n)^3$
$= 64m^3 - 125n^3$

Exercises 3.5

Standard Assignment: Exercises 1, 3, 5, 7, 9, 11, 17, 19, 21, 25, 27, 31, 35, 37, 41, 43, 45, 47, 49, 55, 57, 63, 65, 67, 69

A In Exercises 1–40, express each product as a polynomial. See Examples 1–8.

1. $2x(x + 5)$
2. $3y(y + 15)$
3. $4t(t - 2)$
4. $5k(k - 4)$
5. $6z(2z + 3)$
6. $7y(3y + 8)$
7. $10m^2(3m - 4)$
8. $9x^2(5x - 11)$
9. $-8x^2(2x^2 - y^2)$
10. $-11t^2(7t^2 - 8z^2)$
11. $2x(3x^2 - 6x + 25)$
12. $3k(5k^2 - 9k - 27)$
13. $-6t^2(12 + 10t - 13t^2)$
14. $3m^3(-7 + 21m - 35m^2)$
15. ▦ $5.314x(2.417x^2 - 4.592x + 9.017)$
16. ▦ $3.865y(8.024x - 7.248y - 6.967)$
17. $(z + 1)(z + 2)$
18. $(y + 2)(y + 3)$
19. $(2k + 1)(3k + 2)$
20. $(3r + 5)(4r + 7)$
21. $(4x - 3)(5x + 4)$
22. $(5y - 6)(6y + 9)$
23. $(5t - 9)(11t - 7)$
24. $(10s - 9)(12s - 11)$
25. $(x^2 + 2)(x^2 - 2)$
26. $(y^3 + 5)(y^3 - 5)$
27. $(2r - 1)(2r + 1)$
28. $(3t - 2)(3t + 2)$
29. $\left(\frac{1}{2}x + \frac{3}{2}y\right)\left(\frac{1}{2}x - \frac{3}{2}y\right)$
30. $\left(\frac{4}{5}u + \frac{7}{8}v\right)\left(\frac{4}{5}u - \frac{7}{8}v\right)$
31. $(x + 5)^2$
32. $(u + 10)^2$
33. $(2v - 1)^2$
34. $(3t - 1)^2$
35. $(4k + 5)^2$
36. $(7y + 8)^2$
37. $(5x + 11y)^2$
38. $(9r + 13)^2$
39. $\left(\frac{1}{5}s - \frac{1}{6}t\right)^2$
40. $\left(\frac{1}{4}u - \frac{1}{6}v\right)^2$

3.5 Product of Polynomials

In Exercises 41–60, perform each multiplication of polynomials by the column method. See Example 9.

41. $(x^3 - 2x^2 + 5x - 7)(x - 5)$
42. $(y^3 + 3y^2 - 7y - 9)(y - 4)$
43. $(2t + 3)(4t^2 - 6t + 9)$
44. $(3k + 4)(9k^2 - 12k + 16)$
45. $(4y - 5)(16y^2 + 20y + 25)$
46. $(5y - 7)(25y^2 + 35y + 49)$
47. $(2x^2 + 3x + 1)(3x^2 + 5x + 7)$
48. $(3t^2 + 4t + 2)(4t^2 + 10t + 3)$
49. $(5u^3 - 4u^2 + 3u + 1)(3u^2 - 2u + 5)$
50. $(4v^3 + 3v^2 - 2v - 5)(2v^2 + 3v - 3)$
51. $(7x^5 - 5x^3 - 2x + 8)(x^2 + 3x - 1)$
52. $(8x^4 + 9x^2 - 7x - 3)(2x^3 - x + 3)$
53. ▦ $(0.563x^2 - 2.047x + 7.038)(232x^2 - 305x + 145)$
54. ▦ $(3.54y^2 + 5.23y - 4.03)(117y^2 - 235y - 997)$
55. $(x + 1)^3$ [Hint: $(x + 1)^3 = (x + 1)^2(x + 1)$]
56. $(y - 3)^3$
57. $(2z + 5)^3$
58. $(7 - 9y)^3$
59. $(s + t)^3$
60. $(x - y)^3$

In the following, use the formulas in Equations (3.9) and (3.10) to find each product.

EXAMPLE:

$$(x + y - 5)(x - y - 5) = [(x - 5) + y][(x - 5) - y]$$
$$= (x - 5)^2 - y^2 \quad \text{By Equation (3.9)}$$
$$= (x^2 - 10x + 25) - y^2 \quad \text{By Equation (3.10)}$$
$$= x^2 - y^2 - 10x + 25$$

61. $[(a + b) + 6][(a + b) - 6]$
62. $[(u - v) - 7][(u - v) + 7]$
63. $(3x - y + 9)(3x - y - 9)$
64. $(5x + y - 10)(5x + y + 10)$
65. $(4z^2 + 3z - 10)(4z^2 - 3z - 10)$
66. $(5y^2 + 7y - 9)(5y^2 - 7y - 9)$
67. $(8 + 7x - 3x^2)(8 - 7x + 3x^2)$
68. $(3a + 2b - 5c)(3a - 2b + 5c)$
69. $[(4x + 3y) - 12]^2$
70. $[11 - (7x - 8y)]^2$

B Perform each multiplication. Assume that the variables in the exponents represent integers.

71. $x^n(x^{2n} - 3x^{n-1})$
72. $y^{2n}(y^{3-n} + y)$
73. $(z^n - 1)^2$
74. $(t^n + 5)^2$

75. $5x^{-m}(2x^{2m} - 3x^{-m})$
76. $(7y^{3n} - 5y^{-n})(-6y^{-2n})$
77. $(k^{2r} + 2)(k^{2r} - 2)$
78. $(2m^5 - m)(2m^5 + 5)$
79. $(2z^n + 1)^3$
80. $(2a^r + b)^3$
81. $(3p^n - 2p^{-n})^2$
82. $(5r^n + 3r^{-n})^2$
83. $(4x^m + 7y^n)(4x^m - 7y^n)$
84. $(5z^r + 8w^s)(5z^r - 8w^s)$
85. $(6t^{m-1} + 5t^{n-2})(2t^{m+1} - 3t^{n+2})$
86. $(4s^{k-2} + 3s^{k+1})(2s^{k+2} - s^{k-1})$
87. $(3r^n + 5r^{-n})^3$
88. $(2p^n - 3p^{-n})^3$
89. $(a + b)^z[(a + b)^{-2z} + (a + b)^{-z}]$
90. $(c - d)^{2w}[(c - d)^{-w} - (c - d)^{-2w}]$

3.6 Common Factors; Grouping

To **factor** an expression means to write it as a product of two or more expressions. Factoring reverses the process of multiplication. The formulas used to multiply can also be used to factor.

EXAMPLE 1 Use the distributive property $a(b + c) = ab + ac$ to

(a) Multiply: $3(x + 2) = 3x + 3 \cdot 2 = 3x + 6$
(b) Factor: $3x + 6 = 3x + 3 \cdot 2 = 3(x + 2)$
            ↖ given expression   ↖ written as a product

If there is a factor that is common to every term, then the **common factor** can be "factored out" using the distributive property. This applies to any number of terms. For example, the equation $ax + ay + az + aw = a(x + y + z + w)$ factors out the common factor a.

EXAMPLE 2 Factor out a common factor.

(a) $10x + 15 = 5(2x) + 5(3) = 5(2x + 3)$ 5 is a common factor

(b) $9 + 18y = 9(1) + 9(2y) = 9(1 + 2y)$ 9 is a common factor

(c) $5z + 3$ There is no common factor

(d) $3 + 9x + 12y = 3(1) + 3(3x) + 3(4y)$
$= 3(1 + 3x + 4y)$ 3 is a common factor

(e) $2(a + b) + (a + b)x = (a + b)(2 + x)$ $a + b$ is a common factor

We can check the result of factoring by multiplying. For example, to check (a) $10x + 15 = 5(2x + 3)$,

$$\text{multiply } 5(2x + 3) = 5 \cdot 2x + 5 \cdot 3 = 10x + 15$$

We can factor $3x + 6xy$ in two ways:

$$3x + 6xy = (3)x + (3)2xy = 3(x + 2xy)$$
$$3x + 6xy = (3x)1 + (3x)2y = 3x(1 + 2y)$$

When factoring polynomials, we generally use the common factor that is the product of

(1) the largest integer that divides every term

and

(2) all variables that are common to every term, where we attach to each variable the smallest exponent appearing on that variable in the given expression.

EXAMPLE 3 Factor out the common factor using the preceding steps (1) and (2).

(a) $12x + 18x^2 = (6x)2 + (6x)(3x) = 6x(2 + 3x)$ smallest exponent on x is 1, $(x = x^1)$

(b) $6x^3 + 4x^2 + 12x = (2x)3x^2 + (2x)(2x) + (2x)6$
$\qquad\qquad\qquad\qquad = 2x(3x^2 + 2x + 6)$ smallest exponent on x is 1

(c) $5x^6 - 25x^3 = (5x^3)x^3 - (5x^3)5 = 5x^3(x^3 - 5)$ smallest exponent on x is 3

(d) $8y^3z^3 - 12y^3z^4 + 20y^2z^5 = 4y^2z^3(2y - 3yz + 5z^2)$ smallest exponent on y is 2, on z is 3

Let's check (c) by multiplying factors

$$5x^3(x^3 - 5) = 5x^3(x^3) - 5x^3(5) = 5x^6 - 25x^3 \checkmark$$

We can factor $-3x + 6x^2$ in two ways that are considered to be equally correct.

$$-3x^3 + 6x^2 = -3x^2(x - 2)$$
$$-3x^3 + 6x^2 = 3x^2(-x + 2)$$

The decision to factor out either a positive integer or a negative integer is made as a matter of personal taste or to accommodate a particular use of the factorization. Either method is correct.

By factoring out -1 we can see that $x - y$ and $y - x$ are negatives of each other.

$$\begin{aligned} x - y &= x + (-y) \\ &= (-1)(-x) + (-1)y \\ &= (-1)(-x + y) \\ &= -(y - x) \end{aligned}$$

$$\boxed{x - y = -(y - x)}$$

Sometimes grouping pairs that have common factors results in the discovery of a common factor for the entire expression.

EXAMPLE 4 Factor the following expressions.

(a) $3x^2 + x + 12x + 4$
(b) $cx^2 + cy + dy + dx^2$
(c) $5x + xy + 5y + y^2$

Solution:

(a) $3x^2 + x + 12x + 4 = (3x^2 + x) + (12x + 4)$ Grouping
$ = x(3x + 1) + 4(3x + 1)$
$ = (x + 4)(3x + 1)$ Common factor is $3x + 1$

(b) $cx^2 + cy + dy + dx^2 = (cx^2 + cy) + (dy + dx^2)$ Grouping
$ = c(x^2 + y) + d(y + x^2)$
$ = (c + d)(x^2 + y)$ Common factor is $x^2 + y$

(c) $5x + xy + 5y + y^2 = (5x + xy) + (5y + y^2)$ Grouping
$ = x(5 + y) + y(5 + y)$
$ = (x + y)(5 + y)$ Common factor is $5 + y$

Exercises 3.6

Standard Assignment: Exercises 1, 3, 5, 7, 9, 11, 13, 15, 17, 19, 21, 23, 25, 27, 29, 31, 35, 39, 41, 43, 47, 49, 51, 53

A Factor out the common factor and check by multiplying factors. See examples 2 and 3.

1. $5x + 10$
2. $3x + 9$

3. $27y + 27$
4. $12y - 12$
5. $x^2 + 2x$
6. $2x^2 + x$
7. $5yz - 5$
8. $7 - 7xy$
9. $4x - 2x^2$
10. $5x^2 - 25xy$
11. $33ax + 55x^2$
12. $3by - 2b^2y$
13. $2x^6 - 6x^5 - 14x^4$
14. $3x^3 - 12x^2 + 6x$
15. $10y^2z + 35yz^2 + 5y$
16. $18x^3y + 30x^2y^3$
17. $3y^2 + 3y^3 - 6y^2x$
18. $15xy - 18xy^2 - 3x^2y$
19. $2a^3 - 8a^2 + 6a^5$
20. $7b^3 + 28b^2 - 7b^4$
21. $3a^3b^2 - 18a^5b^3 + 6a^2b^2$
22. $5x^2z + 5x^3z^2 + 5xz^4$
23. $32a^2x^5 - 16a^3x^3 + 8a^5x^4$
24. $3xy^2 - 7z^5x^2 + y^3z^2$
25. $13x^5y^2 + 11xz^4 + 13y^3z^2$
26. $5ax^2 + 7bx^3 + 4ab$
27. $x^3y^3z^3 - 3x^2yz + xz^2$
28. $12a^2b^3y^2 - 4a^3b^2y^3$
29. $5x^2y^2 - 2x^3y^4 + 3x^2y^5 + x^3y^2$
30. $2a^2b^3 - 4a^3b^2 + a^2b - ab^2$
31. $x(2a + b) + y(2a + b)$
32. $m(x + y) - n(x + y)$
33. $3a(2x - b) - 5y(2x - b)$
34. $x(x + y) - y(x + y)$
35. $5x(a + b) - y(a + b)$
36. $a(3x + 2y) - 3b(3x + 2y)$
37. $(3x + y)(a + 2) + (3x + y)(a - 3)$
38. $(x + 3)(y - 1) + (x - 7)(y - 1)$
39. $(2x + 1)(x + 4) - (2x + 1)(x + 1)$
40. $(2y - 3)(a - 2b) + (2y + 3)(a - 2b)$

Supply the missing factor or terms in Exercises 41–46.

EXAMPLE:
(a) $-3x^2 + 2x = -x(? + ?)$ (b) $9x - 3 = (?)(1 - 3x)$

Solution:
(a) $-3x^2 + 2x = -x(3x - 2)$ (b) $9x - 3 = (-3)(1 - 3x)$

41. $10x - 2 = (?)(1 - 5x)$
42. $x - 7 = (?)(7 - x)$
43. $3x^2 - 9xy = -3x(? - ?)$
44. $5a - 5b = (-5)(? - ?)$
45. $a + b - c = (?)(c - b - a)$
46. $2x - 4y + 2z = (-2)(?)$

Factor by grouping. See Example 4.

47. $2x^2 - 3xy - 6xy + 9y^2$
48. $7xy - 7yz - 3x + 3z$
49. $ax + ay - bx - by$
50. $x^2 - xy + xy - y^2$
51. $2xy + xz + 2yw + zw$
52. $2x^2 + 6x - xy - 3y$

53. $x^2 - 2x - 3xy + 6y$
54. $xy - x + 1 - y$
55. $y^3 + 1 + y^2 + y$
56. $ab^2 - a^2b - b^3 + a^3$

B Factor the following expressions, assuming that all variables that are used as exponents represent positive integers.

57. $x^{2n} + x^n$
58. $3x^2 - x^{n+2}$
59. $x^{3n} - x^{6n}$
60. $5x^{4z} + 10x^4$
61. $2p^r - 8p^{2r}$
62. $y^q - y^{2q} + y^{7q}$
63. $y^{p+3} + y^{p+4} - y^{p+5}$
64. $6z^{2p+1} + 3z^{4p} + 9z^{2p+3}$
65. $x'y^p + x'y^{p+1} + z^p + z^p y$
66. $a^p b^q + a^{p+1} b^{q+1} - r^5 - r^5 ab$

3.7 Factoring Techniques

In this section we will investigate how reversing other multiplication formulas results in new factorization techniques.

Differences of Squares

You will recall that $(a + b)(a - b) = a^2 - b^2$. The reverse of this identity, the **difference of squares**, $a^2 - b^2 = (a + b)(a - b)$, is frequently used in factoring.

EXAMPLE 1

Factor:

(a) $x^2 - 25$
(b) $9x^2 - 16y^2$
(c) $9x^2 - 2$
(d) $x^5 y - xy^5$

Solution:
(a) $x^2 - 25 = x^2 - 5^2$
$= (x + 5)(x - 5)$

(b) $9x^2 - 16y^2 = (3x)^2 - (4y)^2$
$= (3x + 4y)(3x - 4y)$

(c) $9x^2 - 2 = (3x)^2 - (\sqrt{2})^2$
$= (3x + \sqrt{2})(3x - \sqrt{2})$

(d) $x^5y - xy^5 = (xy)x^4 - (xy)y^4$
$= xy(x^4 - y^4)$
$= xy[(x^2)^2 - (y^2)^2]$
$= xy(x^2 + y^2)(x^2 - y^2)$
$= xy(x^2 + y^2)(x + y)(x - y)$

In Example 1(c) we used $\sqrt{2}$ as a coefficient. This is quite proper and, as we will see in Chapter 6, quite useful. Sometimes, however, it is desirable to allow only integers as coefficients in the factors. In such cases, this must be clearly stated. Then the expression in part (c) becomes unfactorable (using integer coefficients). That is, it is impossible to find integers a, b, c, and d such that $9x^2 - 2 = (ax + b)(cx + d)$.

Trinomial Factoring—Perfect Squares

Using the reverse of the identities $(a + b)^2 = a^2 + 2ab + b^2$ and $(a - b)^2 = a^2 - 2ab + b^2$, we can write a trinomial as the square of a binomial if two of the terms are **perfect squares** and the square of one half the other term equals the product of these two terms. This is known as **factoring by perfect squares**.

EXAMPLE 2 Factor

(a) $x^2 + 8x + 16$
(b) $49t^2 - 56t + 16$
(c) $-27x^4 + 18\sqrt{2}\, x^3 - 6x^2$
(d) $36k^2 - 24kt + 4t^2$

Solution:
(a) $x^2 + 8x + 16 = x^2 + 2(4x) + 4^2$
$= (x + 4)^2$

(b) $49t^2 - 56t + 16 = (7t)^2 - 2(4)(7t) + 4^2$
$= (7t - 4)^2$

(c) $-27x^4 + 18\sqrt{2}\, x^3 - 6x^2 = -3x^2(9x^2 - 6\sqrt{2}\, x + 2)$
$= -3x^2[(3x)^2 - 2(\sqrt{2} \cdot 3x) + (\sqrt{2})^2]$
$= -3x^2(3x - \sqrt{2})^2$

(d) $36k^2 - 24kt + 4t^2 = (6k)^2 - 2(6k)(2t) + (2t)^2$
$= (6k - 2t)^2$

Trinomial Factoring—Trial and Error

Since $(ax + b)(cx + d) = acx^2 + (ad + bc)x + bd$, in order for a given trinomial $Ax^2 + Bx + C$ to be factored into the form $(ax + b)(cx + d)$, it can be shown that we must have

$$A = ac, \quad C = bd \quad \text{and} \quad B = ad + bc \tag{3.12}$$

There is no general and simple way of choosing the numbers a, b, c, and d that satisfy Equation (3.12). When a, b, c, and d are integers, the easiest way is to follow a modified method of **factoring by trial and error**.

Factoring $x^2 + Bx + C$

When $a \cdot c = A = 1$, the trinomial to be factored has the form $x^2 + Bx + C$. In this case, we can choose $a = 1$ and $c = 1$ in Equation (3.12) and attempt to find two integers b and c whose product is C and whose sum is B. For example, to factor $x^2 - 9x + 14 = (x + b)(x + d)$ we look for two integers b and d such that $b \cdot d = 14$ and $b + d = -9$. The possible ways to factor 14 (ignoring order) as a product of integers are

$$(1)(14), (-1)(-14), (2)(7), \text{ and } (-2)(-7)$$

We see that $(-2)(-7) = 14$ and $(-2) + (-7) = -9$. Consequently, with $b = -2$ and $d = -7$

$$x^2 - 9x + 14 = (x + b)(x + d)$$

becomes

$$x^2 - 9x + 14 = (x + [-2])(x + [-7])$$

or

$$x^2 - 9x + 14 = (x - 2)(x - 7)$$

EXAMPLE 3

Factor

(a) $x^2 + 5x + 6$ (b) $x^2 - x - 12$ (c) $x^2 + 2x + 3$

Solution:
(a) The possible factorizations of the constant term 6 are

$$(1)(6), (-1)(-6), (2)(3), \text{ and } (-2)(-3)$$

We see that $2 + 3 = 5$ and $(2)(3) = 6$ so $b = 2$ and $d = 3$.

$$x^2 + 5x + 6 = (x + 2)(x + 3)$$

(b) The possible factorizations of the constant term -12 are

$$(1)(-12), (-1)(12), (2)(-6), (-2)(6), (3)(-4), \text{ and } (-3)(4)$$

We see that $3 + (-4) = -1$ and $(3)(-4) = -12$, so with $b = 3$ and $d = -4$

$$x^2 - x - 12 = (x + 3)(x + [-4])$$

or

$$x^2 - x - 12 = (x + 3)(x - 4)$$

(c) The possible factorizations of the constant term 3 are

$$(1)(3) \text{ and } (-1)(-3)$$

We see that $1 + 3 \neq 2$ and $(-1) + (-3) \neq 2$ so there are no integers b and d with $b \cdot d = 3$ and $b + d = 2$. Thus $x^2 + 2x + 3$ cannot be factored *using integers*.

A test for determining if a given trinomial with integer coefficients will factor (using integers) is

$Ax^2 + Bx + C$ will factor if and only if there are two integers, m and n, whose product is equal to $A \cdot C$ and whose sum equals B.

EXAMPLE 4 Use the test to determine whether $2x^2 + 5x - 4$ will factor (using integers).

Solution: Here $A = 2$, $B = 5$, and $C = -4$. We want to find integers m and n such that $m \cdot n = (2)(-4) = -8$ and $m + n = 5$.

The possible factorizations of -8 (giving choices for m and n) are

$$(1)(-8), (-1)(8), (2)(-4), \text{ and } (-2)(4)$$

Thus, we see that $1 + (-8) \neq 5$, $(-1) + 8 \neq 5$, $2 + (-4) \neq 5$, and $(-2) + 4 \neq 5$.
Consequently, $2x^2 + 5x - 4$ will not factor.

Factoring $Ax^2 + Bx + C$ when $A \neq 1$

When $Ax^2 + Bx + C = (ax + b)(cx + d)$, an organized method of searching for a, b, c, and d is useful. *When A is positive, it is always possible to pick both a and c to be positive.* For example, suppose we want to perform the factorization

$$3x^2 + 8x + 4 = (ax + b)(cx + d)$$

Here, $A = 3$, $B = 8$, and $C = 4$.
From Equation (3.12) we know that

$$3 = a \cdot c, \; 4 = b \cdot d, \text{ and } 8 = ad + bc$$

Step 1. The possible values for a and c are (positive) factor pairs of 3:

$$(1)(3)$$

Step 2. The possible values of b and d are pairs of factors of 4:

$$(1)(4), (-1)(4), (2)(2), \text{ and } (-2)(-2)$$

Step 3. Try $a = 1$ and $c = 3$ ($a = 3$ and $c = 1$ yields identical results. Why?)

$$3x^2 + 8x + 4 = (1x + b)(3x + d)$$

Step 4. Finally, fill in b and d from the factor pairs of 4

$$3x^2 + 8x + 4 = (x + b)(3x + d)$$

Since $8 = \underset{\underset{\text{outer}}{\uparrow}}{ad} + \underset{\underset{\text{inner}}{\uparrow}}{bc} = (1)d + b(3)$ we may ignore negative factor pairs.

By inspection (or trial) we see $8 = (1) \cdot 2 + 2 \cdot (3)$ with $b = d = 2$. Finally,

$$3x^2 + 8x + 4 = (x + 2)(3x + 2)$$

EXAMPLE 5

Factor $4x^2 - 4x - 3$

Solution: We want $4x^2 - 4x - 3 = (ax + b)(cx + d)$.
We know $4 = ac$, $-3 = b \cdot d$, and $-4 = ad + bc$.

Step 1. The possible values for a and c are (positive) factor pairs of 4, or (1)(4) and (2)(2).

Step 2. The possible values for b and d are pairs of factors of -3, or $(-1)(3)$ and $(1)(-3)$. We have these possible factorizations of $4x^2 - 4x - 3$ (writing $[x + (-1)] = x - 1$, etc.).

$(x - 1)(4x + 3)$ $ad + bc = (1)(3) + (-1)(4) = -1$

$(x + 1)(4x - 3)$ $ad + bc = (1)(-3) + (1)(4) = 1$

$(x - 3)(4x + 1)$ $ad + bc = (1)(1) + (-3)(4) = -11$

$(x + 3)(4x - 1)$ $ad + bc = (1)(-1) + (3)(4) = 11$

$(2x - 1)(2x + 3)$ $ad + bc = (2)(3) + (-1)(2) = 4$

$(2x + 1)(2x - 3)$ $ad + bc = (2)(-3) + (1)(2) = -4$ ✔

The correct factorization is then

$$4x^2 - 4x - 3 = (2x + 1)(2x - 3)$$

Practice will sharpen your skills so that you should not have to try all possible factorizations. However, in the next example we will again go through all possible factorizations before arriving at the correct one. Try to factor $12x^2 + 5x - 2$ yourself before looking at the example.

EXAMPLE 6

Factor $12x^2 + 5x - 2$.

Solution: We want $12x^2 + 5x - 2 = (ax + b)(cx + d)$

Step 1. The possible values for a and c are (positive) factor pairs of 12, or (1)(12), (2)(6), and (3)(4).

Step 2. The possible values of b and d are pairs of factors of -2, or $(-1)(2)$ and $(1)(-2)$.

Step 3. Try $a = 1$ and $c = 12$ ($a = 12$ and $c = 1$ yields identical results).

$$12x^2 + 5x - 2 = (x + ?)(12x + ?)$$

Try all possible factor pairs of -2.

$12x^2 + 5x - 2 \neq [x + (-1)](12x + 2) \quad\quad b = -1, d = 2$

$12x^2 + 5x - 2 \neq (x + 1)[12x + (-2)] \quad\quad b = 1, d = -2$

$12x^2 + 5x - 2 \neq (x + 2)[12x + (-1)] \quad\quad b = 2, d = -1$

$12x^2 + 5x - 2 \neq [x + (-2)](12x + 1) \quad\quad b = -2, d = 1$

Returning to Step 3, try $a = 2$ and $c = 6$. ($a = 6$ and $b = 2$ yields identical results.)

$$12x^2 + 5x - 2 = (2x + ?)(6x + ?)$$

Again try all possible factor pairs of -2.

$12x^2 + 5x - 2 \neq (2x + -1)(6x + 2) \quad\quad b = -1, d = 2$

$12x^2 + 5x - 2 \neq (2x + 1)(6x + -2) \quad\quad b = 1, d = -2$

$12x^2 + 5x - 2 \neq (2x + 2)(6x + -1) \quad\quad b = 2, d = -1$

$12x^2 + 5x - 2 \neq (2x + -1)(6x + 2) \quad\quad b = -1, d = 2$

Now we are back to Step 3 again. Try $a = 3$ and $c = 4$ ($a = 4$ and $c = 3$ yields identical results).

$$12x^2 + 5x - 2 = (3x + ?)(4x + ?)$$

Try $\quad 12x^2 + 5x - 2 = (3x + 2)(4x + [-1]) \quad\quad b = 2, d = -1$

$12x^2 + 5x - 2 = (3x + 2)(4x - 1)$ ✔

Sums or Differences of Cubes

You will recall that

$$(a + b)(a^2 - ab + b^2) = a^3 + b^3$$

and

$$(a - b)(a^2 + ab + b^2) = a^3 - b^3$$

The reverse of these identities gives the following basic factorization formulas, which are termed the **sum of cubes** and the **difference of cubes**, respectively.

$$a^3 + b^3 = (a + b)(a^2 - ab + b^2)$$
$$a^3 - b^3 = (a - b)(a^2 + ab + b^2)$$

EXAMPLE 7

Factor

(a) $x^3 + 27y^3$ (b) $8x^3 - 125y^3$

Solution:
(a) $x^3 + 27y^3 = x^3 + (3y)^3$
$= (x + 3y)[x^2 - x(3y) + (3y)^2]$
$= (x + 3y)(x^2 - 3xy + 9y^2)$

(b) $8x^3 - 125y^3 = (2x)^3 - (5y)^3$
$= (2x - 5y)[(2x)^2 + (2x)(5y) + (5y)^2]$
$= (2x - 5y)(4x^2 + 10xy + 25y^2)$

Exercises 3.7

Standard Assignment: Exercises 1, 3, 5, 7, 9, 11, 13, 15, 23, 25, 27, 29, 31, 33, 35, 37, 39, 41, 43, 45, 53, 55, 57, 59, 61

A In Exercises 1 to 22, factor each binomial. See Example 1.

1. $x^2 - 4$
2. $y^2 - 9$
3. $t^2 - 36$
4. $z^2 - 49$
5. $4k^2 - 1$
6. $9m^2 - 25$
7. $121 - 9x^2$
8. $144 - 25y^2$
9. $y^2 - 9a^2$
10. $z^2 - 36a^2$
11. $4t^2 - 25(s - 1)^2$
12. $9y^2 - 36(x + 3)^2$
13. $k^3n - kn^3$
14. $az^3 - a^3z$
15. $5x^3 - 125x$
16. $4y^3 - 64y$
17. $x^6y^2 - x^2y^6$
18. $x^2y^5 - x^6y$
19. $s^5t^2 - st^6$
20. $x^5y^3 - xy^7$
21. $2x^5 - 32x$
22. $243k - 3k^3$

In Exercises 23–28, factor each trinomial. See Example 2.

23. $y^2 + 14y + 49$

24. $x^2 - 18x + 81$

25. $x^2 + 10x + 25$

26. $y^2 + 12y + 36$

27. $z^2 - z + \dfrac{1}{4}$

28. $r^2 - \dfrac{2}{3}r + \dfrac{1}{9}$

In Exercises 29 to 52, factor each trinomial. See Examples 3–6.

29. $t^2 - 7t - 8$

30. $q^2 - 11q + 24$

31. $9x^2 - 12xy + 4y^2$

32. $25n^2 - 60nt + 36t^2$

33. $(x + y)^2 + 24(x + y) + 144$

34. $(p - q)^2 + 18(p - q) + 81$

35. $4z^2 + 12z + 9$

36. $-9k^2 - 24k - 16$

37. $3x^2 + 14x + 15$

38. $2x^2 + 17x + 21$

39. $-2x^2 + 11x + 21$

40. $3y^2 - 44y - 15$

41. $-3y^2 + 46y - 15$

42. $2t^2 - 17t + 21$

43. $r^2 - 8r + 15$

44. $6s^2 - 5s + 1$

45. $3n^2 - 14n - 5$

46. $-3z^2 + 7z - 2$

B 47. $q^4 - 3q^2 - 4$

48. $s^4 - 8s^2 - 9$

49. $(x + y)^2 - 2(x + y) - 3$

50. $14(2a - 3b)^2 + 3(2a - 3b) - 5$

51. $x^4 + x^2y^2 + y^4$

52. $z^4 - 3z^2t^2 + t^4$

In Exercises 53 to 66, factor each binomial. See Example 7.

53. $r^3 + 1$

54. $z^3 + 8$

55. $y^3 - 27$

56. $t^3 - 1$

57. $27k^3 + \dfrac{1}{8}$

58. $64s^3 + \dfrac{1}{27}$

59. $\dfrac{1}{27}n^3 - 125$

60. $\dfrac{1}{216}t^3 - 1000$

61. $8a^3 + 27b^3$

62. $216y^3 - 343z^3$

63. $27x^4y - 216xy^4$

64. $\dfrac{1}{8}xy^4 - \dfrac{1}{125}x^4y$

65. $s^6 - t^6$

66. $p^6 - 64q^6$

3.8 Diagnosing a Factoring Problem

We have introduced many methods of factoring. Now we will explore how to choose which method to use in factoring a given polynomial. Let us first agree that a polynomial is *factored* if

1. The polynomial is expressed as a product of polynomials with integer coefficients;
2. None of the polynomial factors can be factored further, with the exception that any numerical factor need not be factored;
3. All repeated factors are written as a power of a single factor;
4. The order of presence of the factors is not important.

EXAMPLE 1
(a) According to condition (2), $8x^2 + 24$ should be factored as $8(x^2 + 3)$; it is not necessary to factor it as $2 \cdot 2 \cdot 2 \cdot (x^2 + 3)$.
(b) According to condition (3), $x^5 - 2x^4 + x^3$ should be factored as $x^3(x - 1)^2$, rather than $x \cdot x \cdot x \cdot (x - 1) \cdot (x - 1)$.

The following are steps to follow in factoring polynomials. Each step will be illustrated by an example before we proceed to the next step.

Step I. Factor out any common factor of each term.

Example 2
(a) $50x + 200 = 50(x + 4)$
(b) $3y^4 - 24y^3 + 6y^2 = 3y^2(y^2 - 8y + 2)$
(c) Let m and n be any positive integers. Then
$$4z^m w^{n+1} + 2z^{m+1} w^n + 6z^{m+2} w^{n-1}$$
$$= 2z^m(2w^{n+1} + zw^n + 3z^2 w^{n-1})$$
$$= 2z^m w^{n-1}(2w^2 + zw + 3z^2)$$

Often after factoring out the common factor (Step I), the remaining polynomial may need to be factored further. If it is a binomial, use Step II; if it is a trinomial use Step II'.

Step II. For a binomial, check for the following possibilities:

(a) Difference of squares $\quad a^2 - b^2 = (a + b)(a - b)$
(b) Difference of cubes $\quad a^3 - b^3 = (a - b)(a^2 + ab + b^2)$
(c) Sum of cubes $\quad a^3 + b^3 = (a + b)(a^2 - ab + b^2)$
(d) Sum of squares, $a^2 + b^2$, will generally not factor.

EXAMPLE 3

(a) $49x^2 - 81y^2 = (7x)^2 - (9y)^2$
$= (7x + 9y)(7x - 9y)$

(b) $27z^3 - 125w^3 = (3z)^3 - (5w)^3$
$= (3z - 5w)[(3z)^2 + (3z)(5w) + (5w)^2]$
$= (3z - 5w)(9z^2 + 15zw + 25w^2)$

(c) $8p^3 + 1 = (2p)^3 + 1^3$
$= (2p + 1)[(2p)^2 - (2p) \cdot 1 + 1^2]$
$= (2p + 1)(4p^2 - 2p + 1)$

(d) $36q^2 + 9 = (6q)^2 + 3^2$ is a sum of squares that cannot be factored.

Step II'. For a trinomial, check for the following:

(a) **Perfect Squares**

$$a^2 + 2ab + b^2 = (a + b)^2$$

or

$$a^2 - 2ab + b^2 = (a - b)^2$$

(b) If it is not a perfect square polynomial, the trial and error method of Section 3.7 may be used. See Examples 3–6 in Section 3.7.

EXAMPLE 4

(a) $36x^2 + 12x + 1 = (6x)^2 + 2(6x) 1 + 1^2$
$= (6x + 1)^2$

(b) $49y^2 - 126y + 81 = (7y)^2 - 2(7y) \cdot 9 + 9^2$
$= (7y - 9)^2$

(c) $10z^2 + 13z - 3 = (2z + 3)(5z - 1)$
because, by the trial and error method of Section 3.7, we find $2(-1) + 3 \cdot 5 = 13$.

Step III. If none of the methods in Steps I, II, and II' seem applicable, try **factoring by substitution**, which is introduced in the next example, or **factoring by grouping**.

EXAMPLE 5

Factoring by Substitution
(a) Factor $(x^2 + 5x - 4)^2 + 2(x^2 + 5x - 4) + 1$.
(b) Factor $10(y^2 + 1)^2 + 13(y^2 + 1) - 3$
(c) Factor $x^6 - y^6$.

Solution:
(a) Substituting y for $x^2 + 5x - 4$ in the expression above we have

$$y^2 + 2y + 1 = (y + 1)^2$$

Hence,
$$(x^2 + 5x - 4)^2 + 2(x^2 + 5x - 4) + 1$$
$$= [(x^2 + 5x - 4) + 1]^2 \quad \text{since } y = x^2 + 5x - 4$$
$$= (x^2 + 5x - 3)^2$$

(b) Substituting z for $y^2 + 1$, we have
$$10(y^2 + 1)^2 + 13(y^2 + 1) - 3$$
$$= 10z^2 + 13z - 3$$
$$= (2z + 3)(5z - 1) \quad \text{By trial and error}$$
$$= [2(y^2 + 1) + 3][5(y^2 + 1) - 1] \quad z = y^2 + 1$$
$$= (2y^2 + 5)(5y^2 + 4)$$

(c) Let $a = x^3$ and $b = y^3$. Then $a^2 = x^6$ and $b^2 = y^6$. We have
$$x^6 - y^6 = a^2 - b^2 \quad \text{Substitution}$$
$$= (a + b)(a - b)$$
$$= (x^3 + y^3)(x^3 - y^3) \quad \text{Substitute } a = x^3, b = y^3.$$
$$= (x + y)(x^2 - xy + y^2)(x - y)(x^2 + xy + y^2)$$

If we let $a = x^2$ and $b = y^2$ in Example 5(c), we run into some difficulty, for then we have
$$x^6 - y^6 = a^3 - b^3$$
$$= (a - b)(a^2 + ab + b^2)$$
$$= (x^2 - y^2)(x^4 + x^2y^2 + y^4)$$
$$= (x - y)(x + y)(x^4 + x^2y^2 + y^4)$$

By looking at our previous answer we might guess that
$$x^4 + x^2y^2 + y^4 = (x^2 - xy + y^2)(x^2 + xy + y^2)$$

and we can check that this is correct by direct computation. However, it is unlikely that the "trick" for factoring $x^4 + x^2y^2 + y^4$ would be readily discovered. Here it is: we write $x^4 + x^2y^2 + y^4$ as the difference of squares by adding and subtracting x^2y^2.
$$x^4 + x^2y^2 + y^4 = (x^4 + 2x^2y^2 + y^4) - x^2y^2$$
$$= (x^2 + y^2)^2 - (xy)^2$$
$$= (x^2 + y^2 - xy)(x^2 + y^2 + xy)$$

EXAMPLE 6

Factoring by Grouping

(a) $xy + 5y - 2x - 10 = (x + 5)y - 2(x + 5) \quad$ Grouping
$$= (x + 5)(y - 2)$$

(b) $4y^2z - 5z + 4yz^2 - 5y = (4y^2z - 5y) + (4yz^2 - 5)$ Grouping
$= y(4yz - 5) + z(4yz - 5)$
$= (y + z)(4yz - 5)$

(c) $ax + by + ay + bx = (ax + bx) + (ay + by)$ Grouping
$= (a + b)x + (a + b)y$
$= (a + b)(x + y)$

Exercises 3.8

Standard Assignment: Exercises 1, 5, 7, 9, 13, 17, 19, 21, 25, 27, 29, 31, 33, 37, 39, 41, 43, 47, 49, 51

A Factor each of the following polynomials completely.

1. $60a + 180$
2. $210b - 70$
3. $35x - 70y + 14$
4. $54y + 108z - 27$
5. $27x^2y + 81x$
6. $11xy^2 - 121y$
7. $(a + b)x - (a + b)y$
8. $a(x - y) + b(x - y)$
9. $ax^2y^3 + bx^3y^2 + cx^2y^2$
10. $10z^3w^2 - 15z^2w^3 + 25z^2w^2$
11. $121p^2 - 1$
12. $q^2 - 100$
13. $100m^2 - 121$
14. $225 - 16n^2$
15. $81p^2 - (q + r)^2$
16. $(p + q)^2 - 36r^2$
17. $x^4 - 81$
18. $16 - y^4$
19. $z^3 + 729$
20. $125 + w^3$
21. $216 - p^3$
22. $q^3 - 343$
23. $125x^3 + 343y^3$
24. $216z^3 + 729w^3$
25. $8p^3 - (q + r)^3$
26. $(p - q)^3 - 27r^3$
27. $4x^2 + 4x + 1$
28. $9y^2 + 6y + 1$
29. $y^2 - 12y + 36$
30. $z^2 - 14z + 49$
31. $50a^2b + 40ab + 8b$
32. $27ab^2 + 90ab + 75a$
33. $z^2 - 2z^3 + z^4$
34. $20x^2 - 60xy + 45y^2$
35. $4x^2 - 5x + 1$
36. $5y^2 - 9y - 2$
37. $10y^3 + 70y^2 + 100y$
38. $8x^3 - 10x^2 - 3x$
39. $4x^2 + 6xy - 4y^2$
40. $12z^2 - 15zw + 3w^2$

41. $(z^2 + 6z + 1)^2 + 18(z^2 + 6z + 1) + 81$

42. $(2m + 3n - 4)^2 - 14(2m + 3n - 4) + 49$

43. $4(p + q - 1)^2 - 4(p + q - 1) + 1$

44. $9(r - s + t)^2 - 42(r - s + t) + 49$

45. $16(x + y)^2 - 42(x + y) - 18$

46. $6(s + t)^2 - 8(s + t) + 2$

47. $x^6 + y^6$

48. $x^8 - y^8$

49. $2pq + 3q - 10p - 15$

50. $5pq - 2rq - 5ps + 2rs$

51. $x^3y^2 - x^3 - 5y^2 + 5$

52. $16 - 4z^3 - 4y^2 + y^2z^3$

B In the following, m and n are positive integers.

53. $x^m y^n - x^{m+1} y^n - x^m y^n + x^{m+1} y^{n+1}$

54. $x^{m+3} y^{n+2} + x^{m+3} - 2y^{n+2} - 2$

55. $25x^{2m} - 121y^{2n}$

56. $121y^{2n} - 81z^{2m}$

57. $27z^{3n} + 8w^{3m}$

58. $64p^{3m}r^n - 125q^{3n}r^n$

59. $16x^{2m} - 24x^m y^n + 9y^{2n}$

60. $81y^{2n} + 216y^n z^m + 144z^{2m}$

Chapter 3 Summary

Key Words and Phrases

3.1   Factor
       Exponent
       Base
       Power
3.3   Scientific notation
3.4   Polynomial
       Coefficient
       Degree
       Term
       Ascending order
       Descending order
       Monomial
       Binomial

       Trinomial
       Like terms
3.5   FOIL method
3.6   Factor
3.7   Difference of squares
       Perfect square
       Factoring by perfect square
       Factoring by trial and error
       Sum of cubes
       Difference of cubes
3.8   Factoring by substitution
       Factoring by grouping

Key Concepts and Rules

Exponents satisfy the following rules. For any integers m and n and any real numbers x and y:

Rules of Exponents

Product Rule $\qquad x^m x^n = x^{m+n}$

Quotient Rule $\qquad \dfrac{x^m}{x^n} = x^{m-n} \quad (x \neq 0)$

Power Rules $\qquad (x^m)^n = x^{mn}$

$\qquad\qquad\qquad (xy)^m = x^m y^m$

$\qquad\qquad\qquad \left(\dfrac{x}{y}\right)^n = \dfrac{x^n}{y^n} \quad (y \neq 0)$

Review Exercises

[3.1] Name the base and exponent.

1. 6^5
2. $(-3)^7$
3. -4^{-5}
4. $(a-b)^{-3}$

Evaluate the following expressions.

5. 2^4
6. $(-3)^5$
7. $\left(\dfrac{1}{5}\right)^3$
8. $\left(\dfrac{3}{4}\right)^{-3}$

9. 2.0327×10^4
10. 563174×10^{-5}

11. ▦ $(2.0357)^4$
12. ▦ $(8.5067)^{-3}$

Eliminate negative exponents and simplify. Variable exponents represent natural numbers.

13. $6^9 \cdot 6^{-7}$
14. $5^{-6} \cdot 5^4$
15. $\dfrac{7^{-2}}{7^{-1}}$

16. $\dfrac{10^{k+1}}{10^{k-1}}$
17. $a^2 a^{-4} a^5 \quad (a \neq 0)$
18. $\dfrac{b^{-k} b^{-2k}}{b^{-4k}} \quad (b \neq 0)$

[3.2] 19. $(a^4)^{-3} \quad (a \neq 0)$
20. $\left[\left(\dfrac{3}{4}\right)^{-4}\right]^3$

21. $(5^{-2})^{-3}$
22. $(2^{-3} 3^{-2})^{-3}$

23. $(a^2 b^{-3} c^{-4})^5 \quad (b \neq 0 \neq c)$
24. $\left(\dfrac{3x^{-4} y^{k+1}}{4}\right)^{k-1} \quad (x \neq 0)$

[3.3] Write each number in scientific notation.

25. 214,000,000

26. 131 billion

27. The distance from the earth to the sun is about 93,000,000 miles.

28. The radius of a red corpuscle is about 0.0000375 cm.

29. 0.00000513

30. Calculate $\dfrac{0.0014 \times (8.4 \times 10^{13})}{1.72 \times 10^9}$

Convert from scientific notation to decimal notation.

31. 2.13×10^6

32. -8.19×10^{-7}

Use the scientific notation to compute.

33. $\dfrac{0.012 \times 8 \times 0.0014}{0.032}$

34. $\dfrac{25 \times 8.4 \times 0.0015}{0.002 \times 0.05}$

[3.4] Indicate degree and coefficient.

35. $3x^2$

36. $-\dfrac{1}{3}y^3$

37. 125

Perform the indicated operations.

38. $(2x^3 - 3x^2 + 5x - 6) + (-3x^3 + 4x^2 - 5x - 7)$

39. $(4x^3 + 3x^2 - 6x - 7) - (7x^3 - 2x^2 - 8x - 13)$

[3.5] 40. $(5x^3 - 7x^2 + 3x - 12) \cdot (5x^2 - 13x + 2)$ (Hint: You may use the column method.)

41. $(4x + 3)(5x + 2)$

42. $(2x + 3)(2x - 3)$

43. $(x^2 + 5x + 25)(x - 1)$

44. $(y^2 - 6y + 36)(y + 6)$

[3.6–3.8] Factor, as far as possible, into the product of polynomials.

45. $100x - 25$

46. $3x^2 + 15xy$

47. $(a + b)(3x + 2) - (a + b)(2x - 1)$

48. $(x + 1)x^2 - (x + 1)x$

49. $x^2 - 100$

50. $9x^2 - 49$

51. $9x^2 - 25(y - 3)^2$

52. $(x + 1)^2 - 625(y - 1)^2$

53. $x^3 + 27$

54. $x^3 - 8$

55. $\dfrac{1}{125}x^3 + \dfrac{1}{8}$

56. $64x^6 - y^6$

57. $(2x + 1)^2 - (3y - 1)^2$

58. $16y^4 - 81$

59. $2x^2 + 3x - 2$

60. $3x^2 - 4x - 15$

61. $x^4 - 6x^2y^2 + y^4$

62. $x^4 - 7x^2y^2 + y^4$

63. $3(p+q)^2 - 14(p+q) - 5$

64. $32x^{2m} - 48x^m y^n + 18y^{2n}$

65. $75x^{2n} - 363y^{2m}$

66. $54z^{3m} - 16w^{3n}$

Practice Test (60 minutes)

Evaluate the expressions.

1. $(-5)^3$

2. $\left(-\dfrac{1}{3}\right)^4$

Eliminate negative exponents and simplify.

3. $10^{-5} \cdot 10^3 \cdot 10$

4. $\dfrac{5^{-3}}{5^{-2}}$

5. $(2^3 3^{-2})^{-2}$

6. $\left(\dfrac{4x^{1+k}y^{-2}}{3}\right)^{1-k}$ $x \neq 0 \neq y$, $k > 1$

Write each number in scientific notation.

7. 0.00000006957

8. The national debt, according to one estimate, will reach 246.95 billion dollars.

9. Calculate

$$\dfrac{1.34 \times 10^{12}}{0.0012 \times (5.23 \times 10^{15})}$$

Factor, as far as possible, into the product of polynomials.

10. $21x^2y - 7xy^2 + 28xy$

11. $5x^3 - 3x^2 - 15x + 9$

12. $4x^2 - 81y^2$

13. $81x^4 - 16y^4$

14. $x^3 + 125$

15. $125y^3 - 8$

16. $4x^2 - 4x - 15$

17. $3(x-y)^2 - 4(y-x) + 1$

18. $ax - by + ay - bx$

Extended Applications

Poiseville's Law

Jean Louis Poiseville (1799–1869) was a French physiologist and physician. He discovered, by experimental means, the law of laminar flow, which applies in many circumstances to the flow of blood in an artery or vein.

Suppose we restrict our attention to a piece of an artery of length l that is shaped as a cylindrical tube of constant width, which is 2R centimeters.

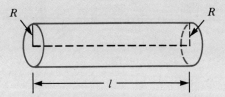

We suppose that blood flows through the tube in such a way that all particles of the blood move parallel to the tube and that the velocity increases regularly from 0 at the wall to a maximum velocity at the center of the tube.

The velocity, v, of a particle of blood at distance r centimeters from the center of the tube is given by the equation

$$v = \frac{P}{4nl}(R^2 - r^2)$$

Here P is the pressure difference (in appropriate units) between the ends of the tube, and n is a number representing the viscosity of the blood, which measures the internal friction of the blood.

Exercises

1. Write the equation for the velocity, v, of human blood flowing through a 3 cm piece of artery with $R = 12 \times 10^{-3}$ cm, $n = 0.027$, and a pressure difference of $P = 3 \times 10^3$ at a distance r from the center of the artery. Find the maximum velocity of blood in the artery.

2. For the artery in Exercise 1, find the velocity of the blood flowing halfway between the center and the wall.

4

Algebraic Fractions

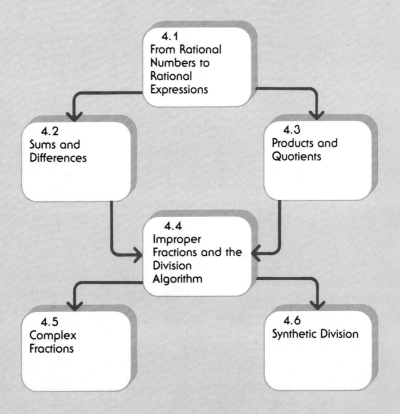

4.1 From Rational Numbers to Rational Expressions

In arithmetic, a rational number a/b is the quotient of any integer a by any nonzero integer b. The rational numbers obey the following basic rules.

Basic Rules of Operation

1. Equality: $\dfrac{a}{b} = \dfrac{c}{d}$ if and only if $ad = bc$

2. Addition: $\dfrac{a}{b} + \dfrac{c}{d} = \dfrac{ad + bc}{bd}$

3. Subtraction: $\dfrac{a}{b} - \dfrac{c}{d} = \dfrac{ad - bc}{bd}$

4. Multiplication: $\dfrac{a}{b} \cdot \dfrac{c}{d} = \dfrac{ac}{bd}$ (4.1)

5. Division: $\dfrac{a}{b} \div \dfrac{c}{d} = \dfrac{a}{b} \cdot \dfrac{d}{c} = \dfrac{ad}{bc}$ $(c \neq 0)$

$$\dfrac{-a}{b} = \dfrac{a}{-b} = -\dfrac{a}{b}$$

6. Reduction: $\dfrac{ca}{cb} = \dfrac{a}{b}$ and $\dfrac{ac}{bc} = \dfrac{a}{b}$
$(c \neq 0)$

Thus, $\dfrac{1}{2} = \dfrac{2}{4}$ because $1 \cdot 4 = 2 \cdot 2$ and $\dfrac{2}{3} = \dfrac{4}{6}$ because $2 \cdot 6 = 3 \cdot 4$. [See Rule 1 in Equations (4.1).] Every integer, whether it is positive, zero, or negative, is a rational number with a denominator of 1. For example,

$$3 = \dfrac{3}{1}, \quad 0 = \dfrac{0}{1}, \quad \text{and} \quad -99 = \dfrac{-99}{1}$$

The set of rational numbers is a "bigger set" that includes the set of integers.

A relation similar to the one between integers and rational numbers in arithmetic exists in algebra. The **rational expression** or **algebraic fraction** p/q is the quotient of any polynomial p by any nonzero polynomial q. Here p is called the *numerator* or *dividend*, and q is called the *denominator* or *divisor*. Thus, for example,

$$\dfrac{x^2 + 2x - 3}{x + 1}, \quad \dfrac{x^2 + 2xy + y^2}{(2 - x)y}, \quad \dfrac{x^2 + 1}{1}, \quad 15 = \dfrac{15}{1}, \quad \text{and} \quad \sqrt{2}$$

are rational expressions.

These examples suggest that the set of rational expressions includes both the set of real numbers (rational and irrational numbers) and the set of polynomials.

Since division by zero is undefined, the values that make a denominator zero are not permitted as replacement values for the variable or variables. In

$$\frac{x^2 + 2x - 3}{x + 1}$$

the number -1 cannot be used as a replacement value for x; and in

$$\frac{x^2 + 2xy + y^2}{(2 - x)y}$$

the number 2 cannot be used as a replacement for x and the number 0 cannot be used as a replacement for y.

EXAMPLE 1

Find the numbers that cannot be used as replacement values for the variable (or variables).

(a) $\dfrac{2x}{(x - 1)(x + 3)}$

(b) $\dfrac{3xy + 5}{xy - 7x}$

(c) $\dfrac{x + y - z}{9}$

Solution:
(a) The numbers that cannot be used to replace x are those that make the denominator $(x - 1)(x + 3)$ zero. We can solve the equation obtained by setting the denominator equal to zero by inspection. Thus, the equation $(x - 1)(x + 3) = 0$ has solutions 1 and -3. The numbers 1 and -3 cannot be used as replacements for x.
(b) If we solve the equation $xy - 7x = 0$ that is obtained by equating the denominator to zero, we have

$$x(y - 7) = 0 \qquad \text{Factor out a common factor of } x$$
$$x = 0 \qquad y = 7$$

Hence, 0 cannot be used as replacement for x and 7 cannot be a replacement for y.
(c) The denominator is constantly 9 and never 0; therefore, no number is excluded from the replacement values.

From now on, in each rational expression all the values that make the denominator zero will be automatically and implicitly excluded.

The basic rules of operations for the rational numbers in Rules (4.1) can each be extended to the rational expressions.

Basic Rules of Operations for the Rational Expressions

Let p, q, r, and s be any polynomials such that q and s are not identically zero. Then

1. Equality: $\dfrac{p}{q} = \dfrac{r}{s}$ if and only if $ps = qr$

2. Addition: $\dfrac{p}{q} + \dfrac{r}{s} = \dfrac{ps + qr}{qs}$

3. Subtraction: $\dfrac{p}{q} - \dfrac{r}{s} = \dfrac{ps - qr}{qs}$ (4.2)

4. Multiplication: $\dfrac{p}{q} \cdot \dfrac{r}{s} = \dfrac{pr}{qs}$

5. Division: $\dfrac{p}{q} \div \dfrac{r}{s} = \dfrac{p}{q} \cdot \dfrac{s}{r} = \dfrac{ps}{qr}$ $(r \neq 0)$

6. Reduction: $\dfrac{rp}{rq} = \dfrac{p}{q}$ and $\dfrac{pr}{qr} = \dfrac{p}{q}$ $(r \neq 0)$

It is often desirable to reduce a fraction such as $\tfrac{8}{12}$ to its lowest terms. In this case, we would divide both the numerator and the denominator by their largest common factor, 4, to obtain $\tfrac{2}{3}$. Similarly, a rational expression can be simplified by dividing out or reducing factors that are common to the numerator and the denominator until no common factors other than 1 (or -1) are present. This requires the use of Rule 6 that is given in Rules (4.2). The resulting rational expression is then said to be in **lowest terms**.

EXAMPLE 2   Reduce the following expressions to lowest terms.

(a) $\dfrac{28xy^2}{35x^2y} = \dfrac{(4y)(7xy)}{(5x)(7xy)}$ Factor out the common factor $7xy$

$= \dfrac{4y}{5x}$ Divide out $7xy$ [Rule 6 in Rules (4.2)]

(b) $\dfrac{12y^2 - 9xy}{8xy - 6x^2} = \dfrac{3y(4y - 3x)}{2x(4y - 3x)}$

$= \dfrac{3y}{2x}$ Reduction [Rule 6 in Rules (4.2)]

(c) $\dfrac{6(x^2 + x + 1)}{9(x^3 - 1)} = \dfrac{3 \cdot 2(x^2 + x + 1)}{3 \cdot 3(x - 1)(x^2 + x + 1)}$ Factor out the common factors

$= \dfrac{2}{3(x - 1)}$ Reduction [Rule 6 in Rules (4.2)]

WARNING $\dfrac{k^2}{2k^2 + x} \neq \dfrac{1}{2 + x}$ since k^2 is not a *factor* of the denominator

The process that is the reverse of reducing to lowest terms is called raising to higher terms. This process is needed when two rational expressions that have different denominators are added or subtracted. (See Section 4.2.) In order to raise a rational expression to higher terms, we simply multiply both the numerator and the denominator by a common factor. [See Rule 6 in Rules (4.2).]

EXAMPLE 3 Raise each rational expression to the higher term with the indicated denominator.

(a) $\dfrac{x - 5}{7}$, $14(x + 5)$

(b) $\dfrac{5}{2x}$, $6x^2y$

(c) $\dfrac{x + y}{2x - 3y}$, $4x^2 - 9y^2$

Solution:

(a) $\dfrac{x - 5}{7} = \dfrac{?}{14(x + 5)}$

Because $14(x + 5) = 7 \cdot 2(x + 5)$, we multiply by $2(x + 5)$ in both the numerator and the denominator of $(x - 5)/7$

$\dfrac{x - 5}{7} = \dfrac{2(x + 5) \cdot (x - 5)}{2(x + 5) \cdot 7}$

$= \dfrac{2(x^2 - 25)}{14(x + 5)}$ Product of sum and difference

(b) $\dfrac{5}{2x} = \dfrac{?}{6x^2y}$

Because $6x^2y = (2x)(3xy)$, we have

$\dfrac{5}{2x} = \dfrac{5(3xy)}{2x(3xy)} = \dfrac{15xy}{6x^2y}$

(c) $\dfrac{x + y}{2x - 3y} = \dfrac{?}{4x^2 - 9y^2}$

Because $4x^2 - 9y^2 = (2x - 3y)(2x + 3y)$, we have

$\dfrac{x + y}{2x - 3y} = \dfrac{(x + y)(2x + 3y)}{(2x - 3y)(2x + 3y)} = \dfrac{2x^2 + 5xy + 3y^2}{4x^2 - 9y^2}$ Use FOIL here

Exercises 4.1

Standard Assignment: Exercises 1, 3, 7, 9, 13, 15, 17, 23, 25, 27, 31, 33, 39, 41, 45, 47, 49, 51, 53, 57

A In Exercises 1–12, find the real numbers that are excluded from the replacement values. See Example 1.

1. $\dfrac{1}{x + 5}$

2. $\dfrac{y}{y - 9}$

3. $\dfrac{3y + 1}{3y - 6}$

4. $\dfrac{5x - 1}{4x - 12}$

5. $\dfrac{x^2 + 1}{x^2 - 1}$

6. $\dfrac{y + 1}{y^2 - 4}$

7. $\dfrac{x + y}{5}$

8. $\dfrac{x^2 - xy}{9}$

9. $\dfrac{2x^2 + x - 1}{x^2 + 1}$

10. $\dfrac{3x^2 - 2xy}{(x - 1)^2 + y^2 + 1}$

11. $\dfrac{1}{(x - 1)(x + 2)}$

12. $\dfrac{x - 2}{(x - 3)(x + 1)}$

In Exercises 13–46, reduce the expressions to lowest terms. See Example 2.

13. $\dfrac{7x}{21y}$

14. $\dfrac{15y}{105x}$

15. $\dfrac{7x^3}{10x^5}$

16. $\dfrac{11x^6}{32x^2}$

17. $\dfrac{4x^2}{20x^4}$

18. $\dfrac{6x^5}{27x^7}$

19. $\dfrac{-125x^2 y}{75xy^2}$

20. $\dfrac{121xy^2}{-22x^2 y}$

21. $\dfrac{56x^4 y^3 z^2}{64x^2 y^4 z^5}$

22. $\dfrac{35x^2 y^4 z^6}{75x^3 y^3 z^7}$

23. $\dfrac{u(u^2 + 1)}{u^2(u^2 + 1)}$

24. $\dfrac{v^2(v - 1)}{v^3(v - 1)^3}$

25. $\dfrac{2x^2 - 2x}{3x - 3}$

26. $\dfrac{3x^2 + 6x}{5x^2 + 10x}$

27. $\dfrac{2x - 1}{1 - 2x}$

28. $\dfrac{2 - 5y}{5y - 2}$

29. $\dfrac{3t + 3}{t^2 - 1}$

30. $\dfrac{10 - 5y}{4 - y^2}$

31. $\dfrac{2x - 6}{9 - x^2}$

32. $\dfrac{15 + 3y}{y^2 - 25}$

33. $\dfrac{10x^2 + 10x}{x^2 + 2x + 1}$

34. $\dfrac{3x^2 + 9x}{x^2 + 6x + 9}$

35. $\dfrac{6x - 3}{4x^2 - 4x + 1}$

36. $\dfrac{9x^2 - 6x + 1}{12x - 4}$

37. $\dfrac{x^3 - y^3}{x - y}$

38. $\dfrac{x^3 + y^3}{x + y}$

39. $\dfrac{(x^2 + x + 1)^2}{x^3 - 1}$

40. $\dfrac{x^3 + y^3}{(x^2 - xy + y^2)^2}$

41. $\dfrac{u^2 - 4}{u^2 + 4u + 4}$

42. $\dfrac{9v^2 - 1}{9v^2 + 6v + 1}$

43. $\dfrac{x^4 + 2x^2y^2 + y^4}{x^4 - y^4}$

44. $\dfrac{x^4 - y^4}{(x - y)^2(x + y)^3(x^2 + y^2)}$

45. $\dfrac{x^2 - 2x - 3}{x^2 - 4x + 3}$

46. $\dfrac{x^2 - 5x + 6}{x^2 - 8x + 15}$

In Exercises 47–62, raise the expressions to higher terms with the indicated denominator. See Example 3.

47. $\dfrac{2}{3x}, \quad 15x^2$

48. $\dfrac{3}{2y}, \quad 10y^3$

49. $\dfrac{3}{2y}, \quad 6xy$

50. $\dfrac{2}{3x}, \quad 6xy$

51. $\dfrac{2xy}{5x^2y}, \quad 15x^2y^2$

52. $\dfrac{7x + y}{3xy^2}, \quad 15x^2y^2$

53. $\dfrac{3x - 2}{x + 1}, \quad x^2 - 1$

54. $\dfrac{8x + 1}{x - 1}, \quad x^2 - 1$

55. $\dfrac{5x - 2}{x^2 + x + 1}, \quad x^3 - 1$

56. $\dfrac{x^2 + x - 1}{x - 1}, \quad x^3 - 1$

57. $\dfrac{3x^2 + 5}{x + 1}$, $x^3 + 1$

58. $\dfrac{3x - 2}{x^2 - x + 1}$, $x^3 + 1$

59. $\dfrac{3}{2u + 5v}$, $4u^2 - 25v^2$

60. $\dfrac{6}{5v - 2u}$, $4u^2 - 25v^2$

61. $\dfrac{5}{5u - 2v}$, $125u^3 - 8v^3$

62. $\dfrac{7}{3a + 5b}$, $27a^3 + 125b^3$

B In Exercises 63–68, insert the appropriate polynomials in the blank spaces.

63. $\dfrac{3x}{5y} = \dfrac{}{5xy - 5y^2}$

64. $\dfrac{4y}{7x} = \dfrac{}{7x^2 - 14xy}$

65. $\dfrac{x + 2}{3x^2} = \dfrac{2x^2 + 9x + 10}{}$

66. $\dfrac{7y}{y - 3} = \dfrac{}{y^2 - 5y + 6}$

67. $\dfrac{x^4 - 2x^2y^2 + y^4}{x^4 - y^4} = \dfrac{x^2 - y^2}{}$

68. $\dfrac{u^4 - v^4}{(u - v)(u + v)^2} = \dfrac{u^2 + v^2}{}$

4.2 Sums and Differences

In order to find the sum or difference of two rational expressions with the same denominators, simply add or subtract their numerators. For example,

$$\dfrac{p}{q} + \dfrac{r}{q} = \dfrac{p + r}{q} \quad \text{and} \quad \dfrac{p}{q} - \dfrac{r}{q} = \dfrac{p - r}{q}$$

EXAMPLE 1 In electronics, when two resistors with resistances R_1 and R_2 are connected parallel to each other, as shown in the diagram, the combined resistance R is found from the equation

$$\dfrac{1}{R} = \dfrac{1}{R_1} + \dfrac{1}{R_2}$$

Find R if $R_1 = 20$(ohms) and $R_2 = 30$(ohms).

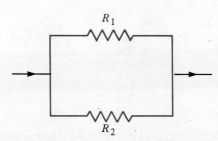

Solution: Substituting 20 for R_1 and 30 for R_2, we get

$$\frac{1}{R} = \frac{1}{20} + \frac{1}{30}$$

Multiply both sides by $60R$.

$$60R \cdot \frac{1}{R} = 60R \cdot \frac{1}{20} + 60R \cdot \frac{1}{30}$$

$$60 = 3R + 2R$$

$$60 = 5R$$

$$12 = R$$

Hence, $R = 12$ (ohms).

EXAMPLE 2

(a) $\dfrac{5x}{7} + \dfrac{3y}{7} = \dfrac{5x + 3y}{7}$

(b) $\dfrac{10x}{z} + \dfrac{5y}{z} - \dfrac{5x + 4y}{z} = \dfrac{10x + 5y - (5x + 4y)}{z}$

$$= \dfrac{10x + 5y - 5x - 4y}{z} = \dfrac{5x + y}{z}$$

(c) $\dfrac{5k - 1}{k + 1} + \dfrac{4k + 10}{k + 1} = \dfrac{5k - 1 + 4k + 10}{k + 1}$

$$= \dfrac{9k + 9}{k + 1}$$

$$= \dfrac{9(k + 1)}{k + 1} = 9$$

EXAMPLE 3

(a) $\dfrac{a - 6}{(a + 1)^2} + \dfrac{a + 8}{(a + 1)^2} = \dfrac{a - 6 + a + 8}{(a + 1)^2}$

$$= \dfrac{2(a + 1)}{(a + 1)^2}$$

$$= \dfrac{2}{a + 1}$$

(b) $\dfrac{x^2}{(x + y)^2} - \dfrac{y^2}{(x + y)^2} = \dfrac{x^2 - y^2}{(x + y)^2}$

$$= \dfrac{(x + y)(x - y)}{(x + y)^2} = \dfrac{x - y}{x + y}$$

(c) $\dfrac{3x - 2}{x^2 - 5x + 6} - \dfrac{2x + 1}{x^2 - 5x + 6} = \dfrac{3x - 2 - (2x + 1)}{x^2 - 5x + 6}$

$= \dfrac{x - 3}{(x - 2)(x - 3)}$

$= \dfrac{1}{x - 2}$

Although the sum or difference of rational expressions with distinct denominators can be found by using Rules (4.2), as in the cases of

$$\dfrac{p}{q} + \dfrac{r}{s} = \dfrac{p \cdot s}{q \cdot s} + \dfrac{r \cdot q}{s \cdot q} = \dfrac{ps + qr}{qs}$$

and

$$\dfrac{p}{q} - \dfrac{r}{s} = \dfrac{ps - qr}{qs}$$

there are two drawbacks. First, these formulas are not directly applicable when the operations involve more than two rational expressions, and second, direct application of the formulas often unnecessarily raises the results to higher terms. For example

$$\dfrac{3}{x^3} + \dfrac{2}{x^4} = \dfrac{3x^4 + 2x^3}{x^3 \cdot x^4} = \dfrac{3x^4 + 2x^3}{x^7}$$

However,

$$\dfrac{3}{x^3} = \dfrac{3x}{x^4}$$

so

$$\dfrac{3}{x^3} + \dfrac{2}{x^4} = \dfrac{3x}{x^4} + \dfrac{2}{x^4} = \dfrac{3x + 2}{x^4}$$

which is now in lowest terms.

In order to remedy these drawbacks, we raise each rational expression to a higher term by using the **least common multiple (LCM)** of the denominators, which is called the **least common denominator (LCD)**, as a common denominator, and then we add or subtract as in Examples 2 and 3. The LCD of a given set of rational expressions with integral coefficients can be found by the following process.

4.2 Sums and Differences

> **To Find the LCD of Any Given Set of Rational Expressions**
> 1. Factor each denominator completely.
> 2. Write a product where each factor of the denominator polynomials appears exactly once.
> 3. Attach to each factor in this product the largest exponent that appears on this factor in any of the denominators.

EXAMPLE 4 Find the LCD.

(a) $\dfrac{1}{50x^2y}$, $\dfrac{x+y}{4xy^2z}$

(b) $\dfrac{1}{t^2 - 4t + 4}$, $\dfrac{2}{t^2 + t - 6}$, $\dfrac{3}{t^2 + 6t + 9}$

Solution:
(a) Factor each denominator.
$$50x^2y = 2 \cdot 5^2 \cdot x^2 \cdot y$$
$$4xy^2z = 2^2 \cdot x \cdot y^2 \cdot z^1$$
Hence, the LCD is $2^2 \cdot 5^2 \cdot x^2y^2z^1$ or $100x^2y^2z$.

(b) Factor each denominator.
$$t^2 - 4t + 4 = (t - 2)^2$$
$$t^2 + t - 6 = (t - 2)(t + 3)$$
$$t^2 + 6t + 9 = (t + 3)^2$$
Hence, the LCD is $(t - 2)^2(t - 3)^2$.

EXAMPLE 5 Raise the expressions to higher terms using the LCD as the common denominator.

(a) $\dfrac{7}{2x^2}$, $\dfrac{5}{4xy}$, $\dfrac{11}{6xy^2}$

(b) $\dfrac{n}{n^2 - 9n + 14}$, $\dfrac{2n}{n^2 - 5n + 6}$

Solution:
(a) In order to find the LCD, factor each denominator.
$$2x^2 = 2 \cdot x^2, \quad 4xy = 2^2 \cdot x \cdot y, \quad \text{and} \quad 6xy^2 = 2 \cdot 3 \cdot x \cdot y^2$$

160   Chap. 4   Algebraic Fractions

Hence, the LCD is $2^2 3 x^2 y^2$. We have

$$\frac{7}{2x^2} = \frac{7 \cdot (2 \cdot 3 \cdot y^2)}{2x^2 \cdot (2 \cdot 3 \cdot y^2)} = \frac{42y^2}{2^2 3 x^2 y^2}$$

$$\frac{5}{4xy} = \frac{5 \cdot (3xy)}{2^2 xy \cdot (3xy)} = \frac{15xy}{2^2 3 x^2 y^2}$$

$$\frac{11}{6xy^2} = \frac{11 \cdot (2x)}{2 \cdot 3xy^2 \cdot (2x)} = \frac{22x}{2^2 3 x^2 y^2}$$

(b) In order to find the LCD, factor each denominator.

$$n^2 - 9n + 14 = (n - 2)(n - 7)$$
$$n^2 - 5n + 6 = (n - 2)(n - 3)$$

Hence, the LCD is $(n - 2)(n - 3)(n - 7)$. We have

$$\frac{n}{n^2 - 9n + 14} = \frac{n(n - 3)}{(n - 2)(n - 7)(n - 3)} = \frac{n^2 - 3n}{(n - 2)(n - 3)(n - 7)}$$

$$\frac{2n}{n^2 - 5n + 6} = \frac{2n \cdot (n - 7)}{(n - 2)(n - 3)(n - 7)} = \frac{2n^2 - 14n}{(n - 2)(n - 3)(n - 7)}$$

EXAMPLE 6   Perform the indicated operation and simplify.

(a) $\dfrac{1}{2y^2} + \dfrac{2}{3y}$

(b) $\dfrac{x^2}{x + 1} - \dfrac{2x}{x^2 - 1}$

Solution:

(a) Because the LCD is $2 \cdot 3y^2$ or $6y^2$, we write

$$\frac{1}{2y^2} + \frac{2}{3y} = \frac{1 \cdot 3}{2y^2 \cdot 3} + \frac{2 \cdot (2y)}{3y \cdot (2y)}$$

$$= \frac{3}{6y^2} + \frac{4y}{6y^2}$$

$$= \frac{3 + 4y}{6y^2}$$

(b) $\dfrac{x^2}{x + 1} - \dfrac{2x}{x^2 - 1}$

$$= \frac{x^2}{x + 1} - \frac{2x}{(x + 1)(x - 1)}$$

Factor the denominator to find the LCD: $(x + 1)(x - 1)$

$$= \frac{x^2 \cdot (x-1)}{(x+1) \cdot (x-1)} - \frac{2x}{(x+1)(x-1)} \qquad \text{Raise to higher terms}$$

$$= \frac{x^3 - x^2 - 2x}{(x+1)(x-1)} \qquad \text{Subtract}$$

$$= \frac{x(x^2 - x - 2)}{(x+1)(x-1)} \qquad \text{Factor out the common factor } x$$

$$= \frac{x(x-2)(x+1)}{(x-1)(x+1)} \qquad \text{Factor } x^2 - x - 2 = (x-2)(x+1)$$

$$= \frac{x(x-2)}{x-1} \qquad \text{Divide out } x+1$$

EXAMPLE 7 Perform the indicated operations and simplify.

$$\frac{2x}{x^2 - 5x + 6} - \frac{x}{x^2 - 9x + 14} + \frac{3}{x^2 - 10x + 21}$$

Solution:

$$\frac{2x}{x^2 - 5x + 6} - \frac{x}{x^2 - 9x + 14} + \frac{3}{x^2 - 10x + 21}$$

$$= \frac{2x}{(x-2)(x-3)} - \frac{x}{(x-2)(x-7)} + \frac{3}{(x-3)(x-7)}$$

It is apparent now that the LCD is $(x-2)(x-3)(x-7)$. Convert each denominator to the LCD.

$$= \frac{2x \cdot (x-7)}{(x-2)(x-3)(x-7)} - \frac{x \cdot (x-3)}{(x-2)(x-3)(x-7)}$$

$$+ \frac{3(x-2)}{(x-2)(x-3)(x-7)}$$

$$= \frac{2x^2 - 14x - (x^2 - 3x) + 3x - 6}{(x-2)(x-3)(x-7)}$$

$$= \frac{x^2 - 8x - 6}{(x-2)(x-3)(x-7)}$$

It is better to leave the denominator of the final result in factored form so that the excluded values of the variables become self-evident. In this case, $x \neq 2, 3, 7$.

Exercises 4.2

Standard Assignment: Exercises 1, 3, 5, 7, 11, 13, 15, 19, 21, 23, 27, 29, 35, 37, 39, 41, 45, 47, 49, 51, 53

A In Exercises 1–8, perform the given operations and simplify. See Examples 2 and 3.

1. $\dfrac{5x-2}{15} + \dfrac{x+5}{15}$

2. $\dfrac{9y+8}{20} + \dfrac{y-3}{20}$

3. $\dfrac{15r-5}{21} - \dfrac{r+2}{21}$

4. $\dfrac{14z+3}{24} - \dfrac{2z-3}{24}$

5. $\dfrac{2t^2-7}{3t} + \dfrac{3t^2+7}{3t}$

6. $\dfrac{3-7y}{5y} + \dfrac{2y-3}{5y}$

7. $\dfrac{3x-5}{(x-1)^2} - \dfrac{x-3}{(x-1)^2}$

8. $\dfrac{5a+1}{(a+2)^2} - \dfrac{2a-5}{(a+2)^2}$

9. The focal length f of a lens, in terms of the object distance d to the lens and its image distance m to the lens, is given by

$$\frac{1}{f} = \frac{1}{d} + \frac{1}{m}$$

Find m if $d = 6$ meters and $f = 2$ meters.

10. A data-processing office has two high-speed card sorters, A and B. Machine A can sort a certain bunch of cards in 20 seconds, while machine B can sort the same bunch of cards in 30 seconds. How long would it take the two sorters to sort this bunch of cards together?

11. A relation between the temperature in degrees Fahrenheit, F, and in degrees Celsius, C, is given by

$$\frac{F-32}{9} = \frac{C}{5}$$

Find F if $C = 40°$

12. In Exercise 11 find C if $F = 50°$.

In Exercises 13–36, only the denominators are given. Find the LCD (least common denominator). See Examples 4 and 5.

13. $6z^2$, $3z$

14. $5t$, $25t^2$

15. $9a^2$, $6ab$

16. $21uv$, $14v^2$

17. $35r$, $14(r-1)$

18. $36(s+1)$, $24ts$

19. $2xy$, $3yz$, $4xz$

20. $4uv$, $6vw$, $8uw$

21. $3r-6$, $4r-8$

22. $7+21z$, $3+9z$

23. $4k^2 - 1$, $(2k + 1)^2$
24. $(3n - 1)^2$, $9n^2 - 1$
25. $16s^2 - 25t^2$, $8s + 10t$
26. $36a^2 - 49b^2$, $18a - 21b$
27. $8k^3 + 27$, $2k + 3$
28. $27 - 64y^3$, $3 - 4y$
29. $x^2 - 7x + 12$, $x^2 - 16$
30. $m^2 + 9m + 20$, $m^2 - 25$
31. $64t^3 + 27$, $16t^2 - 12t + 9$
32. $8y^3 - 27$, $4y^2 + 6y + 9$
33. $z^2 - 3z + 2$, $z^2 - 5z + 6$
34. $y^2 + 4y + 3$, $y^2 + 7y + 12$
35. $r^2 - 5r + 4$, $r^2 - 6r + 5$, $r^2 - 9r + 20$
36. $x^2 + 3x - 4$, $x^2 + 6x - 7$, $x^2 + 11x + 28$

In Exercises 37–60, perform the indicated operation and reduce the result to the lowest terms possible.

37. $\dfrac{1}{5x} + \dfrac{2}{25x^2}$

38. $\dfrac{4}{3y^2} + \dfrac{5}{6y}$

39. $\dfrac{2}{21r^2} - \dfrac{1}{14rs}$

40. $\dfrac{1}{6st} - \dfrac{5}{9t^2}$

41. $\dfrac{a}{24(a + 1)} + \dfrac{1}{36a}$

42. $\dfrac{y}{35(y - 1)} + \dfrac{3}{14y}$

43. $\dfrac{1}{6xy} - \dfrac{1}{8yz} + \dfrac{1}{4xz}$

44. $\dfrac{w}{3uv} + \dfrac{u}{4vw} - \dfrac{v}{2uw}$

45. $\dfrac{2k}{9k + 3} - \dfrac{k}{21k + 7}$

46. $\dfrac{3t}{4t - 8} - \dfrac{2t}{3t - 6}$

47. $\dfrac{2n - 3}{9n^2 - 1} + \dfrac{4n - 1}{(3n - 1)^2}$

48. $\dfrac{3u + 1}{(2u + 1)^2} + \dfrac{u + 3}{4u^2 - 1}$

49. $\dfrac{1}{21a - 18b} - \dfrac{4b}{49a^2 - 36b^2}$

50. $\dfrac{-4t}{25s^2 - 16t^2} - \dfrac{1}{10s + 8t}$

51. $\dfrac{-1}{4x - 3} + \dfrac{16x^2 + 8x + 12}{64x^3 - 27}$

52. $\dfrac{-1}{2k + 3} + \dfrac{4k^2 + 8k + 12}{8k^3 + 27}$

53. $\dfrac{r - 3}{r^2 - 25} - \dfrac{r - 3}{r^2 + 9r + 20}$

54. $\dfrac{2m}{m^2 - 16} - \dfrac{2m - 7}{m^2 - 7m + 12}$

B 55. $\dfrac{-x + 1}{4x^2 + 6x + 9} + \dfrac{2x^2 - 3x}{8x^3 - 27}$

56. $\dfrac{y + 1}{16y^2 - 12y + 9} + \dfrac{-4y^2 + y}{64y^3 + 27}$

57. $\dfrac{5p - 14}{p^2 - 5p + 6} - \dfrac{5p - 6}{p^2 - 3p + 2}$

58. $\dfrac{4v + 12}{v^2 + 7v + 12} - \dfrac{v + 3}{v^2 + 4v + 3}$

59. $\dfrac{-y + 1}{y^2 - 6y + 5} + \dfrac{1}{y^2 - 9y + 20} - \dfrac{-y}{y^2 - 5y + 4}$

60. $\dfrac{5x + 3}{x^2 + 6x - 7} - \dfrac{2x + 2}{x^2 + 11x + 28} - \dfrac{3x + 1}{x^2 + 3x - 4}$

4.3 Products and Quotients

Rational expressions are multiplied in the same manner as rational numbers.

Multiplication of Rational Expressions

If p, q, r, and s are polynomials and $q, s \neq 0$, then

$$\dfrac{p}{q} \cdot \dfrac{r}{s} = \dfrac{pr}{qs} \quad \begin{array}{l}\text{Multiply the numerators} \\ \text{Multiply the denominators}\end{array}$$

EXAMPLE 1 The gravitational force between two masses, m_1 and m_2, that are d units apart is given by the equation

$$F = \dfrac{Gm_1m_2}{d^2}$$

Find m_2 if $m_1 = 5$, $d = 3$, $G = 32$, and $F = 16$.
  By substitution,

$$16 = \dfrac{32 \cdot 5m_2}{3^2}$$

$$16 = \dfrac{160m_2}{9}$$

Multiplying both sides by $\tfrac{9}{160}$, we get

$$\dfrac{9}{160} \cdot 16 = \dfrac{9}{160} \cdot \dfrac{160m_2}{9}$$

$$\dfrac{9}{10} = m_2 \qquad \text{Reduce both sides}$$

Hence, $m_2 = 0.9$.

4.3 Products and Quotients

EXAMPLE 2   Express the products in lowest terms.

(a) $\dfrac{15}{2x^4} \cdot \dfrac{6x^2}{25} = \dfrac{15 \cdot 6x^2}{2x^4 \cdot 25}$ Multiply the numerators
                                       Multiply the denominators

$ = \dfrac{3 \cdot 5 \cdot 2 \cdot 3 \cdot x^2}{5 \cdot 5 \cdot 2x^2 \cdot x^2}$ Factor the numerator and the denominator

$ = \dfrac{3 \cdot 3}{5x^2}$ Reduce to lowest terms

$ = \dfrac{9}{5x^2}$

(b) $\dfrac{y-2}{y^2-9} \cdot \dfrac{y+3}{y^2-4} = \dfrac{y-2}{(y+3)(y-3)} \cdot \dfrac{y+3}{(y+2)(y-2)}$

$ = \dfrac{(y-2)(y+3)}{(y+3)(y-3)(y+2)(y-2)}$

$ = \dfrac{1}{(y-3)(y+2)}$ Reduce to lowest terms

EXAMPLE 3   Compute the products and reduce to lowest terms.

(a) $\dfrac{(xy^2)^3 z}{x^2(y^3 z)^2} \cdot \dfrac{x(yz^2)^2}{(x^2 y)^4 z^4}$

$ = \dfrac{x^3 y^6 z}{x^2 y^6 z^2} \cdot \dfrac{xy^2 z^4}{x^8 y^4 z^4}$ Remove parentheses by using the Power Rule for exponents

$ = \dfrac{x}{z} \cdot \dfrac{1}{x^7 y^2}$ Reduce to lowest terms

$ = \dfrac{x}{x^7 y^2 z}$ Multiply

$ = \dfrac{1}{x^6 y^2 z}$ Reduce again to lowest terms

(b) $\dfrac{a^2 - 9b^2}{9ab - 9b^2} \cdot \dfrac{3a^2 - 3ab}{a^2 - 4ab + 3b^2}$

$ = \dfrac{(a+3b)(a-3b)}{9b(a-b)} \cdot \dfrac{3a(a-b)}{(a-b)(a-3b)}$ Factor

$ = \dfrac{3a(a+3b)(a-3b)(a-b)}{9b(a-b)^2(a-3b)}$ Multiply

$ = \dfrac{a(a+3b)}{9b(a-b)}$ Reduce to lowest terms

EXAMPLE 4 Express the product in lowest terms.

$$\frac{x^2 - 5x + 6}{x^2 - 25} \cdot \frac{x^2 - 10x + 25}{x^2 - 7x + 12} \cdot \frac{x^2 + x - 20}{x^2 - 4}$$

$$= \frac{(x - 2)(x - 3)}{(x + 5)(x - 5)} \cdot \frac{(x - 5)^2}{(x - 3)(x - 4)} \cdot \frac{(x - 4)(x + 5)}{(x + 2)(x - 2)} \qquad \text{Factor}$$

$$= \frac{x - 5}{x + 2} \qquad \text{Multiply and reduce to lowest terms}$$

Quotients

In arithmetic every nonzero rational number a/b has a unique reciprocal b/a such that

$$\frac{a}{b} \cdot \frac{b}{a} = 1$$

Similarly, in algebra every nonzero rational expression r/s has a unique **reciprocal**, s/r, such that

$$\frac{r}{s} \cdot \frac{s}{r} = 1$$

Because division by zero is undefined in the division

$$\frac{p}{q} \div \frac{r}{s}$$

of rational expressions, the divisor r/s cannot be allowed to be zero. Therefore, all replacement values that make the numerator polynomial r zero must be excluded. Consequently, the nonzero divisor r/s always has a unique (nonzero) reciprocal s/r. In deriving a formula for calculating the quotient, let

$$Q = \frac{p}{q} \div \frac{r}{s}$$

As multiplication is the inverse of division, we multiply r/s on both sides of the equality to get

$$Q \cdot \frac{r}{s} = \frac{p}{q}$$

Then, by multiplying the reciprocal of r/s, s/r, on both sides, we have

$$Q = \frac{p}{q} \cdot \frac{s}{r}$$

4.3 Products and Quotients

The Division Formula

$$\frac{p}{q} \div \frac{r}{s} = \frac{p}{q} \cdot \frac{s}{r}$$

The division formula states that the quotient (division) of two rational expressions is the product (multiplication) of the dividend and the reciprocal of the divisor.

EXAMPLE 5 Find the quotient and reduce to lowest terms.

(a) $\dfrac{8r}{15s} \div \dfrac{16r}{5s} = \dfrac{8r}{15s} \cdot \dfrac{5s}{16r}$ Multiply dividend by the reciprocal of the divisor

$= \dfrac{8 \cdot 5rs}{3 \cdot 5 \cdot 8 \cdot 2sr}$ Factor 15 and 16

$= \dfrac{1}{6}$ Reduce to lowest terms

(b) $\dfrac{xy^3z^4}{(x^2y)^2} \div \dfrac{x^5y^2z^{11}}{(yz^2)^3}$

$= \dfrac{xy^3z^4}{(x^2y)^2} \cdot \dfrac{(yz^2)^3}{x^5y^2z^{11}}$ Multiply dividend by the reciprocal of the divisor

$= \dfrac{xy^3z^4}{x^4y^2} \cdot \dfrac{y^3z^6}{x^5y^2z^{11}}$ By the Power Rule for exponents and Rule (3.6)

$= \dfrac{xy^6z^{10}}{x^9y^4z^{11}}$ By the Product Rule of exponents

$= \dfrac{y^2}{x^8z}$ Reduce to lowest terms

EXAMPLE 6 Calculate the quotient and reduce to lowest terms.

$\dfrac{2y^2 - y - 1}{y^2 + 5y + 6} \div \dfrac{y^2 - 5y + 4}{3y + 9}$

$= \dfrac{2y^2 - y - 1}{y^2 + 5y + 6} \cdot \dfrac{3y + 9}{y^2 - 5y + 4}$ Multiply dividend by the reciprocal of the divisor

$= \dfrac{(2y + 1)(y - 1)}{(y + 2)(y + 3)} \cdot \dfrac{3(y + 3)}{(y - 1)(y - 4)}$ Factor

$= \dfrac{3(2y + 1)}{(y + 2)(y - 4)}$ Reduce to lowest terms

As we have seen in Examples 5 and 6, the division of rational expressions is quite similar to the multiplication of rational expressions.

Exercises 4.3

Standard Assignment: Exercises 1, 3, 5, 7, 9, 15, 19, 21, 25, 29, 31, 33, 35, 39, 41, 43, 45, 47, 51, 53, 57

A In Exercises 1–28, find the indicated product and reduce to lowest terms. See Examples 2, 3, and 4.

1. $\dfrac{4x}{27} \cdot \dfrac{9}{8x^2}$

2. $\dfrac{6}{25y^2} \cdot \dfrac{15y}{81}$

3. $\dfrac{7}{12s^2} \cdot \dfrac{3s}{14t}$

4. $\dfrac{4v}{3u} \cdot \dfrac{9u^3}{8uv^3}$

5. $\dfrac{-5a^2}{8b^3} \cdot \dfrac{64b^4}{25a^3}$

6. $\dfrac{-7q}{11p^3} \cdot \dfrac{33p^2}{21q^2}$

7. $\dfrac{2yz^2}{7x^5} \cdot \dfrac{-21x^3}{8y^2z^3}$

8. $\dfrac{3x^2}{8y^3z^4} \cdot \dfrac{-4y^2z^3}{9x^4}$

9. $\dfrac{-3t}{2s} \cdot \dfrac{-4s^2}{6st}$

10. $\dfrac{-5rs}{3r^2} \cdot \dfrac{-6r^2}{10rs}$

11. $\dfrac{(-uv^2)^3 w}{u(vw^2)^2} \cdot \dfrac{(-2v^2w)^2 u}{(u^2v)^3 w}$

12. $\dfrac{(-x^2y)^2 z}{x(y^2z)^3} \cdot \dfrac{(-2yz^2)^3 x}{(x^2y)^2 z}$

13. $\dfrac{(-3x^2y^3)^3}{(2x^3y^2)^2} \cdot \dfrac{(4x^2y)^2}{(6x^3y^4)^3}$

14. $\dfrac{(-2uv^3)^2}{(10u^2v)^4} \cdot \dfrac{(-5uv^2)^3}{(u^3v^2)^2}$

15. $\dfrac{3z}{8z+4} \cdot \dfrac{12z+6}{27z^2}$

16. $\dfrac{25t-5}{9t^2} \cdot \dfrac{27t}{10t-2}$

17. $\dfrac{1-a}{b-1} \cdot \dfrac{1-b}{a-1}$

18. $\dfrac{u-v}{v-w} \cdot \dfrac{w-v}{v-u}$

19. $\dfrac{r^2-1}{15rs^2} \cdot \dfrac{3r^2s}{r+1}$

20. $\dfrac{p-q}{7pq^2} \cdot \dfrac{14p^2q}{p^2-q^2}$

21. $\dfrac{(k+4)^2}{27k^3} \cdot \dfrac{9k^2}{k^2-16}$

22. $\dfrac{(y-5)^2}{49y^4} \cdot \dfrac{7y^2}{y^2-25}$

23. $\dfrac{x^2-5x+6}{x^3-3x^2+2x} \cdot \dfrac{x^2-5x+4}{x^2-7x+12}$

24. $\dfrac{z^2-6z+8}{z^2-5z+4} \cdot \dfrac{z^2-4z+3}{z^3-5z^2+6z}$

25. $\dfrac{y^2-7y+10}{y^3-8} \cdot \dfrac{y^2+2y+4}{y^3-125}$

26. $\dfrac{t^3-27}{t^3-64} \cdot \dfrac{t^2+4t+16}{t^2+3t+9}$

27. $\dfrac{x^2 + 3x + 2}{x^2 + 5x + 6} \cdot \dfrac{x^2 + 4x + 3}{x^2 + 6x + 8} \cdot \dfrac{x^2 + 5x + 4}{x^2 + 4x + 3}$

28. $\dfrac{m^2 - 7m + 12}{m^2 - 5m + 4} \cdot \dfrac{m^2 - 5m + 6}{m^2 - 4m + 3} \cdot \dfrac{m^2 - 3m + 2}{m^2 - 6m + 8}$

In Exercises 29–50, find the indicated quotient and reduce to lowest terms. See Examples 5 and 6.

29. $\dfrac{3x^2}{100} \div \dfrac{27x^2}{150}$

30. $\dfrac{49}{5y^3} \div \dfrac{14}{15y^2}$

31. $\dfrac{81}{(-2t)^3} \div \dfrac{54}{(-4t)^2}$

32. $\dfrac{(-3z)^3}{16} \div \dfrac{(-6z)^2}{81}$

33. $\dfrac{(x^2y)^3}{125} \div \dfrac{(xy^2)^4}{50}$

34. $\dfrac{8xy}{(x^2y^3)^2} \div \dfrac{10x^2y}{(x^3y)^3}$

35. $\dfrac{u - v}{u^2v} \div \dfrac{u^2 - v^2}{uv^2}$

36. $\dfrac{x^2y^3}{x^2 - y^2} \div \dfrac{y^2x^4}{x + y}$

37. $\dfrac{3 - 21k}{4k^2} \div \dfrac{2 - 14k}{3k}$

38. $\dfrac{9xy}{4 + 20x} \div \dfrac{27x^2y}{2 + 10x}$

39. $\dfrac{y^3 - 125}{y^5z} \div \dfrac{y^2 + 5y + 25}{y^3z}$

40. $\dfrac{x^2 - 4x + 16}{x^4y^3} \div \dfrac{x^2 - 64}{x^2y^3}$

41. $\dfrac{2z - 4}{z^2 - 4} \div \dfrac{z + 1}{z + 2}$

42. $\dfrac{2x + 6}{x + 1} \div \dfrac{x^2 - 9}{3x - 9}$

43. $\dfrac{x^2 + 3x}{x^2 + 4x + 3} \div \dfrac{x^2 - 5x}{x + 1}$

44. $\dfrac{t^2 + 5t + 6}{t^2 + 2t} \div \dfrac{t^2 - 9}{t - 1}$

45. $\dfrac{2z^2 + 9z + 4}{z^2 + 6z + 9} \div \dfrac{z^2 + 8z + 16}{2z^2 + 5z - 3}$

46. $\dfrac{3y^2 - 5y - 2}{y^2 - 2y + 1} \div \dfrac{y^2 - 4y + 4}{3y^2 - 2y - 1}$

47. $\dfrac{y^2 - 5y + 6}{y^2 - 9y + 20} \div \dfrac{y^2 - 7y + 12}{y^2 - 7y + 10}$

48. $\dfrac{x^2 + 7x + 12}{x^2 + 11x + 30} \div \dfrac{x^2 + 9x + 18}{x^2 + 9x + 20}$

49. $\dfrac{2u^2 + uv - v^2}{u^2 + uv - 2v^2} \div \dfrac{4u^2 - v^2}{2u^2 + 5uv + 2v^2}$

50. $\dfrac{3u^2 + 2uv - v^2}{3u^2 + 4uv + v^2} \div \dfrac{6u^2 + uv - v^2}{6u^2 + 5uv + v^2}$

In Exercises 51–60, perform each of the indicated operations and reduce the result to lowest terms.

51. $\left(\dfrac{5a^2b}{6} \div \dfrac{15ab^2}{4}\right) \cdot \dfrac{27ab}{(a^2b)^3}$

52. $\left(\dfrac{5a^2b}{6} \cdot \dfrac{15ab^2}{4}\right) \div \dfrac{27ab}{(a^2b)^3}$

53. $\dfrac{5a^2b}{6} \cdot \left(\dfrac{15ab^2}{4} \div \dfrac{27ab}{(a^2b)^3}\right)$

54. $\dfrac{5a^2b}{6} \div \left(\dfrac{15ab^2}{4} \cdot \dfrac{27ab}{(a^2b)^3}\right)$

55. $\left(\dfrac{x+4}{x^2-16} \div \dfrac{x+1}{x^2-7x+12}\right) \cdot \dfrac{x+1}{x^2+4x}$

56. $\dfrac{x+4}{x^2-16} \cdot \left(\dfrac{x+1}{x^2-7x+12} \div \dfrac{x+1}{x^2+4x}\right)$

57. $\left(\dfrac{4x^2-8x+3}{4x^2-8x-5} \div \dfrac{2x^2+5x-3}{2x^2-3x-5}\right) \cdot \dfrac{2x^2+7x+3}{2x^2-x-3}$

58. $\dfrac{4z^2-8z+3}{4z^2-8z-5} \cdot \left(\dfrac{2z^2+5z-3}{2z^2-3z-5} \div \dfrac{2z^2+7z+3}{2z^2-z-3}\right)$

59. $\left(\dfrac{3y^2-2y-1}{3y^2+2y-1} \div \dfrac{y^2-1}{(y+1)^2}\right) \div \dfrac{3y+1}{3y-1}$

60. $\dfrac{3x^2-2x-1}{3x^2+2x-1} \div \left(\dfrac{x^2-1}{(x+1)^2} \div \dfrac{3x+1}{3x-1}\right)$

B 61. Is division an associative operation? (Hint: Compare the results of Exercises 59 and 60.)

62. Is division a commutative operation? Why?

4.4 Improper Fractions and the Division Algorithm

In arithmetic, a fraction a/b is **proper** if $|a| < |b|$; it is an **improper fraction** if $|a| \geq |b|$. Thus, $\frac{1}{2}, -\frac{1}{2}, \frac{1}{3}, -\frac{1}{3}, \frac{2}{3}, -\frac{2}{3}, \ldots$ are proper fractions, and $\frac{2}{2}, -\frac{2}{2}, \frac{3}{2}, -\frac{3}{2}, \frac{4}{3}, -\frac{4}{3}, \ldots$ are improper fractions. In a similar way, we call an algebraic fraction p/s (p and s are polynomials with degree of $s > 0$) proper if the degree of p is less than the degree of s; it is called improper if the *degree* of p is greater than or equal to the *degree* of s. For example,

$$\dfrac{1}{x-2}, \quad \dfrac{2z+1}{z^2}, \quad \dfrac{-x^2+3x-5}{x^3+1}$$

are proper fractions, and

$$\dfrac{2z-3}{z+1}, \quad \dfrac{-y^2+5}{y^2+2}, \quad \dfrac{x^3+2x^2+3x+5}{x^2-1}$$

are improper fractions.

An improper fraction can always be rewritten as a nonzero integer plus (or minus) a proper fraction. For example,

$$\dfrac{11}{3} = 3 + \dfrac{2}{3} \quad \text{and} \quad \dfrac{153}{10} = 15 + \dfrac{3}{10}$$

4.4 Improper Fractions and the Division Algorithm

Every improper algebraic fraction can be rewritten in a unique way as the sum (or difference) of a nonzero polynomial and a proper (algebraic) fraction. (4.3)

EXAMPLE 1 Write each rational expression as the sum of a polynomial and a proper algebraic fraction.

(a) $\dfrac{2x^3 + 3x^2 + 1}{x^2} = \dfrac{x^2(2x + 3) + 1}{x^2}$ Common factor of x^2

$= \dfrac{x^2(2x + 3)}{x^2} + \dfrac{1}{x^2}$ Sum of two expressions

$= (2x + 3) + \dfrac{1}{x^2}$ Divide out x^2

(b) $\dfrac{y^2 + 5}{y^2 + 2} = \dfrac{(y^2 + 2) + 3}{y^2 + 2}$ $5 = 2 + 3$, regroup

$= \dfrac{y^2 + 2}{y^2 + 2} + \dfrac{3}{y^2 + 2}$ Sum of two expressions

$= 1 + \dfrac{3}{y^2 + 2}$ Reduce to lowest terms

(c) $\dfrac{2x^2 + 3x + 5}{x + 1} = \dfrac{(2x^2 + 2x) + (x + 1) + 4}{x + 1}$ Regroup the numerator

$= \dfrac{2x(x + 1) + (x + 1) + 4}{x + 1}$ Factor $2x^2 + 2x$

$= \dfrac{2x(x + 1) + (x + 1)}{x + 1} + \dfrac{4}{x + 1}$ Sum of two expressions

$= \dfrac{(x + 1)(2x + 1)}{x + 1} + \dfrac{4}{x + 1}$ Common factor of $x + 1$

$= 2x + 1 + \dfrac{4}{x + 1}$ Divide out $x + 1$

The reason for the validity of Rule (4.3) is the following Division Algorithm.

Division Algorithm

For any two polynomials $p(x)$ and $d(x)$, with $d(x) \neq 0$, there exists unique polynomials $q(x)$ and $r(x)$ such that

1. $p(x) = d(x) \cdot q(x) + r(x)$ and
2. either $r(x) = 0$ or the degree of $r(x)$ is less than the degree of $d(x)$.

Chap. 4 Algebraic Fractions

The Division Algorithm says that dividing a polynomial $p(x)$ by a nonzero polynomial $d(x)$ results in a polynomial $q(x)$, called the **quotient,** and another polynomial $r(x)$, called the **remainder.** For any improper algebraic fraction $\frac{p(x)}{d(x)}$, we may write

$$\frac{p(x)}{d(x)} = \frac{d(x) \cdot q(x) + r(x)}{d(x)} \qquad \text{By the Division Algorithm}$$

$$= \frac{d(x) \cdot q(x)}{d(x)} + \frac{r(x)}{d(x)} \qquad \text{Sum of two fractions}$$

$$= q(x) + \frac{r(x)}{d(x)} \qquad \text{Divide out } d(x)$$

As the degree of $r(x)$ is less than the degree of $d(x)$, the last expression is a polynomial, $q(x)$, plus a proper fraction, $\frac{r(x)}{d(x)}$, as desired.

In order to find the quotient $q(x)$ and the remainder $r(x)$ for a given improper fraction $\frac{p(x)}{s(x)}$, a long division process similar to that used in arithmetic may be employed. As an illustration of this process, we will use the improper fraction in Example 1(c)

$$\frac{2x^2 + 3x + 5}{x + 1}$$

Step 1. Write the following long division form:

$$x + 1 \overline{\smash{\big)}\, 2x^2 + 3x + 5} \qquad \text{Divisor and dividend are arranged in descending order}$$

Step 2. Write the result of dividing the first term of the dividend, $2x^2$, by the first term of the divisor, x, above $2x^2$.

$$x + 1 \overline{\smash{\big)}\, 2x^2 + 3x + 5}^{\;2x} \qquad \frac{2x^2}{x} = 2x$$

Step 3. Write the product of the divisor, $x + 1$, and the result obtained in Step 2, $2x$, under the dividend.

$$\begin{array}{r} 2x \\ x + 1 \overline{\smash{\big)}\, 2x^2 + 3x + 5} \\ 2x^2 + 2x \end{array} \qquad (x + 1) \cdot 2x = 2x^2 + 2x$$

Step 4. Subtract the results of Step 3, $2x^2 + 2x$, from the line above it and record the result in a line below it.

$$\begin{array}{r} 2x \phantom{{}+2x+5} \\ x+1 \overline{\smash{\big)}\, 2x^2 + 3x + 5} \\ \underline{2x^2 + 2x} \\ x + 5 \end{array}$$ ← Result of subtraction

Step 5. Using the last line of Step 4, $x + 5$, as the new dividend and repeating Steps 2, 3, and 4, we have

$$\begin{array}{r} 2x + 1 \\ x+1 \overline{\smash{\big)}\, 2x^2 + 3x + 5} \\ \underline{2x^2 + 2x} \\ x + 5 \\ \underline{x + 1} \\ 4 \end{array}$$

$\dfrac{x}{x} = 1$ from $x + 1\,\overline{\smash{\big)}\,x+5}$

← The remainder

where the top line, $2x + 1$, is the quotient $q(x)$, and the bottom line, 4, is the remainder r. Therefore,

$$\frac{2x^2 + 3x + 5}{x + 1} = 2x + 1 + \frac{4}{x + 1}$$

EXAMPLE 2 Use the long division process to write

$$\frac{3y^3 - 2y + 7}{y^2 + y + 1}$$

as a polynomial plus a proper fraction.

Solution: In order to follow Steps 1–5 of the long division, we must put $0y^2$ in place of the missing y^2 term in the dividend (numerator):

$\dfrac{3y^3}{y^2} = 3y$

$$\begin{array}{r} 3y - 3 \\ y^2 + y + 1 \overline{\smash{\big)}\, 3y^3 + 0y^2 - 2y + 7} \\ \underline{3y^3 + 3y^2 + 3y} \\ -3y^2 - 5y + 7 \\ \underline{-3y^2 - 3y - 3} \\ -2y + 10 \end{array}$$

← $3y \cdot (y^2 + y + 1)$
← Result of subtraction
← $-3 \cdot (y^2 + y + 1)$
← The remainder

Thus, $$\frac{3y^3 - 2y + 7}{y^2 + y + 1} = 3y - 3 + \frac{-2y + 10}{y^2 + y + 1}$$

EXAMPLE 3

Write the improper fraction $(5t^4 - 4t^3 + 8)/(t^2 + 2)$ as the sum of a polynomial and a proper fraction.

Solution: In order to use long division as in the preceding example, we put $0t$ in place of the missing t term in the divisor, and $0t^2$ and $0t$ in place of the missing terms in the dividend:

$$\frac{5t^4}{t^2} = 5t^2 \quad \text{The quotient}$$

$$
\begin{array}{r}
5t^2 - 4t - 10 \\
t^2 + 0t + 2 \overline{\smash{)}5t^4 - 4t^3 + 0t^2 + 0t + 8} \\
\underline{5t^4 + 0t^3 + 10t^2 } \\
-4t^3 - 10t^2 + 0t \quad \leftarrow \text{Result of subtraction} \\
\underline{-4t^3 + 0t^2 - 8t } \\
-10t^2 + 8t + 8 \quad \leftarrow \text{Result of subtraction} \\
\underline{-10t^2 + 0t - 20} \\
8t + 28 \quad \leftarrow \text{The remainder}
\end{array}
$$

Hence, $\quad \dfrac{5t^4 - 4t^3 + 8}{t^2 + 2} = 5t^2 - 4t - 10 + \dfrac{8t + 28}{t^2 + 2}$

In general, for fractions with shorter denominators, like monomials and binomials, the grouping process illustrated in Example 1 may be used; for fractions with denominators longer than a binomial, we recommend the long division process.

EXAMPLE 4

Rewrite $(z^4 - 3z^3 + 2z^2 + 4z)/(z^2 - 3z + 1)$ using the long division process.

Solution: We put 0 in place of the missing constant term in the dividend. As in the preceding example, follow Steps 1–5 of the long division process:

$$\frac{z^4}{z^2} = z^2 \quad \text{The quotient}$$

$$
\begin{array}{r}
z^2 + 0z + 1 \\
z^2 - 3z + 1 \overline{\smash{)}z^4 - 3z^3 + 2z^2 + 4z + 0} \\
\underline{z^4 - 3z^3 + z^2 } \\
0z^3 + z^2 + 4z \quad \leftarrow \text{Result of subtraction} \\
\underline{0z^3 - 0z^2 + 0z } \\
z^2 + 4z + 0 \quad \leftarrow \text{Result of subtraction} \\
\underline{z^2 - 3z + 1} \\
7z - 1 \quad \text{The remainder}
\end{array}
$$

Hence, $\quad \dfrac{z^4 - 3z^3 + 2z^2 + 4z}{z^2 - 3z + 1} = z^2 + 1 + \dfrac{7z - 1}{z^2 - 3z + 1}$

Exercises 4.4

Standard Assignment: Exercises 1, 3, 5, 7, 9, 13, 15, 17, 19, 21, 23, 27, 29, 31, 33, 35, 37, 41, 43, 47

A In Exercises 1–12, indicate which fractions are proper fractions and which are improper fractions.

1. $\dfrac{99}{100}$
2. $\dfrac{-50}{51}$
3. $\dfrac{-90}{89}$
4. $\dfrac{79}{80}$
5. $\dfrac{x}{x+1}$
6. $\dfrac{5y}{7y-3}$
7. $\dfrac{5z-5}{z^2+10}$
8. $\dfrac{10x-9}{x^2-99}$
9. $\dfrac{-y^2}{5y^2+5}$
10. $\dfrac{-2k^2}{10k^2+1}$
11. $\dfrac{t^2}{1-t-t^2}$
12. $\dfrac{2y^2}{3+y+8y^2}$

In Exercises 13–28, use the grouping process to rewrite each fraction as the sum of a polynomial and a proper fraction. See Example 1.

13. $\dfrac{x^2-x+1}{x}$
14. $\dfrac{3y^2+2y-1}{y}$
15. $\dfrac{21-14t}{7}$
16. $\dfrac{121+22z}{11}$
17. $\dfrac{n^2-2n+2}{n-1}$
18. $\dfrac{y^2+2y+2}{y+1}$
19. $\dfrac{y^2+6y+9}{y+3}$
20. $\dfrac{x^2-8x+16}{x-4}$
21. $\dfrac{3x^2+4x+5}{x^2+x}$
22. $\dfrac{5u^2+3u+4}{u^2-u}$
23. $\dfrac{4u^2+5u+6}{u^2+1}$
24. $\dfrac{7v^2-v-3}{v^2-2}$
25. $\dfrac{9t^2+6t-11}{3t+1}$
26. $\dfrac{12x^2-7x+5}{4x-1}$
27. $\dfrac{21x^2+8}{7x+1}$
28. $\dfrac{18z^2-1}{6z-7}$

In Exercises 29–50, use the long division process to rewrite each fraction as the sum of a polynomial and a proper fraction. See Examples 2 and 3.

29. $\dfrac{6x^2+5x-13}{3x-2}$
30. $\dfrac{8y^2-14y+15}{2y-3}$
31. $\dfrac{8x^4-4x^3+2x^2+x+1}{2x^2+1}$
32. $\dfrac{6x^4+3x^3-2x^2+9x+3}{3x^2-1}$
33. $\dfrac{t^4-3t^3+2t^2+30t-1}{t^2+2t-1}$
34. $\dfrac{x^3-3x^2+4x+7}{x^2-2x+6}$

176 Chap. 4 Algebraic Fractions

35. $\dfrac{5u^2 + 2u^3 + 13 - 3u}{u - u^2 + 3}$

36. $\dfrac{3v^2 + 4v^3 + 5 - 6v}{3v + v^2 - 1}$

37. $\dfrac{4x^4 - 2x^2 + 1}{2x^2 + x - 1}$

38. $\dfrac{6y^4 + 3y^2 - 9}{3y^2 - y - 2}$

39. $\dfrac{15z + 9 - 8z^2 - 24z^4}{3z + 1 + z^2}$

40. $\dfrac{14 + 9x^2 - 25x^4 + 5x}{2x - 4 + x^2}$

41. $\dfrac{2u - 3u^2 + u^3 + 7}{5 - 2u + u^2}$

42. $\dfrac{5v - 4v^2 + 2v^3 + 9}{7 + v + v^2}$

43. $\dfrac{y^4 + 2y^2 - 2y + 10}{y^3 + 2y^2 - y - 2}$

44. $\dfrac{2y^4 + y^3 - 2y + 11}{y^3 + 3y^2 - y - 5}$

45. $\dfrac{13 - t + 3t^2 + 4t^3 + 2t^4}{5 - 3t + 6t^2 + t^3}$

46. $\dfrac{21 + 2x - 5x^2 - 7x^3 + 3x^4}{9 + 6x - 5x^2 + x^3}$

47. $\dfrac{9 - y + 2y^2 - 4y^3 - y^4}{y^3 + 2y^2 - 5}$

48. $\dfrac{14 + y - 3y^2 + 5y^3 - y^4}{y^3 - 3y^2 + 7}$

49. $\dfrac{2x^4 - x^3 + 15x + 20}{10 - 3x - x^3}$

50. $\dfrac{3k^4 + 2k^3 - 11k + 30}{11 - 2k - k^3}$

4.5 Complex Fractions

A **complex fraction** is a fraction that contains either a fraction in the numerator or a fraction in the denominator, or both. For example,

$$\dfrac{\dfrac{1}{2}}{\dfrac{1}{3}}, \quad \dfrac{\dfrac{3}{x} - 2}{5}, \quad \dfrac{4}{7 + \dfrac{1}{y}}, \quad \text{and} \quad \dfrac{\dfrac{1}{x} + \dfrac{1}{y}}{\dfrac{x}{y} - \dfrac{y}{x}}$$

are complex fractions. It is important to recognize that a complex fraction represents another form of division (quotient) in which the numerator is the dividend and the denominator is the divisor. Thus,

$$\dfrac{\dfrac{1}{2}}{\dfrac{1}{3}} = \dfrac{1}{2} \div \dfrac{1}{3}$$

$$= \dfrac{1}{2} \cdot \dfrac{3}{1} \quad \text{Multiply the dividend by the reciprocal of the divisor}$$

4.5 Complex Fractions

There are two effective methods of simplifying complex fractions.

Method 1. Simplify the complex fraction by expressing both the numerator and the denominator as single fractions. Multiply the resulting numerator by the reciprocal of the denominator. It will be helpful to recall that the reciprocal of a/b is b/a.

EXAMPLE 1

In refrigeration, the coefficient of performance formula for the ideal refrigerator is given by $1/[(T_2/T_1) - 1]$. Use Method 1 to simplify this expression.

Solution:

$$\frac{1}{\dfrac{T_2}{T_1} - 1} = \frac{1}{\dfrac{T_2}{T_1} - \dfrac{T_1}{T_1}} \qquad 1 = \frac{T_1}{T_1}$$

$$= \frac{1}{\dfrac{T_2 - T_1}{T_1}} \qquad \text{Express the denominator as a single fraction}$$

$$= 1 \cdot \frac{T_1}{T_2 - T_1} \qquad \text{Multiply the numerator 1 by the reciprocal of the denominator}$$

$$= \frac{T_1}{T_2 - T_1}$$

EXAMPLE 2

Use Method 1 to simplify this expression.

$$\frac{\dfrac{1}{x} + \dfrac{1}{y}}{\dfrac{x}{y} - \dfrac{y}{x}}$$

Solution:

$$\frac{\dfrac{1}{x} + \dfrac{1}{y}}{\dfrac{x}{y} - \dfrac{y}{x}} = \frac{\dfrac{1}{x} \cdot \dfrac{y}{y} + \dfrac{x}{x} \cdot \dfrac{1}{y}}{\dfrac{x}{y} \cdot \dfrac{x}{x} - \dfrac{y}{x} \cdot \dfrac{y}{y}} \qquad \text{Obtain common denominators in the numerator and the denominator using } \dfrac{x}{x} = 1 \text{ and } \dfrac{y}{y} = 1$$

$$= \frac{\dfrac{y}{xy} + \dfrac{x}{xy}}{\dfrac{x^2}{xy} - \dfrac{y^2}{xy}} \qquad \text{Change the numerator and the denominator to single fractions}$$

178   Chap. 4   Algebraic Fractions

$$= \frac{\dfrac{y+x}{xy}}{\dfrac{x^2-y^2}{xy}} = \frac{y+x}{xy} \cdot \frac{xy}{x^2-y^2} \qquad \text{Multiply the numerator by the reciprocal of the denominator}$$

$$= \frac{y+x}{(x+y)(x-y)} \qquad \text{Divide out } xy; \text{ factor } x^2-y^2$$

$$= \frac{1}{x-y} \qquad \text{Divide out } x+y$$

Observe that the complex fraction in this example can be reduced at once to a simple fraction if we multiply both the numerator and the denominator by xy, which is the least common denominator of all the fractions in this complex fraction. That is,

$$\frac{\dfrac{1}{x}+\dfrac{1}{y}}{\dfrac{x}{y}-\dfrac{y}{x}} = \frac{xy\left(\dfrac{1}{x}+\dfrac{1}{y}\right)}{xy\left(\dfrac{x}{y}-\dfrac{y}{x}\right)}$$

$$= \frac{\dfrac{xy}{x}+\dfrac{xy}{y}}{\dfrac{x^2y}{y}-\dfrac{xy^2}{x}} \qquad \text{Use the Distributive Property}$$

$$= \frac{y+x}{x^2-y^2} \qquad \text{Divide out } x \text{ and } y$$

$$= \frac{x+y}{(x+y)(x-y)} \qquad \text{Factor } x^2-y^2$$

$$= \frac{1}{x-y} \qquad \text{Divide out } x+y$$

Method 2. Multiply both the numerator and the denominator of the complex fraction by the least common denominator of all the fractions appearing in the complex fraction.

EXAMPLE 3   Use Method 2 to simplify

(a) $(x^{-1} - y^{-1})^{-1} = \left(\dfrac{1}{x} - \dfrac{1}{y}\right)^{-1}$

$$= \frac{1}{\dfrac{1}{x} - \dfrac{1}{y}} \qquad \text{The least common denominator of all the fractions appearing is } xy$$

$$= \frac{xy \cdot 1}{xy\left(\dfrac{1}{x} - \dfrac{1}{y}\right)} \qquad \text{Multiply the numerator and denominator by } xy$$

$$= \frac{xy}{\dfrac{xy}{x} - \dfrac{xy}{y}}$$

$$= \frac{xy}{y - x} \qquad \text{Divide out } x \text{ and } y$$

(b) $\dfrac{\dfrac{z}{z+1} - \dfrac{z^2}{z^2-1}}{1 + \dfrac{1}{z-1}}$

Solution: We multiply both the numerator and the denominator by the least common denominator of all the fractions appearing, $z^2 - 1$.

$$\frac{(z^2-1)\left(\dfrac{z}{z+1} - \dfrac{z^2}{z^2-1}\right)}{(z^2-1)\left(1 + \dfrac{1}{z-1}\right)} = \frac{\dfrac{(z^2-1)z}{z+1} - \dfrac{(z^2-1)z^2}{z^2-1}}{z^2 - 1 + \dfrac{z^2-1}{z-1}}$$

$$= \frac{\dfrac{(z+1)(z-1)z}{z+1} - z^2}{z^2 - 1 + \dfrac{(z+1)(z-1)}{z-1}} \qquad \text{Factor } z^2 - 1$$

$$= \frac{(z-1)z - z^2}{z^2 - 1 + z + 1} \qquad \text{Divide out common factors}$$

$$= \frac{z^2 - z - z^2}{z^2 + z} \qquad \text{Distribute and simplify}$$

$$= \frac{-z}{z(z+1)} = -\frac{1}{z+1} \qquad \text{Divide out } z$$

EXAMPLE 4   Simplify $\dfrac{k}{2k - \dfrac{7}{3 - \dfrac{1}{2}}}$

Solution: This type of problem needs careful attention. First, rewrite

$$3 - \frac{1}{2} = \frac{6}{2} - \frac{1}{2} = \frac{5}{2}$$

then,

$$\frac{7}{3-\frac{1}{2}} = \frac{7}{\frac{5}{2}} = 7 \cdot \frac{2}{5} = \frac{14}{5}$$

Hence,

$$\frac{k}{2k - \frac{7}{3-\frac{1}{2}}} = \frac{k}{2k - \frac{14}{5}}$$

$$= \frac{5k}{5\left(2k - \frac{14}{5}\right)} \quad \text{Use Method 2}$$

$$= \frac{5k}{10k - 14}$$

Exercises 4.5

Standard Assignment: Exercises 1, 3, 5, 7, 9, 11, 13, 15, 17, 19, 21, 23, 25, 27, 31, 35, 37, 43, 49

A Simplify each of the following complex fractions. See Examples 1–3.

1. $\dfrac{\frac{3}{5}}{\frac{3}{4}}$

2. $\dfrac{\frac{2}{7}}{\frac{3}{7}}$

3. $\dfrac{\frac{1}{3}}{1 + \frac{2}{3}}$

4. $\dfrac{2 - \frac{1}{3}}{\frac{2}{3}}$

5. $\dfrac{\frac{5x}{6}}{\frac{4x}{5}}$

6. $\dfrac{\frac{5}{6y}}{\frac{4}{5y}}$

7. $\dfrac{\frac{5a}{2b}}{\frac{10a^2}{3b^2}}$

8. $\dfrac{\frac{2x^2}{3y}}{\frac{4x}{5y^2}}$

9. $\dfrac{1 - \frac{1}{x}}{3 + \frac{1}{x}}$

10. $\dfrac{\frac{2}{x} + 3}{\frac{3}{x} - 1}$

11. $\dfrac{x + \frac{1}{y}}{x - \frac{2}{y}}$

12. $\dfrac{2t - \frac{3}{s}}{3t + \frac{2}{s}}$

13. $\dfrac{\dfrac{1}{x}+\dfrac{x}{y}}{\dfrac{4}{y}-\dfrac{3}{x}}$

14. $\dfrac{\dfrac{2x}{y}-\dfrac{3}{x}}{\dfrac{1}{x}+\dfrac{4}{y}}$

15. $\dfrac{\dfrac{1}{5}-\dfrac{3}{k}}{\dfrac{5}{2k}+\dfrac{1}{5}}$

16. $\dfrac{\dfrac{4}{2k}+\dfrac{3}{7}}{\dfrac{3}{7}-\dfrac{2}{k}}$

17. $\dfrac{\dfrac{2}{3m}+\dfrac{3}{n}}{\dfrac{1}{3n}-\dfrac{4}{m}}$

18. $\dfrac{\dfrac{5}{m}-\dfrac{6}{7n}}{\dfrac{2}{n}+\dfrac{3}{7m}}$

19. $\dfrac{\dfrac{2}{3x}-\dfrac{x}{2y}}{\dfrac{3}{3y}+\dfrac{5y}{2x}}$

20. $\dfrac{\dfrac{3}{4y}+\dfrac{2y}{5x}}{\dfrac{5y}{4x}-\dfrac{7}{5y}}$

21. $\dfrac{p+\dfrac{2}{p}}{p^2+\dfrac{4}{p^2}}$

22. $\dfrac{q^2-\dfrac{25}{q^2}}{q-\dfrac{5}{q}}$

23. $\dfrac{5+2q^{-1}}{25-4q^{-2}}$

24. $\dfrac{16-25p^{-2}}{4-5p^{-1}}$

25. $(x^{-1}+y^{-1})^{-1}$

26. $(2x^{-1}-y^{-1})^{-1}$

27. $(3x^{-1}+4y^{-1})^{-1}$

28. $(5x^{-1}-6y^{-1})^{-1}$

29. $[(3p)^{-1}+(4q)^{-1}]^{-1}$

30. $[(2a)^{-1}-(5b)^{-1}]^{-1}$

31. $\dfrac{\dfrac{1}{x-1}+\dfrac{1}{x+2}}{4-\dfrac{3}{x+2}}$

32. $\dfrac{\dfrac{3}{x-1}+5}{\dfrac{4}{x+3}-\dfrac{3}{x-1}}$

33. $\dfrac{\dfrac{x^2}{x^2-1}-\dfrac{x}{x-1}}{1+\dfrac{1}{x+1}}$

34. $\dfrac{1-\dfrac{1}{p+3}}{\dfrac{p}{p-3}-\dfrac{p^2}{p^2-9}}$

35. $\dfrac{\dfrac{5}{q^2-q+1}}{\dfrac{3q-2}{q^3+1}}$

36. $\dfrac{\dfrac{2r-5}{r^3-1}}{\dfrac{3}{1+r+r^2}}$

37. $\dfrac{\dfrac{4}{2x-3y}-\dfrac{3}{3y-5x}}{\dfrac{-2}{5x-3y}+\dfrac{5}{3y-2x}}$

38. $\dfrac{\dfrac{2}{a-5b}+\dfrac{-3}{4b-7a}}{\dfrac{6}{7a-4b}-\dfrac{9}{5b-a}}$

39. $\dfrac{\dfrac{5}{x-2}-\dfrac{1+5x}{x^2+4}}{\dfrac{1}{x+2}-\dfrac{x^3}{x^4-16}}$

40. $\dfrac{\dfrac{1-2x}{4x^2+1}+\dfrac{1}{2x+1}}{\dfrac{1}{2x-1}-\dfrac{8x^3}{16x^4+1}}$

41. $\dfrac{\dfrac{p}{p^3 + q^3}}{\dfrac{1}{p+q} + \dfrac{p-q}{p^2 - pq + q^2}}$
42. $\dfrac{\dfrac{n}{m^3 - n^3}}{\dfrac{m+n}{m^2 + mn + n^2}} - \dfrac{1}{m-n}$

Simplify each of the following. See Example 4.

43. $\dfrac{1}{1 - \dfrac{2}{1 - \dfrac{2}{3}}}$
44. $\dfrac{4}{2 - \dfrac{3}{5 - \dfrac{1}{10}}}$
45. $\dfrac{2x}{3x + \dfrac{5}{6 - \dfrac{1}{3}}}$

46. $\dfrac{3y}{2y - \dfrac{8}{10 - \dfrac{1}{5}}}$
47. $\dfrac{3x - 2y}{4x - \dfrac{5y}{7x - \dfrac{x}{y}}}$
48. $\dfrac{p^2 - q^2}{2q + \dfrac{p}{5q - p}}$

49. $y - \dfrac{1}{1 - \dfrac{1}{y - 3}}$
50. $2x - \dfrac{x}{1 - \dfrac{x}{1 - x}}$

B 51. $1 + \dfrac{1}{1 - \dfrac{1}{1 + \dfrac{1}{1 + \dfrac{1}{1 - z}}}}$
52. $1 - \dfrac{1}{1 + \dfrac{1}{1 - \dfrac{1}{1 + t}}}$

53. In electronics, the coupled inductance, with circuits connected when parallel to each other with opposing fields, is given by $\dfrac{1}{\dfrac{1}{L_1 - M} + \dfrac{1}{L_2 - M}}$. Simplify this expression.

54. In studying electronic amplifiers, the expression $\dfrac{\left(\dfrac{m}{m+1}\right)R}{\dfrac{r}{m+1} + R}$ often comes up. Simplify this expression.

4.6 Synthetic Division

We often encounter the division of polynomials, $p(x)/d(x)$, in which the divisor $d(x)$ is of the form $x - c$, where c may be any constant. For this kind of division, there is a simplified procedure, **synthetic division**. In Section 4.4,

4.6 Synthetic Division

we introduced long division for polynomials, which when applied to

$$\frac{x^3 + 3x - 5}{x + 2}$$

results in the following display on the left.

$$
\begin{array}{r}
x^2 - 2x + 7 \\
x + 2 \overline{\smash{\big)}\, x^3 + 0x^2 + 3x - 5} \\
\underline{x^3 + 2x^2} \\
-2x^2 + 3x \\
\underline{-2x^2 - 4x} \\
7x - 5 \\
\underline{7x + 14} \\
-19 \leftarrow \text{Remainder}
\end{array}
$$

$$
\begin{array}{r}
1 - 2 + 7 \\
1 + 2 \overline{\smash{\big)}\, 1 + 0 + 3 - 5} \\
\underline{①+ 2} \\
-2 + ③ \\
\underline{-②- 4} \\
7 - 5 \\
\underline{⑦+ 14} \\
-19 \leftarrow \text{Remainder}
\end{array}
$$

Observe that since the calculations involved only the coefficients of the polynomials, we may as well complete the division without the variable, as shown on the right. Note that a 0 should be inserted for each missing power of x. Observe, next, that in the abbreviated form of division, each circled number is the duplicate of the number directly above it; these encircled duplicates and the leading coefficient 1 in the divisor can be discarded.

$$
\begin{array}{r}
1 - 2 + 7 \\
2 \overline{\smash{\big)}\, 1 + 0 + 3 - 5} \\
2 \\
\underline{} \\
-2 \\
-4 \\
\underline{} \\
7 \\
+14 \\
\underline{} \\
-19 \leftarrow \text{Remainder}
\end{array}
$$

In order to economize space, we can move all numbers close to the numbers directly above them and write the leading coefficient of the dividend, 1, directly below it in the bottom line. The result will look like this.

$$
\begin{array}{r}
1 - 2 + 7 \quad \leftarrow \text{Quotient} \\
2 \overline{\smash{\big)}\, 1 + 0 + 3 - 5} \quad \leftarrow \text{Dividend} \\
\underline{2 - 4 + 14} \\
1 - 2 + 7, -19 \quad \leftarrow \text{Remainder}
\end{array}
$$

There is still room for simplification. In the present form, the top line (quotient) can be omitted since it reappears in the bottom line. The bottom line, which consists of the quotient and the remainder, in the far right, is obtained by subtracting the line above it from the dividend line. We shall turn this subtraction into addition by changing the divisor, 2, into its negative, -2. The resulting form is the famous synthetic division process.

Chap. 4 Algebraic Fractions

$$\begin{array}{c} \text{Divisor} \rightarrow \\ \\ \text{Quotient} \rightarrow \end{array} \quad \begin{array}{r} -2 \underline{\smash{)}\, 1 + 0 + 3 - 5} \\ -2 + 4 - 14 \\ \hline 1 - 2 + 7, -19 \end{array} \quad \begin{array}{l} \leftarrow \text{Dividend} \qquad (1) \\ \qquad\qquad\qquad\qquad (2) \\ \leftarrow \text{Remainder} \quad (3) \end{array}$$

We will summarize the synthetic division process in the following steps.

1. In order to divide by $x + 2$, use -2 (in the upper left) as the *synthetic divisor*.
2. Write down the coefficients of the dividend in descending order, with 0 replacing any missing powers, in line (1).
3. Bring down the first coefficient in line (1) to line (3).
4. As indicated by the arrows, write the product of each coefficient in line (3) and the divisor, -2, in line (2), one place to the right.
5. Write down in line (3) the sum of each number created by Step 4 and the number above it.
6. Repeat Steps 4 and 5 to the end. Separate the last number in line (3), the remainder, by a comma from the coefficients of the quotient.

It is an easy task to write the quotient in polynomial form, complete with powers of the variable. If there are three coefficients such as $1 - 2 + 7$ in line (3), and the variable is x, the degree of the quotient is one less than the degree of the dividend, or $3 - 1 (=2)$. Thus, $1 - 2 + 7$ represents $x^2 - 2x + 7$.

EXAMPLE 1 Use synthetic division to simplify.

(a) $\dfrac{x^5 + 1}{x + 1}$

Convert $x + 1$ to the form $x - c$ by $x + 1 = x - (-1)$. Thus $c = -1$.

$$\begin{array}{r} -1 \underline{\smash{)}\, 1 + 0 + 0 + 0 + 0 + 1} \\ -1 + 1 - 1 + 1 - 1 \\ \hline 1 - 1 + 1 - 1 + 1, \ 0 \end{array}$$

Hence,

$$\frac{x^5 + 1}{x + 1} = x^4 - x^3 + x^2 - x + 1$$

(b) $\dfrac{2x^4 - x^2 + 4}{x - 2}$

$$\begin{array}{r} 2 \underline{\smash{)}\, 2 + 0 - 1 + 0 + 4} \\ 4 + 8 + 14 + 28 \\ \hline 2 + 4 + 7 + 14, \quad 32 \end{array}$$

Hence,

$$\frac{2x^4 - x^2 + 4}{x - 2} = 2x^3 + 4x^2 + 7x + 14 + \frac{32}{x - 2}$$

EXAMPLE 2 Use synthetic division to simplify

$$\frac{9y^3 - 3y^2 + 45y - 12}{3y - 1}$$

Solution: In order to apply synthetic division, we must convert the coefficient of y in the divisor into 1. For this purpose, we multiply both the numerator and the denominator by $\frac{1}{3}$. We now have

$$\frac{9y^3 - 3y^2 + 45y - 12}{3y - 1} = \frac{3y^3 - y^2 + 15y - 4}{y - \frac{1}{3}}$$

Then by synthetic division

$$\begin{array}{r|rrrr} \frac{1}{3} & 3 & -1 & +15 & -4 \\ & & 1 & +0 & +5 \\ \hline & 3 & +0 & +15, & 1 \end{array}$$

Hence,

$$\frac{9y^3 - 3y^2 + 45y - 15}{3y - 1} = 3y^2 + 0 \cdot y + 15 + \frac{1}{y - \frac{1}{3}}$$

$$= 3y^2 + 15 + \frac{3}{3y - 1}$$

The Remainder Theorem

In Example 1(b), when the polynomial $p(x) = 2x^4 - x^2 + 4$ is divided by $x - 2$, the remainder is $r = 32$.

On the other hand, when the variable x in the polynomial $p(x)$ is replaced by 2 (the synthetic divisor) we have

$$p(2) = 2 \cdot 2^4 - 2^2 + 4$$
$$= 2 \cdot 16 - 4 + 4$$
$$= 32$$

In this example, $p(2) = r$ (the remainder). Is this a coincidence? Let us find out by dividing any polynomial $p(x)$ by $x - c$, where c may be any fixed number. According to the Division Algorithm introduced in Section 4.4, we have

$$p(x) = (x - c)q(x) + r$$

where $q(x)$ denotes the quotient and r denotes the remainder after the division. We then replace x by c in the equality

$$p(c) = (c - c)q(c) + r$$
$$= 0 \cdot q(c) + r$$
$$= r$$

We have proven the following theorem.

The Remainder Theorem

When the polynomial $p(x)$ is divided by $x - c$, the remainder r is equal to $p(c)$.

EXAMPLE 3 Find the remainder r when $2x^{100} - 3x^{52} + 13$ is divided by $x + 1$.

Solution: Here $p(x) = 2x^{100} - 3x^{52} + 13$ and $c = -1$. By the Remainder Theorem, the remainder

$$r = p(-1) = 2 \cdot (-1)^{100} - 3 \cdot (-1)^{52} + 13$$
$$= 2 - 3 + 13$$
$$= 12$$

EXAMPLE 4 Let $p(y) = y^4 + 5y^3 - 19y^2 + 44y + 51$. Find $p(-8)$.

Solution: According to the Remainder Theorem, $p(-8) = r$, which is the remainder when $p(y)$ is divided by $y - (-8)$. Using synthetic division

$$\underline{-8 \rvert\, 1 + 5 - 19 + 44 + 51}$$
$$ \underline{- 8 + 24 - 40 - 32}$$
$$ 1 - 3 + 5 + 4, \quad 19 \leftarrow r$$

Hence, $p(-8) = 19$.

Exercises 4.6

Standard Assignment: Exercises 1, 3, 5, 7, 9, 11, 13, 15, 17, 19, 21, 25, 27, 29, 35, 37, 39, 41, 43, 45, 49

A In Exercises 1–24, use synthetic division to divide the first polynomial by the second. Indicate the quotient and the remainder. See Example 1.

1. $x^4 + 1; x + 1$
2. $x^4 - 1; x - 1$
3. $x^5 - 1; x - 1$
4. $x^5 + 32; x + 2$
5. $x^2 - 7x + 13; x - 5$
6. $x^2 + 12x - 6; x - 6$

7. $y^2 + 3y + 6; y + 2$
8. $y^2 - 12y - 10; y + 7$
9. $2z^2 - 9; z - 2$
10. $3z^2 + 8; z - 5$
11. $5w^2 - 2w; w + 3$
12. $4w^2 + 7w; w + 4$
13. $x^3 + 2x - 10; x - 4$
14. $x^3 - 12x + 3; x - 3$
15. $2y^3 - y^2 + 15; y - 3$
16. $3y^3 + 2y^2 - 17; y - 4$
17. $3z^3 - 4z^2 + 8z - 11; z + 5$
18. $4z^3 - 3z^2 + 11z - 7; z + 6$
19. $-4w^3 + 3w^2 - 5w; w - 6$
20. $-5w^3 - 2w^2 + 12w; w - 7$
21. $x^4 - 4x^3 + 13x^2 + 5x - 120; x - 1$
22. $2t^4 - 3t^3 + 5t^2 - 7t + 165; t + 1$
23. $y^5 - y^3 - 15y^2 + 22y + 225; y + 2$
24. $3y^5 - 2y^4 + y^2 - 16y - 132; y - 2$

In Exercises 25–34, use synthetic division to simplify. See Examples 1 and 2.

25. $\dfrac{6x^2 - 3x + 7}{x - 2}$

26. $\dfrac{5y^2 + 4y - 6}{y - 3}$

27. $\dfrac{t^3 - 2t^2 + t - 5}{t - 5}$

28. $\dfrac{2x^3 + 5x^2 - 7x - 8}{x - 4}$

29. $\dfrac{2x^3 - 4x + 8}{2x + 1}$

30. $\dfrac{4z^3 - 2z^2 + 6}{2z - 1}$

31. $\dfrac{9x^3 - 27x + 81}{3x - 2}$

32. $\dfrac{27w^3 - 9w + 243}{3w + 2}$

33. $\dfrac{4k^3 - 8k^2 + 12k - 10}{2k - 3}$

34. $\dfrac{10x^3 - 12x^2 + 8x + 50}{2x + 13}$

In Exercises 35–42, use the Remainder Theorem to find the remainder of each division problem. See Example 3.

35. $(16x^{31} - 13x^{26} + 3) \div (x + 1)$
36. $(7t^{25} + 24t^{18} - 21) \div (t + 1)$
37. $(-9y^{28} + 6y^{17} - 12y^{14} + 17) \div (y - 1)$
38. $(15y^{24} - 13y^{19} + 11y^8 + 20) \div (y - 1)$
39. $(z^5 + z^4 + z^3 + z^2 + z + 1) \div (z + 2)$
40. $(z^5 - z^4 + z^3 - z^2 + z - 1) \div (z + 2)$
41. $(w^4 - 2w^3 + 3w^2 - w + 1) \div (w - 3)$
42. $(k^4 - 3k^3 + 6k^2 - 9k + 12) \div (k - 3)$

In Exercises 43–48, a polynomial $p(x)$ and a constant c are given. Find $p(c)$ by using the Remainder Theorem and synthetic division. See Example 4.

43. $x^3 - 3x^2 + 11x - 29$; $\quad c = 5$
44. $2x^3 + x^2 - 15x - 2$; $\quad c = 7$
45. $x^4 - 2x^2 - 5x + 10$; $\quad c = -7$
46. ▦ $3x^4 - 14x^3 + 7x^2 + 5$; $\quad c = -5$
47. ▦ $x^5 - 64x^3 + 3x - 18$; $\quad c = 8$
48. ▦ $2x^5 + 16x^4 - 15x^2 + 58$; $\quad c = -8$

EXAMPLE: Show that 2 is a solution of the equation $x^3 - 3x^2 + 5x - 6 = 0$.

Solution: Let $p(x) = x^3 - 3x^2 + 5x - 6$. We will show that $p(2) = 0$, or equivalently by the Remainder Theorem, that the remainder $r = 0$. Using synthetic division, we have

$$\underline{2|\,1 - 3 + 5 - 6}$$
$$\,\underline{2 - 2 + 6}$$
$$1 - 1 + 3, \quad 0 = r$$

Hence, 2 is a solution of the given equation.

In Exercises 49–52, an equation and a number are given. Determine whether or not the given number is a solution of the equation.

49. $x^5 + x^4 - 16x - 16 = 0$; 2
50. $2z^3 - z^2 + 12 = 0$; -2
51. $2y^4 + y^3 - 24 = 0$; -2
52. $t^4 + t^2 + 2t + 84 = 0$; 3

Chapter 4 Summary

Key Words and Phrases

4.1 Rational expression (algebraic fraction)
 Lowest terms
4.2 Least Common Multiple (LCM)
 Least Common Denominator (LCD)
4.3 Reciprocal
4.4 Proper Fraction
 Improper Fraction
 Division Algorithm
4.5 Complex Fraction
4.6 Synthetic Division
 Remainder Theorem

Key Concepts and Rules

The quotient of two rational expressions is the product of the dividend and the reciprocal of the divisor. That is,

$$\frac{p}{q} \div \frac{r}{s} = \frac{p}{q} \cdot \frac{s}{r} = \frac{ps}{qr}$$

In order to simplify complex fractions:

Method 1. Express both the numerator and the denominator as single fractions; then multiply the resulting numerator by the reciprocal of the denominator.

Method 2. Multiply both the numerator and the denominator of the complex fraction by the least common denominator (LCD) of all of the fractions appearing in the complex fraction.

Remainder Theorem. When the polynomial $p(x)$ is divided by $x - c$, the remainder r is equal to $p(c)$.

Review Exercises

If you have trouble with any of the problems, look in the section indicated by the numbers in square brackets.

[4.1] Find the real numbers that are excluded from the replacement values.

1. $\dfrac{3}{x - 4}$
2. $\dfrac{2x - 1}{x^2 - 9}$
3. $\dfrac{1}{x^2 - 5x + 6}$

Reduce to the lowest terms.

4. $\dfrac{121x^2}{11x}$
5. $\dfrac{y + 1}{y^2 - 1}$
6. $\dfrac{9z^2 - 8z - 1}{9z^2 - 6z + 1}$

Raise to higher terms with the indicated denominator.

7. $\dfrac{2x + 3}{x - 1}$, $x^2 - 1$
8. $\dfrac{y + 1}{y^2 - 2y + 4}$, $y^3 + 8$

Complete the blank spaces with appropriate polynomials.

9. $\dfrac{x^2}{x + 2} = \dfrac{}{2x^2 + 9x + 10}$
10. $\dfrac{y - 3}{y} = \dfrac{y^2 - 7y + 12}{}$

11. $\dfrac{z^2 + 9}{} = \dfrac{z^4 - 81}{(z - 3)(z + 3)^2}$
12. $\dfrac{}{w^2 - 4} = \dfrac{w^4 + 8w^2 + 16}{w^4 - 16}$

[4.2] Find the LCDs for the indicated denominators.

13. $6x - 3, \quad 8x - 4$

14. $7y + 21, \quad 3y + 9$

15. $16z^2 - 25, \quad 8z + 10$

16. $6w^2 - 5w + 1, \quad 2w^2 - 3w + 1$

Perform the indicated operations.

17. $\dfrac{1}{7x + 21} + \dfrac{2}{3x + 9}$

18. $\dfrac{4y}{4y^2 - 1} - \dfrac{4y + 1}{(2y - 1)^2}$

19. $\dfrac{z - 1}{z^2 - 6z + 5} - \dfrac{z}{z^2 - 5z + 4} - \dfrac{1}{z^2 - 9z + 20}$

[4.3] Perform each of the indicated operations and reduce the result to the lowest terms.

20. $\dfrac{-3s}{8t^3} \cdot \dfrac{4t^2}{9s^3}$

21. $\dfrac{12y - 6}{27y^2} \cdot \dfrac{3y}{8y - 4}$

22. $\dfrac{z^2 + 5z + 25}{z^3w} \div \dfrac{z^3 - 125}{z^5w}$

23. $\dfrac{w - 1}{w^2 - 9} \div \dfrac{w^2 + 2w}{w^2 + 5w + 6}$

24. $\left(\dfrac{2x^2 - x - 3}{2x^2 + 7x + 3} \div \dfrac{2x^2 - 3x - 5}{2x^2 + 5x - 3}\right) \cdot \dfrac{4x^2 - 8x + 3}{4x^2 - 8x - 5}$

[4.4] Rewrite each fraction as the sum of a polynomial and a proper fraction.

25. $\dfrac{8x^2 - 6x + 5}{2x}$

26. $\dfrac{3y^2 + 5y - 6}{y - 1}$

27. $\dfrac{3z^2 - z + 7}{z^2 + 2}$

28. $\dfrac{2w^4 + w^3 - 15w + 19}{w^3 + 3w + 10}$

[4.5] Simplify complex fractions.

29. $\dfrac{2 + \dfrac{1}{t}}{1 - \dfrac{1}{t}}$

30. $\dfrac{4 - \dfrac{3}{x + 2}}{\dfrac{1}{x + 2} - \dfrac{1}{1 - x}}$

31. $\dfrac{\dfrac{z}{y^3 - z^3}}{\dfrac{1}{y - z} + \dfrac{y + z}{y^2 + yz + z^2}}$

32. $\dfrac{4k - \dfrac{3k}{1 - \dfrac{1}{1 + \dfrac{1}{1 - k}}}}{}$

[4.6] Use synthetic division to do the following divisions and indicate the quotient and the remainder.

33. $(x^2 - 9x + 15) \div (x - 5)$
34. $(2y^3 + y - 8) \div (y + 3)$
35. $(z^5 - 2z^4 - 3z^2 + 4z - 5) \div (z - 2)$

Use the Remainder Theorem to find the remainder for each of the following divisions.

36. $(61x^{29} - 19x^{23} + 13x^{11} - 17) \div (x - 1)$
37. $(-8y^{30} + 7y^{29} - 6y^{13} + 10) \div (y + 1)$
38. $(z^5 - z^4 + z^3 - z + 17) \div (z - 2)$

Find $p(c)$ by using the Remainder Theorem and synthetic division.

39. $p(x) = x^4 - 3x^2 - 7x + 15$; $c = -7$
40. $p(y) = 3y^5 - 15y^3 + 12y^2 - 50$; $c = 8$

Practice Test (60 minutes)

1. Raise $\dfrac{3x - 2}{x - 4}$ to the higher term with denominator $x^2 - x - 12$.

2. Fill in the blank space in $\dfrac{x + 2}{x - 3} = \dfrac{}{2x^2 - x - 15}$.

In Problems 3–6 perform the indicated operations and simplify.

3. $\dfrac{2x}{5x - 3} + \dfrac{2}{3x + 1}$

4. $\dfrac{6y}{4y^2 - 9} - \dfrac{5y - 1}{(2y + 3)^2}$

5. $\dfrac{x + 1}{x^2 - 5x} \cdot \dfrac{x^2 + 3x}{x^2 + 4x + 3}$

6. $\dfrac{y^2 + 11y + 30}{y^2 + 7y + 12} \div \dfrac{y^2 + 9y + 20}{y^2 + 9y + 18}$

7. Rewrite $\dfrac{3x^2 + 4x - 9}{x^2 + 1}$ as the sum of a polynomial and a proper fraction.

8. Simplify the complex fraction $\dfrac{\dfrac{1}{x+3} - \dfrac{1}{x-4}}{5 + \dfrac{2}{x+4}}$.

9. Use the Remainder Theorem to find the remainder.
   $(15x^{32} + 14x^{21} - 13x^{17} + 86x^{12} - 100) \div (x + 1)$

10. Find $p(c)$ by using the Remainder Theorem and synthetic division.
    $p(x) = 3x^4 + 5x^3 - 7x + 6; \qquad c = -6$

Extended Applications

Mathematics in the Space Program

On January 31, 1958, the United States launched its first satellite, *Explorer 1*. Suppose that *Explorer 1* was to circle the earth at a fixed height above the surface with only the gravitational attraction of the earth acting on it. Also suppose that the flight data received at the control station showed the gravitational force (per unit mass) for *Explorer 1*'s orbit to be 9.24 m/sec².

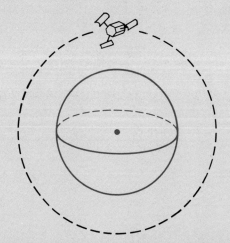

Using Newton's Law of Gravitation, together with the information that the radius of the earth is 6370 km and the gravitational force (per unit mass) at the earth's surface is 9.8 m/sec², it is possible to find

(a) the height above the surface of the earth at which *Explorer 1* was circling;
(b) the speed at which *Explorer 1* was traveling.

According to Newton's Law of Gravitation, the gravitational force at a distance of x km above the earth's surface is given by

$$\frac{G \cdot M}{(x \cdot 10^3 + 6.37 \times 10^6)^2} \quad \text{(in m/sec}^2\text{)} \qquad (*)$$

where

M = the mass of the earth (in kg)

G = the universal gravitational constant, 6.67×10^{-11}

The denominator in Equation (*) represents the square of the distance (in meters) from the center of the earth to a point x km above the earth's surface.

(a) Since the gravitational force (per unit mass) on the earth is 9.8, from Equation (*), we have (since $x = 0$)

$$\frac{6.67 \times 10^{-11} M}{(6.37 \times 10^6)^2} = 9.8$$

or

$$M = \frac{9.8 \times (6.37 \times 10^6)^2}{6.67 \times 10^{-11}} \approx 5.96 \times 10^{24} \text{ (kg)}$$

In order to determine the height in (a), substitute 5.96×10^{24} for M in Equation (*), and equate the result to 9.24. We have

$$\frac{6.67 \times 10^{-11} \times 5.96 \times 10^{24}}{(x \cdot 10^3 + 6.37 \times 10^6)^2} = 9.24$$

or

$$(x \cdot 10^3 + 6.37 \times 10^6)^2 = \frac{6.67 \times 10^{-11} \times 5.96 \times 20^{24}}{9.24}$$

$$\approx 4.3 \times 10^{13}$$

Using a calculator, we find

$$x \cdot 10^3 + 6.37 \times 10^6 \approx 6.56 \times 10^6$$

or

$$x \cdot 10^3 \approx 6.56 \times 10^6 - 6.37 \times 10^6 = 0.19 \times 10^6 \text{ meters}$$

or

$$x \approx 190 \text{ km}$$

Thus, *Explorer 1* was circling approximately 190 km above the earth.

(b) In order to counter a gravitational intensity of 9.24 m/sec^2, *Explorer 1* must maintain a velocity v(m/sec) such that

$$\frac{v^2}{r} = 9.24 \text{ m/sec}^2 \qquad (**)$$

where r is the distance (in meters) from the center of the earth to the *Explorer 1*. We have

$$r = 6370 + 190 \text{ km}$$
$$= 6560 \text{ km}$$
$$= 6.56. \times 10^6 \text{ m}$$

From Equation (**),

$$v^2 = 9.24r = 9.24 \times 6.56 \times 10^6 \approx 6.06 \times 10^7$$

or, again using the calculator,

$$v \approx 7785 \text{ m/sec.}$$

Exercises

1. The moon has mass of 7.36×10^{22} kg and a diameter of approximately 3500 km. Find the lunar gravity on the moon's surface using an equation similar to Equation (*), by replacing the mass and the radius of the earth with that of the moon and reinterpreting x as the distance above the moon's surface.

2. Compute the ratio of the lunar gravity to the earth's gravity.

3. If the gravitational force (per unit mass) for *Explorer 1*'s orbit were measured to be 8.3 m/sec^2, what would be the height and speed of *Explorer 1*?

5

Exponents, Roots, and Radicals

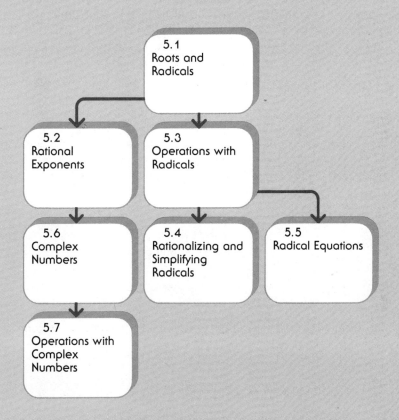

5.1 Roots and Radicals

If b and a are real numbers and n is a positive integer greater than 1, we can say that b is an **nth root** of a if and only if $b^n = a$. When $n = 2$, b is usually called a **square root** of x and when $n = 3$, b is usually called a **cube root** of x. Since $(3)^2 = 9$ and $(-3)^2 = 9$, both 3 and -3 are square roots of 9.

We introduce the notation $\sqrt[n]{a}$ to define a particular nth root of a that is called the **principal nth root of a**. The symbol $\sqrt{}$ is a *radical sign*, n is the *index*, and a is the *radicand*. The expression $\sqrt[n]{a}$ is called a **radical**. If no index is written, the index is understood to be 2. When possible, the principal nth root is chosen to be positive. For example, $\sqrt{4} = 2$ and $\sqrt[4]{81} = 3$. When the index is odd, however, the choice is rather simple. For example, the cube root of a, denoted by $\sqrt[3]{a}$, is the *only* real number whose cube is a. Thus, $\sqrt[3]{8} = 2$ and $\sqrt[3]{-8} = -2$. In general, we have the following definition.

Definition 1

a. If n is an even positive integer and $a \geq 0$, then $\sqrt[n]{a} = b$ if $b \geq 0$ and $b^n = a$.
b. If n is an odd positive integer and a is any (positive, negative, or zero) real number, then $\sqrt[n]{a} = b$ if b is the real number such that $b^n = a$.

Thus,

n even
$(a \geq 0)$ $\sqrt[n]{a} = b$ means $b \geq 0$ and $b^n = a$

n odd
(a real) $\sqrt[n]{a} = b$ means simply $b^n = a$

The number b just described is called the principal nth root of a. When $n = 2$ we write $\sqrt{a}$ for $\sqrt[2]{a}$.

EXAMPLE 1

(a) $\sqrt{49} = 7$, since $7 \geq 0$ and $7^2 = 49$
(b) $\sqrt[4]{16} = 2$, since $2 \geq 0$ and $2^4 = 16$
(c) $\sqrt[n]{0}$, since $0 \geq 0$ and $0^n = 0$ for any positive integer n.
(d) $\sqrt{-4}$ does not exist in the real number system because there is no real number whose square is -4.

In our definition of the nth root of a, $\sqrt[n]{a}$, we require $a \geq 0$ when n is even [Definition 1(a)], because there is no real number whose even number power is negative. Hence, symbols such as $\sqrt[4]{-5}$, $\sqrt[6]{-8}$, and $\sqrt[2k]{-a^2}$ (k is a positive integer) are not defined in the universe of real numbers.

EXAMPLE 2

(a) $\sqrt[3]{125} = 5$, since $5^3 = 125$

(b) $\sqrt[3]{-125} = -5$, since $(-5)^3 = -125$

(c) $\sqrt[5]{-\dfrac{1}{32}} = -\dfrac{1}{2}$, since $\left(-\dfrac{1}{2}\right)^5 = -\dfrac{1}{32}$

(d) $\sqrt{(-4)^2} = \sqrt{16} = 4$

(e) $\sqrt[4]{(-3)^4} = \sqrt[4]{81} = 3$

Because $\sqrt{a^2}$ is the *nonnegative* number, that is, it is either 0 or a positive number, whose square is a^2, we have $\sqrt{a^2} = a$ if a is positive or zero and $\sqrt{a^2} = -a$ if a is negative. But $|a| = a$ if $a \geq 0$ or $-a$ if $a < 0$ so

$$\sqrt{a^2} = |a|$$

In general, for any positive integer n and any real number x,

$$\sqrt[n]{x^n} = \begin{cases} x & \text{if } n \text{ is odd} \\ |x| & \text{if } n \text{ is even} \end{cases} \quad (5.1)$$

For example, $\sqrt[3]{x^3} = x$ and $\sqrt[4]{x^4} = |x|$.

EXAMPLE 3

(a) $\sqrt{a^4} = \sqrt{(a^2)^2} = |a^2| = a^2$ since $a^2 \geq 0$ for any real number a.

(b) $\sqrt{81b^2} = \sqrt{(9b)^2} = |9b| = 9|b|$

(c) $\sqrt[5]{-6^5} = \sqrt[5]{(-6)^5} = -6$, since $-6^5 = (-6)^5$

(d) $-\sqrt[4]{(-7)^4} = -|-7| = -7$

Computations involving radicals obey Equation (5.1) and the following Laws of Radicals, which are true wherever all radicals exist on both sides of the equations.

The Laws of Radicals

(a) $\sqrt[n]{ab} = \sqrt[n]{a} \cdot \sqrt[n]{b}$

(b) $\sqrt[m]{\sqrt[n]{a}} = \sqrt[mn]{a}$ \quad (5.2)

(c) $\sqrt[n]{\dfrac{a}{b}} = \dfrac{\sqrt[n]{a}}{\sqrt[n]{b}}$, \quad $b \neq 0$

EXAMPLE 4 Use Equations (5.1) and (5.2) to simplify the following.
(a) $\sqrt{25 \cdot 36} = \sqrt{25} \cdot \sqrt{36} = 5 \cdot 6 = 30$, by Equation (5.2)(a)
(b) $\sqrt[3]{\sqrt{64}} = \sqrt[3]{8} = \sqrt[3]{2^3} = 2$, by Equation (5.1)
(c) $\sqrt[3]{\dfrac{27}{64}} = \dfrac{\sqrt[3]{27}}{\sqrt[3]{64}} = \dfrac{3}{4}$, by Equation (5.2)(c)
(d) $\sqrt[3]{-p^6} = \sqrt[3]{(-p^2)^3} = -p^2$, by Equation (5.1)
(e) $\sqrt[4]{5x^2} \cdot \sqrt[4]{3xy^2} = \sqrt[4]{15x^3y^2}$, by Equation (5.2) (a)

Although the task at hand dictates the form of radical that is simplest for that task, we will say that a radical is simplified if it conforms to the following specifications.

Simplified Form for Radicals

1. The radicand contains no fractions.
2. No factor of the radicand appears with an exponent greater than or equal to the index of the radical.
3. The index of the radical is as small as possible.

When simplifying the square root of an integer, find the largest perfect square that is a factor of the integer. For example,

$4 = 2^2$ is a perfect square:
$$\sqrt{12} = \sqrt{4 \cdot 3} = \sqrt{4}\sqrt{3} = \sqrt{2^2}\sqrt{3} = 2\sqrt{3}$$

$9 = 3^2$ is a perfect square:
$$\sqrt{45} = \sqrt{9 \cdot 5} = \sqrt{9}\sqrt{5} = \sqrt{3^2}\sqrt{5} = 3\sqrt{5}$$

$121 = 11^2$ is a perfect square:
$$\sqrt{242} = \sqrt{121 \cdot 2} = \sqrt{121}\sqrt{2} = \sqrt{11^2}\sqrt{2} = 11\sqrt{2}$$

Similarly, in order to simplify the nth root of an integer q, find the largest integer p such that $q = p^n \cdot r$, and then take the nth root as is illustrated next. Thus

$$\sqrt[3]{56} = \sqrt[3]{8 \cdot 7} = \sqrt[3]{2^3 \cdot 7} = \sqrt[3]{2^3}\sqrt[3]{7} = 2\sqrt[3]{7}$$
$$\sqrt[5]{352} = \sqrt[5]{32 \cdot 11} = \sqrt[5]{2^5 \cdot 11} = \sqrt[5]{2^5}\sqrt[5]{11} = 2\sqrt[5]{11}$$

When $a \geq 0$, $\sqrt[n]{a}$ is defined and occasionally it is convenient to use

$$\sqrt[n]{a^m} = (\sqrt[n]{a})^m$$

as an alternate method of calculating $\sqrt[n]{a^m}$. For example,
$$\sqrt[3]{8^5} = (\sqrt[3]{8})^5 = 2^5 = 32$$

Exercises 5.1

Standard Assignment: Exercises 1, 5, 9, 11, 13, 15, 21, 23, 25, 27, 29, 31, 33, 35, 37, 39, 41

A Find each indicated root. See Examples 1, 2, and 3.

1. $\sqrt{25}$
2. $\sqrt{64}$
3. $-\sqrt{81}$
4. $-\sqrt{100}$
5. $\sqrt[3]{216}$
6. $\sqrt[3]{64}$
7. $\sqrt[3]{-27}$
8. $\sqrt[3]{-216}$
9. $\sqrt[4]{625}$
10. $\sqrt[4]{10{,}000}$
11. $\sqrt[4]{-16}$
12. $\sqrt[4]{-81}$
13. $\sqrt[5]{32}$
14. $\sqrt[5]{243}$
15. $-\sqrt[5]{-1}$
16. $-\sqrt[5]{-1024}$
17. $-\sqrt[6]{64}$
18. $\sqrt[6]{-64}$
19. $\sqrt[7]{128}$
20. $\sqrt[7]{-1}$
21. $\sqrt{\dfrac{49}{81}}$
22. $\sqrt{\dfrac{121}{144}}$
23. $-\sqrt[3]{-\dfrac{27}{64}}$
24. $\sqrt[3]{\dfrac{125}{1000}}$

In Exercises 25–32, use a calculator with a $\sqrt{\ }$ key; for a fourth root, $\sqrt[4]{\ }$, just push $\sqrt{\ }$ key twice. Remember that $\sqrt[4]{x}$ is the same as $\sqrt{\sqrt{x}}$, according to Equation (5.2)(b). Round your answer to the nearest thousandth.

25. ▦ $\sqrt{988}$
26. ▦ $\sqrt{7821}$
27. ▦ $-\sqrt{0.321}$
28. ▦ $-\sqrt{0.587}$
29. ▦ $\sqrt[4]{7568}$
30. ▦ $\sqrt[4]{30429}$
31. ▦ $\sqrt[4]{3.142}$
32. ▦ $\sqrt[4]{9.173}$

In Exercises 33–56, all variables represent real numbers. Simplify each indicated root. See Examples 3 and 4.

33. $\sqrt{81x^2}$
34. $\sqrt{121y^2}$
35. $-\sqrt{144z^4}$
36. $-\sqrt{100w^4}$
37. $\sqrt{0.16s^2}$
38. $\sqrt{0.25t^2}$
39. $\sqrt[3]{216a^3}$
40. $\sqrt[3]{343b^3}$
41. $\sqrt[3]{-125c^3}$

B 42. $\sqrt[3]{-729d^3}$
43. $\sqrt[3]{1.331e^3}$
44. $\sqrt[3]{-0.216f^3}$

45. $-\sqrt{\dfrac{1}{25}g^4h^2}$
46. $-\sqrt{\dfrac{4}{49}j^2k^4}$
47. $\sqrt{121p^8q^4r^2}$

48. $\sqrt{144p^{10}q^6r^4}$
49. $\sqrt[3]{-27r^3s^6}$
50. $\sqrt[3]{-1000s^6t^9}$
51. $\sqrt[3]{-64u^{12}v^{18}w^3}$
52. $\sqrt[3]{0.125u^3v^6w^{21}}$
53. $\sqrt[4]{0.0016a^8b^{12}c^{16}}$
54. $\sqrt[4]{0.0081d^{16}e^{20}f^4}$
55. $\sqrt[5]{32x^{25}y^{30}z^{35}}$
56. $\sqrt[6]{729p^{12}q^{24}r^{12}}$

In Exercises 57–68, all variables represent nonzero real numbers. Simplify each indicated root. See Example 4.

57. $\sqrt[3]{\sqrt{64a^{12}}}$
58. $\sqrt[3]{-\sqrt{729b^{24}}}$
59. $\sqrt{\sqrt[3]{\dfrac{1}{64}a^{12}b^{18}}}$
60. $\sqrt{-\sqrt[3]{-\dfrac{c^{12}d^{24}}{729}}}$
61. $\sqrt[4]{\sqrt[3]{8^{12}e^{24}f^{36}}}$
62. $\sqrt[3]{\sqrt[4]{10^{24}g^{36}h^{48}}}$
63. $\sqrt{\dfrac{121k^8}{169m^4n^6}}$
64. $\sqrt{\dfrac{49p^2q^6}{144r^4}}$
65. $\sqrt{\dfrac{8100u^2}{v^2-2vw+w^2}},\quad (v\neq w)$
66. $\sqrt{\dfrac{7.29a^4}{b^4+6b^2+9}}$
67. $\sqrt{\dfrac{625(c^4+2c^2+1)}{16d^4e^8f^{10}}}$
68. $\sqrt{\dfrac{x^4+8x^3y+16x^2y^2}{4x^2-12xy+9y^2}},\quad (2x\neq 3y)$

Solve the following problems.

69. In order for an initial investment of P dollars to mature to S dollars after two years when the interest is compounded annually, an annual interest rate of $r = \sqrt{\dfrac{S}{P}} - 1$ is required. Find r if $P = \$1{,}000{,}000$ and $S = \$1{,}210{,}000$.

70. The terminal speed V, of a body that weighs w grams, falling in a cylinder of a particular fluid is given by the equation $V = \sqrt{\dfrac{w}{1.5}}$ cm/sec. Find the terminal speed for a body weighing 6 grams.

71. A spring balance scale attached to a wire that holds a model airplane on a circular path indicates the force F, on the wire. The speed of the plane, V in m/sec, is given by $V = \sqrt{(rF)/m}$ where m is the mass of the plane and r is the length of the wire. What is the top speed of a 2 kg plane on an 18 m wire if the force on the wire at top speed is 49 newtons?

72. The current I (amperes) in a circuit using W watts and having a resistance of R ohms is given by $I = \sqrt{\dfrac{W}{R}}$. Find the current in a circuit using 1058 watts and having 2 ohm resistance.

73. In order to construct a sphere of volume V, we require a radius r, given by $r = \sqrt[3]{\dfrac{3V}{4\pi}}$. Find the radius required for a sphere with a volume of 127.04 cubic units.

5.2 Rational Exponents

In Chapter 3, we defined x^n for integer exponents n. The Power Rule

$$(x^m)^n = x^{mn}$$

already established for integer exponents (see Section 3.2), can be extended to cover fractional powers as well. Our first step is to replace m by $1/n$ in the Power Rule.

In order for $(x^{1/n})^n = x^{(1/n)n} = x$ to be valid, the new expression $x^{1/n}$ must be an nth root of x. This suggests the following definition.

Definition 2

If a is a real number and m and n are integers such that $n \geq 2$, then

a. $a^{1/n} = \sqrt[n]{a}$ provided that $\sqrt[n]{a}$ exists;
b. If the rational number m/n is in lowest terms, then $a^{m/n} = (a^{1/n})^m$ provided that $\sqrt[n]{a}$ exists. (A rational number m/n is in **lowest terms** provided $n > 0$ and m and n have no common divisors other than ± 1.)
c. If p/q is a rational number with $p/q = m/n$, and m/n is in lowest terms, then $a^{p/q} = a^{m/n}$.

It should be noted that Definition 2 extends the definition of exponents to include $a^{p/q}$ for any rational number p/q. If $q = 1$, then $a^{p/q} = a^p$; if q is negative, as in $3/(-5)$ or $(-2)/(-7)$, we use equivalent expressions such as $3/(-5) = (-3)/5$ and $(-2)/(-7) = 2/7$ so that Definition 2 will apply.

WARNING

$(-8)^{2/6} \neq [(-8)^{1/6}]^2$

since $(-8)^{1/6} = \sqrt[6]{-8}$ is undefined

Reducing $\dfrac{2}{6}$ to lowest terms, $\dfrac{1}{3}$

$(-8)^{2/6} = (-8)^{1/3} = \sqrt[3]{-8} = -2$

The laws of exponents discussed in Sections 3.1 and 3.2 apply to rational exponents when only positive bases appear. Consequently, Equation (3.7) yields $a^{m/n} = (a^{1/n})^m = (a^m)^{1/n}$, for positive a. Actually whenever $a^{1/n}$ exists it can be shown that

$$a^{m/n} = (\sqrt[n]{a})^m = \sqrt[n]{a^m}$$

EXAMPLE 1 Simplify: (a) $9^{1/2}$ (b) $(-27)^{1/3}$ (c) $8^{2/3}$ (d) $-16^{5/2}$ (e) $100^{3/2}$ (f) $(-25)^{7/2}$ (g) $125^{2/3} - 16^{-1/4}$ (h) $(a^{2/3}b^{1/2})(a^{1/3}b^{-1/4})$; $a, b > 0$

Solution: Referring to Definition 2:

(a) $9^{1/2} = \sqrt{9} = 3$
(b) $(-27)^{1/3} = \sqrt[3]{-27} = -3$
(c) $8^{2/3} = (\sqrt[3]{8})^2 = 2^2 = 4$
(d) $-16^{5/2} = -(\sqrt{16})^5 = -4^5 = -1024$
(e) $100^{3/2} = (\sqrt{100})^3 = 10^3 = 1000$
(f) $(-25)^{7/2}$ is not a real number since $\sqrt{-25}$ doesn't exist.
(g) $125^{2/3} - 16^{-1/4} = (\sqrt[3]{125})^2 - \dfrac{1}{16^{1/4}} = 5^2 - \dfrac{1}{\sqrt[4]{16}}$

$$= 25 - (1/2) = 24.5$$

(h) $(a^{2/3}b^{1/2})(a^{1/3}b^{-1/4}) = a^{2/3}a^{1/3}b^{1/2}b^{-1/4} = (a^{2/3+1/3})(b^{1/2-1/4}) = ab^{1/4}$

EXAMPLE 2 Rewrite the Laws of Radicals in Equations (5.2) by using rational exponents.

Solution: Using Definition 2, the Laws of Radicals in Equations (5.2) and Equation (5.1), respectively, are translated into the following expressions. From Equation [5.2(a)], $\sqrt[n]{ab} = \sqrt[n]{a} \cdot \sqrt[n]{b}$, we have

$$(ab)^{1/n} = a^{1/n}b^{1/n} \tag{5.3}$$

from Equation [5.2(b)], $\sqrt[m]{\sqrt[n]{a}} = \sqrt[mn]{a}$, we have

$$(a^{1/n})^{1/m} = a^{1/mn} \tag{5.4}$$

and from Equation [5.2(c)], $\sqrt[n]{\dfrac{a}{b}} = \dfrac{\sqrt[n]{a}}{\sqrt[n]{b}}$, $b \neq 0$, we have

$$(a/b)^{1/n} = (a^{1/n})/(b^{1/n}), \; b \neq 0 \tag{5.5}$$

Thus, from Equation (5.1) we have

$$(x^n)^{1/n} = \begin{cases} x & \text{if } n \text{ is odd} \\ |x| & \text{if } n \text{ is even} \end{cases} \tag{5.6}$$

Notice that according to Equation (5.6) we have $[(-5)^2]^{1/2} = |-5| = 5$, and consequently, $[(-5)^2]^{1/2} \neq (-5)^{[2 \cdot (1/2)]} = -5$. This failure of the law $(a^n)^m = a^{mn}$ for rational exponents shows that caution is required in working with negative bases.

If $a < 0$ *and* n *is even, then* $a^{1/n}$ *is undefined*, and so the expression $(a^{1/n})^m = a^{m/n}$ is undefined. However, if m is also even, a^m is positive so that the expression $(a^m)^{1/n}$ is defined. For example, $[(-8)^2]^{1/6} = 64^{1/6} = 2$, whereas $[(-8)^{1/6}]^2$ is undefined. If $a^{1/n}$ is defined as a real number, then:

$$(a^{1/n})^m = (a^m)^{1/n}$$

EXAMPLE 3 Write each radical with a rational exponent and simplify.

(a) $\sqrt[3]{-27a^3b^6} = (-27a^3b^6)^{1/3} = (-27)^{1/3}(a^3b^6)^{1/3}$ Use (5.3)
$= -3a^{3/3}b^{6/3} = -3ab^2$ Simplify exponents

(b) $\sqrt[5]{\dfrac{c^5 d^{10} e^{-15}}{32}} = \left(\dfrac{c^5 d^{10}}{32 e^{15}}\right)^{1/5} = \dfrac{(c^5 d^{10})^{1/5}}{(2^5 e^{15})^{1/5}}$ Use (5.5)

$= \dfrac{c^{5/5} d^{10/5}}{2^{5/5} e^{15/5}} = \dfrac{cd^2}{2e^3}$ Provided $e \neq 0$

EXAMPLE 4 Factor out the indicated common factor. All variables represent positive real numbers.

(a) $3x^{3/2} - 8x^{-1/2}$; $x^{-1/2}$
(b) $7y^{-5/3} + 8y^{-2/3}$; $y^{-5/3}$

Solution:

(a) $3x^{3/2} - 8x^{-1/2} = 3x^{-1/2}x^2 - 8x^{-1/2}$ Since $-\dfrac{1}{2} + 2 = \dfrac{3}{2}$
$= x^{-1/2}(3x^2 - 8)$

(b) $7y^{-5/3} + 8y^{-2/3} = 7y^{-5/3} + 8y^{-5/3}y^{3/3}$
$= y^{-5/3}(7 + 8y)$

Exercises 5.2

Standard Assignment: Exercises 1, 3, 5, 9, 11, 13, 15, 17, 19, 21, 23, 25, 27, 29, 31, 33, 35, 37, 41

In Exercises 1–12, write each radical with a rational exponent.

EXAMPLE: $\sqrt[10]{|x|^5 \cdot |y|^2} = (|x|^5 \cdot |y|^2)^{1/10} = |x|^{5/10}|y|^{2/10} = |x|^{1/2}|y|^{1/5}$

A
1. $\sqrt{345}$
2. $-\sqrt{0.513}$
3. $\sqrt[3]{-606}$
4. $-\sqrt[3]{918}$
5. $\sqrt[3]{a^2}$
6. $\sqrt[4]{|a|^3}$
7. $\sqrt[7]{b^3 c^5}$
8. $\sqrt[6]{|a|^3|e|^2}$
9. $(\sqrt[3]{-11})^2$
10. $(\sqrt[5]{a^2})^3$
11. $\sqrt[5]{\sqrt{|s|^3}}$
12. $\sqrt[3]{\sqrt[5]{t^4}}$

In Exercises 13–24, rewrite each in radical form.

13. $32^{1/2}$
14. $8^{1/2}$
15. $56^{1/3}$
16. $(-99)^{1/3}$
17. $14^{2/3}$
18. $(-10)^{2/3}$
19. $91^{-3/5}$
20. $(-121)^{-4/5}$
21. $\left(\dfrac{7}{8}\right)^{2/3}$
22. $\left(\dfrac{8}{9}\right)^{2/3}$
23. $\left(\dfrac{31}{17}\right)^{-3/5}$
24. $\left(-\dfrac{32}{19}\right)^{-4/5}$

In Exercises 25–44, simplify and express each answer with positive exponents. All variables represent positive real numbers. See Examples 1 and 2.

25. $216^{2/3}$
26. $(-64)^{2/3}$
27. $(-27)^{-1/3}$
28. $(-32)^{-2/5}$
29. $\dfrac{121^{-1/2}}{(-8)^{-1/3}}$
30. $\dfrac{(-216)^{-1/3}}{144^{-1/2}}$
31. $(b^{-4}c^{-12})^{3/4}$
32. $(c^8 d^{-4})^{-3/4}$
33. $(27d^3 c^{-9})^{-2/3}$
34. $(-1000e^{-6}d^{-12})^{-1/3}$
35. $\left(\dfrac{81x^8}{y^{-6}z^2}\right)^{1/2}$
36. $\left(\dfrac{-64y^{12}}{z^{-6}w^9}\right)^{-1/3}$
37. $1000^{2/3} + 16^{-1/4}$
38. $(-343)^{2/3} - 25^{-1/2}$
39. $(8x^3)^{2/3} + (x^{-8})^{-1/4}$
40. $(-27y^6)^{1/3} - (y^{-12})^{-1/6}$
41. $(a^{4/3} b^{-1/4})(a^{-1/3} b^{5/4})$
42. $(c^{-2/3} d^{1/6})(c^{8/3} d^{-7/6})$

B 43. $x^0 \cdot x^{1/2} \cdot x^{1/4} \cdot x^{1/8} \cdot x^{1/16} \cdots x^{1/128}$

44. $y^0 \cdot y^{1/100} \cdot y^{2/100} \cdot y^{3/100} \cdots y^{99/100}$

In exercises 45–52, factor the indicated common factor. Variables x, y, z, and w represent positive numbers, and n is a positive integer. See Example 3.

45. $9x^{4/3} + x^{1/3},\ x^{1/3}$
46. $10y^{6/5} - 3y^{1/5},\ y^{1/5}$
47. $21z^{4/5} - 6z^{-1/5},\ 3z^{-1/5}$
48. $45w^{5/3} - 10w^{2/3},\ 5w^{-1/3}$
49. $7x^{3n/2} + 8x^{5n/2},\ x^{n/2}$
50. $-8y^{7n/3} + 13y^{10n/3},\ -y^{n/3}$
51. $14z^{4n/5} - 21z^{9n/5},\ 7z^{-n/5}$
52. $14z^{4n/5} - 21z^{9n/5},\ 7z^{4n/5}$

In Exercises 53–60, simplify and express all answers with positive exponents. (m and n are positive integers and the other variables represent positive real numbers.)

53. $\left(\dfrac{x^{3n}}{x^{n-2}}\right)^{1/2}$
54. $\left(\dfrac{y^{4n}}{y^{n-3}}\right)^{1/3}$
55. $\left(\dfrac{z^{n-3}}{z^{5n+1}}\right)^{-1/4}$
56. $\left(\dfrac{w^{3-2n}}{w^{3n+13}}\right)^{-1/5}$
57. $\left(\dfrac{125 a^{m+5} b^{2-4n}}{a^{2-2m} b^{2n+8}}\right)^{-1/3}$

58. $\left(\dfrac{c^{3m+4}d^{1-3n}}{81c^{2-5m}d^{n+5}}\right)^{-1/4}$

59. $\left(\dfrac{u^{5m/2}+2u^{3m/2}+u^{m/2}}{u^{m/2}}\right)^{1/2}$

60. $\left(\dfrac{v^{-n/3}}{4v^{5n/3}-12v^{2n/3}+3v^{-n/3}}\right)^{-1/3}$

Solve the following problems.

61. The radius of a sphere of a given volume V, is given by $r = \left(\dfrac{3V}{4\pi}\right)^{1/3}$. If we treat the moon as a sphere and are told that its volume is $2.19 \times 10^{19} \text{m}^3$, find its radius.

62. The atmospheric pressure P, which is measured in lb/in² at altitudes h, of less than 50 miles, is given by the equation $P = 14.7 \left(\dfrac{1}{2}\right)^{h/3.25}$. Compute the atmospheric pressure at an altitude of $16\dfrac{1}{4}$ miles.

63. The side S, of an equilateral triangle of area A, is given by $S = \left(\dfrac{4A}{\sqrt{3}}\right)^{1/2}$. Find the side of an equilateral triangle with an area of 692 cm². (Use the approximation $\sqrt{3} = 1.73$.)

64. The force F, that is exerted on the semicircular end of a trough full of water is approximately $F = 42\, r^{3/2}$ lb, where r is the radius of the semicircular end. Find the force if the radius is 3 feet.

65. ▦ The adiabatic relation $PV^{1.4} = 7200$ gives the relationship between the volume V, in ft³, of a gas and pressure P, in lb/ft², that is exerted on the gas under certain circumstances. If $V = 400$ ft³, find P.

5.3 Operations with Radicals

A **radical expression** is any algebraic expression that contains radicals. For example,

$$5\sqrt{3} - 2\sqrt{2} + 1, \quad \dfrac{\sqrt{5}+1}{\sqrt{2}}, \quad \text{and} \quad (\sqrt{x^2+1} - \sqrt{2})^2 + 6$$

are radical expressions. When adding or subtracting radical expressions, the commutative, associative, and distributive properties will be used whenever they are applicable.

EXAMPLE 1

(a) $3\sqrt{2} + 5\sqrt{2} = (3+5)\sqrt{2}$
$\qquad\qquad\quad\; = 8\sqrt{2}$

(b) $5\sqrt{3} + 7\sqrt{3} - 8\sqrt{3} = (5 + 7 - 8)\sqrt{3}$
$= 4\sqrt{3}$

(c) $(3\sqrt{5} + 10\sqrt{7}) + (6\sqrt{5} - 12\sqrt{7})$
$= (3\sqrt{5} + 6\sqrt{5}) + (10\sqrt{7} - 12\sqrt{7})$
$= 9\sqrt{5} - 2\sqrt{7}$

Often the laws of radicals in Equations (5.1) and (5.2) must be used before adding or subtracting radical expressions.

EXAMPLE 2

(a) $6\sqrt{20} + 2\sqrt{45} = 6\sqrt{4} \cdot \sqrt{5} + 2\sqrt{9} \cdot \sqrt{5}$
$= 6 \cdot 2\sqrt{5} + 2 \cdot 3\sqrt{5}$
$= 12\sqrt{5} + 6\sqrt{5}$
$= 18\sqrt{5}$

(b) $9\sqrt{12x^2} - 5\sqrt{27x^2} = 9 \cdot \sqrt{4x^2} \cdot \sqrt{3} - 5 \cdot \sqrt{9x^2} \cdot \sqrt{3}$
$= 9 \cdot 2|x| \cdot \sqrt{3} - 5 \cdot 3|x| \cdot \sqrt{3}$
$= 18|x|\sqrt{3} - 15|x|\sqrt{3}$
$= 3|x|\sqrt{3}$

(c) $5\sqrt[3]{54} - 21\sqrt[3]{16} = 5\sqrt[3]{27 \cdot 2} - 21\sqrt[3]{8 \cdot 2}$
$= 5\sqrt[3]{27} \cdot \sqrt[3]{2} - 21\sqrt[3]{8} \cdot \sqrt[3]{2}$
$= 5 \cdot 3 \cdot \sqrt[3]{2} - 21 \cdot 2 \cdot \sqrt[3]{2}$
$= (15 - 42)\sqrt[3]{2}$
$= -27\sqrt[3]{2}$

When multiplying radical expressions, the following formulas and identities may be used whenever applicable.

1. Product of sum and difference:

$$(a + b)(a - b) = a^2 - b^2$$

2. Square of sum or difference:

$$(a + b)^2 = a^2 + 2ab + b^2$$
$$(a - b)^2 = a^2 - 2ab + b^2$$

3. FOIL (First - Outer - Inner - Last) method:

$$(a + b)(c + d) = ac + ad + bc + bd$$
$$\quad\quad\quad\quad\quad\quad\quad\uparrow\quad\uparrow\quad\uparrow\quad\uparrow$$
$$\quad\quad\quad\quad\quad\text{First Outer Inner Last}$$

4. Identities:
$$(a + b)(a^2 - ab + b^2) = a^3 + b^3$$
$$(a - b)(a^2 + ab + b^2) = a^3 - b^3$$

EXAMPLE 3 Simplify the following expressions using 1, 2, and 3 from the above list.

(a) Use $(a + b)(a - b) = a^2 - b^2$ where $a = \sqrt{13}$ and $b = \sqrt{7}$.
$$\begin{aligned}(\sqrt{13} + \sqrt{7})(\sqrt{13} - \sqrt{7}) &= (\sqrt{13})^2 - (\sqrt{7})^2 \\ &= 13 - 7 \\ &= 6\end{aligned}$$

(b) Use $(a + b)^2 = a^2 + 2ab + b^2$ where $a = 2\sqrt{3}$ and $b = 3\sqrt{5}$.
$$\begin{aligned}(2\sqrt{3} + 3\sqrt{5})^2 &= (2\sqrt{3})^2 + 2 \cdot 2\sqrt{3} \cdot 3\sqrt{5} + (3\sqrt{5})^2 \\ &= 2^2 (\sqrt{3})^2 + 2 \cdot 2 \cdot 3\sqrt{3} \cdot \sqrt{5} + 3^2 (\sqrt{5})^2 \\ &= 4 \cdot 3 + 12\sqrt{3 \cdot 5} + 9 \cdot 5 \\ &= 12 + 45 + 12\sqrt{15} \\ &= 57 + 12\sqrt{15}\end{aligned}$$

(c) Use $(a - b)^2 = a^2 - 2ab + b^2$ where $a = 5\sqrt{2}$ and $b = 3$.
$$\begin{aligned}(5\sqrt{2} - 3)^2 &= (5\sqrt{2})^2 - 2 \cdot 5\sqrt{2} \cdot 3 + 3^2 \\ &= 5^2 (\sqrt{2})^2 - 2 \cdot 5 \cdot 3 \cdot \sqrt{2} + 9 \\ &= 25 \cdot 2 + 9 - 30\sqrt{2} \\ &= 59 - 30\sqrt{2}\end{aligned}$$

(d) Use FOIL.
$$\begin{aligned}(\sqrt{7} + \sqrt{2})(\sqrt{5} - \sqrt{3}) &= \sqrt{7} \cdot \sqrt{5} - \sqrt{7} \cdot \sqrt{3} + \sqrt{2} \cdot \sqrt{5} - \sqrt{2} \cdot \sqrt{3} \\ &= \sqrt{35} - \sqrt{21} + \sqrt{10} - \sqrt{6}\end{aligned}$$

EXAMPLE 4 (a) Use $(a + b)(a^2 - ab + b^2) = a^3 + b^3$ where $a = \sqrt[3]{10}$ and $b = \sqrt[3]{2}$.
$$\begin{aligned}(\sqrt[3]{10} + \sqrt[3]{2})(\sqrt[3]{100} - \sqrt[3]{20} + \sqrt[3]{4}) &= (\sqrt[3]{10} + \sqrt[3]{2})(\sqrt[3]{10 \cdot 10} - \sqrt[3]{10 \cdot 2} + \sqrt[3]{2 \cdot 2}) \\ &= (\sqrt[3]{10} + \sqrt[3]{2})[(\sqrt[3]{10})^2 - \sqrt[3]{10}\sqrt[3]{2} + (\sqrt[3]{2})^2] \\ &= (\sqrt[3]{10})^3 + (\sqrt[3]{2})^3 = 12\end{aligned}$$

(b) Use $(a - b)(a^2 + ab + b^2) = a^3 - b^3$ where $a = \sqrt[3]{x}$ and $b = \sqrt[3]{y}$.

$$(\sqrt[3]{x} - \sqrt[3]{y})(\sqrt[3]{x^2} + \sqrt[3]{xy} + \sqrt[3]{y^2})$$
$$= (\sqrt[3]{x} - \sqrt[3]{y})\,[(\sqrt[3]{x})^2 + \sqrt[3]{x}\,\sqrt[3]{y} + (\sqrt[3]{y})^2]$$
$$= (\sqrt[3]{x})^3 - (\sqrt[3]{y})^3$$
$$= x - y$$

EXAMPLE 5 Reduce the expressions to the lowest terms.

(a) $\dfrac{-14\sqrt{13} + 21}{-28} = \dfrac{-7 \cdot 2\sqrt{13} + (-7)\cdot(-3)}{(-7)\,4}$ Common factor of -7

$$= \dfrac{-7(2\sqrt{13} - 3)}{-7 \cdot 4}$$

$$= \dfrac{2\sqrt{13} - 3}{4}$$ Divide out -7

(b) $\dfrac{10k + \sqrt[3]{-16k^3}}{14k} = \dfrac{10k + \sqrt[3]{-8k^3}\cdot \sqrt[3]{2}}{14k}$ If $k \neq 0$

$$= \dfrac{10k + \sqrt[3]{(-2k)^3}\cdot \sqrt[3]{2}}{14k}$$

$$= \dfrac{10k - 2k\sqrt[3]{2}}{14k} = \dfrac{2k(5 - \sqrt[3]{2})}{7(2k)}$$

$$= \dfrac{5 - \sqrt[3]{2}}{7}$$ Divide out $2k$

Exercises 5.3

Standard Assignment: Exercises 1, 3, 5, 9, 11, 13, 15, 19, 25, 27, 29, 39, 41, 45, 51, 57, 61, 63

A In Exercises 1–22, simplify each radical expression. All variables represent positive real numbers. See Examples 1 and 2.

1. $\sqrt{32} - \sqrt{8}$
2. $\sqrt{48} - \sqrt{12}$
3. $7\sqrt{5} - (3\sqrt{2} - 3\sqrt{5})$
4. $10\sqrt{5} - (2\sqrt{3} - 5\sqrt{5})$
5. $8\sqrt{3} + (7\sqrt{5} - 2\sqrt{12})$
6. $9\sqrt{2} + (\sqrt{20} - \sqrt{18})$
7. $9\sqrt{25} - \sqrt{50} + \sqrt{2}$
8. $3\sqrt{16} + 2\sqrt{32} - 5\sqrt{18}$

9. $\sqrt[3]{54} - 2\sqrt[3]{16}$
10. $\sqrt[3]{128} + 5\sqrt[3]{64}$
11. $3\sqrt{8x^3} - 5x\sqrt{32x} - x\sqrt{18x}$
12. $3\sqrt{18y^3} - 10y\sqrt{32y} + 2x\sqrt{8x}$
13. $4z\sqrt{32z} - 7\sqrt{8z^2}$
14. $5w\sqrt{48w} - 8\sqrt{75w^3}$
15. $13\sqrt[4]{32} - \sqrt[4]{512}$
16. $7\sqrt[4]{162} - 9\sqrt[4]{32}$
17. $4\sqrt[3]{-128} - \sqrt[3]{16}$
18. $5\sqrt[3]{-16} - 7\sqrt[3]{128}$
19. $\sqrt{50x^2} - 3\sqrt{18x^2} + \sqrt{45y^2}$
20. $\sqrt{50y^2} + 2z\sqrt{32} - 3\sqrt{8z^2}$
21. $4\sqrt{2x^2y} - \sqrt{128x^2y} + 2\sqrt{72x^2y}$
22. $3\sqrt{32zw^2} + 2\sqrt{18zw^2} - \sqrt{50zw^2}$

In Exercises 23–68, perform the indicated multiplications and simplify the expressions. All variables represent positive real numbers. See Examples 3 and 4.

23. $3(2 - \sqrt{2})$
24. $4(\sqrt{3} - 1)$
25. $\sqrt{3}(5 - 2\sqrt{3})$
26. $\sqrt{5}(6\sqrt{5} - 7)$
27. $\sqrt{2}(\sqrt{18} + \sqrt{32})$
28. $\sqrt{7}(\sqrt{28} + 2\sqrt{63})$
29. $(2 + \sqrt{3})(2 - \sqrt{3})$
30. $(3 + \sqrt{2})(3 - \sqrt{2})$
31. $(\sqrt{5} - 1)(\sqrt{5} + 1)$
32. $(\sqrt{7} - 2)(\sqrt{7} + 2)$
33. $(3\sqrt{2} + 2)(3\sqrt{2} - 2)$
34. $(5\sqrt{3} + 4)(5\sqrt{3} - 4)$
35. $(\sqrt{7} + \sqrt{5})(\sqrt{7} - \sqrt{5})$
36. $(\sqrt{5} + \sqrt{2})(\sqrt{5} - \sqrt{2})$
37. $(\sqrt{5} + 4\sqrt{3})(\sqrt{5} - 4\sqrt{3})$
38. $(2\sqrt{7} + \sqrt{2})(2\sqrt{7} - \sqrt{2})$
39. $(\sqrt{m} + \sqrt{n})(\sqrt{m} - \sqrt{n})$
40. $(a\sqrt{x} + \sqrt{y})(a\sqrt{x} - \sqrt{y})$
41. $(3\sqrt{5} + 2\sqrt{7})(3\sqrt{5} - 2\sqrt{7})$
42. $(a\sqrt{x} + b\sqrt{y})(a\sqrt{x} - b\sqrt{y})$
43. $(3 + \sqrt{7})^2$
44. $(5 + \sqrt{2})^2$
45. $(\sqrt{3} - 7)^2$
46. $(\sqrt{5} - 2)^2$
47. $(\sqrt{5} + \sqrt{2})^2$
48. $(\sqrt{3} + \sqrt{7})^2$
49. $(5 - 2\sqrt{7})^2$
50. $(3 - 4\sqrt{5})^2$
51. $(\sqrt{5} + 2)(3 + \sqrt{7})$
52. $(\sqrt{3} + 7)(\sqrt{2} + 5)$
53. $(4 - \sqrt{3})(5 + \sqrt{5})$
54. $(\sqrt{7} - 3)(4 + \sqrt{6})$
55. $(\sqrt{6} + \sqrt{3})(\sqrt{3} + \sqrt{7})$
56. $(\sqrt{7} + \sqrt{5})(\sqrt{7} + \sqrt{8})$
57. $(\sqrt{5} - 3\sqrt{7})(5\sqrt{7} - 1)$
58. $(\sqrt{8} - 4\sqrt{6})(3\sqrt{2} - 5)$
59. $(4\sqrt{7} + 2\sqrt{11})(\sqrt{7} - \sqrt{11})$
60. $(3\sqrt{5} - 4\sqrt{13})(\sqrt{5} + \sqrt{13})$
61. $(3\sqrt{u} + 5)(4\sqrt{u} - 3)$
62. $(\sqrt{2} + 2\sqrt{v})(3\sqrt{2} - \sqrt{v})$
63. $(2\sqrt{v} + 3\sqrt{w})(4\sqrt{v} - 5\sqrt{w})$
64. $(a\sqrt{v} + b\sqrt{w})(c\sqrt{v} + d\sqrt{w})$

B 65. $(\sqrt[3]{3} + \sqrt[3]{7})(\sqrt[3]{9} - \sqrt[3]{21} + \sqrt[3]{49})$ 66. $(\sqrt[3]{5} - \sqrt[3]{3})(\sqrt[3]{25} + \sqrt[3]{15} + \sqrt[3]{9})$

67. $(\sqrt[3]{7} - \sqrt[3]{k})(\sqrt[3]{49} + \sqrt[3]{7k} + \sqrt[3]{k^2})$ 68. $(\sqrt[3]{m} + \sqrt[3]{5})(\sqrt[3]{m^2} - \sqrt[3]{5m} + \sqrt[3]{25})$

In Exercises 69–80, reduce the expressions to the lowest terms. All variables represent positive real numbers. See Example 5.

69. $\dfrac{35 + 21\sqrt{5}}{14}$ 70. $\dfrac{25\sqrt{3} + 15}{10}$

71. $\dfrac{12 - 9\sqrt{7}}{-15}$ 72. $\dfrac{9\sqrt{11} - 15}{-6}$

73. $\dfrac{\sqrt{6} + \sqrt{10}}{3\sqrt{2}}$ 74. $\dfrac{5\sqrt{15} + \sqrt{21}}{4\sqrt{3}}$

75. $\dfrac{k\sqrt{12} - \sqrt{27}}{\sqrt{3}}$ 76. $\dfrac{\sqrt{18} - m\sqrt{50}}{2\sqrt{2}}$

77. $\dfrac{\sqrt{18p^2} - 5p\sqrt{8}}{\sqrt{2p}}$ 78. $\dfrac{2q\sqrt{12} + \sqrt{6q^2}}{\sqrt{3q}}$

79. $\dfrac{s\sqrt[3]{125} + p}{5s + p}$, $(5s + p \ne 0)$ 80. $\dfrac{p\sqrt[3]{16q^3} - 10q}{14q}$

5.4 Rationalizing and Simplifying Radicals

A fraction with a radical denominator, such as $1/\sqrt{2}$, does not immediately give the approximate size of the number. The fraction $1/\sqrt{2} \approx 1/1.414$ requires a lengthy division, whereas in

$$\frac{1}{\sqrt{2}} = \frac{1}{\sqrt{2}} \cdot 1 = \frac{1}{\sqrt{2}} \cdot \frac{\sqrt{2}}{\sqrt{2}} = \frac{\sqrt{2}}{2} \approx \frac{1.414}{2} = 0.707$$

the size of the fraction is quite explicit. The process of removing radicals from a denominator is called **rationalizing the denominator.**

EXAMPLE 1 Rationalize the denominator and simplify

(a) $\dfrac{9}{\sqrt{3}}$ (b) $\sqrt{\dfrac{7}{18}}$

Solution:

(a) $\dfrac{9}{\sqrt{3}} = \dfrac{9}{\sqrt{3}} \cdot 1 = \dfrac{9}{\sqrt{3}} \cdot \dfrac{\sqrt{3}}{\sqrt{3}} = \dfrac{9\sqrt{3}}{3} = 3\sqrt{3}$

(b) $\sqrt{\dfrac{7}{18}} = \dfrac{\sqrt{7}}{\sqrt{18}} = \dfrac{\sqrt{7}}{\sqrt{9 \cdot 2}} = \dfrac{\sqrt{7}}{\sqrt{9}\sqrt{2}} = \dfrac{\sqrt{7}}{3\sqrt{2}} = \dfrac{\sqrt{7}\sqrt{2}}{3\sqrt{2}\sqrt{2}} = \dfrac{\sqrt{7 \cdot 2}}{3 \cdot 2} = \dfrac{\sqrt{14}}{6}$

Thus $\sqrt{\dfrac{7}{18}} = \dfrac{\sqrt{14}}{6}$

Each of the two radicals of the form $a\sqrt{b} + c\sqrt{d}$ and $a\sqrt{b} - c\sqrt{d}$ is called the **conjugate** of the other. Thus $\sqrt{5} - \sqrt{3}$ is the conjugate of $\sqrt{5} + \sqrt{3}$. Conjugates are often used to rationalize a denominator.

WARNING

In general, $\sqrt{4} + \sqrt{9} \neq \sqrt{4 + 9}$ and $\sqrt{4} - \sqrt{9} \neq \sqrt{4 - 9}$
$\sqrt{a} + \sqrt{b} \neq \sqrt{a + b}$ and $\sqrt{a} - \sqrt{b} \neq \sqrt{a - b}$

EXAMPLE 2 Rationalize the denominator and simplify the following.

(a) $\dfrac{1}{\sqrt{3} + \sqrt{2}}$ (b) $\dfrac{6}{\sqrt{5} - \sqrt{3}}$

Solution:

(a) Multiply both the numerator and the denominator by $\sqrt{3} - \sqrt{2}$, and use the identity $(a + b)(a - b) = a^2 - b^2$ where $a = \sqrt{3}$ and $b = \sqrt{2}$.

$$\dfrac{1}{\sqrt{3} + \sqrt{2}} = \dfrac{1}{\sqrt{3} + \sqrt{2}} \cdot \dfrac{\sqrt{3} - \sqrt{2}}{\sqrt{3} - \sqrt{2}}$$

$$= \dfrac{\sqrt{3} - \sqrt{2}}{(\sqrt{3})^2 - (\sqrt{2})^2}$$

$$= \dfrac{\sqrt{3} - \sqrt{2}}{3 - 2}$$

$$= \sqrt{3} - \sqrt{2}$$

(b) Multiply the numerator and denominator by $\sqrt{5} + \sqrt{3}$.

$$\dfrac{6}{\sqrt{5} - \sqrt{3}} = \dfrac{6}{\sqrt{5} - \sqrt{3}} \cdot \dfrac{\sqrt{5} + \sqrt{3}}{\sqrt{5} + \sqrt{3}}$$

$$= \dfrac{6(\sqrt{5} + \sqrt{3})}{(\sqrt{5})^2 - (\sqrt{3})^2}$$

$$= \dfrac{6(\sqrt{5} + \sqrt{3})}{5 - 3}$$

$$= 3(\sqrt{5} + \sqrt{3})$$

EXAMPLE 3 Rationalize the denominator.

(a) $\dfrac{2}{\sqrt[3]{5} - \sqrt[3]{3}}$ (b) $\dfrac{10}{(\sqrt[3]{7})^2 - \sqrt[3]{7}\,\sqrt[3]{3} + (\sqrt[3]{3})^2}$

Solution:

(a) In order to take advantage of the formula $(a - b)(a^2 + ab + b^2) = a^3 - b^3$, we multiply the numerator and denominator by $(\sqrt[3]{5})^2 + (\sqrt[3]{5})(\sqrt[3]{3}) + (\sqrt[3]{3})^2$

$$\dfrac{2}{\sqrt[3]{5} - \sqrt[3]{3}} = \dfrac{2}{\sqrt[3]{5} - \sqrt[3]{3}} \cdot \dfrac{(\sqrt[3]{5})^2 + \sqrt[3]{5} \cdot \sqrt[3]{3} + (\sqrt[3]{3})^2}{(\sqrt[3]{5})^2 + \sqrt[3]{5} \cdot \sqrt[3]{3} + (\sqrt[3]{3})^2}$$

$$= \dfrac{2(\sqrt[3]{25} + \sqrt[3]{15} + \sqrt[3]{9})}{(\sqrt[3]{5})^3 - (\sqrt[3]{3})^3}$$

$$= \dfrac{2(\sqrt[3]{25} + \sqrt[3]{15} + \sqrt[3]{9})}{5 - 3}$$

$$= \sqrt[3]{25} + \sqrt[3]{15} + \sqrt[3]{9}$$

(b) In order to take advantage of $a^3 + b^3 = (a + b)(a^2 - ab + b^2)$, we multiply the numerator and denominator by $\sqrt[3]{7} + \sqrt[3]{3}$.

$$\dfrac{10}{(\sqrt[3]{7})^2 - \sqrt[3]{7} \cdot \sqrt[3]{3} + (\sqrt[3]{3})^2} = \dfrac{\sqrt[3]{7} + \sqrt[3]{3}}{\sqrt[3]{7} + \sqrt[3]{3}} \cdot \dfrac{10}{(\sqrt[3]{7})^2 - \sqrt[3]{7} \cdot \sqrt[3]{3} + (\sqrt[3]{3})^2}$$

$$= \dfrac{(\sqrt[3]{7} + \sqrt[3]{3}) \cdot 10}{(\sqrt[3]{7})^3 + (\sqrt[3]{3})^3}$$

$$= \dfrac{(\sqrt[3]{7} + \sqrt[3]{3}) \cdot 10}{7 + 3} = \sqrt[3]{7} + \sqrt[3]{3}$$

EXAMPLE 4 Rationalize the denominators of the following expressions where y and z are positive, $y \neq z$ and $v \neq w$.

(a) $\dfrac{x}{\sqrt{y} + \sqrt{z}}$ (b) $\dfrac{u}{\sqrt[3]{v} - \sqrt[3]{w}}$ (c) $\dfrac{1}{\sqrt[4]{y} - \sqrt[4]{z}}$

Solution:

(a) Multiply the numerator and denominator by $\sqrt{y} - \sqrt{z}$.

$$\dfrac{x}{\sqrt{y} + \sqrt{z}} = \dfrac{x}{\sqrt{y} + \sqrt{z}} \cdot \dfrac{\sqrt{y} - \sqrt{z}}{\sqrt{y} - \sqrt{z}}$$

$$= \dfrac{x(\sqrt{y} - \sqrt{z})}{(\sqrt{y})^2 - (\sqrt{z})^2}$$

$$= \dfrac{x(\sqrt{y} - \sqrt{z})}{y - z}$$

(b) Multiply the numerator and denominator by $(\sqrt[3]{v})^2 + \sqrt[3]{v} \cdot \sqrt[3]{w} + (\sqrt[3]{w})^2$.

$$\frac{u}{\sqrt[3]{v} - \sqrt[3]{w}} = \frac{u}{\sqrt[3]{v} - \sqrt[3]{w}} \cdot \frac{(\sqrt[3]{v})^2 + \sqrt[3]{v} \cdot \sqrt[3]{w} + (\sqrt[3]{w})^2}{(\sqrt[3]{v})^2 + \sqrt[3]{v} \cdot \sqrt[3]{w} + (\sqrt[3]{w})^2}$$

$$= \frac{u(v^{2/3} + v^{1/3}w^{1/3} + w^{2/3})}{(\sqrt[3]{v})^3 - (\sqrt[3]{w})^3}$$

$$= \frac{u}{v - w}(v^{2/3} + v^{1/3}w^{1/3} + w^{2/3})$$

(c) $\dfrac{1}{\sqrt[4]{y} - \sqrt[4]{z}} = \dfrac{1}{\sqrt[4]{y} - \sqrt[4]{z}} \cdot \dfrac{\sqrt[4]{y} + \sqrt[4]{z}}{\sqrt[4]{y} + \sqrt[4]{z}} = \dfrac{y^{1/4} + z^{1/4}}{(y^{1/4})^2 - (z^{1/4})^2}$

$$= \frac{y^{1/4} + z^{1/4}}{y^{1/2} - z^{1/2}}$$

$$= \frac{y^{1/4} + z^{1/4}}{y^{1/2} - z^{1/2}} \cdot \frac{y^{1/2} + z^{1/2}}{y^{1/2} + z^{1/2}}$$

$$= \frac{(y^{1/4} + z^{1/4})(y^{1/2} + z^{1/2})}{(y^{1/2})^2 - (z^{1/2})^2}$$

$$= \frac{(y^{1/4} + z^{1/4})(y^{1/2} + z^{1/2})}{y - z}$$

Exercises 5.4

Standard Assignment: Exercises 1, 3, 5, 7, 9, 11, 13, 17, 19, 21, 23, 25, 27, 31, 33

A Rationalize and simplify the following radicals.

1. $\dfrac{6}{\sqrt{3}}$
2. $\dfrac{8}{\sqrt{2}}$
3. $\dfrac{1}{\sqrt{5}}$
4. $\dfrac{1}{\sqrt{7}}$
5. $\dfrac{-4}{\sqrt{6}}$
6. $\dfrac{-6}{\sqrt{8}}$
7. $\dfrac{\sqrt{2}}{\sqrt{8}}$
8. $\dfrac{4\sqrt{3}}{\sqrt{6}}$
9. $\dfrac{-\sqrt{3}}{\sqrt{7}}$
10. $\dfrac{-\sqrt{7}}{\sqrt{5}}$
11. $\dfrac{-20\sqrt{3}}{\sqrt{5}}$
12. $\dfrac{-121\sqrt{10}}{\sqrt{11}}$
13. $\sqrt{\dfrac{5}{12}}$
14. $\sqrt{\dfrac{7}{18}}$
15. $\dfrac{4\sqrt{5}}{\sqrt{a}}$
16. $\dfrac{-5\sqrt{2}}{\sqrt{6}}$
17. $\dfrac{3\sqrt{c}}{\sqrt{d}}$
18. $\dfrac{-4\sqrt{d}}{\sqrt{e}}$
19. $\dfrac{1}{\sqrt{2} + m}$
20. $\dfrac{1}{\sqrt{5} + 2n}$
21. $\dfrac{3}{2 - \sqrt{3}}$
22. $\dfrac{-5}{1 - \sqrt{2}}$
23. $\dfrac{1}{\sqrt{3} + \sqrt{2}}$
24. $\dfrac{-2}{\sqrt{5} + \sqrt{7}}$

25. $\dfrac{\sqrt{3}}{\sqrt{5p} - \sqrt{3}}$
26. $\dfrac{\sqrt{5}}{\sqrt{6} - \sqrt{5q}}$
27. $\dfrac{2\sqrt{7}}{3\sqrt{7} - 8x}$
28. $\dfrac{6\sqrt{5}}{4y - 3\sqrt{5}}$

29. $\dfrac{\sqrt{35}}{3\sqrt{5} + 2\sqrt{7y}}$
30. $\dfrac{\sqrt{65}}{2\sqrt{13x} - 3\sqrt{5}}$
31. $\dfrac{\sqrt{b}}{a + \sqrt{b}}$

32. $\dfrac{3\sqrt{c}}{\sqrt{c} - d}$
33. $\dfrac{u - v\sqrt{w}}{u + v\sqrt{w}}$
34. $\dfrac{a\sqrt{x} + 3}{a\sqrt{x} - 3}$

B 35. $\dfrac{1}{a\sqrt{x} + b\sqrt{y}}$
36. $\dfrac{\sqrt{xy}}{a\sqrt{x} - b\sqrt{y}}$
37. $\dfrac{1}{\sqrt[3]{2} + 1}$

38. $\dfrac{1}{\sqrt[3]{3} + 2}$
39. $\dfrac{1}{2\sqrt[3]{3} - 3}$
40. $\dfrac{1}{4\sqrt[3]{2} - 1}$

41. $\dfrac{\sqrt[3]{15}}{\sqrt[3]{5} + \sqrt[3]{3}}$
42. $\dfrac{\sqrt[3]{14}}{\sqrt[3]{7} - \sqrt[3]{2}}$
43. $\dfrac{1}{\sqrt[3]{x} + 1}$

44. $\dfrac{1}{\sqrt[3]{x} - b}$
45. $\dfrac{c}{a\sqrt[3]{x} - b}$
46. $\dfrac{\sqrt[3]{x}}{a\sqrt[3]{x} + b}$

47. $\dfrac{1}{a\sqrt[3]{x} + \sqrt[3]{y}}$
48. $\dfrac{1}{\sqrt[3]{x} - b\sqrt[3]{y}}$
49. $\dfrac{\sqrt[3]{uv}}{c\sqrt[3]{u} - d\sqrt[3]{v}}$

50. $\dfrac{\sqrt[3]{v}}{c\sqrt[3]{u} + d\sqrt[3]{v}}$
51. $\dfrac{1}{\sqrt[4]{5} + 1}$
52. $\dfrac{1}{2 - \sqrt[4]{3}}$

53. $\dfrac{\sqrt[4]{3}}{\sqrt[4]{3} - \sqrt[4]{2}}$
54. $\dfrac{\sqrt[4]{5}}{\sqrt[4]{7} + \sqrt[4]{5}}$

55. $\dfrac{\sqrt[4]{15}}{2\sqrt[4]{3} + \sqrt[4]{5}}$
56. $\dfrac{\sqrt[4]{21}}{\sqrt[4]{3} - 3\sqrt[4]{7}}$

5.5 Radical Equations

An equation in which the variable occurs under a radical sign is called a **radical equation**. For example,

$$\sqrt{3x - 2} = 9, \qquad \sqrt[3]{2x + 7} = 3, \quad \text{and} \quad \dfrac{3}{\sqrt{x + 5}} = 7x - 1$$

are radical equations.

To solve radical equations, we use the following rule.

If $A = B$ then $A^n = B^n$ for any positive integer n.

5.5 Radical Equations

EXAMPLE 1

Solve $\sqrt[3]{2x + 7} = 3$

Solution: Because the equation contains a cubic root, we cube both sides.

$$(\sqrt[3]{2x + 7})^3 = 3^3$$
$$2x + 7 = 27$$
$$2x = 20$$
$$x = 10$$

In order to be sure that $x = 10$ is the solution, we substitute it into the given equation.

Does $\sqrt[3]{2 \cdot 10 + 7} = 3$?   Yes! The solution set is $\{10\}$.

Unfortunately the converse of the previously mentioned rule need not be true. The converse is: "If $A^n = B^n$ then $A = B$," which is not necessarily true since $(-2)^2 = 2^2$, but $-2 \neq 2$. Therefore, the rule gives only the information that

All solutions of $A = B$ are solutions of $A^n = B^n$, but some solutions of $A^n = B^n$ may not be solutions of $A = B$.

Hence, the final checking of the solutions of $A^n = B^n$ in the original equation is essential.

EXAMPLE 2

Solve $\sqrt{x^2 + 3x - 9} - 1 + x = 0$.

Solution: Add $1 - x$ to both sides to isolate the radical on one side. We have

$$\sqrt{x^2 + 3x - 9} = 1 - x$$

Then square both sides.

$$(\sqrt{x^2 + 3x - 9})^2 = (1 - x)^2$$

Using $(a - b)^2 = a^2 - 2ab + b^2$ where $a = 1$ and $b = x$, we get

$$x^2 + 3x - 9 = 1 - 2x + x^2$$

Adding $2x - x^2$ to both sides, we get

$$5x - 9 = 1 \quad \text{or} \quad 5x = 10$$

Dividing both sides of $5x = 10$ by 5, we conclude that $x = 2$ is a possible solution.

Now check by substitution.

Check: Let $x = 2$. Does $\sqrt{2^2 + 3 \cdot 2 - 9} - 1 + 2 = 0$?

Does $\sqrt{4 + 6 - 9} + 1 = 0$?   No.

The only possible solution, $x = 2$, did not satisfy the given radical equation. Therefore, the equation has no solution, or we may say that the solution set is the empty set, $\emptyset$.

In general, to solve a radical equation we follow the following steps.

To Solve a Radical Equation

1. Relocate radicals if necessary, so that one radical is alone on one side of the equal sign.
2. Raise each side of the equation to a power so that at least one radical sign disappears.
3. Repeat steps 1 and 2, if necessary, until all radical signs disappear.
4. All possible solutions must be checked in the original equation.

EXAMPLE 3

Solve $\sqrt{x - 5} + \sqrt{x} = 5$.

Solution: Subtract $\sqrt{x}$ from both sides to relocate one radical on each side. We have

$$\sqrt{x - 5} = 5 - \sqrt{x}$$
$$(\sqrt{x - 5})^2 = (5 - \sqrt{x})^2 \qquad \text{Square both sides}$$
$$x - 5 = 25 - 10\sqrt{x} + x \qquad \text{Use } (a - b)^2 = a^2 - 2ab + b^2$$
$$10\sqrt{x} = 30 \qquad \text{Combine terms}$$
$$\sqrt{x} = 3 \qquad \text{Divide by 10}$$
$$x = 9 \qquad \text{Square both sides}$$

Check: Let $x = 9$. Does $\sqrt{9 - 5} + \sqrt{9} = 5$?
$$\sqrt{4} + \sqrt{9} = 5?$$
$$2 + 3 = 5? \qquad \text{Yes.}$$

Hence, $\{9\}$ is the solution set of the given equation.

EXAMPLE 4

Solve $\sqrt{x + 6\sqrt{x}} = \sqrt{x} + 2$.

Solution:

$$(\sqrt{x + 6\sqrt{x}})^2 = (\sqrt{x} + 2)^2 \qquad \text{Square both sides}$$
$$x + 6\sqrt{x} = x + 4\sqrt{x} + 4 \qquad \text{Use } (a + b)^2 = a^2 + 2ab + b^2$$
$$2\sqrt{x} = 4 \qquad \text{Isolate the radical}$$
$$\sqrt{x} = 2 \qquad \text{Divide by 2}$$
$$x = 4 \qquad \text{Square both sides}$$

Hence, $x = 4$ is a possible solution.

Check: Let $x = 4$.
$$\sqrt{4 + 6\sqrt{4}} = \sqrt{4} + 2$$
$$\sqrt{4 + 6 \cdot 2} = 2 + 2$$
$$\sqrt{4 + 12} = 4 \quad \checkmark$$

Therefore, the solution set is $\{4\}$.

EXAMPLE 5

Solve $\sqrt[4]{k^2 + 5k - 1} = \sqrt{k + 2}$.

Solution: Squaring both sides, we have
$$\sqrt{k^2 + 5k - 1} = k + 2$$

Again, squaring both sides, we get
$$k^2 + 5k - 1 = k^2 + 4k + 4$$

Now, by isolating the variable k on the left side and the constants on the right side, we have
$$k = 4 + 1$$
$$k = 5$$

Check: Let $k = 5$.
$$\sqrt[4]{5^2 + 5 \cdot 5 - 1} = \sqrt{5 + 2}$$
$$\sqrt[4]{25 + 25 - 1} = \sqrt{7}$$
$$\sqrt[4]{49} = \sqrt{7}$$
$$\sqrt[4]{7^2} = \sqrt{7}$$
$$7^{2/4} = \sqrt{7}$$
$$7^{1/2} = \sqrt{7} \quad \checkmark$$

Therefore, the solution set is $\{5\}$.

Exercises 5.5

Standard Assignment: Exercises 1, 3, 7, 13, 17, 19, 21, 23, 27, 29, 31, 35, 37, 39, 41, 43, 45, 47

A Solve the following radical equations. See Examples 1–4.

1. $\sqrt{x - 1} = 3$
2. $\sqrt{y - 2} = 4$
3. $\sqrt{2z + 1} = 5$
4. $\sqrt{3z + 1} = 7$
5. $\sqrt{5y - 4} = 9$
6. $\sqrt{7w - 10} = 5$
7. $\sqrt{7s - 5} + 2 = 0$
8. $\sqrt{8 - 5t} + 1 = 0$
9. $\sqrt{9 - 5t} - 2 = 0$
10. $\sqrt{7s - 5} - 3 = 0$
11. $\sqrt{6u + 1} - \sqrt{13} = 0$
12. $\sqrt{5u + 2} - \sqrt{17} = 0$

13. $\sqrt[3]{2k-1} = 5$
14. $\sqrt[3]{3k+4} = 4$
15. $\sqrt[3]{5p+12} + 2 = 0$
16. $\sqrt[3]{4p-11} + 3 = 0$
17. $\sqrt[3]{1-2q} + 7 = 0$
18. $\sqrt[3]{14-3q} + 4 = 0$
19. $\sqrt[5]{7-r} - 1 = 0$
20. $\sqrt[5]{12-2r} - 2 = 0$
21. $\sqrt{3s-7} = \sqrt{2s+1}$
22. $\sqrt{5s+21} = \sqrt{3s+93}$
23. $\sqrt{13-2t} + \sqrt{4t-8} = 0$
24. $\sqrt{31-5t} + \sqrt{6t+7} = 0$
25. $\sqrt{5-u} + \sqrt{3u-15} = 0$
26. $\sqrt{3u-21} + \sqrt{7-u} = 0$
27. $\sqrt{31-5t} - \sqrt{6t+7} = 0$
28. $\sqrt{4t-8} - \sqrt{13-2t} = 0$
29. $2\sqrt{x} - \sqrt{5x-14} = 0$
30. $\sqrt{3x+17} - 2\sqrt{x+1} = 0$
31. $y = \sqrt{y^2 - 3y + 21}$
32. $2y = \sqrt{4y^2 - 5y + 125}$
33. $\sqrt{y^2 - 2y + 14} - y - 4 = 0$
34. $\sqrt{4y^2 - 3y - 26} + 2y - 1 = 0$
35. $\sqrt{9z^2 - z + 2} + 3z - 1 = 0$
36. $\sqrt{z^2 + 2z + 1} + z + 5 = 0$
37. $\sqrt{x+3} = \sqrt{x} + \sqrt{3}$
38. $\sqrt{x+5} = \sqrt{x} + \sqrt{5}$
39. $\sqrt{y+5} + \sqrt{y+3} = 4$
40. $\sqrt{y+6} + \sqrt{y+5} = 7$
41. $\sqrt{5+z} = \sqrt{z} - 1$
42. $\sqrt{z+13} = 1 + \sqrt{z}$
43. $\sqrt{x + 2\sqrt{x}} = 1 + \sqrt{x}$
44. $\sqrt{x - 6\sqrt{x}} = \sqrt{x} - 5$
45. $\sqrt[4]{k^2 + 3k - 5} = \sqrt{k+1}$
46. $\sqrt[4]{k^2 - k + 10} = \sqrt{k-1}$
47. $\sqrt{x+5} \cdot \sqrt{x} = \sqrt{6+x^2}$
48. $\sqrt{x+3} \cdot \sqrt{x-2} = \sqrt{14+x^2}$

5.6 Complex Numbers

In this book, we have confined ourselves so far to the *real number system*. It has been adequate in many aspects, but simple quadratic equations, such as $x^2 + 1 = 0$, remain unsolvable when only real numbers are used. It is possible to enlarge the real number system to include new numbers that will satisfy equations such as $x^2 + 1 = 0$. For now, we will assume the existence of a number, denoted by i, with $i^2 + 1 = 0$. We also have $i^2 = -1$ and $i = \sqrt{-1}$. Now $x^2 + 1 = 0$ has two solutions, $\pm i$.

Definition of *i*:

$i = \sqrt{-1}$ and $i^2 = -1$

5.6 Complex Numbers

This new number ($i = \sqrt{-1}$) will be combined with the real numbers through addition and multiplication, with the requirement that the basic properties of the real numbers, including the associative, commutative, and distributive properties, as well as the power and radical laws, must be preserved. For example,

$$\sqrt{-4} = \sqrt{-1 \cdot 4} = \sqrt{-1}\sqrt{4} = i \cdot 2 = 2i$$

$$i + (-1)i = [1 + (-1)]i = 0 \cdot i = 0$$

$$i^3 = i^2 \cdot i = (-1)i = -i$$

$$(-3) \cdot 4i = (-3 \cdot 4)i = -12i$$

$$\sqrt{\frac{-4}{3}} = \frac{\sqrt{-4}}{\sqrt{3}} = \frac{2i}{\sqrt{3}}$$

$$\frac{2i}{\sqrt{3}} = \frac{2i}{\sqrt{3}} \cdot \frac{\sqrt{3}}{\sqrt{3}} = \frac{2\sqrt{3}}{3}i \qquad \text{Rationalize the denominator}$$

EXAMPLE 1 Solve the equation $3x^2 + 4 = 0$.

Solution: $3x^2 + 4 - 4 = 0 - 4$ Subtract 4 from both sides

$$3x^2 = -4$$

$$x^2 = \frac{-4}{3} \qquad \text{Divide both sides by 3}$$

$$x = \pm\sqrt{\frac{-4}{3}} \qquad \text{Find the square roots of both sides}$$

$$= \pm\frac{2\sqrt{3}}{3}i$$

The solution set is $\left\{\dfrac{2\sqrt{3}}{3}i, -\dfrac{2\sqrt{3}}{3}i\right\}$.

We shall call any nonzero multiple of i a **pure imaginary number**.

Definition of Complex Numbers

A **complex number** is a number of the form

$$a + bi$$

where a and b can be any real numbers.

The number a is called the **real part** and b is called the **imaginary part** of $a + bi$. Note that despite the name "imaginary part," the number b is a real number.

EXAMPLE 2

(a) $5 - 4i$ is a complex number; the real part is 5 and the imaginary part is -4.

(b) Any real number a can be thought of as a complex number, $a + 0i$, in which the real part is a and the imaginary part is 0.

(c) Any pure imaginary number bi ($b \neq 0$) is a complex number, $0 + bi$, whose real part is 0.

We call the complex number $a + bi$ an **imaginary number** if $b \neq 0$. A pure imaginary number is an imaginary number $a + bi$ such that $a = 0$. A complex number $a + bi$ is a real number if $b = 0$.

The set C of complex numbers is the largest set of numbers thus far introduced. It contains the set of natural numbers, the set of integers, the set of rational numbers, and the set of real numbers as subsets.

Number System

N Natural numbers
J Integers
Q Rational numbers
R Real numbers
C Complex numbers

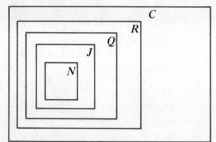

$$N \subset J \subset Q \subset R \subset C$$

Two complex numbers such as $a + bi$ and $c + di$ are said to be equal,

$$a + bi = c + di$$

exactly when $a = c$ and $b = d$.

If $x + yi = -6 - \sqrt[3]{2}\, i$, then $x = -6$ and $y = -\sqrt[3]{2}$.

The addition of two complex numbers is accomplished by adding the real parts and adding the imaginary parts. That is,

$$(a + bi) + (c + di) = (a + c) + (b + d)i$$

Similarly,

$$(a + bi) - (c + di) = (a - c) + (b - d)i$$

EXAMPLE 3

(a) $(6 + i) + (3 + 2i) = (6 + 3) + (1 + 2)i$
$\qquad = 9 + 3i$

(b) $(16 + 21i) - (13 + 121i) = (16 - 13) + (21 - 121)i$
$\qquad = 3 + (-100)i$
$\qquad = 3 - 100i$

(c) $a + bi + (-a - bi) = [a + (-a)] + [b + (-b)]i$
$= 0 + 0i$
$= 0$

Thus, $-(a + bi) = -a - bi$.

EXAMPLE 4

Solve the equation $2x + 7i = 6 + (3y - 2)i$.

Solution: Two complex numbers are equal only when their real parts are equal and their imaginary parts are equal. We have $2x = 6$ and $7 = 3y - 2$. Hence, $x = 3$ and $y = 3$.

Exercises 5.6

Standard Assignment: Exercises 1, 5, 9, 11, 12, 15, 17, 19, 23, 29, 31, 33

In each of Exercises 1–16, identify the real part and the imaginary part. See Example 2.

A
1. $3 + 4i$
2. $6 + 7i$
3. $5 - 2i$
4. $10 - 3i$
5. $-7 - i$
6. $-9 - 8i$
7. $5i - 11$
8. $i - 1$
9. i
10. $-i$
11. -1
12. 0
13. i^2
14. $10i^2 - 1$
15. $\sqrt{3} + \sqrt{-3}$
16. $\sqrt{-9}$

In Exercises 17–35, perform the addition or subtraction as indicated. All variables represent any real numbers. See Example 3.

17. $(3 + 4i) + (4 + 5i)$
18. $(4 + 3i) + (7 + 6i)$
19. $(6 - 5i) + (10 - 3i)$
20. $(5 - 6i) + (5 - 2i)$
21. $ai + bi$
22. $ci + (-di)$
23. $ci - di$
24. $ai - bi$
25. $(u + vi) + w$
26. $u - (v + wi)$
27. $pi + (q - ri)$
28. $ri - (s - ti)$
29. $(5 + i) + (2 - 3i) + (3 + i)$
30. $(-6 + 3i) + (5 - i) + (7 - 2i)$
31. $(8 - 7i) - (4 - 9i) + 10i$
32. $(9 - 10i) - (5 - 8i) - (-13i)$
33. $(a + 2bi) + (a - bi) - (c + bi)$
34. $(2m - 2ni) - (m - ni) + (3 - ni)$

B
35. Is the set of complex numbers closed under addition? In other words, is the sum of any two complex numbers always a complex number?

36. Is the set of imaginary numbers closed under addition?

37. What is the additive identity of the complex number system?

38. What is the additive inverse of $a + bi$?
39. Can the sum of a real number and an imaginary number be a real number?
40. Can the sum of a real number and an imaginary number be an imaginary number?
41. Are all imaginary numbers complex numbers?
42. Are all real numbers complex numbers?

Solve the following equations. See Examples 1 and 4.

43. $x^2 + 4 = 0$
44. $y^2 + 9 = 0$
45. $-2y^2 - 9 = 0$
46. $-5x^2 - 4 = 0$
47. $(x - 1)^2 + 3 = 0$
48. $(y - 2)^2 + 5 = 0$
49. $(2 - y)^2 + 7 = 0$
50. $(1 - x)^2 + 11 = 0$
51. $x^4 - 81 = 0$
52. $x^4 - 16 = 0$

Solve the following equations where all variables represent real numbers. See Example 4.

53. $(2x - 1) + 3i = 5 + (1 - 2y)i$
54. $(3x + 5) - 6i = 1 + (3y - 3)i$
55. $(u + v)i = 3u + (2v - 1)i$
56. $(s - 2) + (3 - 2t)i = (s - t)i$

5.7 Operations with Complex Numbers

In Section 5.6 we introduced the complex number system as the set of all $a + bi$, where a and b are any real numbers and $i = \sqrt{-1}$. We also showed how to add and subtract complex numbers. In this section, we will study the behavior of i^n and the multiplication and division of the complex numbers.

First, let us observe the behavior of i^n for any integer power n. We know that $i^2 = -1$ and i^n can be derived successively

$$i^3 = i^2 \cdot i = (-1)i = -i \qquad i^4 = i^2 \cdot i^2 = (-1)(-1) = 1$$
$$i^5 = i^4 \cdot i = i \qquad i^6 = i^4 \cdot i^2 = i^2 = -1$$
$$i^7 = i^4 \cdot i^3 = -i \qquad i^8 = i^4 \cdot i^4 = 1 \cdot 1 = 1$$

You can see the general pattern in the values of i^n that powers of i follow a repeating pattern of values: $i, -1, -i, 1$. The value of any power of i can be determined by using this pattern.

5.7 Operations with Complex Numbers

EXAMPLE 1   Identify each of the following as 1, -1, i, or $-i$ by using the fact that $i^4 = 1$.

(a) i^{101}
(b) i^{242}
(c) i^{-360}
(d) i^{-401}

Solution:

(a) Since $i^{101} = i^{4 \cdot 25 + 1} = (i^4)^{25} \cdot i^1$, we have
$$i^{101} = i^1 = i$$

(b) Since $i^{242} = i^{4 \cdot 60 + 2} = (i^4)^{60} \cdot i^2$, we have
$$i^{242} = i^2 = -1$$

(c) Since $i^{-360} = i^{4 \cdot (-90) + 0} = (i^4)^{-90} \cdot i^0$, we have
$$i^{-360} = i^0 = 1$$

(d) Since $i^{-401} = i^{4 \cdot (-101) + 3} = (i^4)^{-101} \cdot i^3$, we have
$$i^{-401} = i^3 = -i$$

In order to multiply complex numbers, use the FOIL method.

EXAMPLE 2   (a) Multiply $3 + 4i$ and $6 + 7i$.

Solution: Using the FOIL method, we have
$$(3 + 4i)(6 + 7i) = 3 \cdot 6 + 3 \cdot 7i + 4i \cdot 6 + 4i \cdot 7i$$
$$= 18 + 21i + 24i + 28i^2$$
$$= (18 - 28) + (21 + 24)i$$
$$= -10 + 45i$$

(b) $(3 - 2i)(7 + 4i) = 3 \cdot 7 + 3 \cdot 4i + (-2i) \cdot 7 + (-2i) \cdot 4i$
$$= (21 - 8i^2) + (12 - 14)i$$
$$= 21 - 8(-1) + (-2)i$$
$$= 29 - 2i$$

(c) $(5 + 6i)(5 - 6i) = 5^2 - (6i)^2$ Use $(x + y)(x - y) = x^2 - y^2$ where $x = 5$ and $y = 6i$
$$= 5^2 - 6^2 i^2$$
$$= 5^2 - 6^2(-1)$$
$$= 5^2 + 6^2$$
$$= 61$$

In general, if a, b, c, and d are any real numbers, then

$$(a + bi)(c + di) = ac + (adi + cbi) + bi \cdot di$$
$$= ac + bdi^2 + (ad + cb)i$$
$$= ac + bd(-1) + (ad + bc)i$$
$$= (ac - bd) + (ad + bc)i$$

and

$$(a + bi)(a - bi) = a^2 - (bi)^2 \quad \text{See Example 2(c)}$$
$$= a^2 - b^2 i^2$$
$$= a^2 - b^2(-1)$$
$$= a^2 + b^2$$

Each of the two complex numbers of the form $a + bi$ and $a - bi$ is called the **complex conjugate** of the other. Example 2(c) and the previous calculations show

The product of a complex number and its conjugate results in a real number.

Conjugates are the most important keys to the division of complex numbers.

EXAMPLE 3 Find $(5 + 6i)/(4 + 3i)$, giving your answer in the form $a + bi$.

Solution: We always multiply both the numerator and the denominator by the conjugate of the denominator; this process converts the denominator into a real number.

$$\frac{5 + 6i}{4 + 3i} = \frac{5 + 6i}{4 + 3i} \cdot \frac{4 - 3i}{4 - 3i}$$

$$= \frac{20 - 15i + 24i - 18i^2}{16^2 - (3i)^2} \quad \text{FOIL method}$$

$$= \frac{20 + 9i - 18 \cdot (-1)}{16 + 9}$$

$$= \frac{38 + 9i}{25} \quad \text{or} \quad \frac{38}{25} + \frac{9}{25}i$$

In general, if a, b, c, and d are real numbers, and $c^2 + d^2 \neq 0$, then

$$\frac{a + bi}{c + di} = \frac{(a + bi)(c - di)}{(c + di)(c - di)}$$

$$= \frac{ac - adi + bci - bdi^2}{c^2 + d^2}$$

$$= \frac{(ac + bd) + (bc - ad)i}{c^2 + d^2}$$

$$= \frac{ac + bd}{c^2 + d^2} + \frac{bc - ad}{c^2 + d^2}i$$

The last result need not be memorized; the student should only understand the process leading to the result and be able to apply it.

Exercises 5.7

Standard Assignment: Exercises 1, 5, 9, 13, 17, 23, 27, 31, 35, 37, 39, 41

Simplify each of the following. See Example 1.

A
1. i^{16}
2. i^{20}
3. i^{81}
4. i^{85}
5. i^{-78}
6. i^{-62}
7. i^{-39}
8. i^{-99}
9. $-10i^{106}$
10. $-13i^{203}$
11. $-15i^{-111}$
12. $-18i^{-223}$

Perform each of the following operations as indicated and give your answer in the form $a + bi$ where a and b are real numbers. See Examples 2 and 3.

13. $7i \cdot (-3)$
14. $-8i \cdot 9$
15. $(-5) \cdot (-6i)$
16. $(-8i) \cdot (-9)$
17. $i(3 + 4i)$
18. $2i(4 - 5i)$
19. $(-7)(-5 + i)$
20. $(2 - 10i)(-3)$
21. $(1 + i)(1 - i)$
22. $(\sqrt{2} + i)(\sqrt{2} - i)$
23. $(1 - 5i)(1 + 5i)$
24. $(6 - 7i)(6 + 7i)$
25. $(29 + 17i)(13 + 12i)$
26. $(19 + 12i)(11 + 13i)$
27. $(\sqrt{3} + 6i)(\sqrt{2} - 3i)$
28. $(\sqrt{5} - \sqrt{2}i)(\sqrt{2} + \sqrt{5}i)$
29. $(\sqrt{7} - 3i)(\sqrt{2} - 5i)$
30. $(\sqrt{6} - 7i)(\sqrt{6} - 12i)$
31. $(-\sqrt{5} + 13i)^2$
32. $(\sqrt{3} - 12i)^2$
33. $(\sqrt{18} + \sqrt{72}i)^2$
34. $(\sqrt{27} - \sqrt{48}i)^2$
35. $\dfrac{2 + \sqrt{3}i}{i}$
36. $\dfrac{\sqrt{5} - 4i}{i}$
37. $\dfrac{-1}{1 + i}$
38. $\dfrac{1}{2 - i}$
39. $\dfrac{3i}{2 - i}$
40. $\dfrac{5i}{1 + i}$
41. $\dfrac{-7}{i + 5}$
42. $\dfrac{-13}{2i - 3}$
43. $\dfrac{2 + 3i}{1 + i}$
44. $\dfrac{3 + 5i}{4 + i}$
45. $\dfrac{1 - i}{3 + 2i}$
46. $\dfrac{2 - 5i}{4 - 7i}$
47. $\dfrac{7i - 6}{5i - 8}$
48. $\dfrac{8i - 3}{2i - 11}$

49. ▦ $(\sqrt{391.235} + \sqrt{72.336i})(\sqrt{77.567} + \sqrt{0.915i})$

50. ▦ $(321.155 - \sqrt{62.246i})^2$

51. ▦ $(125.025 + 0.935i)(125.025 - 0.935i)$

52. ▦ $\dfrac{415.13 + 152.47i}{213.17 + 171.23i}$

53. ▦ $\dfrac{-702.43 + 516.97i}{612.57 - 203.09i}$

54. ▦ $\dfrac{1}{(113.14 + 215.23i)^2}$

B 55. Is the set of all complex numbers closed under multiplication? In other words, is the product of any two complex numbers always a complex number?

56. Can the product of two complex numbers be a real number?

57. If a and b are real numbers and $a^2 + b^2 \neq 0$, what is the multiplicative inverse of $a + bi$?

58. Find the multiplicative inverse of $4 - 3i$.

59. Find all real values of x such that $(3 + xi)/(1 - i) = 1 + 2i$.

60. Find all real values of x such that $(x^2 + 3i)/(1 + i) = 6 - 3i$.

61. Find all real values of x and y such that $(x + yi)/i = 5 - 7i$.

62. Find all real values of x and y such that $(5x + yi)/(2 - i) = 2 + i$.

63. Let $z = -3 + 2i$. Find the value of $z^2 + 6z + 13$. Is $-3 + 2i$ a solution of the equation $z^2 + 6z + 13 = 0$? How about $-3 - 2i$?

64. Let $z = 5 - 3i$. Find the value of $z^2 - 10z + 34$. Is $5 - 3i$ a solution of the equation $x^2 - 10x + 34 = 0$? How about $5 + 3i$?

Chapter 5 Summary

Key Words and Phrases

5.1	nth root	5.5	Radical equation
	Square root	5.6	Pure imaginary numbers
	Cube root		Complex numbers
	Principal nth root		Real part
	Radical		Imaginary part
5.2	Lowest terms		Imaginary numbers
5.3	Radical expression	5.7	Complex conjugate

Key Concepts and Rules

The Laws of Radicals:

$\sqrt[n]{a^m} = (\sqrt[n]{a})^m$

$\sqrt[n]{ab} = \sqrt[n]{a}\,\sqrt[n]{b}$

$\sqrt[m]{\sqrt[n]{a}} = \sqrt[mn]{a}$

$\sqrt[n]{\dfrac{a}{b}} = \dfrac{\sqrt[n]{a}}{\sqrt[n]{b}}$

Review Exercises

If you have difficulty with any of the following exercises, look in the section indicated by the numbers in square brackets.

[5.1] Find or simplify each indicated root. All variables represent nonzero real numbers.

1. $\sqrt{36}$
2. $\sqrt{49}$
3. $-\sqrt{121}$
4. $-\sqrt{144}$
5. $\sqrt[3]{27}$
6. $-\sqrt[3]{-64}$
7. $-\sqrt[3]{\dfrac{80^3}{1000}}$
8. $\sqrt[3]{\dfrac{27b^3}{125c^3}}$
9. $\sqrt{81x^4}$
10. $\sqrt[3]{0.216y^3}$
11. $\sqrt[4]{81z^8}$
12. $\sqrt[6]{64s^6 t^{12}}$
13. $\sqrt{\dfrac{144k^4}{k^4 + 8k^2 + 16}}$
14. $\sqrt{\dfrac{625(d^4 + 10d^2 + 25)}{49a^8 b^4 c^{12}}}$

[5.2] Simplify and express each answer with positive exponents. All variables represent positive real numbers.

15. $(-64)^{1/3}$
16. $(-27)^{2/3}$
17. $(-8)^{-1/3}$
18. $(-32k^{15})^{1/5}$
19. $\dfrac{(-8)^{-1/3}}{144^{-1/2}}$
20. $(a^7 u^6 v^{-12})^{-2/3}$
21. $(64m^{-12})^{-1/6}$
22. $(n^8)^{1/4} + (8n^{-3})^{-2/3}$
23. $a^0 \cdot a^{1/3} \cdot a^{1/9} \cdot a^{1/27} \cdot a^{1/81} \cdot a^{1/243}$
24. $b^0 \cdot b^{1/10} \cdot b^{2/10} \cdot b^{3/10} \ldots b^{10/10}$

[5.3] Simplify each radical expression. All variables represent positive real numbers.

25. $\sqrt[3]{64} - 4\sqrt[3]{128}$
26. $7\sqrt[3]{16} - 2\sqrt[3]{-128}$
27. $4k\sqrt{32k^2} - 3\sqrt{49k^4}$
28. $4\sqrt{50ab^2} - 3\sqrt{18bc^2} + 2\sqrt{32ca^2}$

Perform each indicated multiplication and simplify. All variables represent positive real numbers.

29. $\sqrt{5}(2\sqrt{5} - 6)$
30. $\sqrt{2}(3\sqrt{18} - 4\sqrt{32})$
31. $(2\sqrt{7} + 3)(2\sqrt{7} - 3)$
32. $(\sqrt{s} + \sqrt{t})(\sqrt{s} - \sqrt{t})$
33. $(\sqrt{3} - \sqrt{7})^2$
34. $(2\sqrt{u} + 3\sqrt{v})(4\sqrt{u} - 5\sqrt{v})$
35. $(a\sqrt{2} + b\sqrt{3})(c\sqrt{2} + d\sqrt{3})$

Reduce to lowest terms. All variables represent positive real numbers.

36. $\dfrac{3\sqrt{18} - k\sqrt{50}}{2\sqrt{2}}$
37. $\dfrac{3\sqrt{18t^2} - 4t\sqrt{8}}{\sqrt{2}\,t}$
38. $\dfrac{u\sqrt[3]{16v^3} - 50v}{15v}$

[5.4] Rationalize the denominators and simplify. All denominators represent positive real numbers.

39. $\dfrac{-17\sqrt{5}}{\sqrt{3}}$
40. $\dfrac{6\sqrt{m}}{18\sqrt{n}}$
41. $\dfrac{2\sqrt{2}}{4\sqrt{3} + 5\sqrt{2}}$
42. $\dfrac{a\sqrt{x} + b}{a\sqrt{x} - b}$
43. $\dfrac{1}{\sqrt[3]{y} + c}$
44. $\dfrac{1}{1 - \sqrt[4]{5}}$
45. $\dfrac{1}{a\sqrt[4]{x} + 1}$
46. $\dfrac{\sqrt[3]{st}}{a\sqrt[3]{s} + b\sqrt[3]{t}}$

[5.5] Solve the following radical equations.

47. $\sqrt{x^2 + 2} - x + 4 = 0$
48. $\sqrt{y} - \sqrt{y - 5} + 1 = 0$
49. $\sqrt{13 + \sqrt{s}} = 4$
50. $\sqrt[4]{t^2 - t - 5} = \sqrt{t - 1}$
51. $\sqrt{\sqrt{u} - 2}\,\sqrt{\sqrt{u} - 3} = \sqrt{56 - 5\sqrt{u}}$
52. $5v - 5\sqrt{v^2 + 6} = 2$
53. $2\sqrt{x + 3} - \sqrt{4x + 3} = 0$
54. $-4\sqrt[4]{z - 3} + 4 = 0$

[5.6] Identify the real and the imaginary part of each of the following expressions.

55. $-3 - 8i$
56. $\sqrt{-5}$
57. $\sqrt{5} - \sqrt{-5}$
58. 0
59. i^2
60. $50i^2 - i$
61. $-i$
62. -1

Perform the addition or subtraction as indicated. All variables represent real numbers.

63. $(13 - 7i) + (15 + 11i)$
64. $(16 + 17i) - (25 - 19i)$
65. $(4m + 5mi) - (m - 2ni) + (3m - 7ni)$
66. Is the sum of any two complex numbers always a complex number?
67. Is the sum of any two imaginary numbers always a complex number?
68. Solve $(x - 2)^2 + 7 = 0$.
69. Solve $16y^4 - 1 = 0$.

70. Let s and t be real numbers. Solve the following equation for s and t:
$(3 - s) + (5 - 2t)i = 1 + (s - t)i$

[5.7] 71. Simplify $3i^{48} - 5i^{65} + 2i^{82}$.

72. Simplify $i^{-82} + 2i^{-65} - 3i^{-48}$

Perform each operation and simplify.

73. $(-6i)(-3 + 2i)$

74. $(2 - 5i)(2 + 5i)$

75. $(-\sqrt{7} + 6i)^2$

76. $(\sqrt{5} + \sqrt{2}\,i)(\sqrt{2} + \sqrt{5}\,i)$

77. $\dfrac{4 - \sqrt{5}\,i}{2i}$

78. $\dfrac{1}{2 + 3i}$

79. $\dfrac{8 + 5i}{2i + 11}$

80. $\dfrac{1}{(7 + 9i)^2}$

81. Find the multiplicative inverse of $9 - 2i$.

82. Find all real values of x and y such that
$$\dfrac{x^2 - 3i}{1 + i} = \dfrac{1}{2} + yi$$

83. Let $z = -3 - 5i$. Find the value of $z^2 + 6z + 34$.

84. Is $-3 + 5i$ a solution of the equation $x^2 + 6x + 34 = 0$?

Practice Test (60 minutes)

1. Simplify each root, where a, b, and c represent nonzero real numbers.

   (a) $\sqrt[3]{-125}$ (b) $\sqrt[4]{81a^8}$ (c) $\sqrt{\dfrac{121b^8}{c^4 + 10c^2 + 25}}$

2. Simplify and express each result with positive exponents. All variables represent positive real numbers.
   (a) $(-27)^{-2/3}$
   (b) $(-8s^{-12}t^{-6})^{-2/3}$
   (c) $(k^{-1/4})^{-8} + (2^{-3}k^{-6})^{-1/3}$

3. Simplify. $7\sqrt[3]{-125} - 4\sqrt[3]{-128}$.

4. Find $(5\sqrt{2} + 6\sqrt{3})(4\sqrt{2} + 7\sqrt{3})$.

5. Reduce $(10\sqrt{30} - 9\sqrt{54})/(6\sqrt{6})$ to lowest terms.

6. Rationalize the denominator and simplify.

$$\dfrac{2\sqrt{5}}{3\sqrt{7} - 6\sqrt{8}}$$

7. Solve $\sqrt{2x+6} - \sqrt{2x} + 3 = 0$.
8. Solve $\sqrt{11 + 2\sqrt{y}} - \sqrt{y} = 1$.
9. Simplify $5i^{49} - 4i^{101} + 2i^{205}$.
10. Find all real values of u and v such that
$$\frac{3 + u^2 i}{1 + i} = -v + \frac{i}{2}$$

Extended Applications

Mathematics in Biology

The body surface area of an animal is an important factor that is considered by physiologists in studies involving heat dispersion from the animal's body. It is a boring and tedious task to measure the body surface area of the animals in an experiment. The difficulty has been greatly reduced by an approximation formula in which the surface area is related to two more easily measured factors. In the case of dogs, an empirical formula for the approximate surface area, A, was established by G. R. Cowgill and D. L. Drabkin.

$$A \approx (4.381)V^{0.425}L^{0.725} \qquad (*)$$

where

V = the volume of the dog in cubic centimeters,

L = the length of the dog measured in centimeters, and

A is given in square centimeters.

When calculating the surface area of a dog by using Formula (*), a scientific calculator is recommended.

EXAMPLE: Calculate the body surface area of a dog of volume 7350 cm³ and $L = 44$ cm.

Solution: Using Formula (*) and a scientific calculator, the area

$$A \approx 4.381 \times (7350)^{0.425} \times (44)^{0.725}$$
$$\approx 4.381 \times 43.97 \times 15.54$$
$$\approx 2993.51 \text{ cm}^2$$

In the case of a human, an empirical formula for the body surface area was devised by DuBois and DuBois:

$$A \approx (0.0071184)W^{0.425}H^{0.725}$$

Where

W = the weight of the body in kilograms,

H = the height in centimeters, and

A is given in square meters.

Exercises

1. Find the body surface area of Miss America if her weight is 59 kilograms and her height is 1.68 meters.

2. If Miss America gained 2 kg, by how much would her surface area (in sq cm) increase?

3. A teenager weighing 55 kg increased in height from 5'2" to 5'5" in one year without increasing in weight. How much of an increase in surface area is there?

6

Second-Degree Equations and Inequalities

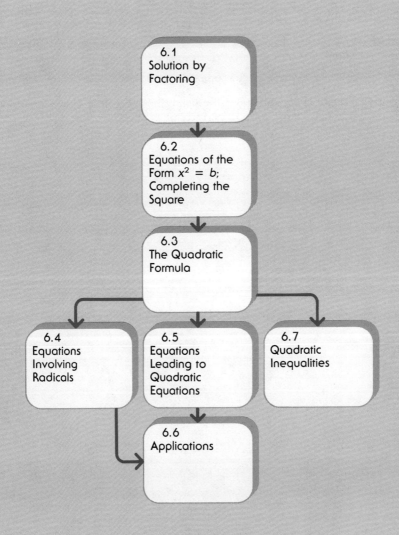

6.1 Solution by Factoring

In Chapter 2 we solved first-degree equations in one variable. We now turn our attention to second-degree equations in one variable; these are called **quadratic equations** in the variable. As with linear equations, there is a **standard form** for such equations, namely

$$ax^2 + bx + c = 0$$

where $a \neq 0$, and a, b, and c are constants representing real or complex numbers.

In Section 2.1 we solved linear equations by performing elementary operations that resulted in equivalent equations. These techniques apply to quadratic and higher degree equations as well. For example, $2x^2 + 3 = -5x$ is equivalent to $2x^2 + 5x + 3 = 0$, which is in standard form, where $a = 2$, $b = 5$, and $c = 3$.

If the left-hand side of a quadratic equation is factorable, we can use the following property to solve the equation. If a and b are real (or complex) numbers,

$$a \cdot b = 0 \quad \text{if and only if} \quad a = 0 \text{ or } b = 0$$

This property is easily established. If either $a = 0$ or $b = 0$, then $a \cdot b = 0$ by the zero factor property (Section 1.3). On the other hand, if $a \cdot b = 0$ and $a \neq 0$ we can show that $b = 0$ as follows. If $a \neq 0$, then $1/a$ exists:

$$a \cdot b = 0$$

$$\frac{1}{a} \cdot a \cdot b = \frac{1}{a} \cdot 0 \qquad \text{Multiply by } \frac{1}{a}$$

$$b = 0 \qquad \frac{1}{a} \cdot a = 1$$

as desired.

It should now be clear that a product involving any number of factors equals zero if and only if at least one of the factors is zero.

EXAMPLE 1 Solve $(x + 3)(x - 2) = 0$.

Solution: Because the product is zero, one of the factors must be zero, and thus either $x + 3 = 0$ or $x - 2 = 0$. In the first case, $x = -3$; in the second case, $x = 2$. Thus, both 2 and -3 are solutions of the original equation and the solution set is $\{-3, 2\}$.

EXAMPLE 2

Solve $x^2 + 9x + 20 = 0$.

Solution: First we factor
$$x^2 + 9x + 20 = (x + 4)(x + 5)$$

Now
$$(x + 4)(x + 5) = 0$$

if and only if $\quad x + 4 = 0 \quad$ or $\quad x + 5 = 0$

$\qquad\qquad\qquad\quad x = -4 \quad$ or $\quad x = -5$

The solution set is $\{-4, -5\}$.

EXAMPLE 3

Solve $2y^2 + 13y - 7 = 0$.

Solution: First we factor
$$2y^2 + 13y - 7 = (2y - 1)(y + 7)$$

Now
$$(2y - 1)(y + 7) = 0$$

if and only if $\quad 2y - 1 = 0 \quad$ or $\quad y + 7 = 0$

$\qquad\qquad\qquad\quad 2y = 1 \quad$ or $\quad y = -7$

$\qquad\qquad\qquad\quad y = \dfrac{1}{2} \quad$ or $\quad y = -7$

The solution set is $\left\{\dfrac{1}{2}, -7\right\}$.

EXAMPLE 4

Solve $3k^2 - 18k = 0$.

Solution: First notice that this is in standard form: $ak^2 + bk + c = 0$ where the constant term $c = 0$.
Factor
$$3k^2 - 18k = 3k(k - 6)$$

Now
$$3k(k - 6) = 0$$

if and only if $\quad 3k = 0 \quad$ or $\quad k - 6 = 0$

$\qquad\qquad\qquad\quad k = 0 \quad$ or $\quad k = 6$

The solution set is $\{0, 6\}$.

EXAMPLE 5

Solve $2t(t - 4) - t^2 = (2t - 5)^2 - 16$.

Solution: First expand both sides to obtain the standard form.

$$2t^2 - 8t - t^2 = 4t^2 - 20t + 25 - 16$$
$$t^2 - 8t = 4t^2 - 20t + 9 \qquad \text{Simplify}$$
$$-3t^2 + 12t - 9 = 0 \qquad \text{Standard form}$$

Dividing by -3

$$t^2 - 4t + 3 = 0 \qquad \text{Standard form}$$
$$(t - 1)(t - 3) = 0 \qquad \text{Factor}$$

Thus

$$t - 1 = 0 \quad \text{or} \quad t - 3 = 0$$
$$t = 1 \quad \text{or} \quad t = 3$$

The solution set is $\{1, 3\}$.

EXAMPLE 6 Solve $x^2 + 2x + 1 = 0$.

Solution: First we factor.

$$x^2 + 2x + 1 = 0$$

Now

$$(x + 1)(x + 1) = 0$$

if and only if

$$x + 1 = 0 \quad \text{or} \quad x + 1 = 0$$
$$x = -1 \quad \text{or} \quad x = -1$$

The solution set is $\{-1\}$.

When a quadratic equation has only one root, as in Example 6, we say that the root is of **multiplicity two.**

Let us note that the equation $(x - r_1)(x - r_2) = 0$ has $\{r_1, r_2\}$ as its solution set. If we were asked to find a quadratic equation having the numbers r_1 and r_2 as solutions, we would write the previous equation. Expanding the left-hand side will then convert the equation to standard form. The symbol r_1 is read "r sub one," or "r one"; r_2 is read "r sub two" or "r two."

EXAMPLE 7 Find a quadratic equation having the given solution set.

(a) Solution set $\{2, -3\}$ (b) Solution set $\left\{-\dfrac{1}{2}, \dfrac{2}{3}\right\}$

Solution:
(a) Write

$$(x - 2)(x - [-3]) = 0 \qquad r_1 = 2,\ r_2 = -3$$
$$(x - 2)(x + 3) = 0$$
$$x^2 + x - 6 = 0 \qquad \text{Standard form}$$

(b) Write

$$\left(x - \left[-\frac{1}{2}\right]\right)\left(x - \frac{2}{3}\right) = 0 \qquad r_1 = -\frac{1}{2}, \, r_2 = \frac{2}{3}$$

$$\left(x + \frac{1}{2}\right)\left(x - \frac{2}{3}\right) = 0$$

$$2 \cdot 3\left(x + \frac{1}{2}\right)\left(x - \frac{2}{3}\right) = 2 \cdot 3 \cdot 0 \qquad \text{Multiply by } 2 \cdot 3 \text{ to clear fractions}$$

$$2\left(x + \frac{1}{2}\right)3\left(x - \frac{2}{3}\right) = 0 \qquad \text{Rearrange factors}$$

$$(2x + 1)(3x - 2) = 0 \qquad \text{Distribute}$$

$$6x^2 - x - 2 = 0 \qquad \text{Standard form}$$

EXAMPLE 8   A box-shaped refrigeration unit (see Figure 6.1) houses vaccines in an upper compartment and medical specimens in a lower compartment. It is three times as high as it is wide and is 3 feet deep. The compartment divider is 2 feet below the top of the unit and the volume of the lower level compartment is 24 cubic feet. Find the dimensions of the unit.

Figure 6.1

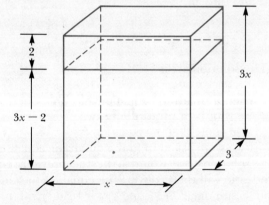

Solution:
Let x = width, $3x$ = height of whole unit
Volume of lower compartment = (height) × (width) × (depth)
$$= (3x - 2) \cdot x \cdot 3$$

(a) $(3x - 2) \cdot x \cdot 3 = 24$
$(3x - 2) \cdot x = 8$
$3x^2 - 2x - 8 = 0$
$(3x + 4)(x - 2) = 0$

(b) $3x + 4 = 0$ or $x - 2 = 0$
$x = -\dfrac{4}{3}$ or $x = 2$

As $x = -\dfrac{4}{3}$ does not make sense in this problem, we must have $x = 2$. Consequently, the width is 2 ft, the height is $3(2) = 6$ ft, and the depth is 3 ft.

(c) *Check:* Does height = $3 \cdot$ (width)? Does $6 = 3 \cdot 2$? Yes!
Lower compartment height = $6 - 2 = 4$.
Does the lower compartment volume = 24 cubic feet?
Does $4 \cdot 2 \cdot 3 = 24$? Yes!

EXAMPLE 9

A deck of uniform width is constructed around a swimming pool that is 25 meters long and 20 meters wide. If the deck area is 550 square meters, find its width. (See Figure 6.2.)

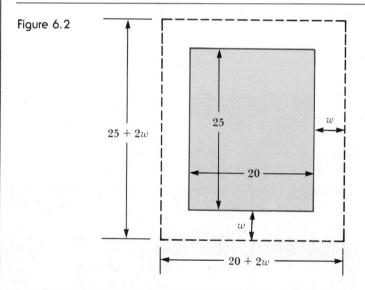

Figure 6.2

Solution:
Let w = width of deck
Area of deck = $(20 + 2w)(25 + 2w) - 20 \cdot 25$

(a) $\qquad 550 = (20 + 2w)(25 + 2w) - 20 \cdot 25$
$\qquad\qquad 550 = 20 \cdot 25 + 90w + 4w^2 - 20 \cdot 25$
$4w^2 + 90w - 550 = 0$
$(4w + 110)(w - 5) = 0$
$4w + 110 = 0$ or $w - 5 = 0$

(b) $\quad w = -\dfrac{110}{4} \quad$ or $\quad w = 5$

The negative solution $-\dfrac{110}{4}$ does not make sense in this problem and is discarded. The width is 5 meters.

(c) *Check:* Length of pool plus deck = 25 + 10 = 35
Width of pool plus deck = 20 + 10 = 30
Deck Area = 30 · 35 − 20 · 25 = 1050 − 500 = 550 sq m

Exercises 6.1

Standard Assignment: Exercises 3, 5, 9, 11, 13, 15, 17, 21, 25, 31, 37, 43, 45, 51, 53, 57, 61, 63

Solve the following equations by inspection or as in Example 1.

A
1. $(x + 2)(x + 5) = 0$
2. $(y - 3)(y + 7) = 0$
3. $(z + 1)(2z - 3) = 0$
4. $(x - 3)(x - 3) = 0$
5. $(2t + 5)(1 - 3t) = 0$
6. $x \cdot x = 0$
7. $k(k + 4) = 0$
8. $-y(2y + 1) = 0$
9. $5(r + 2)(2r - 9) = 0$
10. $-3(4x + 2)(3x + 1) = 0$
11. $z(z + 1)(z - 2) = 0$
12. $-3y(y + 6)(3y - 7) = 0$

Solve the following equations by factoring (after first putting the equation in standard form). See Examples 2–6.

13. $x^2 - x - 2 = 0$
14. $y^2 - 2y - 3 = 0$
15. $t^2 + 4t - 5 = 0$
16. $z^2 + 12 = 7z$
17. $m^2 = 5m - 6$
18. $p^2 + 8p = -15$
19. $x^2 - 11x + 10 = 0$
20. $u^2 - 8u + 16 = 0$
21. $9z + 18 = -z^2$
22. $t^2 + 6t - 7 = 0$
23. $2y^2 - 3y + 1 = 0$
24. $6m^2 + m = 1$
25. $3x^2 - x - 2 = 0$
26. $2z^2 + 6 = 7z$
27. $2k^2 + 7k - 4 = 0$
28. $7r^2 + 10 = 19r$
29. $12p^2 + 28p - 5 = 0$
30. $8 + 6k - 5k^2 = 0$
31. $5 + 13m = 6m^2$
32. $5z^2 - 12z + 4 = 0$
33. $3t^2 + 10t + 7 = 0$
34. $-4 - 5x = -6x^2$
35. $3a^2 - 9 = 0$
36. $12y^2 - 11y + 2 = 0$

37. $u^2 - 7u = 0$
38. $m^2 = -5m$
39. $k(4k - 1) = 5$
40. $-3y^2 + 27 = 0$
41. $5z^2 + 25z = 0$
42. $r^2 = 16$
43. $3x^2 + 6x - 14 = 5x - x^2$
44. $3y(y + 10) - 9y = 2y(y - 1)$
45. $(z + 1)^2 - 4 = 5(z - 1)$
46. $(2r + 5)(r + 1) = 9$
47. $-3(t - 3)(t + 1) = t(7 - 2t) - 3$
48. $(4k + 2)(k - 1) - 3k = 4k(4 - k) + 7$
49. $x - \dfrac{4}{x} = 0$
50. $\dfrac{5}{2y} - \dfrac{1}{y^2} = 1$

Find a quadratic equation (in standard form) that has the indicated solution set. See Example 7.

51. $\{2, 3\}$
52. $\{4, -1\}$
53. $\left\{-2, -\dfrac{1}{3}\right\}$
54. $\left\{\dfrac{1}{2}, 0\right\}$
55. $\{3\}$
56. $\{-5\}$
57. $\{-3, 0\}$
58. $\left\{-\dfrac{1}{2}, \dfrac{2}{3}\right\}$
59. $\left\{-7, \dfrac{3}{4}\right\}$

The following exercises involve geometric considerations similar to those in Examples 8 and 9.

60. A fence divides a large rectangular plot of land into two smaller plots. The rear section is a grazing area of 34,500 square feet. The depth of the plot is twice the width and the dividing fence is 70 feet from the front of the lot. Find the length and width of the plot of land.

61. A border of uniform width is constructed around a picture 75 cm high and 65 cm wide. If the area of the border is 5100 sq cm, find its width.

62. A topless box is to be made from a square piece of cardboard by cutting 2-inch squares from each of the four corners. If the box is to have a volume of 200 cubic inches, find the size of the piece of cardboard that must be used.

63. The area of a rectangle is 240 sq cm. If the length is increased by 2 cm and the width by 4 cm, the new area is 336 sq cm. Find the length and width of the original rectangle. (There are *two* such rectangles.)

B Solve each of the following equations for x in terms of a and b.

EXAMPLE: Solve $2x^2 = -bx + 3b^2$ for x.

Solution: $2x^2 + bx - 3b^2 = 0$ Rewrite in standard form

$(2x + 3b)(x - b) = 0$ Factor

$$2x + 3b = 0 \quad \text{or} \quad x - b = 0$$
$$x = \frac{-3b}{2} \quad \text{or} \quad x = b$$

The solution set is $\left\{b, \dfrac{-3b}{2}\right\}$.

64. $x^2 = 25a^2$

65. $4x^2 - 49b^2 = 0$

66. $x^2 - 5bx = -6b^2$

67. $x^2 - 8ax + 16a^2 = 0$

68. $3x^2 = -7bx - 4b^2$

69. $12x^2 - 11bx + 2b^2 = 0$

70. $x^2 - (2 + b)x + 2b = 0$

71. $x^2 - (a + b)x + ab = 0$

6.2 Solution of Equations of the Form $x^2 = b$; Completing the Square

We can solve any quadratic equation of the form $x^2 = b$ by the **square root method**. Using the fact that $b = (\sqrt{b})^2$, if $b \geq 0$

$$x^2 = b$$
$$x^2 - b = 0$$
$$x^2 - (\sqrt{b})^2 = 0$$
$$(x - \sqrt{b})(x + \sqrt{b}) = 0$$
$$x - \sqrt{b} = 0 \quad \text{or} \quad x + \sqrt{b} = 0$$
$$x = \sqrt{b} \quad \text{or} \quad x = -\sqrt{b}$$

The solution set is $\{-\sqrt{b}, \sqrt{b}\}$. Thus, $x^2 = b$ if and only if $x = \pm\sqrt{b}$.

EXAMPLE 1 Solve (a) $x^2 = 7$ (b) $6y^2 - 102 = 0$ (c) $16z^2 = 0$

Solution:

(a) $x^2 = 7$ if and only if $x = \pm\sqrt{7}$. The solution set is $\{-\sqrt{7}, \sqrt{7}\}$.

(b) $6y^2 - 102 = 0$
$6y^2 = 102$
$y^2 = 17$
$y = \pm\sqrt{17}$

The solution set is $\{-\sqrt{17}, \sqrt{17}\}$.

(c) $16z^2 = 0$
$z^2 = 0$
$z = \pm\sqrt{0} = 0$

The solution set is $\{0\}$.

When $b < 0$ we write the solutions using $\sqrt{-1} = i$.

6.2 Solution of Equations of the Form $x^2 = b$; Completing the Square

EXAMPLE 2 Solve (a) $x^2 = -5$ (b) $2y^2 + 1 = 0$

Solution:

(a) $x^2 = -5$

$x = \pm\sqrt{-5}$

$x = -\sqrt{5}\,i$

The solution set is $\{-\sqrt{5}\,i, \sqrt{5}\,i\}$.

(b) $2y^2 + 1 = 0$

$2y^2 = -1$

$y^2 = -\dfrac{1}{2}$

$y = \pm\sqrt{-\dfrac{1}{2}} = \pm\dfrac{1}{\sqrt{2}}\,i = \pm\dfrac{\sqrt{2}}{2}\,i$

The solution set is $\left\{-\dfrac{\sqrt{2}}{2}\,i, \dfrac{\sqrt{2}}{2}\,i\right\}$.

The previous method may also be easily extended to solve equations of the form $(x - a)^2 = b$ or $(ax - b)^2 = c$.

EXAMPLE 3 Solve (a) $(x + 5)^2 = 27$ (b) $(4k + 1)^2 = 3$ (c) $(y - 3)^2 = -49$

Solution: In order to solve (a), we could substitute z for $x + 5$.

$z^2 = 27$ Substitute z for $x + 5$

$z = \sqrt{27}$ or $z = -\sqrt{27}$

that is,

$x + 5 = \sqrt{27}$ or $x + 5 = -\sqrt{27}$

$x = -5 + \sqrt{27}$ or $x = -5 - \sqrt{27}$

The solution set is $\{-5 + \sqrt{27}, -5 - \sqrt{27}\}$. Because $27 = 9 \cdot 3$, $\sqrt{27} = \sqrt{9 \cdot 3} = \sqrt{9}\sqrt{3} = 3\sqrt{3}$, so we can also write

$\{-5 + 3\sqrt{3}, -5 - 3\sqrt{3}\}$

for the solution set.

Usually the substitution made in (a) is done mentally. Thus for (b)

$(4k + 1)^2 = 3$

$4k + 1 = \sqrt{3}$ or $4k + 1 = -\sqrt{3}$

$4k = -1 + \sqrt{3}$ or $4k = -1 - \sqrt{3}$

$k = \dfrac{-1 + \sqrt{3}}{4}$ or $k = \dfrac{-1 - \sqrt{3}}{4}$

The solution set is $\{(-1 + \sqrt{3})/4, (-1 - \sqrt{3})/4\}$.

242   Chap. 6 Second-Degree Equations and Inequalities

For (c)
$$(y - 3)^2 = -49$$
$$y - 3 = \sqrt{49}\, i \quad \text{or} \quad y - 3 = -\sqrt{49}\, i$$
$$y - 3 = 7i \quad \text{or} \quad y - 3 = -7i$$
$$y = 3 + 7i \quad \text{or} \quad y = 3 - 7i$$

The solution set is $\{3 + 7i,\ 3 - 7i\}$.

EXAMPLE 4 Suppose $20,000 is invested at an annual interest rate of r (decimal rate) and interest is compounded annually. After 2 years, the investment has earned $4200 in interest. Find r.

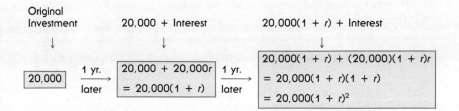

The diagram shows us that the original investment of $20,000 becomes $20,000(1 + r)$ after 1 year and at the end of 2 years, the original investment has grown to $20,000(1 + r)^2$. Thus

(a) $20,000(1 + r)^2 = 20,000 + 4200$

$$(1 + r)^2 = \frac{24,200}{20,000}$$

$$(1 + r)^2 = 1.21$$

$$1 + r = \pm\sqrt{1.21}$$

$$1 + r = \pm 1.1$$

$$r = -1 \pm 1.1$$

(b) $r = -2.1 \quad$ or $\quad r = 0.1$

As -2.1 does not make sense as an interest rate, $r = 0.1$ or 10%.

(c) *Check.* $20,000(1 + r)^2 = 20,000(1.1)^2$
$$= 20,000(1.21)$$
$$= 24,200$$

Interest $= 24,200 - 20,000 = 4200$ as given

Any quadratic equation can be solved first by rewriting the equation in the form $(x - p)^2 = q$ and then by using the square root method.

6.2 Solution of Equations of the Form $x^2 = b$; Completing the Square

In order to rewrite $x^2 - 6x - 5 = 0$ in the form $(x - p)^2 = q$, we proceed as follows.

$$x^2 - 6x - 5 = 0$$
$$x^2 - 6x = 5$$
$$x^2 - 6x + 9 = 5 + 9 \quad \text{Add 9 to both sides}$$
$$(x - 3)^2 = 14$$
$$x - 3 = \sqrt{14} \quad \text{or} \quad x - 3 = -\sqrt{14}$$
$$x = 3 + \sqrt{14} \quad \text{or} \quad x = 3 - \sqrt{14}$$

The solution set is $\{3 + \sqrt{14}, 3 - \sqrt{14}\}$. How was 9 chosen? From the expansion of $(x - p)^2$ we see that $(x - p)^2 = x^2 - 2px + p^2$. There is a relation between the coefficient, $-2p$, of the "x term" and the constant p^2. Namely, $p^2 = $ the square of half of $-2p$.

$$\left[\frac{1}{2}(-2p)\right]^2 = (-p)^2 = p^2$$

Starting with $x^2 - 6x$, we **complete the square** by adding the square of half of -6, the coefficient of the "x term." Since $[\frac{1}{2}(-6)]^2 = (-3)^2 = 9$, we added 9 to $x^2 - 6x$ to obtain $(x - 3)^2$; the same number 9 must be added to the other side so that the equation remains equivalent.

EXAMPLE 5 Solve $k^2 + 10k + 8 = 0$.

Solution: First, isolate the constant on one side of the equation.

$$k^2 + 10k + 8 = 0$$
$$k^2 + 10k = -8$$

Next, add $[\frac{1}{2}(10)]^2 = 5^2 = 25$ to both sides.

$$k^2 + 10k + 25 = -8 + 25$$
$$(k + 5)^2 = 17$$
$$k + 5 = \sqrt{17} \quad \text{or} \quad k + 5 = -\sqrt{17}$$
$$k = -5 + \sqrt{17} \quad \text{or} \quad k = -5 - \sqrt{17}$$

The solution set is $\{-5 + \sqrt{17}, -5 - \sqrt{17}\}$.

EXAMPLE 6 Solve $t^2 - 4t + 4 = 0$.

Solution: Caution: $[\frac{1}{2}(-4)]^2 = (-2)^2 = 4$, so the left side is already a perfect square.

$$t^2 - 4t + 4 = 0$$
$$(t - 2)^2 = 0$$
$$t - 2 = 0$$
$$t = 2$$

The solution set is $\{2\}$.

The technique of completing the square will also work in this example but it adds unnecessary steps.

If we want to complete the square on a quadratic equation where the coefficient of the "square" term is not 1, we must first divide both sides by this coefficient. We will illustrate and summarize this technique next.

Completing the Square	Illustration
Given: $ax^2 + bx + c = 0$	$2x^2 - 9x + 6 = 0$
1. $a \neq 0$ Divide both sides by a. $x^2 + \frac{b}{a}x + \frac{c}{a} = 0$	1. $x^2 - \frac{9}{2}x + 3 = 0$
2. Rewrite the equation by isolating the constant on the right side: $x^2 + \frac{b}{a}x = -\frac{c}{a}$	2. $x^2 - \frac{9}{2}x = -3$
3. Square half the coefficient of x: $\left(\frac{1}{2} \cdot \frac{b}{a}\right)^2 = \left(\frac{b}{2a}\right)^2$	3. $\left[\frac{1}{2}\left(-\frac{9}{2}\right)\right]^2 = \frac{81}{16}$
4. Add the square to both sides. $x^2 + \frac{b}{a}x + \left(\frac{b}{2a}\right)^2 = -\frac{c}{a} + \left(\frac{b}{2a}\right)^2$	4. $x^2 - \frac{9}{2}x + \frac{81}{16} = -3 + \frac{81}{16}$ $= \frac{-48}{16} + \frac{81}{16}$ $= \frac{33}{16}$
5. The left side is the perfect square. $\left(x + \frac{b}{2a}\right)^2 = \frac{b^2 - 4ac}{(2a)^2}$	5. $\left(x + \frac{-9}{4}\right)^2 = \frac{33}{16}$ $\left(x - \frac{9}{4}\right)^2 = \frac{33}{16}$
6. Complete the solution by the square root method.	6. $x - \frac{9}{4} = \sqrt{\frac{33}{16}}$ or $x - \frac{9}{4} = -\sqrt{\frac{33}{16}}$ $x - \frac{9}{4} = \frac{\sqrt{33}}{4}$ or $x - \frac{9}{4} = -\frac{\sqrt{33}}{4}$ $x = \frac{9}{4} + \frac{\sqrt{33}}{4}$ or $x = \frac{9}{4} - \frac{\sqrt{33}}{4}$ $x = \frac{9 + \sqrt{33}}{4}$ or $x = \frac{9 - \sqrt{33}}{4}$

6.2 Solution of Equations of the Form $x^2 = b$; Completing the Square

Exercises 6.2

Standard Assignment: Exercises 3, 5, 7, 9, 11, 13, 15, 17, 19, 23, 25, 27, 33, 39, 45, 47

A Solve each of the following equations. See Examples 1–3.

1. $y^2 = 25$
2. $k^2 = 10{,}000$
3. $x^2 = \dfrac{1}{16}$
4. $t^2 = \dfrac{4}{81}$
5. $m^2 = -4$
6. $z^2 = 14$
7. $q^2 + 1 = 0$
8. $y^2 + 4 = 0$
9. $a^2 - 4 = 0$
10. $(y - 1)^2 = 1$
11. $(k + 2)^2 = 1$
12. $(t - 1)^2 = 2$
13. $(m - 5)^2 = 3$
14. $(z - 5)^2 = -3$
15. $(a - 7)^2 = -4$
16. $(2x - 4)^2 = 4$
17. $(-2y - 4)^2 = 4$
18. $(2x - 3)^2 = -36$

Solve the following equations by completing the square. See Examples 5 and 6.

19. $x^2 - 4x + 5 = 0$
20. $y^2 + 6y = -7$
21. $12t^2 + t = 6$
22. $4k - 1 = 2k^2$
23. $25m^2 + 1 = 10m$
24. $4z^2 - 17z + 15 = 0$
25. $q^2 - 10q + 25 = 0$
26. $2y^2 - 12y + 8 = 0$
27. $3a^2 + 3a = 18$
28. $2r^2 + 3r = 9$
29. $3k^2 - 5k + 1 = 0$
30. $z^2 - 2z + 2 = 0$
31. $t^2 = 2t - 17$
32. $3x^2 - x - 3 = -1$
33. $16q^2 + 40q + 9 = 0$
34. $3m^2 + 2m + 1 = 0$
35. $3r^2 + 3r + 1 = 0$
36. $4y^2 + 4y + 5 = 0$
37. $2r^2 - 7r - 9 = 0$
38. $x^2 - 6x + 11 = 0$
39. $7z^2 - 2z - 2 = 0$
40. $2k^2 - k + 1 = 0$
41. $3y^2 = 4y + 1$
42. $2t^2 = 3t + 1$
43. $3r^2 + 2r - 7 = 0$
44. $2x^2 + 5x = 15 - 2x$
45. $5a^2 - 6a = 4a^2 + 6a - 3$
46. $x^2 + 7x + 5 = x - x^2$

Solve the following problems.

47. Follow Example 4 closely and show that if a principal of P dollars is invested at a (decimal) interest rate r, and interest is compounded annually, then $P(1 + r)^2$ dollars will be accumulated after 2 years.

48. How much interest will \$40,000 earn over 2 years if it is invested at 10% interest that is compounded annually?

B Solve for x in terms of a and/or b, where a and b are positive.

49. $2x^2 - b = 0$
50. $(x - b)^2 = 0$
51. $(x - b)^2 = 25$
52. $4x^2 - 9b = 0$
53. $(3x + a)^2 = -1$
54. $(2x + a)^2 = 16b$
55. $x^2 + a^2 = 1$
56. $x^2 + 16 = b^2$
57. $x^2 + a^2 = b^2$
58. $a^2 + 9 = x^2$
59. $a^2 + b^2 = x^2$
60. $(1 + x)^2 = a$

6.3 The Quadratic Formula

It was demonstrated in Section 6.2 that the method of completing the square can be used to solve any quadratic equation. In this section we will use this method to solve the general quadratic equation and produce a formula that can then be used to find a solution to *any* given quadratic equation. The formula is much easier and faster to use than the method of completing the square.

Given the equation

$$ax^2 + bx + c = 0 \quad (a \neq 0)$$

We solve the equation by completing the square.

1. Divide by a.

$$x^2 + \frac{b}{a}x + \frac{c}{a} = 0$$

2. Isolate the constant on the right-hand side.

$$x^2 + \frac{b}{a}x = -\frac{c}{a}$$

3. Add the square of half the coefficient of x to both sides.

$$x^2 + \frac{b}{a}x + \left(\frac{b}{2a}\right)^2 = \left(\frac{b}{2a}\right)^2 - \frac{c}{a}$$

4. The left side is a perfect square.

$$\left(x + \frac{b}{2a}\right)^2 = \frac{b^2}{4a^2} - \frac{c}{a}$$

$$\left(x + \frac{b}{2a}\right)^2 = \frac{b^2 - 4ac}{4a^2}$$

5. Complete the solution by the square root method.

$$x + \frac{b}{2a} = \pm\sqrt{\frac{b^2 - 4ac}{4a^2}}$$

Thus $\quad x = \dfrac{-b}{2a} + \dfrac{\sqrt{b^2 - 4ac}}{2a} \quad$ or $\quad x = \dfrac{-b}{2a} - \dfrac{\sqrt{b^2 - 4ac}}{2a}$

We can abbreviate the two equations by using the sign $\pm$ and writing the solution as a single fraction. The solutions of the equation $ax^2 + bx + c = 0$ are given by

> **The Quadratic Formula**
> $$x = \frac{-b \pm \sqrt{b^2 - 4ac}}{2a}$$

In order to use the **quadratic formula**, the given quadratic equation must first be put into *standard form*, $ax^2 + bx + c = 0$.

EXAMPLE 1

Solve $3x^2 = 5x + 2$ by the quadratic formula.

Solution: Rewrite

$$\begin{array}{ccc} ax^2 + & bx + & c = 0 \\ \downarrow & \downarrow & \downarrow \\ 3x^2 - & 5x - & 2 = 0 \end{array}$$ Standard form

$3x^2 + (-5)x + (-2) = 0$

$a = 3, b = -5,$ and $c = -2$ In the quadratic formula

$$x = \frac{-b \pm \sqrt{b^2 - 4ac}}{2a}$$

$$= \frac{-(-5) \pm \sqrt{(-5)^2 - 4(3) \cdot (-2)}}{2(3)}$$

$$= \frac{5 \pm \sqrt{25 + 24}}{6}$$

$$= \frac{5 \pm \sqrt{49}}{6}$$

$$= \frac{5 \pm 7}{6}$$

So $x = \dfrac{5 + 7}{6} = \dfrac{12}{6} = 2$ or $x = \dfrac{5 - 7}{6} = \dfrac{-2}{6} = -\dfrac{1}{3}$

The solution set is $\left\{-\dfrac{1}{3}, 2\right\}$.

EXAMPLE 2 Solve $k^2 - 2k = 2$.

Solution: Rewrite

$$k^2 - 2k - 2 = 0 \qquad \text{Standard form}$$

$a = 1$, $b = -2$, and $c = -2$

$$k = \frac{-b \pm \sqrt{b^2 - 4ac}}{2a}$$

$$= \frac{-(-2) \pm \sqrt{(-2)^2 - 4(1) \cdot (-2)}}{2(1)}$$

$$= \frac{2 \pm \sqrt{4 + 8}}{2}$$

$$= \frac{2 \pm \sqrt{12}}{2}$$

$$= \frac{2 \pm \sqrt{4 \cdot 3}}{2}$$

$$= \frac{2 \pm 2\sqrt{3}}{2}$$

So $\qquad k = \dfrac{2 + 2\sqrt{3}}{2} = \dfrac{2(1 + \sqrt{3})}{2} = 1 + \sqrt{3}$

or $\qquad k = \dfrac{2 - 2\sqrt{3}}{2} = \dfrac{2(1 - \sqrt{3})}{2} = 1 - \sqrt{3}$

The solution set is $\{1 + \sqrt{3},\ 1 - \sqrt{3}\}$.

EXAMPLE 3 The area of a square metal plate doubles and its sides increase in length by 1 cm when the plate is heated. Find the length of a side of the (unheated) square.

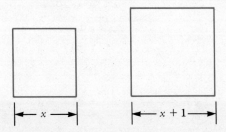

Solution:

Let $\qquad x = $ side length before heating (in cm)

$x + 1 = $ side length of heated square (in cm)

$$x^2 = \text{area of unheated square}$$
$$2x^2 = \text{double the area of the unheated square}$$
$$(x + 1)^2 = \text{area of the heated square}$$

Thus
$$(x + 1)^2 = 2x^2$$
$$x^2 + 2x + 1 = 2x^2$$
$$x^2 - 2x - 1 = 0$$

$$x = \frac{-(-2) \pm \sqrt{(-2)^2 - 4(1)(-1)}}{2(1)}$$

$$= \frac{2 \pm \sqrt{4 + 4}}{2}$$

$$= \frac{2 \pm \sqrt{8}}{2}$$

$$= \frac{2 \pm 2\sqrt{2}}{2}$$

$$= 1 \pm \sqrt{2}$$

Negative length is nonsense, so we discard $x = 1 - \sqrt{2}$. Thus $x = 1 + \sqrt{2} \approx 2.41421$ (from a calculator).

Check: The equation for doubling the area of the unheated square is
$$2x^2 = 2(1 + \sqrt{2})^2 = 2(1 + 2\sqrt{2} + 2) = 6 + 4\sqrt{2} \text{ cm}^2$$

The equation for the area of the heated square is
$$(x + 1)^2 = (1 + \sqrt{2} + 1)^2 = (2 + \sqrt{2})^2$$
$$= 4 + 4\sqrt{2} + 2 = 6 + 4\sqrt{2} \text{ cm}^2$$

Therefore the areas are equal.

EXAMPLE 4

Solve $2r(r + 6) = -19$.

Solution: Rewrite
$$2r^2 + 12r + 19 = 0 \quad \text{Standard form}$$
where $a = 2, b = 12, \text{ and } c = 19$

$$r = \frac{-b \pm \sqrt{b^2 - 4ac}}{2a}$$

$$= \frac{-12 \pm \sqrt{(12)^2 - 4(2) \cdot (19)}}{2(2)}$$

$$= \frac{-12 \pm \sqrt{144 - 152}}{4}$$

$$= \frac{-12 \pm \sqrt{-8}}{4}$$

$$= \frac{-12 \pm 2\sqrt{2}\,i}{4} \qquad \sqrt{-8} = \sqrt{4}\sqrt{2}\sqrt{-1}$$

$$= \frac{2(-6 \pm \sqrt{2}\,i)}{4}$$

$$= \frac{-6 \pm \sqrt{2}\,i}{2}$$

$$r = \frac{-6 + \sqrt{2}\,i}{2} \qquad \text{or} \qquad r = \frac{-6 - \sqrt{2}\,i}{2}$$

The solution set is $\{(-6 + \sqrt{2}\,i)/2,\ (-6 - \sqrt{2}\,i)/2\}$.

We prefer to express complex numbers in the standard complex form $a + bi$, so the solution set is given as $\left\{ -3 + \dfrac{\sqrt{2}}{2}i,\ -3 - \dfrac{\sqrt{2}}{2}i \right\}$.

EXAMPLE 5 Solve $3iy^2 - y + 2i = 0$.

Solution: The equation is already in standard form, where $a = 3i$, $b = -1$, and $c = 2i$

$$y = \frac{-b \pm \sqrt{b^2 - 4ac}}{2a}$$

$$= \frac{1 \pm \sqrt{(-1)^2 - 4(3i)\cdot(2i)}}{2(3i)}$$

$$= \frac{1 \pm \sqrt{1 - 24i^2}}{6i}$$

$$= \frac{1 \pm \sqrt{25}}{6i}$$

$$= \frac{1 \pm 5}{6i}$$

Thus,

$$y = \frac{1 + 5}{6i} = \frac{6}{6i} = \frac{1}{i} \qquad \text{or} \qquad y = \frac{1 - 5}{6i} = \frac{-4}{6i} = \frac{-2}{3i}$$

and

$$\frac{1}{i} = \frac{1}{i} \cdot \frac{(-i)}{(-i)} = \frac{-i}{1} = -i \qquad \text{and} \qquad \frac{-2}{3i} = \frac{-2}{3i} \cdot \frac{(-i)}{(-i)} = \frac{2i}{3} = \frac{2}{3}i$$

The solution set is $\left\{ -i,\ \dfrac{2}{3}i \right\}$, with roots given in the form $a + bi$.

If a, b, and c are real numbers, we can predict the number and type (real or imaginary) of solutions that the equation $ax^2 + bx + c = 0$ will have. The number $b^2 - 4ac$, which is called the **discriminant** of the equation, reveals the nature of the solution as shown in the following table.

Discriminant	Solutions
$b^2 - 4ac = 0$	one real root
$b^2 - 4ac > 0$	two unequal real roots
$b^2 - 4ac < 0$	two unequal imaginary roots

In the last solution, remember that a complex number $a + bi$ is imaginary when $b \neq 0$.

Examination of the form of the solutions will help to explain these results. We have

$$\frac{-b}{2a} + \frac{\sqrt{b^2 - 4ac}}{2a} \quad \text{and} \quad \frac{-b}{2a} - \frac{\sqrt{b^2 - 4ac}}{2a}$$

If $b^2 - 4ac = 0$, then $\sqrt{b^2 - 4ac} = 0$ and both expressions reduce to $\frac{-b}{2a}$, which is the only root.

If $b^2 - 4ac > 0$, then $\sqrt{b^2 - 4ac}$ is a nonzero real number, and we produce two different roots by first adding (and then subtracting) the real number $\frac{\sqrt{b^2 - 4ac}}{2a}$ to $\frac{-b}{2a}$.

If $b^2 - 4ac < 0$, then $\sqrt{b^2 - 4ac}$ is imaginary and, as in the previous example, it again produces two distinct roots, which are imaginary this time.

EXAMPLE 6   Examine the equations in Examples 2, 4, and 5, using the discriminant.

Example	Equation	$b^2 - 4ac$	Conclusion
2	$k^2 - 2k - 2 = 0$	$(-2)^2 - 4(1)(-2)$ $= 12 > 0$	Two unequal real roots
4	$2r^2 + 12r + 19 = 0$	$(12)^2 - 4(2)(19)$ $= -8 < 0$	Two unequal imaginary roots
5	$3iy^2 - y + 2i = 0$	The coefficients are not all real; therefore, no information is obtained from the discriminant.	

EXAMPLE 7   Use the discriminant to determine whether the equation $4x^2 + 4x + 1 = 0$ has one real root, two real unequal roots, or two unequal imaginary roots.

Solution:
Because $a = 4$, $b = 4$, and $c = 1$, $b^2 - 4ac = 4^2 - 4 \cdot (4) \cdot (1)$
$$= 16 - 16$$
$$= 0$$
Thus there is only one root, namely $\dfrac{-b}{2a} = \dfrac{-1}{2}$.

Some additional facts about the solutions of $ax^2 + bx + c = 0$ will be stated using the notation,
$$r_1 = \frac{-b}{2a} + \frac{\sqrt{b^2 - 4ac}}{2a}, \quad \text{and} \quad r_2 = \frac{-b}{2a} - \frac{\sqrt{b^2 - 4ac}}{2a}$$

It is always true that

$$r_1 + r_2 = \frac{-b}{a} \quad \text{and} \quad r_1 \cdot r_2 = \frac{c}{a}$$

and

$$ax^2 + bx + c = a(x - r_1)(x - r_2)$$

Verification is straightforward and will be left to the student. (Exercises 61–63.)

Exercises 6.3

Standard Assignment: Exercises 1, 3, 7, 11, 13, 17, 23, 25, 27, 29, 31, 33, 41, 43, 45, 47, 51, 53, 55

A Solve each of the following equations by using the quadratic formula. See Examples 1, 2, 4, and 5.

1. $y^2 + 2y - 4 = 0$
2. $m^2 + 2m + 2 = 0$
3. $6x^2 = 7x + 5$
4. $2r^2 + 3r - 9 = 0$
5. $k^2 + 3k + 5 = 0$
6. $2q^2 - 4q + 5 = 0$
7. $t^2 + 7 = 4t$
8. $3z^2 - 2z = 7$
9. $6y^2 + 11y = 10$

10. $3p^2 + 8p + 4 = 0$ 11. $8(x^2 - x) = x^2 - 3$ 12. $6r^2 + 5r = -1$
13. $7k^2 - 2k + 4 = 2k^2 + 6k$ 14. $10t^2 = 11t + 6$
15. $2p^2 + 7p - 4 = 0$ 16. $(3x - 2)^2 = 6x$ 17. $2(r^2 + 1) = -3r$
18. $4y^2 + 6y - 1 = 0$ 19. $5m = 2m^2 + 1$ 20. $2k^2 - 5k + 4 = 0$
21. $2t^2 - 49 = 0$ 22. $p^2 = 5(p - 1)$
23. $y^2 + 8y + 5 = y(2 - y)$ 24. $(z + 13)(z + 5) = -2$
25. $9k^2 + 25 = 0$ 26. $2r^2 + 3r = -3$
27. $3m^2 + 2m + 2 = 0$ 28. $5x^2 - 4x + 3 = 0$
29. $t(t + 1) = 3t^2 + 1$ 30. $3(x^2 + 1) = 2x^2 + 4x + 1$
31. $z^2 + 6z - 2 = 0$ 32. $x^2 - 2ix + 3 = 0$
33. $k^2 - 4ik - 1 = 0$ 34. $iy^2 - y + 2i = 0$
35. $z^2 = (2i + 2)z + (7 - 2i)$ 36. $it^2 + (10 - 6i)t = 10$

Find the solution to each of the following equations and round it to the nearest thousandth. You will need a calculator with a square root key.

37. $3w^2 - 5w - 4 = 0$ 38. $y^2 - 4y - 15 = 0$
39. $4t^2 + 3t - 2 = 0$ 40. $5z^2 + 11z + 3 = 0$
41. $4r^2 - 63.7 = 0$ 42. $8.91k^2 + 2.3k - 1 = 0$

Use the discriminant to determine whether each equation has one real root, two unequal roots, or two unequal imaginary roots. See Example 7.

43. $-3x^2 + 21 = 0$ 44. $7p^2 - 2p = 0$ 45. $6y^2 - 7y + 3 = 0$
46. $3k^2 + 7k + 6 = 0$ 47. $2z^2 + 12z + 8 = 0$ 48. $r^2 - 6r = 9$
49. $4t^2 - 17t + 15 = 0$ 50. $p^2 + 2 = 4p$ 51. $x^2 + 2x + \dfrac{1}{2} = 0$

Solve the following word problems. See Example 3.

52. The surface area A, of a cylinder with height h, and radius r, is given by the equation $A = 2\pi rh + 2\pi r^2$. Find the radius of a cylinder with height 2 cm and area 4π cm^2.

53. The diagonal of a square tile is 2 cm longer than its side. Find the length of the side. (Hint: Use the fact that the hypotenuse c, of a right triangle with legs a and b, satisfies the equation $c^2 = a^2 + b^2$.)

54. Six square photographs of identical size will cover the same area as nine square photographs of identical size. If the side of the smaller photograph is 4 inches less than the side of the larger photograph, what is the length of a side of the larger photograph?

B Use the factorization $ax^2 + bx + c = a(x - r_1)(x - r_2)$, where r_1 and r_2 are determined by the quadratic formula, to factor the given expression.

EXAMPLE: Factor $6x^2 + 13x - 5$ as $a(x - r_1)(x - r_2)$.

Solution: Because $a = 6$, $b = 13$, and $c = -5$ we have

$$\sqrt{b^2 - 4ac} = \sqrt{(13)^2 - 4(6)(-5)} = \sqrt{169 + 120} = \sqrt{289} = 17$$

$$r_1 = \frac{-b}{2a} + \frac{\sqrt{b^2 - 4ac}}{2a} = \frac{-13}{2(6)} + \frac{17}{2(6)} = \frac{-13 + 17}{12} = \frac{4}{12} = \frac{1}{3}$$

$$r_2 = \frac{-b}{2a} - \frac{\sqrt{b^2 - 4ac}}{2a} = \frac{-13}{2(6)} - \frac{17}{2(6)} = \frac{-13 - 17}{12} = \frac{-30}{12} = \frac{-5}{2}$$

Thus

$$6x^2 + 13x - 5 = 6\left(x - \frac{1}{3}\right)\left[x - \left(\frac{-5}{2}\right)\right]$$

$$= 6\left(x - \frac{1}{3}\right)\left(x + \frac{5}{2}\right)$$

55. $2x^2 - 11x - 12$ 56. $k^2 + 81$ 57. $10y^2 + 3y - 1$

58. $t^2 - 6t + 13$ 59. $z^2 + 8z + 5$ 60. $\frac{1}{2}r^2 + 5r - 13$

61. Verify by direct calculation that $r_1 + r_2 = -b/a$.

62. Verify by direct calculation that $r_1 \cdot r_2 = c/a$.

63. Verify the factorization $ax^2 + bx + c = a(x - r_1)(x - r_2)$ by using the results of problems 61 and 62.

6.4 Equations Involving Radicals

We introduced equations involving radicals in Section 5.5. This section will focus on radical equations whose solution entails quadratic equations. You will recall from Section 5.5 that

All solutions of $A = B$ are also solutions of $A^n = B^n$, but some solutions of $A^n = B^n$ may not be solutions of $A = B$.

Our present interest in this principle is for $n = 2$.

EXAMPLE 1   Solve $x - 1 - \sqrt{1 - x} = 0$.

Solution: In order to isolate the radical on one side, add $\sqrt{1 - x}$ to both sides. We have

$$x - 1 = \sqrt{1 - x}$$

Now square both sides.

$$(x - 1)^2 = (\sqrt{1 - x})^2$$
$$x^2 - 2x + 1 = 1 - x \qquad \text{Use } (a - b)^2 = a^2 - 2ab + b^2$$
$$x^2 - x = 0 \qquad \text{Add } x - 1 \text{ to both sides}$$
$$x(x - 1) = 0 \qquad \text{Factor}$$

Thus, $x = 0$ or $x = 1$

Before we can determine the solution set, we must check these possible solutions by substitution into the *original* equation.

Check: Let $x = 0$. Let $x = 1$.
Does $0 - 1 - \sqrt{1 - 0} = 0$? Does $1 - 1 - \sqrt{1 - 1} = 0$?
$-1 - 1 = 0$? $0 - 0 = 0$?
No! Yes!

The solution set is $\{1\}$. The number 0 is not a root of the original equation; it is discarded as an **extraneous root**.

It is sometimes necessary to use the principle from Section 5.5 more than once when solving a radical equation.

EXAMPLE 2   Solve $\sqrt{5 + \sqrt{t}} = \sqrt{t} - 1$.

Solution:
$$(\sqrt{5 + \sqrt{t}})^2 = (\sqrt{t} - 1)^2 \qquad \text{Square both sides}$$
$$5 + \sqrt{t} = t - 2\sqrt{t} + 1$$
$$4 - t = -3\sqrt{t} \qquad \text{Isolate the radical}$$
$$(4 - t)^2 = (-3\sqrt{t})^2 \qquad \text{Square both sides}$$
$$16 - 8t + t^2 = 9t$$
$$t^2 - 17t + 16 = 0$$
$$(t - 16)(t - 1) = 0 \qquad \text{Factor}$$

Hence, $t = 16$ and $t = 1$ are the possible solutions.

Check: Let $t = 16$. Let $t = 1$.
Does $\sqrt{5 + \sqrt{16}} = \sqrt{16} - 1$? Does $\sqrt{5 + \sqrt{1}} = \sqrt{1} - 1$?
$\sqrt{5 + 4} = 4 - 1$? $\sqrt{5 + 1} = 1 - 1$?
$\sqrt{9} = 3$? $\sqrt{6} = 0$?
Yes! No!

The solution set is $\{16\}$; the extraneous root 1 is rejected.

EXAMPLE 3 Solve $\sqrt{3y+4} - \sqrt{y-3} = 3$.

Solution: First rewrite the equation with one radical on each side.

$$\sqrt{3y+4} = \sqrt{y-3} + 3$$
$$(\sqrt{3y+4})^2 = (\sqrt{y-3} + 3)^2 \qquad \text{Square both sides}$$
$$3y + 4 = y - 3 + 6\sqrt{y-3} + 9$$
$$2y - 2 = 6\sqrt{y-3} \qquad \text{Isolate the radical}$$
$$y - 1 = 3\sqrt{y-3} \qquad \text{Divide each term by 2}$$
$$(y-1)^2 = (3\sqrt{y-3})^2 \qquad \text{Square both sides}$$
$$y^2 - 2y + 1 = 9(y-3)$$
$$y^2 - 11y + 28 = 0$$
$$(y-4)(y-7) = 0 \qquad \text{Factor}$$

The possible roots are $y = 4$ and $y = 7$.

Check: Let $y = 4$.   Let $y = 7$.
Does $\sqrt{3(4)+4} - \sqrt{4-3} = 3$?   Does $\sqrt{3(7)+4} - \sqrt{7-3} = 3$?
$\sqrt{16} - \sqrt{1} = 3$? $\sqrt{25} - \sqrt{4} = 3$?
$4 - 1 = 3$? $5 - 2 = 3$?
Yes!   Yes!

The solution set is $\{4, 7\}$.

Exercises 6.4

Standard Assignment: Exercises 1, 3, 5, 9, 11, 13, 15, 19, 23, 25, 31, 37, 39

A Solve each of the following equations. See Examples 1–3.

1. $\sqrt{x^2 - 4} = 0$
2. $\sqrt{y+6} = y$
3. $\sqrt{2}\,t = \sqrt{4-7t}$
4. $p - 2\sqrt{p} + 1 = 0$
5. $p + 2\sqrt{p} + 1 = 0$
6. $q - 5\sqrt{q} + 4 = 0$
7. $q - 10\sqrt{q} + 10 = 0$
8. $r + 8\sqrt{r} - 9 = 0$
9. $r + 5\sqrt{r} - 36 = 0$
10. $x + 1 = 5\sqrt{x-5}$
11. $x + 11 = 6\sqrt{x+3}$
12. $y - 8 + 7\sqrt{y+2} = 0$
13. $y - 26 = -2\sqrt{y-2}$
14. $r - \sqrt{3r+6} = -2$
15. $2 + \sqrt{3r-2} = r$
16. $p - 15 = 2\sqrt{p}$
17. $\sqrt{q+13} = \sqrt{q} + \sqrt{13}$
18. $\sqrt{5k^2 - 10k + 9} = 2k - 1$

19. $2\sqrt{t^2 - 1} = t + 1$
20. $\sqrt{5m^2 + 282} = m + 17$
21. $\sqrt{16y + 121} = y + 10$
22. $\sqrt{7y - 7} - \sqrt{5y - 4} = 1$
23. $\sqrt{k - 3} = \sqrt{2k - 5} - 1$
24. $x + \sqrt{x + 1} = 5$
25. $\sqrt{2r + 9} = 2 + \sqrt{r + 1}$
26. $\sqrt{m - 1} = \sqrt{m - 5}$
27. $\sqrt{7z + 1} - \sqrt{5z - 4} = 1$
28. $\sqrt{3q + 1} - \sqrt{q - 1} = 2$
29. $\sqrt{2 + t} - \sqrt{3 - t} = 3$
30. $\sqrt{r + 2} = \sqrt{3r + 4}$
31. $\sqrt{z^2 + 2} - \sqrt{z - 4} = 0$
32. $\sqrt{p + 7} = \sqrt{2p - 3} + 2$
33. $\sqrt{p} = \sqrt{5p + 4} - 2$
34. $\sqrt{y + 5} = \sqrt{3 - 3y} + 1$
35. $\sqrt{2m - 1} - \sqrt{m - 4} = 2$
36. $\sqrt{k + 8} - 2 = \sqrt{k + 24}$
37. $\sqrt{y + 3} - \sqrt{y + 5} = 2$
38. $\sqrt{x + 7} - \sqrt{x - 2} = 3$
39. $\sqrt{t + 20} - 2\sqrt{t + 7} = 4$
40. $\sqrt{q + 28} - \sqrt{9q + 19} = 5$

6.5 Equations Leading to Quadratic Equations

Sometimes equations that are not quadratic equations can be converted into quadratic equations. A substitution may convert an equation into a quadratic with standard form

$$au^2 + bu + c = 0$$

when u is substituted for an expression involving another variable. Some examples will illustrate this procedure.

EXAMPLE 1

Given Equation	Substitute	Resulting Quadratic
$5(x^2 - 1)^2 - (x^2 - 1) - 3 = 0$	$u = x^2 - 1$	$5u^2 - u - 3 = 0$
$3\left(\dfrac{1}{4 - y}\right)^2 + 1 = 5\left(\dfrac{1}{4 - y}\right)$	$u = \dfrac{1}{4 - y}$	$3u^2 + 1 = 5u$
$7(z + 1)^2 + 2 = 8(z + 1)$	$u = z + 1$	$7u^2 + 2 = 8u$
$3t^4 + 4t^2 - 5 = 0$	$u = t^2$	$3u^2 + 4u - 5 = 0$

We can solve the resulting quadratic equation for

$$u = r_1 \quad \text{and} \quad u = r_2$$

replace u with the expression for which it was substituted, and solve these equations for the desired solutions.

Chap. 6 Second-Degree Equations and Inequalities

EXAMPLE 2 Solve $x^{2/3} + 3x^{1/3} - 4 = 0$.

Solution: Rewrite

$$[x^{1/3}]^2 + 3x^{1/3} - 4 = 0$$
$$u^2 + 3u - 4 = 0 \quad \text{Substitute } u = x^{1/3}$$
$$(u + 4)(u - 1) = 0 \quad \text{Factor}$$
$$u = -4 \quad \text{or} \quad u = 1$$
$$x^{1/3} = -4 \quad \text{or} \quad x^{1/3} = 1 \quad \text{Replace } u \text{ by } x^{1/3}$$
$$[x^{1/3}]^3 = (-4)^3 \quad \text{or} \quad [x^{1/3}]^3 = 1^3 \quad \text{Cube both sides}$$
$$x = -64 \quad \text{or} \quad x = 1$$

The solution set is $\{-64, 1\}$. We leave it to the student to check these solutions in the original equation.

EXAMPLE 3 Solve $\dfrac{1}{(1-y)^2} - \dfrac{7}{1-y} + 10 = 0$.

Solution: Rewrite

$$\left[\frac{1}{1-y}\right]^2 - 7\left[\frac{1}{1-y}\right] + 10 = 0$$
$$u^2 - 7u + 10 = 0 \quad \text{Substitute } u = \frac{1}{1-y}$$
$$(u - 5)(u - 2) = 0 \quad \text{Factor}$$
$$u = 5 \quad \text{or} \quad u = 2$$
$$\frac{1}{1-y} = 5 \quad \text{or} \quad \frac{1}{1-y} = 2 \quad \text{Replace } u \text{ by } \frac{1}{1-y}$$
$$1 = 5(1 - y) \quad \text{or} \quad 1 = 2(1 - y)$$
$$1 = 5 - 5y \quad \text{or} \quad 1 = 2 - 2y$$
$$5y = 4 \quad \quad\quad\quad\quad 2y = 1$$
$$y = \frac{4}{5} \quad\quad\quad\quad\quad y = \frac{1}{2}$$

We multiplied both sides of the equation by an expression containing a variable, namely $1 - y$, and so we must check to make sure that $\frac{4}{5}$ and $\frac{1}{2}$ are *not* roots of $1 - y = 0$.

Check:

$$y = \frac{4}{5} \qquad\qquad y = \frac{1}{2}$$

$$1 - y = \frac{1}{5} \neq 0 \qquad 1 - y = \frac{1}{2} \neq 0$$

The solution set is $\left\{\dfrac{4}{5}, \dfrac{1}{2}\right\}$.

EXAMPLE 4 Solve $4t^4 - 37t^2 + 9 = 0$

Solution: Rewrite

$$\begin{aligned} 4[t^2]^2 - 37t^2 + 9 &= 0 & \\ 4u^2 - 37u + 9 &= 0 & \text{Substitute } u = t^2 \\ (u - 9)(4u - 1) &= 0 & \text{Factor} \end{aligned}$$

$$u - 9 = 0 \quad \text{or} \quad 4u - 1 = 0$$

$$u = 9 \quad \text{or} \quad u = \frac{1}{4}$$

$$t^2 = 9 \quad \text{or} \quad t^2 = \frac{1}{4} \qquad \text{Replace } u \text{ by } t^2$$

$$t = \pm 3 \quad \text{or} \quad t = \pm \frac{1}{2}$$

The solution set is $\{-3, 3, -\tfrac{1}{2}, \tfrac{1}{2}\}$. This can be verified by substitution into the original equation. (See Exercise 41 in Exercises 6.5.)

EXAMPLE 5 Solve $z - 6\sqrt{z} = 7$.

Solution: Rewrite

$$\begin{aligned} [\sqrt{z}]^2 - 6\sqrt{z} - 7 &= 0 & \text{Since } [\sqrt{z}]^2 = z \\ u^2 - 6u - 7 &= 0 & \text{Substitute } u = \sqrt{z} \\ (u + 1)(u - 7) &= 0 & \text{Factor} \end{aligned}$$

$$u = -1 \quad \text{or} \quad u = 7$$

$$\sqrt{z} = -1 \quad \text{or} \quad \sqrt{z} = 7 \qquad \text{Replace } u \text{ by } \sqrt{z}$$

$$[\sqrt{z}]^2 = (-1)^2 \quad \text{or} \quad [\sqrt{z}]^2 = 7^2$$

$$z = 1 \quad \text{or} \quad z = 49$$

We squared both sides of the equation, and so we must check the solution set by substitution into the original equation.

Check: Let $z = 1$.   Let $z = 49$.
         Does $1 - 6\sqrt{1} = 7$?   Does $49 - 6\sqrt{49} = 7$?
              $1 - 6 = 7$? $49 - 6 \cdot 7 = 7$?
                  No! $49 - 42 = 7$?
                                                    Yes!

The solution set is {49}; we discard the extraneous root 1.

Exercises 6.5

Standard Assignment: Exercises 1, 3, 5, 7, 9, 11, 15, 17, 19, 21, 23, 25, 31, 33, 35, 41

Solve each of the following equations. See Examples 1–4.

A 1. $x^4 - 2x^2 + 1 = 0$
   2. $81y^4 + 1 = 18y^2$
   3. $13z^2 - 36 = z^4$

4. $m^4 - 23m^2 = -112$
   5. $2t^4 + t^2 - 1 = 0$
   6. $r^4 - 36 = 5r^2$

7. $p^{-2} - 10p^{-1} + 9 = 0$
   8. $40x^{-2} + 1 = 14x^{-1}$

9. $9y^{-2} + 8y^{-1} = -1$
   10. $9t^{-4} + 8t^{-2} - 1 = 0$

11. $m - 5\sqrt{m} + 6 = 0$
   12. $2k - 15\sqrt{k} = -7$

13. $y + 44 = 15\sqrt{y}$
   14. $k - 3\sqrt{k} - 10 = 0$

15. $p^2 - 3 + 4\sqrt{p^2 - 3} - 5 = 0$
   16. $x^2 - 3 - 4\sqrt{x^2 - 3} - 12 = 0$

17. $2y^2 - 1 - 12\sqrt{2y^2 - 1} + 35 = 0$
   18. $(3t + 1)^2 - 2(3t + 1) + 2 = 0$

19. $(7z + 5)^2 + 2(7z + 5) - 15 = 0$
   20. $6(2q - 3)^2 - 5(2q - 3) + 1 = 0$

21. $r^{2/3} + r^{1/3} - 6 = 0$
   22. $2k^{2/3} - 3k^{1/3} - 2 = 0$

23. $3m^{2/3} - m^{1/3} = 2$

24. $(\sqrt{y} + 5)^2 - 9(\sqrt{y} + 5) + 20 = 0$

25. $(2\sqrt{t} + 1)^2 - 2(2\sqrt{t} + 1) - 3 = 0$

26. $\dfrac{11}{(p + 7)^2} = \dfrac{12}{p + 7} - 1$
   27. $\dfrac{-1}{(3x + 1)^2} = 66 - \dfrac{17}{3x + 1}$

28. $\dfrac{6}{(2 - r)^2} = \dfrac{1}{2 - r} + 1$

29. $2\left(\dfrac{k - 1}{6}\right)^2 + 3\left(\dfrac{k - 1}{6}\right) + 2 = 10\left(\dfrac{k - 1}{6}\right) - \left(\dfrac{k - 1}{6}\right)^2$

30. $\dfrac{1}{(2m + 1)^2} - \dfrac{1}{2m + 1} = 6$
   31. $\left(5y - \dfrac{1}{y}\right)^2 = 2\left(5y - \dfrac{1}{y}\right)$

32. $6\left(2x - \dfrac{3}{x}\right) + 5 = -\left(2x - \dfrac{3}{x}\right)^2$ 33. $6\left(\dfrac{1}{z-1}\right)^4 - 5\left(\dfrac{1}{z-1}\right)^2 + 1 = 0$

34. $3\sqrt{t} + 2 = \dfrac{1}{\sqrt{t}}$ 35. $2\sqrt{p} + 5 = \dfrac{3}{\sqrt{p}}$ 36. $\left(\dfrac{r^2}{r+1}\right)^2 = \dfrac{r^2}{r+1}$

37. $\left(\dfrac{7m}{m+1}\right)^2 - 3\left(\dfrac{7m}{m+1}\right) = 18$ 38. $\left(\dfrac{4x^2 - 3}{x}\right)^2 = 1$

39. $6\left(\dfrac{k^2 - 1}{k}\right)^2 = 7\left(\dfrac{k^2 - 1}{k}\right) + 24$ 40. $\left(\dfrac{2t^2 + 3}{t}\right)^2 + 35 = 12\left(\dfrac{2t^2 + 3}{t}\right)$

41. Verify that each number in $\left\{-3, 3, -\dfrac{1}{2}, \dfrac{1}{2}\right\}$ satisfies the equation

$$4t^4 - 37t^2 + 9 = 0$$

in Example 4.

6.6 Applications

Quadratic equations have a wide range of applications. Word problems whose solutions involve constructing and solving appropriate quadratic equations surface in surprisingly varied circumstances. We will encounter several such applications in this section.

The methods of solving the problems in this section are identical to those of Section 2.3. This is the right time to review the discussion and summary that are presented in Section 2.3.

In the following exercises we will state a sample problem and then explain how to solve it. Similar problems will then be given as exercises for you to solve. After a suitable variety of types of applications have been presented, you will be asked to work a set of exercises where you must determine the "type" of problem and then provide a solution.

Exercises 6.6

Standard Assignment: All odd-numbered exercises

In each of the following problems:

    a. set up an equation representing the conditions of the problem,

    b. solve the equation and interpret the result,

    c. check the solution in the original problem,

NUMERICAL PROBLEMS

EXAMPLE 1: The numerator of a fraction is 5 less than the denominator. The sum of the fraction and 6 times its reciprocal is $\frac{25}{2}$. Find the numerator and the denominator if each is an integer.

Solution:

Let
$$x = \text{the numerator}$$
$$x + 5 = \text{the denominator}$$

(a) $\quad \dfrac{x}{x+5} + 6\left(\dfrac{x+5}{x}\right) = \dfrac{25}{2}$

$\quad 2x^2 + 12(x+5)^2 = 25x(x+5) \qquad$ Multiply by $2x(x+5)$

$\quad 2x^2 + 12(x^2 + 10x + 25) = 25x^2 + 125x$

$\quad 2x^2 + 12x^2 + 120x + 300 = 25x^2 + 125x$

$\quad 11x^2 + 5x - 300 = 0$

$\quad (11x + 60)(x - 5) = 0$

(b) $\quad x = \dfrac{-60}{11} \qquad \text{or} \qquad x = 5$

We are interested in integer solutions and so we disregard $\dfrac{-60}{11}$.

(c) *Check:* $x = 5 =$ the numerator

$\quad x + 5 = 10 =$ the denominator

The numerator $5 = 10 - 5$ is then 5 less than the denominator. The fraction is $\dfrac{5}{10} = \dfrac{1}{2}$. The reciprocal of the fraction is $\dfrac{2}{1} = 2$. The sum of the fraction and 6 times its reciprocal is

$$\dfrac{1}{2} + 6(2) = \dfrac{1}{2} + 12$$

$$= \dfrac{1}{2} + \dfrac{24}{2}$$

$$= \dfrac{25}{2} \qquad \text{As desired.}$$

Solve the following Exercises. See Example 1.

1. The numerator of a fraction is 2 less than the denominator. The sum of the fraction and its reciprocal is $\dfrac{25}{12}$. Find the numerator and the denominator.

2. The sum of the squares of two consecutive positive integers is 113. Find the integers.

3. Two positive integers differ by 6 and have a product of 91. Find the integers.

GEOMETRY PROBLEMS

EXAMPLE 2: A triangle with area 120 sq cm has a base that is 16 cm shorter than 4 times its altitude. Find the base and altitude.

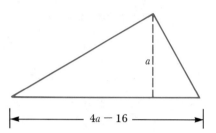

Solution: Let a = the altitude of the triangle, then the base $b = 4a - 16$. Since the area of the triangle $= \frac{1}{2}$ (altitude) · (base) we have

(a)
$$120 = \frac{1}{2} a(4a - 16)$$
$$120 = 2a^2 - 8a$$
$$2a^2 - 8a - 120 = 0$$
$$a^2 - 4a - 60 = 0$$
$$(a - 10)(a + 6) = 0$$

(b) $a = 10$ or $a = -6$

The solution $a = -6$ does not make sense in this problem and is discarded. Thus, $a = 10$, and $b = 4(10) - 16 = 24$ cm.

(c) *Check:* Does $120 = \frac{1}{2}(10) \cdot (24)$? Yes!

The following exercises involve geometric considerations similar to those of Example 2.

4. A triangle of area 4 sq m has a base that is 2 m shorter than 3 times the altitude. Find the base and altitude.

5. One leg is 14 in longer than the other leg in a right triangle whose hypotenuse is 26 in. Find the length of the two legs.

BUSINESS

EXAMPLE 3: A civic club charters a yacht for one day for a cost of $575. When two more people join the group, each person's cost decreases by $2. How many people were in the original group?

Solution:

Let
$$n = \text{number of people in the original group}$$
$$n + 2 = \text{number of people in the final group}$$
$$\frac{575}{n} = \text{original cost per person}$$
$$\frac{575}{n} - 2 = \text{final cost per person}$$

(number of people going) × (cost per person) = 575

(a) $\quad (n + 2)\left(\dfrac{575}{n} - 2\right) = 575$

$575 - 2n + \dfrac{2 \cdot 575}{n} - 4 = 575 \qquad$ Expand the left side by FOIL

$-2n + \dfrac{2 \cdot 575}{n} - 4 = 0 \qquad$ Subtract 575

$2n^2 + 4n - 2 \cdot 575 = 0 \qquad$ Multiply by $-n$

$n^2 + 2n - 575 = 0 \qquad$ Divide by 2

$(n - 23)(n + 25) = 0 \qquad$ Factor

(b) $n = 23 \quad$ or $\quad n = -25$

In this context -25 makes no sense, and so we have $n = 23$ people.

(c) *Check:* $n = 23$ people in the original group gives $\dfrac{575}{23} = \$25$ per person as the original cost. $n + 2 = 23 + 2 = 25$ people in the final group gives $\dfrac{575}{25} = \$23$ per person as the final cost. Since $\$23 = \$(25 - 2)$ each person's cost has indeed decreased by \$2.

Solve the following exercises.

6. A stockbroker buys some stock for \$1800. If the price of each share of the stock had been \$18 less, she could have bought 5 more shares for the same \$1800. How many shares of stock did she buy? How much did she pay for each share?

7. A speculator bought a small coin collection for \$200. He later sold some of the coins, making a profit of \$40 on each coin sold and regaining his initial \$200 investment. If he has 16 coins left, how many coins were in the original collection?

UNIFORM RATE PROBLEMS

EXAMPLE 4: On a still day, Gail can bicycle at an average rate of 24 mph. Yesterday it took her half an hour more to travel 5 miles against a steady wind than it took her to travel the same 5 miles with the wind. How fast was the wind blowing?

Solution:

Let
$$r = \text{rate at which the wind was blowing (in mph)}$$
$$24 - r = \text{Gail's bicycling rate against the wind}$$
$$24 + r = \text{Gail's bicycling rate with the wind}$$

From the formula $d = rt$ (distance = rate · time), we have $d/r = t$.

Now time traveled against the wind = ($\frac{1}{2}$ hr) + (time with the wind) or

(a) $\dfrac{5}{24 - r} = \dfrac{1}{2} + \dfrac{5}{24 + r}$

$$2 \cdot 5(24 + r) = 24^2 - r^2 + 2 \cdot 5(24 - r) \quad \text{Multiply by } 2(24^2 - r^2)$$
$$240 + 10r = 576 - r^2 + 240 - 10r \quad \text{Expand}$$
$$r^2 + 20r - 576 = 0 \quad \text{Simplify}$$
$$(r + 36)(r - 16) = 0 \quad \text{Factor}$$

(b) $r = -36 \quad \text{or} \quad r = 16$

We discard -36 as nonsense, so $r = 16$ mph.

(c) *Check:* If the wind is blowing at 16 mph then Gail travels $24 - 16 = 8$ mph against the wind, and Gail travels $24 + 16 = 40$ mph with the wind. Gail travels 5 miles against the wind in $\dfrac{5}{8}$ of an hour. Gail travels 5 miles with the wind in $\dfrac{5}{40} = \dfrac{1}{8}$ of an hour.

Does $\dfrac{5}{8}$ of an hour $= \dfrac{1}{2}$ hour more than $\dfrac{1}{8}$ of an hour? Does $\dfrac{5}{8} = \dfrac{1}{2} + \dfrac{1}{8}$? Yes!

Solve the following exercises. See Example 4.

8. Jim can row at a rate of 5 mph in still water. If it takes Jim 20 minutes longer to row 4 miles upstream than it takes for him to row the same 4 miles downstream, what is the rate of the current?

9. Karen lives 3 miles from her dentist's office. She averaged 8 mph faster and took 6 minutes less time bicycling home than she did bicycling to her dentist's office. How long did the return trip take?

10. A runner averages 4 mph less than his normal 1-mile rate on an 8-mile uphill section of a 12-mile course. On the 4-mile downhill section of the course he averages 2 mph faster than his normal 1-mile rate. If he runs the entire course in 1 hour and 40 minutes, what is his normal 1-mile rate?

11. A saleswoman noticed that she prolonged her 110-mile trip by 12 minutes as a result of decreasing her normal speed by 5 mph. What is her normal speed?

WORK PROBLEMS

These problems assume that work performed in the accomplishment of a task is done at a steady rate. For example, if a man mows a lawn in 5 hours, we assume he mows $\frac{1}{5}$ of the lawn in 1 hour.

EXAMPLE 5: A swimming pool can be filled by two hoses together in 1 hour and 12 minutes. The larger hose alone will fill the pool in 1 hour less than the smaller one. How long does it take the smaller hose to fill the pool?

Solution: Let t = time required for the smaller hose to fill the pool (in hours); then $t - 1$ = time required for the larger hose to fill the pool (in hours). We know

The portion of the pool filled by the smaller hose in one hour	+	The portion of the pool filled by the larger hose in one hour	=	The portion of the pool filled by both hoses together in one hour.

Now $\quad$ 1 hr 12 min $= \left(1 + \frac{12}{60}\right)$ hr $= \left(1 + \frac{1}{5}\right)$ hr $= \frac{6}{5}$ hr

The portion of the pool filled by both hoses together in 1 hour is

$$\frac{1}{6/5} = \frac{5}{6}$$

The equation becomes

(a) $\quad \dfrac{1}{t} + \dfrac{1}{t-1} = \dfrac{5}{6}$

$\qquad 6(t-1) + 6t = 5t(t-1) \qquad$ Multiply by $6t(t-1)$

$\qquad 12t - 6 = 5t^2 - 5t$

$\quad 5t^2 - 17t + 6 = 0$

$\quad (5t - 2)(t - 3) = 0$

$\quad 5t - 2 = 0 \quad$ or $\quad t - 3 = 0$

(b) $\; t = \dfrac{2}{5} \quad$ or $\quad t = 3$

(c) *Check:* Let $t = \dfrac{2}{5} \quad$ then $\quad t - 1 = \dfrac{-3}{5}$

Since $-\dfrac{3}{5}$ does not make sense in the problem $t = \dfrac{2}{5}$ it is discarded.

Let $t = 3 \quad$ then $\quad t - 1 = 2$

In one hour $\dfrac{1}{3} + \dfrac{1}{2} = \dfrac{2+3}{6} = \dfrac{5}{6}$ of the pool will be filled, as given.

The smaller hose alone will fill the pool in 3 hours.

6.6 Applications

Solve each of the following exercises.

12. An experienced worker can clean an air-conditioning unit in 2 hours, whereas a beginner requires 3 hours to clean the same unit. If a beginner teams up with an experienced worker, how long will it take for them to clean a unit?

13. A computer currently being used to process a company payroll does the job in half the time the old computer took. A new enhancement will shorten the time it now takes to do the job by 2 hours. If the old computer and the newly enhanced system could work together, the job would be done in 3 hours. How long does it take for the current computer to do the job?

14. A boy can clear the stumps from a plot of land in 6 hours less than the time it takes his dad to do the same job. If the father and son work together, they can clear the land in 7 hours and 12 minutes. How long does it take the son to do the job if he works alone?

The skills developed by working the previous exercises should enable you to solve the remaining exercises. Most of these exercises are quite similar to those that appeared earlier. Word problems, however, are always a challenge.

15. A percentage of $100,000 of a certain type of corporate revenue is taken off for state taxes. The same percentage is taken off the amount remaining for federal taxes. After both taxes had been paid, the amount remaining was $96,040. What was the tax rate?

16. A candy company decides to make its popular "Big Bar" even bigger. The present bar is 14 cm long, 6 cm wide, and 2 cm thick. The new bar will have the same thickness but the length and width will each be increased by an identical amount. If the new bar is 25% heavier than the old bar, what are the new dimensions?

17. Three consecutive even integers are the lengths of the sides of a right triangle. What are the integers?

18. A topless box is made from a square of tin by cutting four 4 cm squares from the corners and turning up the extending sides. If the box has a volume of 576 cm^3, find the length of the side of the original tin square.

19. A rectangular garden is 10 yards by 15 yards. There are enough pansy seeds for a 54 square yard area. If a pansy border of uniform width is put around the garden, what will the width be?

20. The product of two consecutive integers is 6 times their sum plus 6. Find the integers.

21. John bought several old Mercedes automobiles for $60,000. After selling all but 8 at an average increase of $10,000 per car, he had regained his original $60,000. How many Mercedes automobiles had he originally bought?

22. The Jackson twins sold some comic books for a total of $520. If they had sold the books for $1 less per book, they would have needed to sell 26 more books to receive the same total. How many books did they sell?

23. In a race on Field Day, Susan skated a quarter of a mile, jumped on a bike and raced the quarter mile back to finish the race. It took her 1 minute less time to bike the second quarter than it did to skate the first and she averaged $\frac{1}{6}$ of a mile per minute faster on the second quarter. How long did it take Susan to complete the race?

24. The larger of two positive numbers is 5 times the smaller. The square of the larger number minus the square of the smaller number is 4 times one more than the smaller number. Find the numbers.

25. Working alone, Sal can sort the mail from the midmorning collection in 2 hours less time than Ron. Working together, they can sort the mail in 1 hour and 48 minutes. How long does it take Sal to sort the mail working alone?

26. A guy wire is staked to the ground 24 feet from the base of a pole. The wire is fastened to the top of the pole and is 4 feet longer than 3 times the height of the pole. How high is the pole?

27. A freight train requires $2\frac{1}{2}$ hours longer to make a 300-mile journey than an express train does. If the express averages 20 mph faster than the freight train, how long does it take the express train to make the trip?

28. Jan is 5 years older than her brother. The product of their present ages minus 6 times the sum of their present ages is 48. How old is Jan?

29. The tens' digit of a two-digit number is one less than the units' digit. The sum of the squares of the digits is 4 less than the number itself. What is the number?

30. The area of a right triangle is 30 sq in. If one leg is 7 in longer than the other, how long is the hypotenuse?

31. The numerator of a fraction is 21 more than the denominator. The fraction minus 9 times its reciprocal equals $\frac{5}{2}$. Find the numerator and the denominator.

32. Find two consecutive odd integers such that the square of the larger integer minus the square of the smaller integer is six more than twice the smaller integer.

33. A club is going to spend $216 to have a picnic. When 6 new members join the club, each member's cost is reduced by $3. How much would it have cost each member before the new members joined the club?

34. ▦ You will recall that if P dollars is invested at a (decimal) interest rate r that is compounded annually, at the end of 2 years the return is $P(1 + r)^2$. If a firm invests $1,000,000 now and gets a return of $1,276,900 in two years, what was the interest rate?

35. ▦ A tank can be filled in 5 hours by two pipes working together. If the smaller pipe is used alone, it requires 2 more hours to fill the tank than the larger one does. How many hours does it take the larger pipe to fill the tank alone?

6.7 Quadratic Inequalities

In Section 2.6 we learned to solve linear inequalities such as $2x - 7 < 2$. Now we will investigate methods for solving quadratic inequalities such as $x^2 + x < 6$. A **quadratic inequality** is any inequality that is equivalent to either $ax^2 + bx + c < 0$ or $ax^2 + bx + c \le 0$, where a, b, and c are real numbers and $a \ne 0$. Note that $3x^2 - 7 \ge 0$ is equivalent to $-3x^2 + 7 \le 0$. Although we will again use the "Addition Property" and "Multiplication Property" of inequalities, additional procedures are required to determine the solution sets of quadratic inequalities.

EXAMPLE 1   Solve $x^2 + x < 6$.

Solution: First rewrite

$$x^2 + x - 6 < 0 \qquad \text{Isolate 0 on the right}$$

Next, factor

$$(x + 3)(x - 2) < 0 \qquad \text{A negative product}$$

For a product of two factors to be negative, the factors must have opposite signs, $(+) \cdot (-) = (-)$, or $(-) \cdot (+) = (-)$. In order for $(x + 3)(x - 2)$ to be negative, the two factors $x + 3$ and $x - 2$ must have opposite signs. (If one is positive, the other must be negative.)

We can determine when these conditions are true by using a **sign chart** as is shown in Figure 6.3. Since $x + 3 > 0$ exactly when $x > -3$, we put $+$ signs to the right of -3 and $-$ signs to the left of -3 over the number line. Since $x - 2 > 0$ exactly when $x > 2$, we put $+$ signs to the right of 2 and $-$ signs to the left of 2.

Figure 6.3

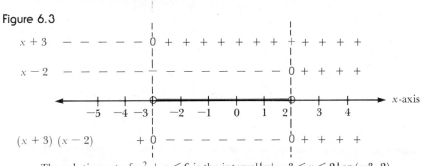

The solution set of $x^2 + x < 6$ is the interval $\{x \mid -3 < x < 2\}$ or $(-3, 2)$.

The dashed vertical lines through 2 and −3, together with the open circles at these points, indicate that 2 and −3 are not solutions of $x^2 + x < 6$.

The chart also shows that unlike signs occur for these two factors only when $-3 < x < 2$, resulting in $(x + 3)(x + 2)$ being negative. The graph of the solution set is shaded on the number line in the chart.

If a quadratic inequality involves the signs ≤ or ≥, then the roots of the linear factors are solutions, and solid circles and solid lines are used on the sign chart.

EXAMPLE 2 Solve $y^2 \geq 2$.

Solution: Rewrite

$$y^2 - 2 \geq 0 \qquad \text{Isolate 0 on the right}$$

$$(y - \sqrt{2})(y + \sqrt{2}) \geq 0 \qquad \text{Factor}$$

In order to determine where the product is positive, make a sign chart for the factors $y - \sqrt{2}$ and $y + \sqrt{2}$ (see Figure 6.4).

Figure 6.4

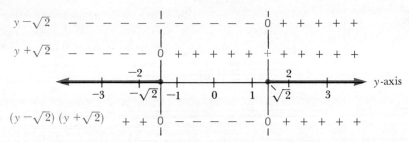

The solution set for $y^2 \geq 2$ is $(-\infty, -\sqrt{2}] \cup [\sqrt{2}, \infty)$ or $\{y | y \leq -\sqrt{2} \text{ or } y > \sqrt{2}\}$.

Now $y - \sqrt{2} \geq 0$ exactly when $y \geq \sqrt{2}$, and $y + \sqrt{2} \geq 0$ exactly when $y \geq -\sqrt{2}$.

EXAMPLE 3 Solve $k^2 - 2k - 2 > 0$.

Solution: Because 0 is already isolated on the right, you can factor $k^2 - 2k - 2 = (k - r_1)(k - r_2)$ by the quadratic formula, where r_1 and r_2 are the roots of $k^2 - 2k - 2 = 0$.

$$k = \frac{-b \pm \sqrt{b^2 - 4ac}}{2a} = \frac{-(-2) \pm \sqrt{(-2)^2 - 4(1)(-2)}}{2}$$

$$= \frac{2 \pm \sqrt{4 + 8}}{2}$$

$$= \frac{2 \pm \sqrt{12}}{2}$$

$$= \frac{2 \pm 2\sqrt{3}}{2}$$

$$= 1 \pm \sqrt{3}$$

Thus, $r_1 = 1 - \sqrt{3}$ and $r_2 = 1 + \sqrt{3}$.

$$k^2 - 2k - 2 = (k - [1 - \sqrt{3}])(k - [1 + \sqrt{3}]) > 0.$$

In order to determine where the product is positive, make a sign chart for the factors $k - [1 - \sqrt{3}]$ and $k - [1 + \sqrt{3}]$ (see Figure 6.5).

Figure 6.5

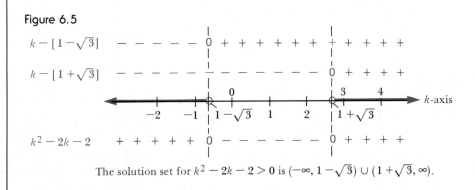

The solution set for $k^2 - 2k - 2 > 0$ is $(-\infty, 1 - \sqrt{3}) \cup (1 + \sqrt{3}, \infty)$.

When we are estimating the position of $1 - \sqrt{3}$ and $1 + \sqrt{3}$ on the graph without a calculator (which gives -0.73205 and 2.73205 respectively) we can either use a table of square roots or proceed in the following manner. Because $1 < 3 < 4$, we have $\sqrt{1} < \sqrt{3} < \sqrt{4}$, or $1 < \sqrt{3} < 2$. Multiplying by -1 yields $-2 < -\sqrt{3} < -1$. If we now add 1 to each of the two preceding inequalities we get $2 < 1 + \sqrt{3} < 3$ and $-1 < 1 - \sqrt{3} < 0$. This will be of some help when graphing the inequalities.

Life is somewhat simpler if, when the inequality sign in a quadratic inequality is replaced by an equal sign, the resulting equation has no real roots. We simply substitute zero or any other real number for the variable in the given inequality. If a true statement results, the solution set is the set of real numbers; if a false statement results, the solution set is ∅. (See Exercise 59 in Exercises 6.7.)

EXAMPLE 4 Solve $2z^2 + 3z + 2 \leq 0$.

Solution: Now $a = 2$, $b = 3$, and $c = 2$ in $2z^2 + 3z + 2 = 0$, and the discriminant $b^2 - 4ac = 3^2 - 4(2)(2) = 9 - 16 = -7 < 0$. Thus, $2z^2 + 3z + 2 = 0$ has no real roots.

Substituting $z = 0$, $2(0)^2 + 3(0) + 2 < 0$?
$$2 < 0 ?$$
No!

The solution set is $\emptyset$.

The sign chart method can also be used for inequalities that involve rational expressions. The factors of both the numerator and the denominator are used in the chart. It is important to note that the roots of the factors of the denominator will *never* be solutions of the given inequality because division by 0 is undefined.

EXAMPLE 5 Solve $\dfrac{3}{t-1} \geq 1$.

Solution: Rewrite

$$\frac{3}{t-1} - 1 \geq 0 \qquad \text{Isolate 0 on the right side}$$

$$\frac{3 - (t-1)}{t-1} \geq 0 \qquad \text{Use } t - 1 \text{ as a common denominator}$$

$$\frac{4-t}{t-1} \geq 0$$

In order to determine where the quotient is positive, make up a sign chart for the factors $4 - t$ and $t - 1$ (see Figure 6.6). Because 1 is a root of a denominator factor, it will *not* be a solution of the given inequality.

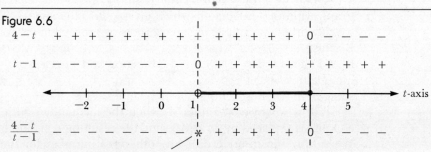

Figure 6.6

Denominator 0, quotient undefined

The solution set for $\dfrac{3}{t-1} \geq 1$ is $(1, 4]$.

6.7 Quadratic Inequalities

EXAMPLE 6

Solve $x^3 - 2x^2 - 8x \le 0$.

Solution: You will recall that any polynomial is a rational expression with denominator 1.

$$x^3 - 2x^2 - 8x = x(x^2 - 2x - 8) \quad \text{Factor}$$
$$= x(x + 2)(x - 4)$$

In order to determine when this product is negative, make a sign chart for the factors x, $x + 2$, and $x - 4$ (see Figure 6.7). Because the inequality sign is $\le$, the roots of the factors are solutions of the given inequality.

Figure 6.7

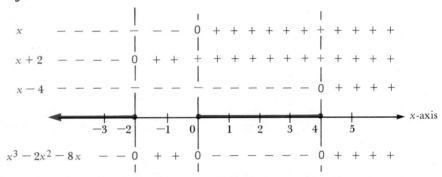

The solution set for $x^3 - 2x^2 - 8x \le 0$ is $(-\infty, -2] \cup [0, 4]$.

EXAMPLE 7

University Apartments consists of 50 apartments. At $300 per month, all of the apartments can be rented. However, for each increase of $20 per month, two vacancies are created with no chance of renting them. The owner wants to be sure that the rental income from the apartments is at least $15,640 per month. What rental rate can be charged?

Solution:

Rental income = (rent per apartment) · (number of apartments rented).

Let
$$n = \text{number of } \$20 \text{ increases}$$
$$20n = \text{increase in rent per apartment}$$
$$2n = \text{number of vacancies}$$
$$50 - 2n = \text{number of apartments rented}$$
$$\text{rental income} = (300 + 20n)(50 - 2n)$$

We want
$$\text{rental income} \ge \$15{,}640$$

274   Chap. 6   Second-Degree Equations and Inequalities

(a) $(300 + 20n)(50 - 2n) \geq 15{,}640$
$15{,}000 + 400n - 40n^2 \geq 15{,}640$
$-640 + 400n - 40n^2 \geq 0$
$40n^2 - 400n + 640 \leq 0$
$n^2 - 10n + 16 \leq 0$
$(n - 8)(n - 2) \leq 0$

Using the sign chart,

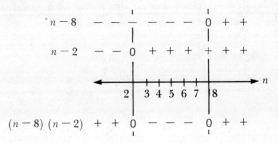

(b) We see that n can be 2, 3, 4, 5, 6, 7, or 8 because we are only interested in integer values of n. The rent could be any amount from $300 + 2(20) = \$340$ up to $300 + 8(20) = \$460$ per month. Considerations other than income, such as maintenance expense, parking facilities, and other use of the vacant apartments, may be used to determine the actual rent charged.

(c) *Check:* We will check \$340 and \$460 and leave the remaining values for you to check.

$$\text{rent} = \$340, \quad n = 2 \qquad\qquad \text{rent} = \$460, \quad n = 8$$
$$\text{rental income} = (340)(50 - 2 \cdot 2) \qquad \text{rental income} = (460)(50 - 2 \cdot 8)$$
$$= (340)(46) \qquad\qquad\qquad\qquad = (460)(34)$$
$$= 15{,}640 \qquad\qquad\qquad\qquad\; = 15{,}640$$

Exercises 6.7

Standard Assignment: Exercises 1, 3, 5, 7, 9, 13, 15, 17, 19, 25, 31, 33, 35, 37, 39

Solve the following inequalities.

A   1. $(x + 4)(x - 1) > 0$ 2. $(y - 2)(y + 5) \geq 0$ 3. $t(t - 3) < 0$

4. $p(p + 2) \leq 0$ 5. $(2q - 1)(q + 3) < 0$ 6. $(4r + 7)(3r - 1) > 0$

7. $m^2 - 3m + 2 \geq 0$ 8. $x^2 + x - 2 \leq 0$ 9. $z^2 - 4z - 21 < 0$

10. $y^2 + 2y - 15 > 0$ 11. $t^2 - t < 6$ 12. $r^2 + 8 \geq 6r$

13. $5k^2 + 7k \leq 6$
14. $6p + 8 \geq 5p^2$
15. $4m^2 + 6 > 11m$
16. $4q^2 + 9 < 12q$
17. $y^2 \geq 3$
18. $x^2 < 7$
19. $(2t + 1)^2 < 4$
20. $(4r - 9)^2 > 4$
21. $5m^2 + 13m - 6 \geq 0$
22. $k^2 + 6 \leq 5k$
23. $z^2 + 8 \geq 4z$
24. $q^2 < -7$
25. $3y^2 - 2 \leq y$
26. $4m + 13 > -m^2$
27. $6p^2 - 11p \geq 35$
28. $x^2 - 8x > -10$
29. $2k^2 + 3 \leq 7k$
30. $z^2 + 5 < 2z$
31. $p^2 - 10p < -3$
32. $t^2 + 11 \leq 6t$
33. $q^2 - 8q - 4 > 0$
34. $y^2 - 8y \geq -16$
35. $r^2 + 4r + 2 < 0$
36. $m^3 + 2m^2 - 3m \geq 0$

Solve the following word problems. See Example 7.

37. An import firm pays a tax of $10 on each radio they import. In addition to this tax, there is a penalty tax that must be paid if more than 1000 radios are imported. The penalty tax is computed by multiplying 5 cents times the number of radios imported in excess of 1000. This tax must be paid on each and every radio imported. If 1006 radios are imported, the penalty tax is 6 · 5 = 30 cents, and is paid on each of the 1006 radios. If the firm wants to spend no more than a total of $640,000 on import taxes, how many radios can they import?

38. A TV quiz program pays a contestant $100 for each correct answer for 10 questions. If all 10 questions are answered correctly, bonus questions are asked. The reward for every correct answer is increased by $50 for each bonus question that is correctly answered. Any incorrect answer ends the game. If 2 bonus questions are answered correctly, the contestant receives $200 *for each of the 12 questions*. If a contestant won more than $3500, how many questions must have been answered correctly?

39. A firm wants a return of at least $125,440 two years from now. They can invest at a 12% interest rate compounded annually. How much must they invest? [You will recall that if P dollars is invested at a (decimal) rate of r compounded annually, then $P(1 + r)^2$ has accrued after 2 years.]

40. The length of a rectangle is 5 m longer than the width. Find all possible widths so that the area of the rectangle will be at least 204 sq m.

Solve the following inequalities.

B 41. $x^3 - 4x < 0$
42. $p^3 - 2p^2 - 15p \leq 0$
43. $(k - 2)(k + 3)(k + 5) > 0$
44. $(2t - 1)(t + 2)(3t + 2) < 0$
45. $\dfrac{1}{2z + 5} > 0$
46. $\dfrac{3}{1 - 2m} \leq 0$
47. $\dfrac{-3}{5y - 2} \leq 0$
48. $\dfrac{r}{r - 4} > 5$
49. $\dfrac{q}{q - 2} \leq 3$
50. $1 - \dfrac{1}{1 + p} < 0$
51. $2 - \dfrac{2x}{3x - 4} > 0$
52. $1 + \dfrac{1}{2 - z} \geq 0$
53. $2 + \dfrac{6}{t - 1} > 0$

54. $\dfrac{k-2}{2k+1} < -1$ 55. $3 \le \dfrac{2r+6}{2r+1}$ 56. $\dfrac{x(x-1)}{x+2} > 0$

57. $\dfrac{p^2+1}{p} \le 2$ 58. $\dfrac{-5t}{t-3} \ge 3t$

59. Use the fact that $ax^2 + bx + c < 0$ $(a > 0)$ is equivalent, by completing the square, to $[x + (b/2a)]^2 < [(b^2 - 4ac)/4a^2]$ to show that if $ax^2 + bx + c = 0$ $(a > 0)$ has no real roots, then the solution set for the original inequality is $\emptyset$. (Hint: If the equation has no real roots, we have $b^2 - 4ac < 0$, while $[x + (b/2a)]^2$ is always ≥ 0.)

Chapter 6 Summary

Key Words and Phrases

6.1 Quadratic equation
     Standard form
6.2 Square root method
     Completing the square
6.3 Quadratic formula
     Discriminant
6.4 Extraneous root
6.7 Quadratic inequality
     Sign chart

Key Concepts and Rules

In order to solve a quadratic equation by *factoring*, we

1. Rewrite the equation in standard form;
2. Factor the quadratic expression as a product of two linear factors;
3. Use the fact that $A \cdot B = 0$ if and only if $A = 0$ or $B = 0$ to complete the solution by equating each linear factor to zero and solving the two resulting equations. These solutions are the roots of the quadratic equation.

In order to solve a quadratic equation by the *square root method*, we use the fact that $x^2 = b$ if and only if $x = \sqrt{b}$ or $x = -\sqrt{b}$. If $b < 0$, we use the imaginary number $i = \sqrt{-1}$ when writing the solutions. This method is applicable to equations in the form $(ax + d)^2 = b$.

The *quadratic formula* gives the solutions for any quadratic equation in standard form $ax^2 + bx + c = 0$ with $a \neq 0$ by

$$x = \frac{-b - \sqrt{b^2 - 4ac}}{2a} \quad \text{or} \quad x = \frac{-b + \sqrt{b^2 - 4ac}}{2a}$$

If a, b, and c are real numbers, the real number $b^2 - 4ac$ is called the *discriminant* of the equation. If $b^2 - 4ac = 0$, there is only one solution to the equation, $-b/2a$. If $b^2 - 4ac \neq 0$, there are two (unequal) roots. The roots are both real if $b^2 - 4ac > 0$ and both are imaginary if $b^2 - 4ac < 0$.

Review Exercises

If you have difficulty with any of these exercises, look in the section indicated by the numbers in square brackets.

Solve the following equations by factoring.

[6.1]
1. $(x + 1)^2 = 0$
2. $(y - 3)(y + 7) = 0$
3. $(z - i)(z + i) = 0$
4. $2t^2 - 6t = 0$
5. $7k - k^2 = 0$
6. $r^2 - 9 = 0$
7. $16x^2 - 25 = 0$
8. $y^2 + 12y + 36 = 0$
9. $k^2 + k - 30 = 0$
10. $2z^2 + 17z = -21$
11. $3t^2 = 44t + 15$
12. $2r^2 - 12rm + 18m^2 = 0$
13. $3y^2 = 19ky - 20k^2$
14. $3p^2 + 14pq + 15q^2 = 0$

Find a quadratic equation in standard form with integer coefficients having the following solution sets.

15. $\{5, -2\}$
16. $\{7\}$
17. $\left\{0, \frac{1}{2}\right\}$
18. $\left\{-\frac{2}{3}, \frac{5}{6}\right\}$

[6.2] Solve the following equations by taking square roots or completing the squares.

19. $(x + 5)^2 = 16$
20. $(4a - 1)^2 = 25$
21. $(7 - 2k)^2 = 12$
22. $(t + 4)^2 = -9$
23. $(3r - 1)^2 = -5$
24. $2m^2 = 1 - m$
25. $3y^2 + 13y = 10$
26. $4r^2 + 5 = 4r$
27. $z^2 - 6z + 6 = 0$
28. $x^2 + 2 = 3x$
29. $2p(p - 2) = 5$
30. $t^2 - 2t + 1 = 0$

[6.3] Solve each of the following equations by the quadratic formula.

31. $k^2 + 1 = 0$
32. $t^2 + 20 = 10t$
33. $(y + 6)(y - 1) = 1$

34. $5z(z - 1) = 2(1 - z)$
35. $(2x + 1)(x - 2) = -4$
36. $r^2 + 4r + 6 = 0$
37. $m^2 + 10m + 25 = 0$
38. $y^2 + 4y + 9 = 0$
39. $2t^2 - 12t + 37 = 0$
40. $(p + 3)(p - 1) = 1$
41. $k^2 - 4k + 5 = 0$
42. $2a^2 + 1 = 5a$
43. $4x^2 - 3ix = 0$
44. $2z^2 - iz + 1 = 0$
45. $3im^2 - 7m + 6i = 0$

In each of the following equations, use the discriminant to determine whether the equation has one real, two distinct real, or two distinct imaginary solutions.

46. $y^2 - y + 13 = 0$
47. $-2k^2 + 24k = 72$
48. $2r^2 - r + 2 = 0$
49. $x^2 + 7 = 0$
50. $m^2 + \sqrt{7}\,m + 3 = 0$
51. $4t^2 + 4t + 1 = 0$

[6.4] Solve each of the following equations.

52. $\sqrt{4z - 4} = z$
53. $10p - 13\sqrt{p} = 3$
54. $x + 5 = \sqrt{6x + 61}$
55. $\sqrt{m} - 1 = \sqrt{m - 7}$
56. $\sqrt{2k - 5} = \sqrt{k + 2}$
57. $\sqrt{2t + 5} - \sqrt{10t + 5} = 2$

[6.5] Solve each of the following equations.

58. $8a^4 + 19a^2 - 27 = 0$
59. $\dfrac{9}{x^2} - \dfrac{30}{x} + 25 = 0$
60. $8r^{2/3} + 7r^{1/3} = 1$
61. $(2\sqrt{t} - 3)^2 - 8(2\sqrt{t} - 3) + 7 = 0$
62. $4\sqrt{m} + 1 = \dfrac{3}{\sqrt{m}}$
63. $6\left(\dfrac{1}{t - 1}\right)^2 + \left(\dfrac{1}{t - 1}\right) - 1 = 0$

[6.6] Solve each of the following inequalities.

64. $x^2 + x < 2$
65. $2y^2 \geq 5y + 3$
66. $5k^2 \leq 12k$
67. $t^2 > 2t + 11$
68. $m^2 + 13 < m$
69. $2(a^2 + 1) \geq a$
70. $2z^3 + z^2 - 6z \leq 0$
71. $\dfrac{3x + 1}{4x + 8} < 0$
72. $\dfrac{x + 4}{3x - 2} \geq 1$

[6.7] Solve the following word problems.

73. A 15 ft ladder leans against a building, resting on the ledge of a window 12 ft above the ground. How far is the foot of the ladder from the base of the building?

74. The numerator of a fraction is one more than the denominator. The sum of the fraction and one minus the reciprocal of the fraction is $\dfrac{11}{6}$. Find the numerator and the denominator.

75. A frame of uniform width is to be constructed about a painting with dimensions of 3 ft by 2 ft. How wide will the frame be if the area of the frame is 6 sq ft?

76. A rectangular box of width 4 ft is constructed so that its length is 1 ft less than 3 times its height. If the height were increased by 1 ft and the length shortened by 5 ft, the resulting box would have half the original volume. What are the dimensions of the original box?

77. A small plane flies 348 miles with the wind in 15 minutes less time than it takes to fly 210 miles against the wind. If the plane flies 40 mph in still air, find the rate at which the wind is blowing.

78. A mixture of two types of coffee contains $27 worth of the cheaper blend and $20.25 worth of the more expensive blend. The cheaper blend sells for $1.35 per kg less than the expensive blend. The mixture contains 5 kg less of the expensive blend than it does of the cheaper blend. How many kg of each blend does the mixture contain?

Practice Test (60 minutes)

Solve the following equation by completing the square.

1. $3x^2 + 5x - 2 = 0$

Solve the following equations by any method.

2. $(2m + 4)^2 = -1$

3. $a^2 + 16 = 10a$

4. $7 - 3z + 2z^2 = 0$

5. $\sqrt{p + 15} - p = 3$

6. $\left(\dfrac{2}{k}\right)^2 - 14\left(\dfrac{2}{k}\right) + 49 = 0$

7. $x^2(3 + x^2) = 4$

8. $\sqrt{t + 7} + \sqrt{3 - t} = 4$

9. A woman invests $1650 to purchase some shares of stock. She later retains 16 shares, but sells the others at a profit of $10 per share. If she recovers her original investment of $1650 by the sale, how many shares of stock were originally purchased?

10. The sum of the squares of two consecutive even intergers equals 20 times one less than the smaller integer. Find the integers.

Solve the following inequalities.

11. $2x^2 \leq 3x + 14$

12. $\dfrac{y + 10}{3y + 4} > 2$

Extended Applications

Rescue Mission

Climatic conditions made it impossible to immediately rescue a climbing party that was trapped and isolated in a gorge high on Mount McKinley in Alaska. A plane was to drop a package of medical and food supplies to the stranded climbers to enable them to survive until a rescue could be accomplished. The plane was to approach the gorge at 110 mph at an elevation of 144 feet above the top of the gorge. The pilot's problem was to determine when to drop the supplies from the plane so that they would drop into the gorge. We will see that dropping the supplies when the plane is directly over the gorge will not work!

Two facts are needed to solve the pilot's problem.

(1) An object will fall a distance of d feet in t seconds with
$$d = 16t^2 + v_o t,$$
where v_o is the initial downward velocity of the object.

(2) The dropped object will continue to move in the horizontal direction, at the same rate that the plane was traveling when the object was dropped, until it hits the mountain.

These two statements ignore air resistance to the object.

Now because the supply package is *dropped*, the initial downward velocity of the package is $0 (= v_o)$. Since the package will enter the top of the gorge when it has fallen 144 feet, we have from (1)

$$144 = 16t^2 + 0 \cdot t$$
$$144 = 16t^2$$
$$9 = t^2$$
$$\pm 3 = t$$

Of course, $t = -3$ is nonsense. The package will drop for 3 seconds before entering the gorge.

The package is traveling 110 mph horizontally when it is dropped and will still be traveling at this rate when it enters the gorge after a 3-second drop. The formula $d = r \cdot t$ can be used to determine how far in the horizontal direction the package will travel in 3 seconds. First, we convert 110 miles per hour to feet per second.

$$110 \text{ miles} = 110 \text{ miles} \times 5280 \frac{\text{feet}}{\text{mile}} = 580{,}800 \text{ feet}$$

and $$110 \frac{\text{miles}}{\text{hr}} \times \frac{1 \text{ hr}}{60 \text{ min}} \times \frac{1 \text{ min}}{60 \text{ sec}} = \frac{110}{60 \cdot 60} \frac{\text{miles}}{\text{sec}}$$

so $$110 \frac{\text{miles}}{\text{hr}} = \frac{110}{60 \cdot 60} \frac{\text{miles}}{\text{sec}} = \frac{580{,}800}{60 \cdot 60} \frac{\text{feet}}{\text{sec}} = 161\frac{1}{3} \text{ ft/sec}$$

The horizontal distance, d, that is traveled by the package in 3 seconds, is determined by $d = r \cdot t$; thus, $d = (161\frac{1}{3}) \cdot (3) = 484$ feet.

The supplies should be dropped when the plane is still 484 feet away from the top of the gorge.

Exercises

1. Suppose the supplies were dropped from the plane in the situation previously described at an initial downward velocity of 12 feet per second. How far away from the gorge should the supplies be released?

2. ▦ The Vehicle Assembly Building at the Kennedy Space Center is 525 feet high. If a wrench is dropped from the top of the building, how long will it take to reach the ground?

3. ▦ If the wrench in Exercise 2 is thrown down from the top with an initial velocity of 80 ft/sec, how long will it take to reach the ground?

7

Linear Equations, Inequalities, Functions

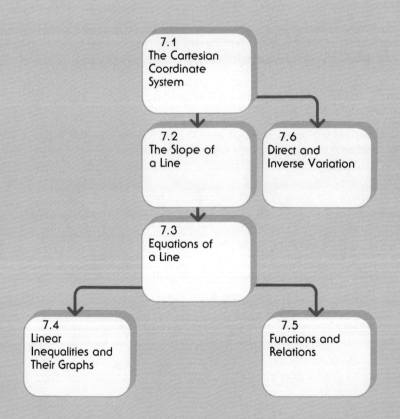

7.1 The Cartesian Coordinate System

A pair of real numbers that is enclosed in parentheses and separated by a comma is called an **ordered pair** of real numbers. Thus, (0, 2), (−5, 7), and ($\frac{1}{2}$, −π) are ordered pairs of real numbers. In this section, we will describe a method for establishing a **Cartesian Coordinate System** in a plane so that each point in the plane will correspond to a unique ordered pair of real numbers, and vice versa.

We select two real number lines in the plane, one vertical and one horizontal, which intersect at their zero points. Usually the positive numbers are chosen to be on the right and above the zero point. The vertical line is called the *y-axis* and the horizontal line is called the *x-axis*, and their common zero point is called the **origin.** Each of the numbers in an ordered pair is called a **component** of the ordered pair. The first component is called the **x-coordinate** of the point and the second component is called the **y-coordinate** of the point. The axes divide the plane into four regions called **quadrants,** which are numbered as shown in Figure 7.1(a). The points on the axes themselves do not belong to any of the quadrants.

The point (1, 2) is located by traveling one unit on the *x*-axis to the right from zero and then going straight upward 2 units (as measured on the *y*-axis). In Figure 7.1(b), the point labeled *A* corresponds to the ordered pair (1, 2) and is called the **graph of the ordered pair** (1, 2) in the coordinate system. Similarly, the points labeled *B, C, D, E,* and *F* in Figure 7.1(b) are the graphs of (5, 0), (0, 5), (0, 0), (−3, 4), and (−3, −2) respectively. We frequently ignore the distinction between an ordered pair and its graph.

Figure 7.1

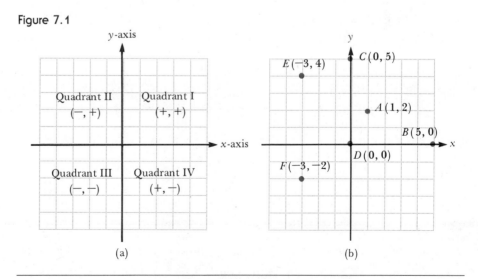

284   Chap. 7   Linear Equations, Inequalities, Functions

EXAMPLE 1

Graph $(-3, 0)$, $(0, -3)$, $(4, 2)$, $(-1, -4)$, $(-2, 3)$, and $(3, -1)$.

Solution: $(-3, 0)$ is 3 units to the left of the origin on the *x*-axis.

$(0, -3)$ is 3 units down from the origin on the *y*-axis.

$(4, 2)$ is 4 units to the right of the origin on the *x*-axis and up 2 units.

$(-1, -4)$ is 1 unit to the left of the origin on the *x*-axis and down 4 units.

$(-2, 3)$ is 2 units to the left of the origin on the *x*-axis and up 3 units.

$(3, -1)$ is 3 units to the right of the origin on the *x*-axis and down 1 unit.

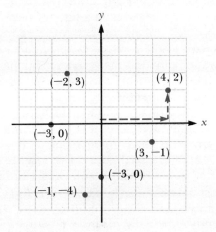

Equations in two variables have ordered pairs of real numbers as solutions. Although any two letters may be chosen as variables, we usually choose *x* and *y*. If we replace *x* by 3 and *y* by 1 in the equation

$$y = 4 - x$$

we have

$$1 = 4 - 3$$

which is true. We say that the ordered pair $(3, 1)$ is a solution of the equation $y = 4 - x$. In general, we say that the ordered pair (a, b) is a **solution** of an equation or inequality in two variables *x* and *y* if replacing *x* by *a* and *y* by *b* results in a true statement.

EXAMPLE 2

Determine by substitution whether each of the following points, $(0, 5)$, $(0, 7)$, $(3, 1)$, $(1, 3)$, and $(1, 5)$ are solutions of the equation $2x + y = 7$.

Solution: Replace *x* by the 1st component and *y* by the 2nd component. Now:

$2(0) + 5 \neq 7$; the point $(0, 5)$ is *not* a solution.

$2(0) + 7 = 7$; the point $(0, 7)$ *is* a solution.

$2(3) + 1 = 7$; the point (3, 1) *is* a solution.
$2(1) + 3 \neq 7$; the point (1, 3) is *not* a solution.
$2(1) + 5 = 7$; the point (1, 5) *is* a solution.

In order to find solutions for a given equation in two variables, we simply substitute *any* number for one of the variables, and then solve the equation for the other variable. For example, to obtain a solution for the equation

$$2x + y = 7,$$

we arbitrarily choose $x = 2$, and then substitute

$$2(2) + y = 7$$
$$y = 7 - 4$$
$$y = 3$$

When $x = 2$, $y = 3$; thus (2, 3) is a solution of $2x + y = 7$.

EXAMPLE 3 Complete the ordered pair (, -1) so that it is a solution of the equation $2x + y = 7$.

Solution: From

$$2x + y = 7$$
$$2x + (-1) = 7 \qquad \text{Substitute } -1 \text{ for } y$$
$$2x = 7 + 1$$
$$2x = 8$$
$$x = 4$$

Thus, $(4, -1)$ is a solution of $2x + y = 7$.

The **graph of an equation** in two variables is the graph of the solution set of the equation, or the graph of all solutions of the equation.

EXAMPLE 4 Graph $2x + y = 7$.

Solution: In Examples 2 and 3 we saw that (0, 7), (1, 5), (3, 1), and $(4, -1)$ are solutions of $2x + y = 7$. These points are graphed in Figure 7.2(a). Note that we chose points with integer components for ease of computation. However, $(-1.3, 9.6)$, $(\sqrt{2}, 7 - 2\sqrt{2})$, $\left(\frac{1}{3}, \frac{19}{3}\right)$, and countless other points are also on the graph.

If we were to graph all the solutions of $2x + y = 7$, they would lie on a straight line in the plane. See Figure 7.2(b). In fact, *all first degree equations in two variables have straight lines as their graphs.* The line in Figure 7.2(b) is the graph of $2x + y = 7$.

286   Chap. 7 Linear Equations, Inequalities, Functions

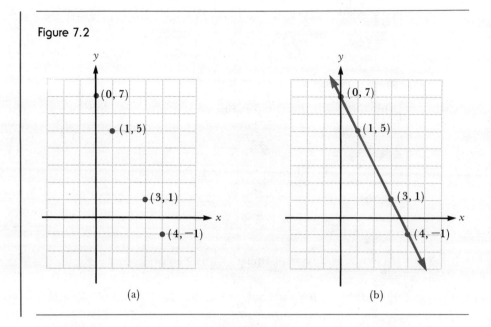

Figure 7.2

(a) (b)

Equivalent equations have identical solution sets and consequently identical graphs. We frequently use the fact that

Any equation that is equivalent to one of the form

$$ax + by = c$$

has a straight line for its graph.

Any equation that is equivalent to $ax + by = c$ is called a **linear equation** because its graph is a line. Two points determine a line, and so we need only find two points on the graph of a first-degree equation; then we can use a ruler to finish the graph.

EXAMPLE 5   Graph $x + 3y = 6$.

Solution: First we find two points on the line that is the graph of $x + 3y = 6$.

Let $x = 0$, $0 + 3y = 6$ Let $y = 0$, $x + 3(0) = 6$
              $3y = 6$ $x = 6$
              $y = 2$

(0, 2) is on the graph.   (6, 0) is on the graph.

Now graph (0, 2) and (6, 0) and draw a line through these two points.

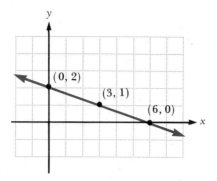

It is a good idea to check a third point on the graph as a safeguard against error. Letting $x = 3$ we have

$$3 + 3y = 6$$
$$3y = 3$$
$$y = 1$$

Thus, $(3, 1)$ should be on the graph. Inspection of the graph above shows that $(3, 1)$ is indeed on the line.

In Example 5, the graph of $x + 3y = 6$ crossed the x-axis at $(6, 0)$ and the y-axis at $(0, 2)$. In general, the solutions $(a, 0)$ and $(0, b)$ of an equation in two unknowns are the points where the graph crosses the x-axis and y-axis respectively. The number a is called the **x-intercept** and b is called the **y-intercept** of the graph. It is easy to find the x- and y-intercepts for the graph of an equation, if such points exist.

1. In order to find the x-intercept, substitute 0 for y and solve for x.
2. In order to find the y-intercept, substitute 0 for x and solve for y.

EXAMPLE 6

Find the x- and y-intercepts of $x = 2$ and sketch the graph.

Solution: Notice that $x = 2$ can be rewritten as $x + 0 \cdot y = 2$ so that both x and y appear in the equation.

Let $y = 0$, $x + 0 \cdot 0 = 2$ Let $y = 3$, $x + 0 \cdot 3 = 2$
                $x = 2$ $x = 2$

$(2, 0)$ is on the line. $(2, 3)$ is on the line.

If we let $x = 0$, we get $0 + 0 \cdot y = 2$ or $0 = 2$! Absurd! This indicates that *no* point $(0, y)$ is on the line and thus, there is no y-intercept. The line is vertical (parallel to the y-axis) and has x-intercept 2.

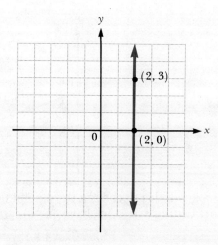

EXAMPLE 7   Graph $4x + 3y = 12$.

Solution:

Let $x = 0$, $4(0) + 3y = 12$ Let $y = 0$, $4x + 3(0) = 12$
                      $3y = 12$ $4x = 12$
                       $y = 4$ $x = 3$

(0, 4) is on the line.   (3, 0) is on the line.
4 is the y-intercept.   3 is the x-intercept.

Graph the points (0, 4) and (3, 0) and use a ruler to finish the sketch.

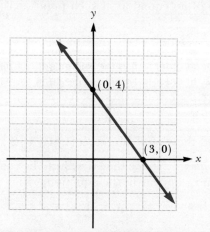

EXAMPLE 8   Graph $y = 2$.

Solution: Rewriting $y = 2$ as $0 \cdot x + y = 2$ produces an equivalent equation containing both x and y.

Let $x = 0$, $\quad 0 \cdot 0 + y = 2$ $\qquad$ Let $y = 0$, $\quad 0 \cdot x + 0 = 2$
$\qquad\qquad\qquad\qquad y = 2$ $\qquad\qquad\qquad\qquad\qquad\qquad 0 = 2$
$\qquad\qquad\qquad\qquad\qquad\qquad\qquad\qquad\qquad\qquad$ Nonsense!

(0, 2) is on the line. $\qquad\qquad$ No point $(x, 0)$ is on the line.
2 is the y-intercept. $\qquad\qquad$ The line is parallel to the
$\qquad\qquad\qquad\qquad\qquad\qquad\qquad$ x-axis.

A second point on the line can be obtained by choosing a second value for x.

Let $x = 3$, $\quad 0 \cdot 3 + y = 2$
$\qquad\qquad\qquad y = 2$

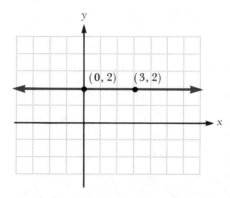

(3, 2) is on the line.

As Examples 6 and 8 suggest, if x appears alone in a linear equation, the graph is a vertical line; if y appears alone, the graph is a horizontal line. To summarize, for any real number k,

Equation	Graph	x-intercept	y-intercept
$x = k$	Vertical line	k	none
$y = k$	Horizontal line	none	k

If x and y both appear with nonzero coefficients in a linear equation, the graph is a line that is neither horizontal nor vertical.

If a line goes through the origin, then both the x-intercept and the y-intercept are 0. We need a point other than an intercept to graph such a line.

Exercises 7.1

Standard Assignment: Exercises 1, 3, 7, 11, 13, 19, 21, 25, 27, 29, 31, 33, 35, 39, 43, 45, 47

A State which quadrant, if any, contains each of the following points.

1. (2, 2) 2. (3, −1) 3. (−1, 0) 4. (−2, −5)
5. (0, 0) 6. (−7, 4) 7. (0, 3) 8. (−4, 2)

Graph each of the following on a Cartesian Coordinate System. See Example 1.

9. (2, 2) 10. (−2, 2) 11. (−2, −2) 12. (2, −2)
13. (0, 6) 14. (−6, 0) 15. (6, 0) 16. (0, −6)
17. (4, 1) 18. (−7, 3) 19. (0.5, −3) 20. (−4, −3.5)

Determine by substitution whether each of the following points is on the given line. See Example 2.

21. $5x - 2y = 3$ (0, 0), (1, 1), (3, 6), (−1, 7)
22. $x + y = 13$ (15, −2), (0, 13), (−1, −12), (−20, 6)
23. $2x - y = 0$ (0, 0), (2, 4), (2, −4), (−3, 6)
24. $3x = 2y + 1$ (0, 1), (1, 0), (2, 3), (3, 4)
25. $x = 4y$ (0, 0), (2, 8), (8, 2), $\left(1, \dfrac{1}{4}\right)$

Complete each ordered pair so that it is a solution of the given equation. See Example 3.

26. $y = 3x + 6$ (0,), (, 0), (3,), (, 9)
27. $x - y = 4$ (0,), (, 0), (4,), (, 1)
28. $2x + 3y = 5$ (0,), (, 0), (1,), (, 3)
29. $x - 2y = 8$ (0,), (, 0), (2,), (, −2)
30. $2x + 6y = 6$ (0,), (, 0), (−6,), (, −1)

Find the x-intercept and y-intercept (if any) of each of the following equations and sketch the graph of the equation.

31. $y = 2x + 3$ 32. $2x - y = 5$
33. $3x - y = 4$ 34. $x + 2y - 3 = 0$
35. $3x + 2y = 6$ 36. $4x + y = 0$
37. $x = 4$ 38. $y = -3$
39. $\dfrac{y}{6} = 1$ 40. $\left(\dfrac{2}{7}\right)x = 6$

41. Graph $y = 2x$, $y = 3x$, and $y = 4x$ on the same coordinate system.
42. Graph $y = 2x$, $y = 2x + 1$, and $y = 2x + 2$ on the same coordinate system.

43. Find m if the x-intercept for $y = mx + 3$ is 1.
44. Find b if the y-intercept for $y = 2x + b$ is 5.
45. Find m if the line $y = mx + 1$ passes through the point $(1, 2)$.
46. Find b if the line $y = 3x + b$ passes through the point $(-1, 1)$.
47. The value V, of a tractor that is purchased for \$14,000 and depreciates linearly at a rate of 10% per year is $V = 14{,}000 - 1400\,t$ where t represents the number of years since the purchase. Find the value of the tractor after
    (a) 2 years   (b) 6 years
    When will the tractor have no value?
48. The number y, of photocopying machines that a company can purchase during a given year is related to the number x, of personal computers that it will purchase during the year by $2x + y = 50$. Graph the equation. How many photocopying machines can be purchased if 10 personal computers are bought that year?
49. The cost C, of producing x line printers is given by $c = 12x + 1000$. Graph the equation with C represented on the vertical axis. What is the cost for producing
    (a) 100 printers?   (b) 101 printers?   (c) 300 printers?   (d) 301 printers?

B 50. Find m and b if the line $y = mx + b$ passes through the point $(1, 4)$ and the origin.

51. Find m and b if the line $y = mx + b$ has the x-intercept 2 and the y-intercept -8.
52. Graph $y = |x|$. (Hint: $|x| = \begin{cases} x & \text{if } x \geq 0 \\ -x & \text{if } x < 0 \end{cases}$ so the graph $y = |x|$ coincides with the graph of the line $y = x$ for $x \geq 0$ and with the line $y = -x$ for $x < 0$.)
53. Graph $y = -|x|$.

7.2 The Slope of a Line

We are accustomed to talking about the "steepness" of a hill, or of the ascent of an airplane. In this section we will see how a number called the slope can be assigned to a line as a measure of the steepness of the line.

If (x_1, y_1) and (x_2, y_2) are points on a line L, the **slope** of the line, which is denoted by the letter m, is the fraction

$$\text{slope} = m = \frac{y_2 - y_1}{x_2 - x_1} \quad \text{for } x_1 \neq x_2$$

If $x_1 = x_2$, the denominator of this fraction is 0; in this case L is a vertical line and the slope is undefined.

Consider the line that passes through $(-2, -3)$ and $(1, 3)$ in Figure 7.3. Letting $(x_1, y_1) = (-2, -3)$ and $(x_2, y_2) = (1, 3)$, we find the slope

$$m = \frac{3 - (-3)}{1 - (-2)} = \frac{6}{3} = 2$$

The numerator 6 is the number of units that a point on the line rises as that point moves from $(-2, -3)$ to $(1, 3)$. The denominator 3 is the number of units that a point moves across, or "runs" above or below the x-axis as it moves from $(-2, -3)$ to $(1, 3)$ on the line. See the dashed arrows in Figure 7.3.

It is sometimes helpful to formulate the equation for the slope of a line containing the points (x_1, y_1) and (x_2, y_2) as

$$m = \frac{\text{change in } y}{\text{change in } x} = \frac{\text{rise}}{\text{run}}$$

where the "rise" is the change in y-coordinates between the points (x_1, y_1) and (x_2, y_2) and the "run" is the corresponding change in the x-coordinates. A positive run indicates movement to the right; a negative run indicates movement to the left. Similarly, a positive rise indicates upward movement; a negative rise indicates downward movement. If we let $\Delta y =$ change in y and $\Delta x =$ change in x then

$$m = \frac{\Delta y}{\Delta x}$$

Figure 7.3

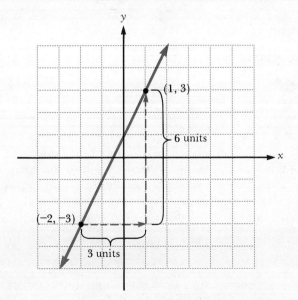

EXAMPLE 1 | Find the slope of the line containing the points $(-1, 4)$ and $(3, 2)$. Graph the line.

Solution:

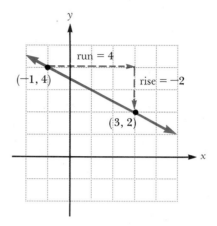

Letting $(x_1, y_1) = (-1, 4)$ and $(x_2, y_2) = (3, 2)$ we have

$$m = \frac{y_2 - y_1}{x_2 - x_1} = \frac{2 - 4}{3 - (-1)} = \frac{-2}{4}$$

The dashed arrows in the figure represent the rise and the run. The negative number -2 obtained for the rise indicates that there is actually a drop in height when going from $(-1, 4)$ to $(3, 2)$ on the line.

It is important to recognize that, when using the slope formula, the result does not depend on which point is called (x_1, y_1) and which is called (x_2, y_2). In Example 1, if we had set $(x_1, y_1) = (3, 2)$ and $(x_2, y_2) = (-1, 4)$, we would still obtain

$$m = \frac{4 - 2}{-1 - 3} = \frac{-2}{4}$$

as before. Similar triangles can be used to show that the same number is obtained for the slope regardless of which two points on the line are chosen.

Given the slope of a line and a point on the line, we can graph the line by using the rise and run to obtain a second point. We express the slope as a fraction and interpret the numerator as the rise and the denominator as the run.

EXAMPLE 2 | Graph the line that passes through (x_1, y_1) and has a slope of m.

(a) $(x_1, y_1) = (-2, -3)$, $m = 3$.

(b) $(x_1, y_1) = (0, 3)$, $m = -\frac{2}{3} = \frac{-2}{3}$.

Solution:

(a) First write $m = 3 = \frac{3}{1}$ and interpret 3 as the rise and 1 as the run. See Figure 7.4(a). Then we move to the right 1 unit and up 3 units from $(-2, -3)$ to find a second point $(-1, 0)$ on the line. This second point is then $(x_1 + \text{run}, y_1 + \text{rise})$ with $(x_1, y_1) = (-2, -3)$.

Figure 7.4

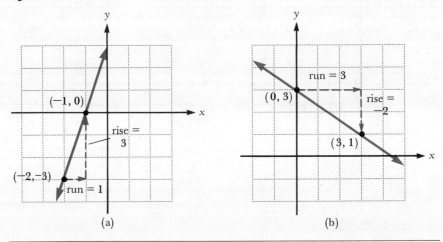

(b) Here the rise is -2 and the run is 3. We move to the right 3 units and drop down 2 units from $(0, 3)$ to find a second point $(3, 1)$ on the line. Again, the second point is $(x_1 + \text{run}, y_1 + \text{rise})$. See Figure 7.4(b).

Lines in a plane can be classified relative to a coordinate system as being one of four kinds:

1. rising—from left to right;
2. falling—from left to right;
3. horizontal, or
4. vertical.

Each type is characterized by its slope:

1. positive slope;
2. negative slope;
3. zero slope, and
4. undefined.

See Figure 7.5. You should remember that a horizontal line has slope zero; slope for a vertical line is undefined.

Figure 7.5

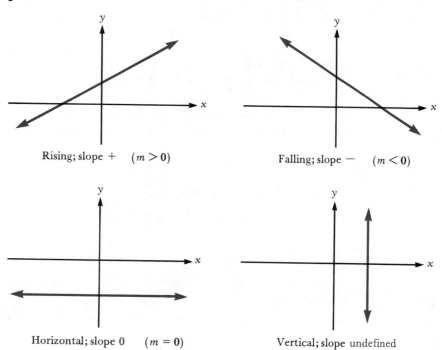

Rising; slope + ($m > 0$)

Falling; slope − ($m < 0$)

Horizontal; slope 0 ($m = 0$)

Vertical; slope undefined

EXAMPLE 3   Find the slope m, of each of the following lines and classify the line as rising, falling, vertical, or horizontal.

(a) $3x + 2y = -12$
(b) $x = 4$
(c) $3y = 7$

Solution:

(a) First we find two points on the line.

Let $x = 0$, $3(0) + 2y = -12$ Let $y = 0$, $3x + 2(0) = -12$
$$2y = -12$$
$$y = -6$$
$$3x = -12$$
$$x = -4$$

$(0, -6)$ is on the line. $(-4, 0)$ is on the line.

Thus, $m = \dfrac{-6 - 0}{0 - (-4)} = \dfrac{-6}{4} = \dfrac{-3}{2}$

Since m is negative, the line is falling.

(b) We could find two points here, but we have already discussed the equation $x = k$. The graph is a vertical line since y is absent from the equation. The slope for this line is undefined.

(c) Rewriting $3y = 7$ as $y = \dfrac{7}{3}$, we see that this is a horizontal line and consequently has slope 0. We could find two points and do the calculation for m, but there is no need to do so.

It is possible to decide whether two lines are parallel, perpendicular, or neither by comparing their slopes. The test is described next.

Parallel and Perpendicular Lines

Let L_1 and L_2 be lines having slopes m_1 and m_2 respectively.

L_1 is parallel to L_2 if and only if $m_1 = m_2$

L_1 is perpendicular to L_2 if and only if $m_1 \cdot m_2 = -1$

Any two vertical lines are parallel, and any horizontal line is perpendicular to any vertical one.

EXAMPLE 4

Decide whether the following pairs of lines are parallel, perpendicular, or neither.

(a) L_1: $x + 2y = 17$ and L_2: $x = -2$.
(b) L_1: $y = x + 2$ and L_2: $3y - 3x = 0$.
(c) L_1: $y = 2x + 7$ and L_2: $2y + x = 0$.

Solution:

(a) As L_2 is a vertical line, we need only decide whether L_1 is vertical, horizontal, or neither. Both x and y appear in the equation for L_1 with nonzero coefficients, and so L_1 is neither horizontal nor vertical. Thus, L_1 and L_2 are neither parallel nor perpendicular.

(b) In order to find the slope of each line, we first find two points on the line. The points $(0, 2)$ and $(-2, 0)$ are easily seen to be on L_1. Thus,

$$m_1 = \text{slope } L_1 = \frac{0 - 2}{-2 - 0} = \frac{-2}{-2} = 1$$

The points $(0, 0)$ and $(1, 1)$ are on L_2. Thus,

$$m_2 = \text{slope } L_2 = \frac{1 - 0}{1 - 0} = \frac{1}{1} = 1$$

Consequently, L_1 and L_2 are parallel.

(c) To find the slope of each line, we first find two points on the line. The points (0, 7) and (1, 9) are on L_1, and hence

$$m_1 = \frac{9 - 7}{1 - 0} = \frac{2}{1} = 2$$

The points (0, 0) and (−2, 1) are on L_2, and hence

$$m_2 = \frac{1 - 0}{-2 - 0} = \frac{1}{-2} = -\frac{1}{2}$$

Since $m_1 \cdot m_2 = 2\left(-\frac{1}{2}\right) = -1$, L_1 and L_2 are perpendicular.

Another technique for solving this type of problem will be introduced in the next section.

Exercises 7.2

Standard Assignment: Exercises 3, 5, 7, 9, 13, 15, 17, 19, 21, 23, 25, 27, 29, 35

A Find the slope of the line containing each of the following pairs of points. Graph the line. See Example 1.

1. (0, 4) and (2, 0)
2. (1, 0) and (2, 3)
3. (4, 6) and (−2, 6)
4. (−3, 7) and (−3, −4)
5. (7, 8) and (−2, −1)
6. (3, 2) and (2, 3)
7. $\left(1, \frac{1}{2}\right)$ and (0, 2)
8. (−1, 2) and (6, 9)

Graph the line that passes through the given point and has a slope of m by using the rise and run method. See Example 2.

9. $m = -3$; passes through (1, 1)
10. $m = 1$; passes through (0, 0)
11. $m = \frac{1}{2}$; passes through (0, 4)
12. $m = -\frac{1}{2}$; passes through (0, 4)
13. $m = 0$; passes through (3, 5)
14. $m = \frac{3}{2}$; passes through (2, 1)
15. $m = -\frac{3}{2}$; passes through (2, 1)
16. $m = \frac{2}{5}$; passes through (−1, 0)

Find the slope m, of each of the given lines and classify the lines as rising, falling, vertical, or horizontal. See Example 3.

17. $2x - y = 7$
18. $y = -3x + 9$

19. $y = \frac{1}{2}x + 1$

20. $4x + 2y = 3$

21. $7x = 21$

22. $3y - 2 = 10$

23. $x = \frac{2}{3}y - 1$

24. $7y - 2x = 4$

Determine whether the following pairs of lines are parallel, perpendicular, or neither.

25. $x = -1$ and $2x + 7 = 9$

26. $x = 0$ and $y = 0$

27. $2x + 7y = 5$ and $y = 2$

28. $y = 3x + 1$ and $6y + 2x = 0$

29. $10x + 2y = 3$ and $y + 1 = -5x$

30. $4x + 3y = 1$ and $3 + y = 2x$

31. $y = x$ and $x + y = 12$

32. $y = x$ and $y = 1$

33. $3x + 8y = 7$ and $5x - 7y = 0$

34. $2y + 7 = 0$ and $9 - 5x = 13$

35. $7y = 3x + 14$ and $6x = 14y + 14$

36. $x = 4y + 8$ and $y = -4x + 1$

B 37. If the slope of the line is $\frac{3}{2}$, find the change in y, $y_2 - y_1$, that is associated with the points (x, y_1) and $(x + 1, y_2)$ on this line.

38. If the slope of a line is m, find the change in y, $y_2 - y_1$, that is associated with the points (x, y_1) and $(x + 1, y_2)$ on the line.

39. Suppose that an air conditioner with an initial cost of $8000 has a value $v = 8000 - 800t$ after t years.
    (a) What is the value of the air conditioner after 2 years?
    (b) What is its value after 3 years?
    (c) How much did it depreciate from the 2nd to the 3rd year?
    (d) What is the slope of the line with the equation $v = 8000 - 800t$?

40. The amount owed A, after t years on an initial loan of $10,000 at a simple interest rate of 12%, is given by the equation: $A = 10,000 + 1200t$ (dollars). How much is owed after
    (a) 3 years?
    (b) 4 years?
    (c) What is the slope of the given equation?

7.3 Equations of a Line

In Section 7.1 we saw that any equation that is equivalent to the linear equation $ax + by = c$ has a straight line for its graph. In this section we will learn how to find an equation for a line.

Figure 7.6

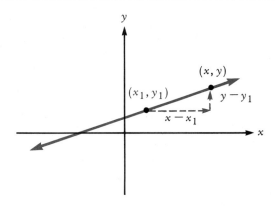

If L is a line with slope m that contains the fixed point (x_1, y_1), we can determine an equation for L as follows: All points (x, y) on L other than (x_1, y_1) must satisfy the slope formula in Equation (7.1). See Figure 7.6.

$$\frac{y - y_1}{x - x_1} = m \qquad (7.1)$$

Multiplying both sides by $x - x_1$ yields

$$y - y_1 = m(x - x_1) \qquad (7.2)$$

The left side of Equation (7.1) is undefined for $x = x_1$, and consequently, (x_1, y_1) is missing from the graph of Equation (7.1). Equation (7.2), however, describes every point on L, including (x_1, y_1). Substituting $x = x_1$, and $y = y_1$ gives $y_1 - y_1 = m(x_1 - x_1)$ or $0 = 0$. Thus, (x_1, y_1) is on the graph of Equation (7.2). It is called the **point-slope form** of the linear equation and has L for its graph.

EXAMPLE 1   Find an equation for the line that passes through the point $(3, 2)$ and has a slope of $m = -2$.

Solution: In order to use the point-slope formula in Equation (7.2), we need only to determine the three numbers x_1, y_1, and m from the given information. The given point is $(3, 2) = (x_1, y_1)$, and so we have $x_1 = 3$ and $y_1 = 2$. The slope is $-2 = m$. Thus

$$y - y_1 = m(x - x_1) \qquad \text{Point-slope form}$$
$$y - 2 = -2(x - 3) \qquad \text{(We could stop here.)}$$

Equivalently,

$$y - 2 = -2x + 6$$

or

$$2x + y = 8$$

The form $ax + by = c$ is called the **general form** of the linear equation.

EXAMPLE 2 Find an equation for the line that passes through the point (0, 3) and has a slope of $m = \frac{1}{3}$. Graph the line.

Solution: We use the point-slope formula in Equation (7.2) as in Example 1. Here $x_1 = 0$, $y_1 = 3$, and $m = \frac{1}{3}$. Thus

$$y - y_1 = m(x - x_1) \quad \text{Point-slope equation}$$

$$y - 3 = \frac{1}{3}(x - 0)$$

$$y = \frac{1}{3}x + 3$$

The graph is sketched in Figure 7.7.

Figure 7.7

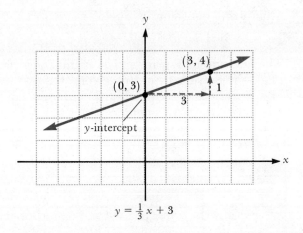

$y = \frac{1}{3}x + 3$

If we want to find an equation of a line having a y-intercept of b and a slope of m, we can proceed as in Example 2. The point $(0, b)$ is on the line, according to Equation (7.2), with $(x_1, y_1) = (0, b)$,

$$y - y_1 = m(x - x_1)$$
$$y - b = m(x - 0)$$

The desired line equation is

$$y = mx + b \tag{7.3}$$

7.3 Equations of a Line

The equation $y = mx + b$ is called the **slope-intercept form** of the linear equation. Note that when the equation is written in this form, the coefficient of the x term is the slope of the line (m), and the constant (b) is the y-intercept. If we are given any linear equation, we can isolate y on one side of the equation and rewrite the other side as $mx + b$ in order to determine the slope and y-intercept of the line. This is called solving explicitly for y in terms of x. We will demonstrate this in Example 3.

EXAMPLE 3 Rewrite the equation $6x + 2y = -8$ in slope-intercept form and find the slope and y-intercept.

Solution:
$$6x + 2y = -8$$
$$2y = -6x - 8$$
$$y = -3x - 4$$
$$y = mx + b \quad \text{With } m = -3 \text{ and } b = -4$$

Thus, the slope is -3 and the y-intercept is -4.

WARNING When finding the slope of $6x + 2y = -8$, notice that $m \neq 6$. You must solve for y as in Example 3.

EXAMPLE 4 Find an equation for the line that has a slope of $m = \frac{1}{2}$ and the y-intercept is $b = 5$.

Solution: Simple substitution into the slope-intercept equation is all that is required.

$$y = mx + b$$
$$\downarrow \quad \downarrow$$
$$y = \frac{1}{2}x + 5$$

Equivalently
$$2y = x + 10$$
$$-x + 2y = 10$$

We can also find an equation of a line that passes through two given points. Use the two points to compute the slope of the line and then use either point in the point-slope form of the line equation.

EXAMPLE 5 Find an equation for the line that passes through $(-1, 2)$ and $(3, 3)$.

Solution: First we find the slope of the line.

$$m = \frac{3 - 2}{3 - [-1]} = \frac{1}{4}$$

Next use

$$y - y_1 = m(x - x_1)$$
$$\downarrow \quad \downarrow \quad \quad \downarrow$$
$$y - 3 = \frac{1}{4}(x - 3) \quad \text{With } (x_1, y_1) = (3, 3)$$

or
$$4y - 12 = x - 3$$
$$-x + 4y = 9$$

Choosing $(x_1, y_1) = (-1, 2)$ yields an equivalent equation.

EXAMPLE 6 Find an equation of the line that passes through the point $(1, 4)$ and is parallel to the line $3y + 6x = 9$.

Solution: We are given that the point $(1, 4)$ is on the line whose equation we want to find; we must know the slope of the line in order to find the equation. However, from Section 7.2 we know that parallel lines have the same slope and we can find the slope of $3y + 6x = 9$ as follows.

$$3y + 6x = 9$$
$$3y = -6x + 9$$
$$y = -2x + 3 \quad \text{Slope-intercept form}$$

Thus, $m = -2$ is the slope of $3y + 6x = 9$ and also of the parallel line that passes through the point $(1, 4)$ whose equation we are seeking. Consequently, using the point-slope formula

$$y - y_1 = m(x - x_1)$$
$$\downarrow \quad \downarrow \quad \quad \downarrow$$
$$y - 4 = -2(x - 1) \quad \text{With } (x_1, y_1) = (1, 4)$$
$$y = -2x + 6$$

EXAMPLE 7 Find an equation of the line that passes through $(-1, 3)$ and is perpendicular to the line $y = \frac{1}{7}x + 1$.

Solution: We know the point $(-1, 3)$ is on the line whose equation we seek. If we can determine the slope m, we can use the point-slope formula. The line $y = \frac{1}{7}x + 1$ has a slope of $\frac{1}{7}$ and is perpendicular to the line in question. From Section 7.2 we know that the product of the slopes is -1.

$$\frac{1}{7}m = -1$$

or
$$m = -7$$

Consequently, using

$$y - y_1 = m(x - x_1)$$
$$y - 3 = -7[x - (-1)] \quad \text{With } (x_1, y_1) = (-1, 3)$$
$$y - 3 = -7(x + 1)$$
$$y = -7x - 4$$

Next we will summarize the different forms for equations whose graphs are lines.

Form	Equation	Slope	Other Information
General	$ax + by = c$ $a \neq 0, b \neq 0$	$-a/b$	x-intercept is c/a y-intercept is c/b
Point-Slope	$y - y_1 = m(x - x_1)$	m	line passes through (x_1, y_1)
Slope-Intercept	$y = mx + b$	m	y-intercept is b x-intercept is $-b/m$ if $m \neq 0$
Vertical Line	$x = k$	undefined	x-intercept is k no y-intercept
Horizontal Line	$y = k$	0	no x-intercept y-intercept is k

Exercises 7.3

Standard Assignment: Exercises 3, 5, 9, 11, 13, 15, 17, 21, 23, 25, 29, 35, 37, 39, 43, 45, 47, 51, 57.

A Find an equation for each of the following lines. See Examples 1 and 2.

1. $m = 0$; passes through $(-9, 3)$
2. $m = 1$; passes through $(3, -2)$
3. $m = \dfrac{1}{7}$; passes through $(-1, 1)$
4. $m = -3$; passes through $(5, 1)$
5. $m = -\dfrac{1}{3}$; passes through $(0, 4)$
6. $m = \dfrac{3}{2}$; passes through $(3, 0)$
7. $m = \dfrac{5}{2}$; passes through $(4, 0)$
8. $m = -\dfrac{2}{5}$; passes through $(-2, -2)$

9. m is undefined; passes through $\left(-\frac{1}{2}, 6\right)$

10. $m = 0$; passes through $(23, 5)$

11. A vertical line that passes through $(5, 1.072)$.

12. A horizontal line that passes through $(1.414, -1)$.

13. A horizontal line that passes through $(0, 0)$.

14. A vertical line that passes through $(0, 0)$.

Rewrite each of the following linear equations in slope-intercept form. Find the slope and y-intercept. See Example 3.

15. $x - y = 0$
16. $x + y = 0$
17. $x + 2y = 1$
18. $2x + y = -1$
19. $-2x + y = 6$
20. $-x + 2y = 4$
21. $3x + 6y = -12$
22. $7x + 4y = 3$
23. $2x - 5y = 11$
24. $8x + 3y = -4$

Find an equation of the line that passes through the given pair of points. See Example 5.

25. $(0, 1)$ and $(1, 0)$
26. $(0, 1)$ and $(1, 3)$
27. $(3, 4)$ and $(5, 7)$
28. $(3, 4)$ and $(5, 1)$
29. $(-1, 3)$ and $(3, 3)$
30. $(-2, -1)$ and $(1, 1)$
31. $(-5, 2)$ and $(2, 7)$
32. $(0, 5)$ and $\left(1, \frac{1}{2}\right)$
33. $(8, 5)$ and $(-1, 4)$
34. $(-1, -3)$ and $(6, -9)$
35. $\left(\frac{1}{2}, \frac{1}{4}\right)$ and $(0, 2)$
36. $(4, -7)$ and $(4, 3)$

Find an equation for each of the following lines. See Examples 4, 6, and 7.

37. $m = 0$; y-intercept 14
38. $m = 2$; y-intercept 5
39. $m = -\frac{1}{2}$; y-intercept 0
40. $m = -\frac{2}{3}$; y-intercept -4
41. $m = \frac{5}{3}$; y-intercept 2
42. $m = -6$; y-intercept -3
43. $m = -\frac{2}{7}$; y-intercept -1

44. parallel to $y = 3$; passes through $(4, 7)$

45. perpendicular to $x = -4$; passes through $(-3, -8)$
46. parallel to $x + y = 1$; passes through $(1, 1)$
47. parallel to $y = 6x + 5$; y-intercept -2
48. perpendicular to $3x - 9y = 18$; passes through $(-2, 4)$
49. perpendicular to $-2x + y = 14$; passes through $(0, 0)$
50. perpendicular to $y = 6x + 5$; y-intercept 5
51. parallel to $x = 3$; passes through $(14, 21)$
52. perpendicular to $2y - 8 = 0$; passes through $(11, -1)$
53. parallel to $-2x + 3y - 7 = 0$; passes through $(1, 0)$
54. perpendicular to $-5x + y = 17$; passes through $(-3, 0)$
55. perpendicular to $\frac{1}{2}y = 7$; passes through $(-2, -5)$

B In the following exercises assume that the situation can be approximated by using a linear equation to describe the relationship between the variables.

EXAMPLE: In 1980, testing showed that each 1000 liters of water from Lake Heron contained 7 milligrams of a polluting mercury compound. One year later, test results showed that 7.5 milligrams of the compound were contained in each 1000 liters of the lake's water. Assuming that the increase of pollutants is linear, find a linear equation relating the year the test was given to the number of milligrams of pollutants per 1000 liters of lake water. What is the pollutant content per 1000 liters of water predicted to be in 1988?

Solution: We assume that the number of milligrams per 1000 liters of water, y, is related to the year the test is run, x, by a linear equation, $y = mx + b$. The points $(1980, 7)$ and $(1981, 7.5)$ are on the graph of this equation according to the two tests described earlier. Thus

$$m = \frac{7.5 - 7}{1981 - 1980}$$

$$= \frac{0.5}{1} = 0.5$$

and we have

$$y = 0.5x + b$$

Because $(1980, 7)$ is on the line

$$7 = 1980(0.5) + b$$
$$7 = 990 + b$$
$$-983 = b$$

Finally,

$$y = 0.5x - 983$$

In 1988 $y = 0.5(1988) - 983$
or $y = 994 - 983$
           $y = 11$

There will be 11 milligrams of the pollutant in each 1000 liters of water in Lake Heron in 1988 at the projected rate of increase.

56. The Forever Toaster Company sold its most expensive toaster for $146 in 1980 and for $158 in 1982. Assuming the price increases linearly over the next six years, find the price of the toaster in 1988.

57. A car purchased for $8400 is worth $6300 after three years. Assuming the value of the car decreases linearly over the next several years, what will be its value four years after its purchase? Six years after its purchase?

58. Four months after a simple interest loan is obtained, the amount due is $6522.67. The 20-year payment would be $13,760. What is the amount due after ten years? What is the simple interest rate?

7.4 Linear Inequalities and Their Graphs

A **linear inequality** occurs when we replace the equal sign in the expression $ax + by = c$, where either a or b is not zero, with one of the inequality signs, $<, \leq, >,$ or $\geq$.

In order to investigate the graph of a linear inequality such as $-2x + y \leq 4$, we first write the inequality with y isolated on one side.

$$-2x + y \leq 4$$
$$y \leq 2x + 4$$

A solution for this inequality is any ordered pair (x, y) for which either $y = 2x + 4$ or $y < 2x + 4$. The graph of $y = 2x + 4$ is a straight line. Every point on the line is a solution of the inequality and is on the graph of the inequality. This line is graphed in Figure 7.8(a). The graph of $y < 2x + 4$ consists of all the points (x, y) below the line $y = 2x + 4$. This is easily visualized by considering any vertical line $x = x_0$. See Figure 7.8(a). The point (x_0, y_1) is on the line so that $y_1 = 2x_0 + 4$; however, for the point (x_0, y_2) below the line, we see that $y_2 < y_1 (= 2x_0 + 4)$, so (x_0, y_2) satisfies $y < 2x + 4$. Thus, (x_0, y_2) is on the graph of $y < 2x + 4$.

On the other hand, for the point (x_0, y_0) above the line, we have $y_0 > y_1$ $(= 2x_0 + 4)$, so that (x_0, y_0) does not satisfy $y < 2x + 4$ and is not on the graph of $y < 2x + 4$. The graph of $y \leq 2x + 4$ consists of all the points below

7.4 Linear Inequalities and Their Graphs

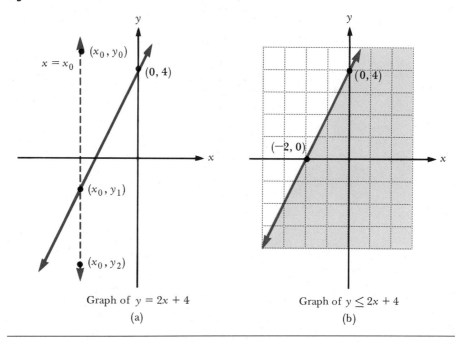

Figure 7.8

Graph of $y = 2x + 4$
(a)

Graph of $y \leq 2x + 4$
(b)

the line $y = 2x + 4$ together with all the points on the line $y = 2x + 4$. See Figure 7.8(b).

In general, if either a or b is not zero, the graph of each of $ax + by < c$, $ax + by > c$, $ax + by \leq c$, and $ax + by \geq c$ is a *half-plane*. This half-plane is either above or below the line $ax + by = c$ (if $b \neq 0$). If $b = 0$, the graph consists of either the half-plane on the left or the half-plane on the right of the vertical line $ax = c$. The line is included in the graph of a linear inequality involving the signs $\leq$ or $\geq$, and is excluded otherwise. We indicate the exclusion of a line from the graph by graphing it as a broken or dashed line. This is illustrated in Example 1.

EXAMPLE 1 Graph (a) $2x + 3y \geq 6$ (b) $2x + 3y > 6$

Solution:

(a) First we graph the line $2x + 3y = 6$. It is included in the graph because the sign $\geq$ is involved, and consequently, it is graphed as a *solid* line. See Figure 7.9(a). In order to determine whether to shade the half-plane above or below the line, we use any point that is not on the line $2x + 3y = 6$ as a **test point**. The point $(0, 0)$ should be used as a test point whenever it is not on the line.

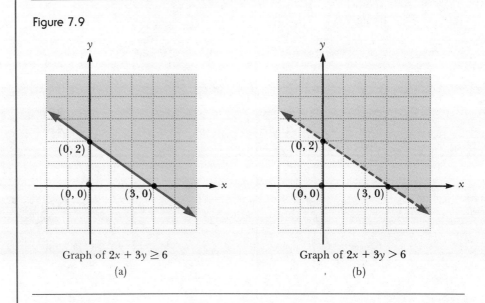

Figure 7.9

Graph of $2x + 3y \geq 6$
(a)

Graph of $2x + 3y > 6$
(b)

Testing $(0, 0)$ in the original inequality $2x + 3y \geq 6$, we have

$$2(0) + 3(0) \geq 6$$
$$0 \geq 6 \quad \text{False!}$$

The test point $(0, 0)$ is not a solution of the inequality $2x + 3y \geq 6$. The point $(0, 0)$ is on the half-plane below the line, and so this half-plane is not part of the graph. The graph consists of the half-plane above the line $2x + 3y = 6$, together with the line $2x + 3y = 6$. See Figure 7.9(a).

(b) We have only one alteration to make in the previous graph. The equation $2x + 3y > 6$ involves the $>$ sign, and so we should graph the line $2x + 3y = 6$ as a broken line to indicate that it is excluded from the graph of $2x + 3y > 6$. The rest of the procedure is identical to part (a). The graph is shown in Figure 7.9(b).

The procedure for graphing linear inequalities is summarized in the following box.

Graphing a Linear Inequality

1. *Graph* the line $ax + by = c$, which results when the inequality sign is replaced by an equal sign in the given inequality:

   as a *dashed line* if the inequality sign is $<$ or $>$; or

   as a *solid line* if the inequality sign is $\leq$ or $\geq$.

2. *Test Point.* Choose any point not on the line $ax + by = c$ and substitute the coordinates into the original inequality. One of the points (0, 0), (1, 0), or (0, 1) can always be chosen to make calculations easy.
3. *Shade* the half-plane containing the test point if the test point is a solution of the original inequality; shade the half-plane not containing the test point if the test point is not a solution of the original inequality.

EXAMPLE 2

(a) Graph $x \geq -2$ and $2x - y < 0$.
(b) Graph the intersection of the graphs of $x \geq -2$ and $2x - y < 0$.
(c) Graph the union of the graph of $x \geq -2$ and $2x - y < 0$.

Solution:

(a) In order to graph $x \geq -2$, graph $x = -2$ as a *solid* line.
Test point: (0, 0) in $x \geq -2$
$0 \geq -2$ True!
Shade the half-plane containing (0, 0). See Figure 7.10(a).

In order to graph $2x - y < 0$, graph $2x - y = 0$ as a *dashed* line. Test point: (1, 0) in $2x - y < 0$
$2(1) - 0 < 0$
$2 < 0$ False!
Shade the half-plane not containing (1, 0). See Figure 7.10(b).

Figure 7.10

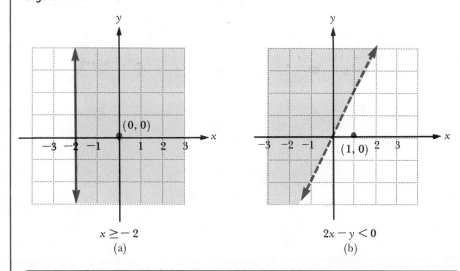

$x \geq -2$
(a)

$2x - y < 0$
(b)

(b) The intersection of the graphs consists of the points in the plane that were shaded in both graphs. See Figure 7.11(a).

Figure 7.11

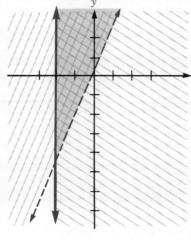

$x \geq -2$ intersect $2x - y < 0$
(a)

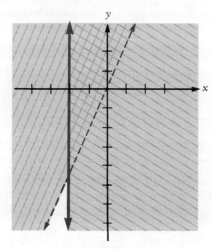

$x \geq -2$ union $2x - y < 0$
(b)

(c) The union of the graphs consists of the points in the plane that were shaded in either graph. See Figure 7.11(b).

Exercises 7.4

Standard Assignment: Exercises 1, 3, 7, 9, 15, 17, 19, 21, 25, 29, 35

A Graph each of the following inequalities. See Example 1.

1. $x + y > 0$
2. $x - y \leq 0$
3. $x + 2y \geq 8$
4. $3x + 4y < -12$
5. $2x - 5y > 10$
6. $2x \geq 6$
7. $3y - 9 \leq 2$
8. $4x + 15 < 7$
9. $3 \leq y - 2$
10. $2y - 3x - 6 < 0$
11. $4y - 5x \geq 20$
12. $x - 2y \leq 1$
13. $y - 3 \geq -2x$
14. $x + 2y - 20 > 0$
15. $9x > 2y - 10$

Graph the intersection of the graphs of the following pairs of inequalities. See Example 2.

16. $x + y \geq 5$; $x + 2y \leq 8$
17. $x - 3y \geq 0$; $x \leq 4$
18. $2y - x \leq 6$; $y > -4$
19. $2x + 3y \leq 6$; $3x + 2y > 6$

20. $2x - 3y \leq 12; y + x + 1 < 0$ 21. $y < 6 - 4x; y + 2 > 0$
22. $3x + 5y \geq 2; y - 3x > 4$ 23. $3x - 4y - 12 > 0; y \leq 2x + 3$
24. $x - 9y < 11; 7x + 2y \geq 12$ 25. $-2x + y \geq 4; 2x + y \geq 2$
26. $y \leq 3x + 3; y + 1 > 3x$ 27. $2y - 7 \geq 5; y \leq -1$

Graph the union of the graphs of the following pairs of inequalities. See Example 2.

28. $x - y \leq 1; x - 3 > 0$ 29. $2x + y > 5; x - y < 0$
30. $2x + 7 \leq 3; y \leq 4$ 31. $-3x + 2y > 6; 2x + y < 2$
32. $y < x + 1; x < y + 1$ 33. $2y + 5 \leq 3; y - 7 \leq 4$
34. $x - 1 \geq 5; 2x < 0$ 35. $5y + 2x - 4 < 0; 7x + 5 > 3y$

7.5 Functions and Relations

The correspondences and relationships that are familiar to us in our everyday lives contain the basic idea for the most fundamental and important concept in mathematics. Consider the following correspondences:

1. To each actor in a play, there corresponds one or more roles;
2. To each worker in a business, there corresponds a social security number;
3. To each professional basketball player, there corresponds one or more positions he plays; and
4. To each person ordering concert tickets, there corresponds the number of tickets ordered.

Each of the previous correspondences can be represented as a set of ordered pairs. Pairs such as: (Tom C. Head, 223 41 6007), (Jane L. Pitts, 346 20 7141), (Mary B. Good, 442 77 0625), ... can represent the correspondence in the second example. Pairs such as ("Magic" Johnson, Forward), ("Magic" Johnson, Center), ("Magic" Johnson, Guard), (Kareem Abdul Jabaar, Center), (Julius Erving, Forward), (Julius Erving, Guard), ... can represent the correspondence in the third example.

Mathematicians have found that ordered pairs are an exceptionally useful method for displaying and organizing correspondences. Any set of ordered pairs is called a **relation**.

The set of the first components of the ordered pairs in a relation is called the **domain** of the relation. The set of the second components of the ordered pairs is called the **range** of the relation.

We are interested most frequently in relations that can be represented as ordered pairs of numbers. There is usually some useful way to code any given correspondence so that it can be represented by ordered pairs of numbers. In this book, both the domain and the range of every relation is assumed to be a subset of the set of real numbers, unless otherwise stated.

EXAMPLE 1 | Given the relation $R = \{(-2, 2), (-1, 4), (0, 0), (0, 2), (3, 4)\}$, find the domain and the range.

Solution: Domain: $\{-2, -1, 0, 3\}$, the set of first components
Range: $\{0, 2, 4\}$, the set of second components

A type of relation with which we are especially concerned is called a function. A **function** is a relation in which no two ordered pairs have the same first components.

The relation R in Example 1 is *not* a function because the two ordered pairs $(0, 0)$ and $(0, 2)$ in R have the same first component, 0. The second relation in our list of examples of correspondences on page 311 *is* a function because each employee has only one social security number.

The solution sets of equations in two variables provide a rich supply of relations. If the variables are x and y and the ordered pairs are written (x, y), then the solution set is a function if no value of x is paired with more than one value of y.

EXAMPLE 2 | Determine whether the solution sets of the following equations are functions.

(a) $y^2 = x$ (b) $y = 2x + 1$

Solution:
(a) Substituting $x = 1$ in (a) yields $y^2 = 1$. However, $(1)^2 = 1$ and $(-1)^2 = 1$, so $y = \pm 1$. Thus, $(1, 1)$ and $(1, -1)$ are pairs in the solution set for $y^2 = x$. As 1 is paired with two different numbers, 1 and -1, the solution set for $y^2 = x$ *is not* a function.
(b) Regardless of what number we substitute for x in the equation $y = 2x + 1$, only one value of y results. For $x = 1$, we get $y = 2(1) + 1 = 3$; for $x = -\frac{1}{2}$, $y = 2(-\frac{1}{2}) + 1 = 0$, and so on. Each value of x is paired with only one value of y; thus the solution set for $y = 2x + 1$ *is* a function.

When the solution set of an equation is a function, we say the function is *defined* by the equation. The domain of such a function is the set of real numbers, which when substituted for x in the equation, determine y as a real number.

EXAMPLE 3 | Find the domain of the functions defined by the following equations.

(a) $y = \dfrac{1}{1 - x^2}$ (b) $y = \sqrt{x}$ (c) $y = \dfrac{1}{\sqrt{x - 1}}$ (d) $y = 2x + 1$

Solution:
(a) Division by zero is undefined, and so we cannot calculate $y = 1/(1 - x^2)$ if $1 - x^2 = 0$. But $1 - x^2 = 0$ only if $x = \pm 1$. These values of x are not in the domain. The domain is $\{x \mid x \text{ is real}, x \neq \pm 1\}$. We usually consider the "x is real" statement as understood and simply write $\{x \mid x \neq \pm 1\}$.

(b) The square root of a negative number is undefined. In order to calculate $y = \sqrt{x}$, we must have $x \geq 0$. The domain is then $\{x \mid x \geq 0\}$.

(c) The equation $y = 1/(\sqrt{x - 1})$ has a double restriction. We cannot take the square root of a negative number, so we must have $x - 1 \geq 0$ in order to calculate $\sqrt{x - 1}$. However, we cannot allow $x - 1 = 0$, for then $0 = \sqrt{x - 1}$ occurs in the denominator. We must have $x - 1 > 0$, or $x > 1$. The domain is $\{x \mid x > 1\}$.

(d) When any real number is substituted for x in $y = 2x + 1$, a unique real number y is determined. The domain is $\{x \mid x \text{ is real}\}$.

When finding the domain of a function remember that

1. Division by zero is undefined, and
2. The square root (or any even root) of a negative number is undefined.

If we are given a graph in the coordinate plane, we can easily decide if the ordered pairs of the graph constitute a function. If two ordered pairs have the same first component, such as (x_0, y_1), (x_0, y_2), then the vertical line $x = x_0$ passes through both points of the graph. This observation can be expanded to the following test.

Vertical Line Test

The graph of a relation represents a function if and only if no vertical line passes through more than one point of the graph.

The graph in Figure 7.12(a), on the next page, is the graph of a function because no vertical line passes through more than one graph point. Figure 7.12(b) shows a graph that is not the graph of a function because the vertical line $x = x_0$ (among others) passes through more than one point of the graph.

Usually a single letter such as f, g, P, or Q is used to name the function. If we write $f = \{(x, y) \mid y = \sqrt{x}\}$ then f is the name of the function. The domain of f is $\{x \mid x \geq 0\}$ (see Example 3) and $(1, 1)$, $(4, 2)$, and $(81, 9)$ are some of the ordered pairs of f. If (x, y) is an ordered pair in f, we write

$$f(x) = y \quad \text{or} \quad y = f(x)$$

314   Chap. 7 Linear Equations, Inequalities, Functions

and say that "f of x equals y" or that "the value of f at x is y." Thus, because $(4, 2)$ is an ordered pair of $f = \{(x, y) \mid y = \sqrt{x}\}$, we write $f(4) = 2$ and say that the value of f at 4 is 2, or that $f(x) = 2$ when $x = 4$. Since $y = \sqrt{x}$ we can write the general expression for $f(x)$ as $f(x) = \sqrt{x}$.

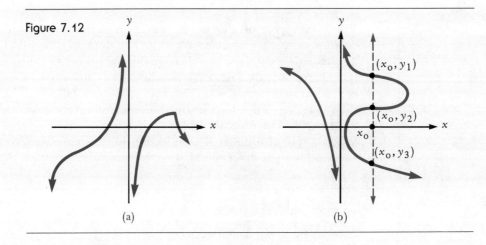

Figure 7.12

(a)   (b)

EXAMPLE 4

Let f be defined by the equation $x + 2y = 3$. Find

(a) an expression for $f(x)$
(b) the value of f at 2
(c) $f(-3)$

Solution:
(a) Since
$$x + 2y = 3$$
$$2y = 3 - x$$
$$y = \frac{(3 - x)}{2}$$
$$f(x) = \frac{(3 - x)}{2}$$

(b) Substituting $x = 2$ in the equation for f yields
$$f(2) = \frac{(3 - 2)}{2}$$
$$f(2) = \frac{1}{2}$$

The value of f at 2 is $\frac{1}{2}$.

(c) Substituting $x = -3$ in

$$f(x) = \frac{(3-x)}{2}$$

$$f(-3) = \frac{(3-[-3])}{2}$$

$$f(-3) = \frac{6}{2}$$

$$f(-3) = 3$$

EXAMPLE 5 Let f be defined by the equation $f(x) = 2x^2 - 7$. Find
(a) $f(1)$
(b) $f(0)$
(c) $f[f(1)]$
(d) $f\left(\dfrac{1}{a}\right)$

Solution:
(a) $f(1) = 2(1)^2 - 7 = -5$
(b) $f(0) = 2(0)^2 - 7 = -7$
(c) Replacing x by $f(1)$ in $f(x) = 2x^2 - 7$ and using part (a), $f[f(1)] = 2[f(1)]^2 - 7 = 2[-5]^2 - 7 = 43$
(d) Replacing x by $1/a$ in $f(x) = 2x^2 - 7$ gives

$$f\left(\frac{1}{a}\right) = 2\left(\frac{1}{a}\right)^2 - 7 = \frac{2}{a^2} - 7$$

EXAMPLE 6 Let f be defined by the equation $f(x) = 1 - x^2$. Find
(a) $f(x + h)$
(b) $\dfrac{f(x + h) - f(x)}{h}$

Solution:
(a) Replacing x by $x + h$ in $f(x) = 1 - x^2$

$f(x + h) = 1 - (x + h)^2 = 1 - (x^2 + 2xh + h^2) = 1 - x^2 - 2xh - h^2$

(b) $f(x + h) - f(x) = (1 - x^2 - 2xh - h^2) - (1 - x^2) = -2xh - h^2$

$$\frac{f(x + h) - f(x)}{h} = \frac{-2xh - h^2}{h} = -2x - h \qquad \text{for } h \neq 0$$

We learned in Section 7.3 that any non-vertical line is the graph of an equation of the form $y = mx + b$. By the vertical line test or by direct inspection of the equation, we see that this equation defines a function. Any function that is defined by an equation of the form $f(x) = mx + b$ is called a **linear function.**

By the vertical line test, the graph shown in Figure 7.13 is the graph of a function, $y = f(x)$. It is important to realize that in the coordinate plane

1. x is on the x-axis.
2. $y = f(x)$ is on the y-axis.
3. (x, y) or $(x, f(x))$ is a point on the graph.
4. $f(x)$ can be represented by the directed length from the point x to the point $(x, f(x))$ on the graph.

Figure 7.13

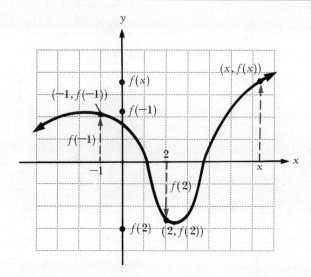

EXAMPLE 7

Graph the linear function defined by $f(x) = -2x + 1$ and represent $f(2)$ by sketching the directed length from 2 to $(2, f(2))$.

Solution: In order to find two points on the linear graph of $f(x) = -2x + 1$, we substitute two values for x into the equation.

Let $x = 0$
$f(0) = -2(0) + 1$
$f(0) = 1$
Thus $(0, 1)$ is on the graph.

Let $x = 2$
$f(2) = -2(2) + 1$
$f(2) = -3$
Thus $(2, -3)$ is on the graph.

The graph and required representation of $f(2)$ are shown in Figure 7.14.

Figure 7.14

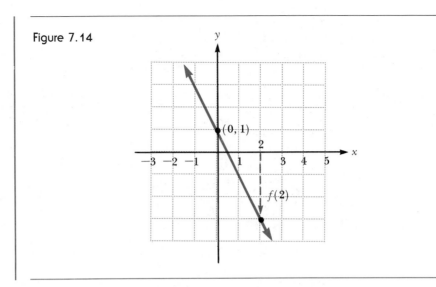

Functions that are defined by equations in two variables, such as x and y, can be thought of as "number processors" in the following way. Given the equation $f(x) = 3x + 4$, we think of x as an *input* that is "processed" by first multiplying x by 3, and then adding 4 to the result of the multiplication. The *output* is the result of this processing, or, $f(x) = 3x + 4$. For an input of $x = 2$, the output is $f(2) = 3(2) + 4 = 10$.

We can think of a number processing function as a one button calculator: Key in x, push the button f; $f(x)$ appears on the display. If x is not in the domain of f, the calculator should display the word "error."

A second machine-like representation is shown in Figure 7.15. Here the input, x, is dropped into an entry slot and processed; it exits as the output, $f(x)$.

Figure 7.15

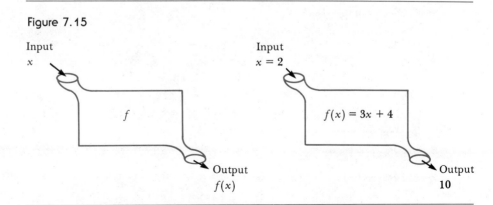

Exercises 7.5

Standard Assignment: Exercises 1, 3, 5, 9, 11, 13, 23, 25, 27, 29, 31, 33, 37, 41, 43, 45, 47, 49, 51

A Determine the domain and range of each of the following relations. State whether the given relation is a function. See Example 1.

1. $\{(-2, 2), (0, 0), (2, 2)\}$
2. $\{(3, 1), (4, 1), (5, 1), (6, 1)\}$
3. $\{(-2, -2), (0, 0), (1, 1) (2, 2)\}$
4. $\{(1, 3), (1, 4), (1, 5), (1, 6)\}$
5. $\{(4, 2), (9, 3), (4, -2), (9, -3)\}$
6. $\{(0, 3), (-1, 1), (-2, -1), (2, 7)\}$

Find the domain of each of the following relations and determine which are functions. See Examples 2 and 3.

7. $x + y = 2$
8. $x = y$
9. $x = \sqrt{y}$
10. $y = \dfrac{1}{x}$
11. $xy = 1$
12. $y = \dfrac{1}{|x - 1|}$
13. $y = \dfrac{1}{\sqrt{x^2 - 1}}$
14. $x = |y|$
15. $y = \sqrt{2x + 1}$
16. $3x - 5y = 15$
17. $y = |x|$
18. $y = \dfrac{1}{x^2 + 1}$
19. $y = \sqrt{x^2 + x + 1}$
20. $y = \dfrac{3x}{2x - 1}$
21. $y = \dfrac{\sqrt{x}}{x^2 + 2}$

Find an expression for $f(x)$ where f is defined by the given equation and find $f(4)$. See Example 4.

22. $3x - 5y = 15$
23. $x = \dfrac{y}{y - 1}$
24. $x = \dfrac{2}{y - 4}$
25. $xy - 3 = 2y$
26. $(x^2 + 1)y + x = 2$
27. $yx^2 - \sqrt{x} = -2y$

Let $f(x) = x^2 - 1$, $g(x) = 2/\sqrt{x}$, and $h(x) = \sqrt{2 - x}$. Find each of the following. See Example 5.

28. $f(-1)$
29. $g(4)$
30. $f(0)$
31. $h(-2)$
32. $h[h(2)]$
33. $g[g(1)]$
34. $h[g(1)]$
35. $g[h(1)]$
36. ▦ $f(36.812)$
37. ▦ $f(-3.4112)$
38. ▦ $g(362.075)$
39. ▦ $h(-4.1705)$

Use the vertical line test to determine whether each of the following graphs represent functions.

40.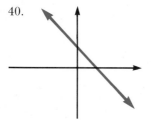

41.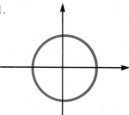

42.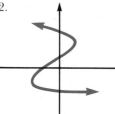

43.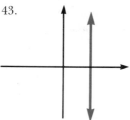

44.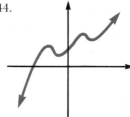

45.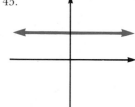

B   For each of the functions in Exercises 46–49 (See Example 6), find

(a) $f(x + h)$ (b) $f(x + h) - f(x)$ (c) $\dfrac{f(x + h) - f(x)}{h}$

46. $y = x$ 47. $y = 2x + 1$ 48. $y = x^2$ 49. $y = x^2 - 3x$

50. Graph $f(x) = x + 2$ and represent $f(1)$ by sketching the directed length from 1 to $(1, f(1))$.

51. Find the slope of the line joining the points $(x_1, f(x_1))$ and $(x_2, f(x_2))$ on the graph of $f(x) = x^2$. Simplify your answer if possible.

52. Solve Exercise 51 for
    (a) $x_1 = 0$ and $x_2 = 1$
    (b) $x_1 = -1$ and $x_2 = 2$
    (c) $x_1 = 3$ and $x_2 = 7$

53. Find the slope of the line joining the points $(a, f(a))$ and $(b, f(b))$ on the graph of $f(x) = 2x^2 - 3x + 5$. Simplify your answer if possible.

54. Solve Problem 53 for $f(x) = \sqrt{x}$, where a and b are two positive real numbers.

7.6 Direct and Inverse Variation

There are two types of functional relationships that occur so frequently they are given special names. Direct variation describes the relationship between two quantities where any increase (or decrease) in one causes a proportional

Chap. 7 Linear Equations, Inequalities, Functions

increase (or decrease) in the other. Inverse variation describes the relationship between two quantities where an increase (or decrease) in one causes a proportional decrease (or increase) in the other.

The following is an example of direct variation. Assuming a sales tax of 5%, the sales tax on a $1000 purchase is $50. If we *double* the purchase price to $2000, the sales tax also *doubles* to $100. If we *reduce* the purchase price by half to $500, the sales tax is also *reduced* by half to $25. Because of this relationship, we say that the sales tax varies directly as the purchase price. We know that

$$\text{Sales tax} = (0.05)(\text{Purchase price})$$

In general, y is said to **vary directly** as x if there is a positive constant k such that $y = kx$. The constant k is called the **constant of variation.** In the sales tax illustration, the constant of variation is $k = 0.05$.

EXAMPLE 1

Assume that y varies directly as x, and $y = 36$ when $x = 24$. Find the value of k and the equation relating x and y.

Solution: Since y varies directly as x

$$y = kx \qquad \text{For some constant } k$$

When $x = 24$, we know $y = 36$

$$36 = k \cdot 24 \qquad \text{By substitution}$$

$$\frac{36}{24} = k$$

$$\frac{3}{2} = k$$

Consequently, $y = \frac{3}{2}x$ is the desired equation.

EXAMPLE 2

The current in a circuit that is connected to a 220-volt battery is 50 amperes. If the current in this circuit varies directly as the voltage of the attached battery, what voltage battery is required to produce a current of 75 amperes?

Solution: Let I = current in amperes and v = battery voltage so that $I = kv$.

$$50 = k \cdot 220$$

$$\frac{50}{220} = k$$

$$\frac{5}{22} = k$$

Thus
$$I = \frac{5}{22}v$$

If we want $I = 75$ amperes, then

$$75 = \frac{5}{22}v$$

$$\frac{22}{5} \cdot 75 = v$$

$$330 = v$$

A battery of 330 volts will produce the required 75 amperes current.

The following is an example of inverse variation. If a plane takes 3 hours to fly from Atlanta to Dallas at an average rate of 200 mph, *increasing* the average rate to 400 mph (twice the previous rate) *decreases* the flight time to $1\frac{1}{2}$ hours (half the previous time). For the Atlanta-Dallas trip, time *varies inversely* as the rate. We can write the formula

$$\text{time} = \frac{\text{distance}}{\text{rate}}$$

In general, y is said to **vary inversely** as x if there is a positive constant k such that

$$y = \frac{k}{x}$$

As before, k is called the constant of the variation.

EXAMPLE 3 Assume that y varies inversely as x and that $y = 35$ when $x = 11$.

(a) Find the constant of variation, k.
(b) Find y if $x = 55$.

Solution:
(a) We know

$$y = \frac{k}{x}$$

So

$$35 = \frac{k}{11} \qquad y = 35 \text{ when } x = 11$$

$$35 \cdot 11 = k$$

$$385 = k$$

(b) From part (a) we know that

$$y = \frac{385}{x}$$

Thus, if $x = 55$,

$$y = \frac{385}{55}$$

$$y = 7$$

There is a natural extension of the language of variation when one quantity varies directly or inversely as a *power* of another quantity.

In particular, y is said to vary directly as the nth power of x if there is a positive constant k such that $y = kx^n$. Also, y is said to vary inversely as the nth power of x if there is a positive constant k such that $y = k/x^n$.

EXAMPLE 4
(a) The area, A, of a circle of radius r varies directly as the square of r. The constant of variation is π since $A = \pi r^2$.
(b) The volume, V, of a cube of side s varies directly as the cube of s. The constant of variation is 1 since $V = s^3$ (or, $V = 1 \cdot s^3$).
(c) The gravitational intensity, G, that is felt by an object near the earth varies inversely as the square of the distance, r, between the object and the earth's center. Thus we have $G = k/r^2$; the constant variation is $k = 39.35 \times 10^{13}$ meters/sec^2.

There are also many combinations in which one variable varies as the product of two or more other variables or powers of variables. We refer to these relationships as joint variation. If y **varies jointly** as t and the square of u, we have $y = ktu^2$.

We can combine joint variation and inverse variation in an obvious way. If T varies jointly as q and the square root of v, and inversely as the square of r, we have $T = (kq\sqrt{v})/(r^2)$.

EXAMPLE 5
Suppose that z varies directly as the square of u and inversely as the square root of t.

(a) If $z = 8$ when $u = 6$ and $t = 9$, find k.
(b) Find z when $u = 12$ and $t = 4$.

Solution:
(a) We have

$$z = \frac{ku^2}{\sqrt{t}}$$

thus
$$8 = \frac{k(6)^2}{\sqrt{9}}$$

$$8 = \frac{k36}{3}$$

$$8 = k \cdot 12$$

$$\frac{2}{3} = k$$

(b) Since
$$z = \frac{ku^2}{\sqrt{t}}$$

from (a)
$$z = \frac{(2/3)u^2}{\sqrt{t}}$$

For $u = 12$ and $t = 4$ we have
$$z = \frac{(2/3)(12)^2}{\sqrt{4}}$$

$$= \frac{2 \cdot 144}{3 \cdot 2} = 48$$

The expression "y is **directly proportional** to x" is sometimes used in place of "y varies directly with x." The phrase "proportional to" replaces "varies with" in a similar way when you are describing the other relationships presented in this section. Thus "y is inversely proportional to x" has the same meaning as "y varies inversely with x," and so on. (See Exercises 13, 14, and 15.)

We will now summarize the steps involved in solving problems with direct and inverse variations.

---

1. Write the equation that models the problem.
   (a) y varies directly with x, $y = kx$
   (b) y varies inversely with x, $y = \dfrac{k}{x}$
   (c) z varies jointly with x and y, $z = kxy$
   (d) z varies directly with x and inversely with y, $z = \dfrac{kx}{y}$
2. Substitute the given values into (1) to solve for k.
3. Rewrite the equation in (1) with the value of k.
4. Solve the resulting equation.

Exercises 7.6

Standard Assignment: Exercises 1, 3, 5, 7, 9, 11, 13, 15, 17

A Solve the following exercises for the constant of variation, k, the equation relating the variables, and the requested value of the variable. See Examples 1, 3, and 5.

1. x varies directly as y, and $y = 30$ when $x = 15$. Find y if $x = 28$.

2. t varies directly as s, and $t = 15$ when $s = 30$. Find t if $s = \frac{13}{2}$.

3. r varies inversely as u, and $r = 3$ when $u = 11$. Find r if $u = \frac{1}{3}$.

4. y varies inversely as z, and $y = 24$ when $z = \frac{1}{6}$. Find y if $z = 1$.

5. m varies directly as q and inversely as p, and $m = \frac{1}{2}$ when $q = 13$ and $p = 26$. Find m when $q = 7$ and $p = 14$.

6. z varies directly as the square of x, and $z = 32$ when $x = 4$. Find z if $x = 5$.

7. u varies inversely as the cube of t, and $u = 9$ when $t = 2$. Find u if $t = 6$.

8. z varies jointly as x and y, and $z = 42$ when $x = 2$ and $y = 7$. Find z when $x = 15$ and $y = 3$.

9. P varies jointly as T and the square of Q, and $P = 36$ when $T = 17$ and $Q = 6$. Find P when $T = 4$ and $Q = 9$.

10. a varies jointly as b and the square root of c, and inversely as d; $a = 9$ when $b = 13$, $c = 81$, and $d = 3$. Find a when $b = 5$, $c = 9$, and $d = 14$.

11. ▦ z varies jointly as u and the cube of v and inversely as the square of w; $z = 29.471$ when $u = 5.34027$, $v = 9.7744$, and $w = 26.9972$. Find z when $u = 35.717$, $v = 21.2412$, and $w = 417.296$.

12. ▦ w varies jointly as y and the square of x and varies inversely as the cube of z; $w = 13.002$ when $x = 47.294$, $y = 217.009$, and $z = 14.7298$. Find w when $x = 22.771$, $y = 42.337$, and $z = 20.20$.

Solve each of the following without determining the constant of variation, k. Use the fact that: If $x_1 = ky_1$, and $x_2 = ky_2$, then $\frac{x_1}{y_1}$ is the same proportion as $\frac{x_2}{y_2}$, and so $\frac{x_1}{y_1} = \frac{x_2}{y_2}$.

> EXAMPLE: If y is directly proportional to x and $y = 12$ when $x = 16$, find y when $x = 8$.

Solution: Using the proportion

$$\frac{12}{16} = \frac{y}{8}$$

we have

$$\frac{3}{4} = \frac{y}{8}$$

$$6 = y$$

13. If z varies directly with w and $z = 17$ when $w = 22$, find z when $w = 110$.

14. The quantity of solid obtained by evaporating seawater is directly proportional to the quantity of seawater evaporated. If 3.5 kg of solid results from 100 kg of seawater, how many kg of solid can be obtained from 200 kg of water?

15. If r is directly proportional to s and $r = 100$ for the value of s_0 of s, find r when s_0 is doubled, that is, when $s = 2s_0$.

Solve each of the following. See Example 2.

16. The quantity p, of protein obtained from dried soybeans is directly proportional to the quantity s, of dried soybeans used. If 20 grams of dried soybeans produces 7 grams of protein, how much protein can be produced from 100 grams of soybeans?

17. The distance d, that an object falls in t seconds varies directly as the square of t. If an object falls 64 feet in 2 seconds, how far does it fall in 7 seconds?

B 18. Boyle's Law says that the pressure P, of a compressed gas is inversely proportional to the volume V. If the pressure is 20 pounds per square inch when the volume of the gas is 300 cubic inches, what is the pressure when the gas is compressed to 100 cubic inches?

19. Hook's Law says that the force F, that is required to stretch an elastic spring d units is directly proportional to the length d. If $F = 10$ newtons when $d = \frac{1}{2}$ meter, what force is required to stretch the spring $\frac{3}{5}$ of a meter?

20. The simple interest returned on an investment for a fixed period of time varies jointly as the principal and the rate. If a $1000 investment earned $600 at 12%, how much interest would be earned on $500 invested at 6% for the same period of time?

21. The intensity of illumination, I, from a light is inversely proportional to the square of the distance d, from the light. If the intensity is 320 candlepower at a distance of 10 feet from the light, what is the intensity 5 feet from the light?

22. The pressure exerted on a given point by a liquid varies directly as the depth of the point beneath the surface of the liquid. If the pressure 15 meters below the surface is 35 newtons, what is the pressure at 70 meters?

23. The distance d, that a ball rolls down an inclined plane varies directly as the

square of the time t, traveled. If the ball rolls 9 feet in $1\frac{1}{2}$ seconds, how far will it roll in 2 seconds?

24. The volume V, of a sphere varies directly as the cube of its radius r. If $V = 972\pi$ when $r = 9$, find V where $r = 6$.

25. For a certain species of snake, the weight of the snake, W, is directly proportional to its length, L. If a snake weighing 57.4 grams has a length of 0.47 meters, find the length of a snake weighing 44.8 grams.

26. The amount A, of protein that can be broken down by x grams of pepsin varies inversely as the square root of x. If 1.539 grams of protein are digested in 43.62 grams of pepsin, how many grams of protein are digested in 84.44 grams of pepsin?

27. The electrical resistance R, of a wire varies directly as the length of the wire and inversely as the square of its diameter d. If 14.7 meters of wire of diameter 1.24 millimeters has a resistance of 14 ohms, find the resistance of 16.3 meters of wire with diameter 0.743.

Chapter 7 Summary

Key Words and Phrases

7.1 Ordered pair
Cartesian Coordinate System
x-coordinate
y-coordinate
Quadrant
Graph of an ordered pair
Graph of an equation
Linear equation
x-intercept
y-intercept
7.2 Slope

7.3 Point-slope form
Slope-intercept form
7.4 Linear inequality
7.5 Relation
Domain
Range
Function
7.6 Varies directly
Varies inversely
Varies jointly

Key Concepts and Rules

If (x_1, y_1) and (x_2, y_2) are two points on a line, $x_1 \neq x_2$, then the *slope* of the line is given by

$$m = \frac{y_2 - y_1}{x_2 - x_1}$$

Two lines L_1 and L_2, with respective slopes of m_1 and m_2, are parallel if $m_1 = m_2$ and perpendicular if $m_1 \cdot m_2 = -1$.

If the slope m, and the y-intercept b, of a line are known, then an equation of the line is $y = mx + b$. This is the *slope-intercept form* of the equation of the line.

If the slope m, and a point, (x_1, y_1) of the line are known, then the *point-slope form* of the equation of the line is $y - y_1 = m(x - x_1)$.

We say that y *varies directly* as x if there is a positive number k such that $y = kx$.

We say that y *varies inversely* as x if there is a positive number k such that $y = k/x$.

We say that z *varies jointly* as x *and* y if there is a positive number k such that $z = kxy$.

Review Exercises

If you have difficulty with any of the exercises, look in the section indicated by the number in brackets.

[7.1] 1. Determine by substitution which of the points $(0, 0)$, $(2, 1)$, $(3, 7)$, and $(-4, 17)$ are on the line $3x + y = 5$.

2. Complete each of the ordered pairs $(0, \ )$, $(1, \ )$, $(\ , 0)$, and $(\ , 7)$ so that it is on the line with equation $4x - 6y = 12$.

Find the x- and y-intercepts (if any) of the following equations and sketch the graphs.

3. $5x + 2y + 10 = 0$ 4. $3x + 2 = 0$ 5. $y - 3 = 7$

6. $y = 2x - 4$ 7. $5x - 3y = 12$

[7.2] Find the slope of the line that passes through each pair of points and graph the line.

8. $(1, 3)$ and $(3, 5)$ 9. $(3, 3)$ and $(2, 7)$

10. $(4, -2)$ and $-5, -2)$

Graph the line that passes through the given point and has a slope of m.

11. $(-2, 3)$, $m = -2$ 12. $(1, -4)$, $m = \dfrac{2}{5}$ 13. $(0, 1)$, $m = \dfrac{9}{7}$

Find the slope of the given line and determine whether it is rising, falling, horizontal, or vertical.

14. $y = -2x + 7$ 15. $5x + 8y = 2$ 16. $3x - 2 = 7$

Decide whether the following pairs of lines are parallel, perpendicular, or neither.

17. $x + y = 3$ and $x - y = 10$ 18. $x + 2y = 4$ and $3x + 6y = -1$

[7.3] Find an equation for each of the following lines. (m = slope)

19. $m = 4$; passes through $(0, -1)$
20. $m = -\frac{2}{3}$; passes through $(-3, -2)$
21. $m = \frac{3}{5}$; passes through $(0, 0)$
22. $m = -2$; passes through $(4, -5)$
23. passes through $(-2, -2)$ and $(1, 3)$
24. passes through $(-2, 4)$ and $(7, 4)$
25. passes through $(1, -2)$ and $(1, 7)$
26. passes through $(-5, 0)$ and vertical
27. parallel to $2x - 8y = 12$ and passes through $(2, 1)$
28. perpendicular to $x - 3y = 1$ and passes through $(4, -3)$

[7.4] Graph the following inequalities.

29. $2x + y < 5$
30. $x \geq 2$
31. $2x + 3y \geq 6$
32. $x + 4y > 12$
33. $5x - 3y \leq 9$
34. $y < 1$

[7.5] Determine which of the following relations are functions and state the domain of each relation.

35. $y^2 - x = 1$
36. $3x - 9y + 9 = 0$
37. $y = |x|$
38. $y = \dfrac{2}{|x - 1|}$
39. $y = \sqrt{2x + 3}$
40. $y = \dfrac{1}{x^2 - 4}$

For $f(x) = 2x^2 + x - 1$, find

41. $f(0)$
42. $f(1)$
43. $f(-1)$
44. $f(1 + h)$
45. $f[f(-1)]$
46. $f\left(\dfrac{1}{x}\right)$

[7.6] 47. Suppose y varies directly as x and inversely as t. If $y = 12$ when $x = 3$ and $t = 2$, find y when $x = 700$ and $t = 56$.

48. Suppose P varies directly as x and the square of t and inversely as the cube of z. If $P = 100$ when $x = 3$, $z = 2$, and $t = 1$, find P when $x = 6$, $z = 4$, and $t = 5$.

49. The frequency f, of a radio wave is inversely proportional to the wave length L. If the frequency is 150 megacycles per second for a wave 2 meters long, find the frequency for a wave 5 meters long.

50. When a given amount of natural gas is heated with twice as much steam, the number of moles of hydrogen produced is directly proportional to the number of moles of natural gas that is heated. When 4.609 moles of natural gas is heated in 9.218 moles of steam, 12.37 moles of hydrogen is produced. How much hydrogen is produced from 13.477 moles of gas and twice as much steam?

Practice Test (60 minutes)

Find the slope, the x- and y-intercepts (if any), and sketch the graph of each of the following equations.

1. $5y - 3x = 1$
2. $7x + y = 5$
3. Graph the line that passes through $(1, 1)$ and has a slope of 3.

Find an equation for each of the following lines.

4. passes through $(0, 2)$ and $(2, 3)$
5. $m = -\frac{2}{7}$; passes through $(-3, 4)$
6. parallel to $x = 7$; passes through $(9, -4)$
7. perpendicular to $4x + 12y = 3$; passes through $(5, 0)$

Graph each of the following inequalities.

8. $y \leq -4$
9. $2x + y > 1$

Determine the domain of each of the following relations and state which are functions.

10. $y^2 = x^2$
11. $xy = 1$
12. $y = \dfrac{1}{\sqrt{x - 1}}$

For the function $f(x) = \dfrac{1}{2x - 1}$, find

13. $f(0)$
14. $-f[f(0)]$
15. If y varies directly as x and inversely as the square of t, and $y = 1$ when $x = 15$ and $t = 3$, find y if $x = 1$ and $t = \dfrac{1}{5}$.

Extended Applications

Radar

We are all familiar with the fact that radar is used in such commercial activities as aircraft guidance at airports and navigation. The Department of Defense also uses radar to detect objects invading U.S. airspace.

Chap. 7 Linear Equations, Inequalities, Functions

The basic principle behind the radar device is simple. Radio waves that are traveling at the speed of light, which is approximately 186,300 miles per second, are transmitted from a directional antenna. If the radiated waves strike an object such as a plane or missile, some of the waves may be reflected back to the radar antenna. These waves induce current that flows through a circuit that is similar to a television receiver. A bright spot on the screen of the radar scope indicates the position of the object.

Simple mathematics is all that is required to do what the indicator on the radar scope does—convert the time between transmission and reception of the waves into units of distance.

Let

d = distance between the radar and the object
c = speed of light (approximately 186,300 mps)
t_1 = time (in seconds) for a pulse of radio waves to travel to the object
t_2 = time (in seconds) for radio waves to travel back from the object after being reflected
t = time (round trip) read by the radar unit

The waves are reflected instantaneously from the object and travel at the same speed (c) both going and returning. We have

$$t_1 = t_2$$

and the round trip time

$$t = t_1 + t_2 = 2t_1$$

$$\frac{1}{2}t = t_1$$

Thus, t_1 is half the time read by the radar unit.

From the formula Distance = Rate × Time we have

$$d = ct_1$$

so (approximately)

$$d = (186,300)\left(\frac{1}{2}t\right)$$

or

$$d = 93,150t \text{ miles}$$

EXAMPLE: If a Greenland radar station receives a reflection 0.02 seconds after transmission, how far away is the object that reflects the waves?

Solution: In the equation

$$d = 93{,}150t \text{ (miles)}$$

we have $t = 0.02$;

$$d = 93{,}150(0.02)$$
$$d = 1863 \text{ miles}$$

Exercises

A radar station in Louisiana receives a reflection from the space shuttle 0.01 seconds after transmission.

1. How far from the station is the shuttle?

2. How long did it take for the transmitted waves to travel from the station to the shuttle?

3. Three minutes later, a reflection is returned from the shuttle 0.0125 seconds after transmission. How far did the shuttle travel during those three minutes?

4. Under the previously stated conditions, what was the average speed of the shuttle over the three-minute interval?

8

Functions and Conic Sections

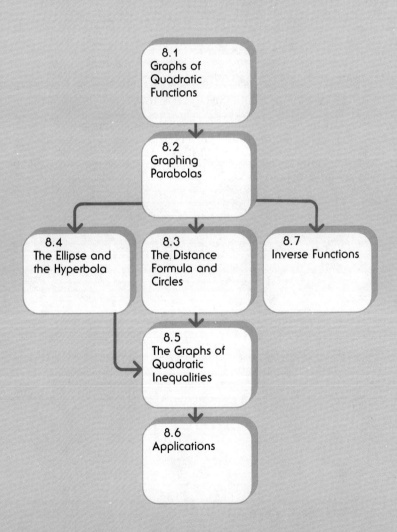

8.1 Graphs of Quadratic Functions

A function that can be written in the form

$$y = ax^2 + bx + c \quad \text{(with } a \neq 0\text{)}$$

is called a **quadratic function**. In order to understand the graph of a quadratic function, we will start with one of the simplest quadratic functions, $y = x^2$, in which $a = 1$ and $b = c = 0$. First, a few points whose coordinates satisfy $y = x^2$ are plotted; the coordinates of these points are obtained by arbitrarily assigning values to x and computing the corresponding values for y, just as we did for the linear equations in Section 7.1. The points $(0, 0)$, $(-1, 1)$, $(1, 1)$, $(2, 4)$, and $(-2, 4)$ lie on the graph. If we plot these points in a coordinate system and connect them by a smooth curve, we will obtain the graph shown in Figure 8.1(a). This graph is known as a **parabola**. The parabola with equation $y = -x^2$ ($a = -1$ and $b = c = 0$) is shown in Figure 8.1(b).

Figure 8.1

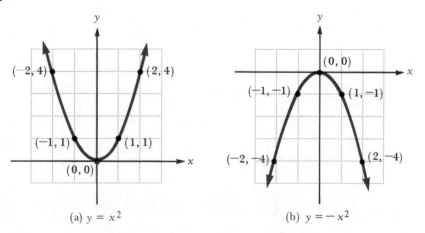

(a) $y = x^2$ (b) $y = -x^2$

We say that the parabola in Figure 8.1(b) **opens downward** and that the parabola in Figure 8.1(a) **opens upward**. The coefficient of x^2, the number a, determines whether the parabola opens upward or downward. If $a > 0$, the parabola opens upward; if $a < 0$, it opens downward.

For a parabola that opens upward, the lowest point is called the **vertex**. If the parabola opens downward, the highest point is called the **vertex**. The vertex in Figure 8.1(a) is $(0, 0)$. Can you identify the vertex of the parabola in Figure 8.1(b)? As you can see in the graph, it is also $(0, 0)$.

334   Chap. 8   Functions and Conic Sections

Parabolas represent, among other things, the path of a ball thrown upward at an angle (if wind resistance is neglected) and a cross section of the reflector of an automobile's headlight.

EXAMPLE 1   Sketch the graphs of (a) $y = x^2 - 5$ (b) $y = -x^2 + 9$

Solution:
(a) For any given value of x, the corresponding y-value in $y = x^2 - 5$ is 5 units less than the corresponding y-value in $y = x^2$. The graph of $y = x^2 - 5$ is shown in Figure 8.2(a). This graph is the same shape as the graph of $y = x^2$ in Figure 8.1(a), but is *pulled downward* by 5 units.
(b) Compare the equation $y = -x^2 + 9$ with the equation $y = -x^2$, whose graph appeared in Figure 8.1(b). Notice that for any given x-value, the corresponding y-value in $y = -x^2 + 9$ is always 9 units greater than the corresponding y-value in $y = -x^2$. We conclude that the graph is the same shape as the graph of $y = -x^2$ that is shown in Figure 8.1(b), but is *lifted upward* by 9 units, as shown in Figure 8.2(b). The vertex is located at (0, 9).

Figure 8.2

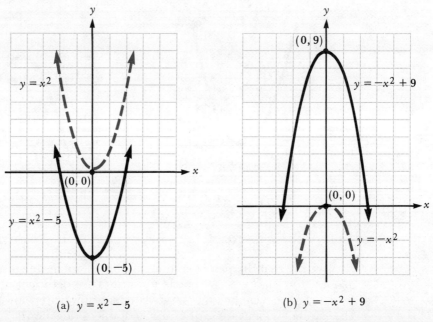

(a) $y = x^2 - 5$ (b) $y = -x^2 + 9$

In general, the graph of $y = ax^2 + c$, (with $a \neq 0$), is the same as the graph of $y = ax^2$ except that it is lifted upward by c units if $c > 0$, and is pulled downward by c units if $|c| < 0$.

EXAMPLE 2 Graph $y = (x - 1)^2$.

Solution: The following table of sample values shows the relation between the graphs of $y = (x - 1)^2$ and $y = x^2$:

x	-2	-1	0	1	2	3	4
$y = x^2$		4	1	0	1	4	9	16
$y = (x - 1)^2$	9	4	1	0	1	4	9

Notice that the arrows in the table indicate that $y = (x - 1)^2$ and $y = x^2$ take on the same y-values; that is, the y-value of $y = x^2$ at $x = v$ is the same as the y-value of $y = (x - 1)^2$ at $x = v + 1$. The graph of $y = (x - 1)^2$ is shown in Figure 8.3; it is the same shape as the graph of $y = x^2$, but is shifted one unit to the right. The vertex is located at (1, 0).

Figure 8.3

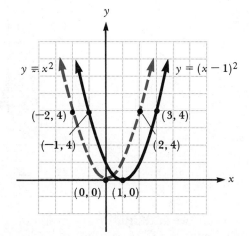

> In general, $y = \pm(x - h)^2$ has the same shape graph as the graph of $y = \pm x^2$, but is shifted h units to the right if h is positive, or $|h|$ units to the left if h is negative. For example, the graph of $y = (x + 3)^2$ is the graph of $y = x^2$ shifted 3 units to the left. This is because $y = (x + 3)^2 = [x - (-3)]^2$ has the form $y = (x - h)^2$ where $h = -3$.

EXAMPLE 3 Graph $y = -x^2 - 4x - 1$ and locate the vertex.

Solution: In order to locate the vertex, rewrite $-x^2 - 4x - 1$ in the form $-(x - h)^2$ by completing the square.

$$\begin{aligned}-x^2 - 4x - 1 &= -(x^2 + 4x) - 1 \\ &= -(x^2 + 4x + 4) + 4 - 1 \quad \text{Subtract 4 and add 4} \\ &= -(x + 2)^2 + 3\end{aligned}$$

The given function is now expressed as

$$y = -[x - (-2)]^2 + 3 \qquad (h = -2)$$

whose graph is the graph of $y = -x^2$ shifted 2 units to the left and lifted 3 units upward, as shown in Figure 8.4. The vertex is located at $(-2, 3)$.

Figure 8.4

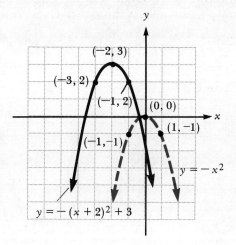

Thus far, we have studied the graph of $y = ax^2 + bx + c$ with $a = 1$ or -1. The following examples illustrate what happens if $a \neq \pm 1$.

EXAMPLE 4

Graph $y = 2x^2$.

Solution: The following table shows that for each x-value, the y-value of $y = 2x^2$ is twice the y-value of $y = x^2$.

x	-2	-1	0	1	2
$y = 2x^2$	8	2	0	2	8
$y = x^2$	4	1	0	1	4

The graph of $y = 2x^2$, as shown in Figure 8.5, has a narrower span than that of $y = x^2$. Both graphs have vertices at the origin.

Figure 8.5

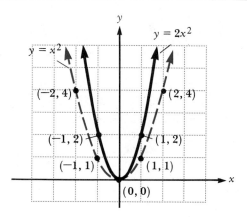

EXAMPLE 5

Graph $y = \left(-\dfrac{1}{2}\right)x^2$.

Solution: As the following table shows, the y-values of this function are one-half of the corresponding y-values of $y = -x^2$.

x	-2	-1	0	1	2
$y = \left(-\dfrac{1}{2}\right)x^2$	-2	$-\dfrac{1}{2}$	0	$-\dfrac{1}{2}$	-2
$y = -x^2$	-4	-1	0	-1	-4

The graph of $y = \left(-\dfrac{1}{2}\right)x^2$, which is sketched in Figure 8.6, has a wider span than the graph of $y = -x^2$; both graphs have vertices at the origin and both open downward.

Figure 8.6

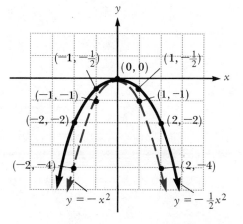

In general, the graph of $y = ax^2 + bx + c$, with $a \neq 0$, can be obtained from the graph of $y = x^2$ or $y = -x^2$ by moving upward or downward, shifting to the right or to the left, and narrowing or widening the span. We will summarize the process next.

1. Rewrite the equation $y = ax^2 + bx + c$ (by completing the square) as $y = a(x - h)^2 + k$. The vertex is located at (h, k).
2. The parabola opens upward if $a > 0$ and downward if $a < 0$.
3. The span is narrower than the spans of $y = \pm x^2$ if $|a| > 1$, and is wider if $a < 1$. The span is the same as $y = \pm x^2$ if $|a| = 1$.
4. The number k is the maximum value of y if $a < 0$, and is the minimum value of y if $a > 0$.

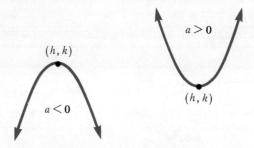

Exercises 8.1

Standard Assignment: Exercises 1, 3, 5, 7, 9, 11, 13, 17, 19, 21, 23, 25, 27, 29, 31, 35, 39

A Sketch each of the following parabolas and indicate the vertex of each parabola. See Examples 1 and 2.

1. $y = -x^2 - 3$
2. $y = -x^2 + 1$
3. $y = -x^2 + 2$
4. $y = -x^2 - 1$
5. $y = -(x - 1)^2$
6. $y = -(x - 2)^2$
7. $y = -(-x + 1)^2$
8. $y = -(x + 2)^2$
9. $y = x^2 - 2$
10. $y = x^2 + 3$
11. $y = (x - 1)^2 - 3$
12. $y = (x - 2)^2$
13. $y = -(x - 1)^2 + 2$
14. $y = -(x - 2)^2 + 1$
15. $y = (x - 2)^2 + 3$
16. $y = (x - 1)^2 - 4$
17. $y = 3x^2$
18. $y = 4x^2$

19. $y = \frac{1}{2}x^2$

20. $y = -\frac{1}{3}x^2$

21. $y = 2(x - 1)^2 + 3$

22. $y = -4(x + 3)^2 - 2$

EXAMPLE: Nancy runs a hot dog stand. From years of experience, she has found that the daily profit P (in dollars) depends on the number x, of hot dogs sold in approximately the following way.

$$P = -x^2 + 120x - 3570$$

How many hot dogs should be sold to yield the maximum profit?

Solution: By the method of completing the square, as shown in Example 3, we get

$$\begin{aligned} P &= -x^2 + 120x - 3570 \\ &= -(x^2 - 120x \quad) - 3570 \\ &= -(x^2 - 120x + 3600) + 3600 - 3570 \\ &= -(x - 60)^2 + 30 \end{aligned}$$

The graph is the parabola opening downward with the vertex located at (60, 30) as shown. Hence, 60 hot dogs should be sold daily in order to yield the maximum daily profit of $30.

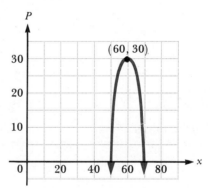

23. Mary runs a day care center. From past results, she summarizes her profits in the equation $P = -x^2 + 60x - 700$, where P denotes the profit in dollars and x is the number of kids she takes. How many kids should she take to make the maximum profit?

24. Paul wants to fence off a rectangular area along a wall with 60 meters of fencing. The side on the wall need not be fenced. What is the maximum area that he could enclose?

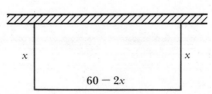

25. The power W in watts delivered by a 220-volt circuit with 22 ohms of resistance depends upon the current I, in amperes, in the following way:

$$W = 220I - 22I^2$$

How many watts of power could this circuit deliver?

26. What is the maximum area possible in a rectangle whose perimeter is 80 cm?

Sketch each of the following parabolas. See Examples 4 and 5.

27. $y = 4x^2$
28. $y = 3x^2 + 1$
29. $y = \frac{1}{2}x^2 - 2$
30. $y = \left(\frac{1}{3}\right)x^2 - 1$
31. $y = 2(x - 1)^2 - 3$
32. $y = 4(x + 3)^2 - 2$

Sketch each of the following parabolas by first completing the square. See Example 3.

33. $y = x^2 + 2x$
34. $y = x^2 - 2x$
35. $y = x^2 - 2x + 5$
36. $y = x^2 + 2x - 4$
37. $y = -x^2 + 10x$
38. $y = -x^2 - 8x$
39. $y = -x^2 - 8x + 10$
40. $y = -x^2 + 10x - 8$
41. $y = 2x^2 + 3x$
42. $y = 3x^2 + 6x - 1$

8.2 Graphing Parabolas

In the previous section, you were given an overview of the parabolas that are graphs of quadratic functions. In order to graph a parabola more accurately, you should know, in addition to the vertex, the x- and y-intercepts and the axis of symmetry. The **axis of symmetry** of a parabola that opens upward or downward is the vertical line that passes through the vertex of the parabola. When a parabola is folded along the axis of symmetry, the two halves of the parabola coincide. See Figure 8.8.

Intercepts

The y-intercept of the graph of $y = ax^2 + bx + c$ is obtained by substituting 0 for x

$$y = a(0)^2 + b(0) + c = c$$

which is the y-intercept. In order to get the x-intercepts, substitute 0 for y

$$0 = ax^2 + bx + c$$

and then solve the equation for x. As we have seen in Section 6.3, a quadratic equation can have one, two, or no real roots (= two imaginary roots); correspondingly, a parabola can have one, two, or no x-intercepts, as is illustrated in Figure 8.7.

Figure 8.7

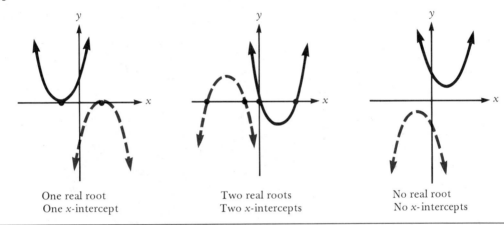

One real root
One x-intercept

Two real roots
Two x-intercepts

No real root
No x-intercepts

EXAMPLE 1

Find the vertex, the axis of symmetry, x- and y-intercepts, and sketch the graph of $y = x^2 - 4x - 5$.

Solution: Here $a = 1$, $b = -4$, and $c = -5$; since $a > 0$, the parabola opens upward. To find the vertex, we complete the square on $x^2 - 4x$

$$\begin{aligned} y &= x^2 - 4x - 5 \\ &= (x^2 - 4x \quad) - 5 \\ &= (x^2 - 4x + 4) - 4 - 5 \qquad \text{Add 4 and subtract 4} \\ &= (x - 2)^2 - 9 \end{aligned}$$

The equation is now in the form of $y = a(x - h)^2 + k$, where $a = 1$, $h = 2$, and $k = -9$. The vertex is at $(2, -9)$ and the axis of symmetry is $x = 2$, the vertical line through $(2, -9)$.

In order to find the x-intercepts, we set $y = 0$ and solve for x.

$$\begin{aligned} 0 &= x^2 - 4x - 5 \\ &= (x - 5)(x + 1) \end{aligned}$$

The equation has two real roots: 5 and -1. Thus, the x-intercepts are 5 and -1. The y intercept is -5 since $c = -5$.

The graph of $y = x^2 - 4x - 5$ is shown in Figure 8.8.

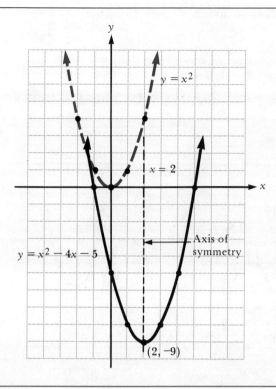

Figure 8.8

In general, when we complete the square as we did in Sections 6.2 and 6.3 on

$$y = ax^2 + bx + c, \text{ with } a \neq 0$$

we get

$$y = a(x + b/2a)^2 + (4ac - b^2)/4a$$

from which we conclude that, if $x = -b/2a$, the term $a(x + b/2a)^2$ vanishes and $y = (4ac - b^2)/4a$ is the maximum value if $a < 0$ (the parabola opens downward) or the minimum value if $a > 0$ (the parabola opens upward). In either case, the point

$$\left(-\frac{b}{2a}, \frac{4ac - b^2}{4a}\right)$$

is the vertex, and the equation of the axis of symmetry is $x = -b/2a$.

EXAMPLE 2 Graph $y = -x^2 + 2x - 1$.

Solution: Here, $a = -1$, $b = 2$, $c = -1$. Since $a < 0$, the parabola opens downward. The x-coordinate of the vertex is $-b/2a = -2/2(-1) = 1$. The y-coordinate of the vertex can be calculated by substituting 1 (the

x-coordinate of the vertex) for x in the original equation to get $y = -(1)^2 + 2(1) - 1 = 0$. Hence, the vertex is located at $(1, 0)$ and the axis of symmetry is $x = 1$. The y-intercept is -1 (the constant term c). The graph is shown in Figure 8.9.

Figure 8.9

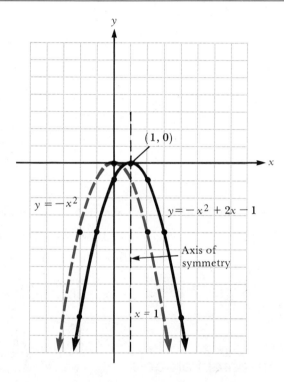

Remark: Setting $y = 0$ we find that the equation $0 = -x^2 + 2x - 1$ has only one root, $x = 1$, and thus the parabola has only one x-intercept, 1.

A summary of the most important features of the parabola $y = ax^2 + bx + c$ follows.

The Parabola $y = ax^2 + bx + c$

1. Vertex = (x_0, y_0) where

$$\begin{cases} x_0 = -b/2a \\ y_0 = a(x_0)^2 + bx_0 + c \end{cases}$$

2. *x*-intercepts: the solutions of $ax^2 + bx + c = 0$ (if any)
3. Axis of symmetry: $x = -b/2a$
4. *y*-intercept is c
5. Graphs:

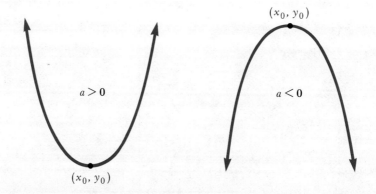

In Example 1 we investigated the graph of $y = x^2 - 4x - 5$. Using the preceding information with $a = 1$, $b = -4$, and $c = -5$, we see that

$$x_0 = -b/2a = -(-4)/(2 \cdot 1) = 2$$
$$y_0 = (2)^2 - 4(2) - 5 = -9$$

The vertex is $(2, -9)$, the axis of symmetry is $x = 2$, and the y intercept is -5. The x-intercepts are determined exactly as before in Example 1 and the graph appears in Figure 8.8.

Horizontal Parabolas

As in the case of $y = ax^2 + bx + c$, the graph of an equation of the form

$$x = ay^2 + by + c \qquad (a \neq 0)$$

is also a parabola. In this case, to plot a skeleton of the graph, we assign values of y and calculate the corresponding values of x.

EXAMPLE 3

Sketch the graph of $x = y^2 - 6y + 5$.

Solution: By assigning $y = 0, 1, 2, 3, 4, 5,$ and 6 we find that $x = 5, 0, -3, -4, -3, 0,$ and 5, respectively. Hence the points with coordinates $(5, 0), (0, 1), (-3, 2), (-4, 3), (-3, 4), (0, 5),$ and $(5, 6)$ form a skeleton of the graph [as shown in Figure 8.10(a)].

An approximate graph is sketched by joining those points in the skeleton of the graph by a smooth curve as shown in Figure 8.10(b). The parabola in Figure 8.10 is said to **open to the right**.

8.2 Graphing Parabolas

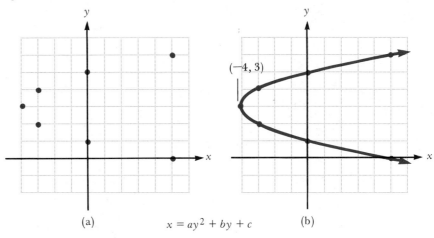

Figure 8.10

$x = ay^2 + by + c$

(a)   (b)

In general, the parabola

$$x = ay^2 + by + c \quad (a \neq 0)$$

opens to the right if $a > 0$ and opens to the left if $a < 0$. See Figure 8.11. Obviously, this relation is *not* a function, since it fails the vertical line test.

The farthest point to the left on a parabola opening to the right is called the **vertex**; the point farthest to the right on a parabola opening to the left is also called the **vertex**. The **axis of symmetry** is the horizontal line through the vertex.

EXAMPLE 4

(a) Find the vertex of the parabola given by the equation $x = -y^2 + 4y$.
(b) Sketch a graph of the parabola described in (a).

Solution:

(a) As in Example 1, we shall complete the square on the quadratic expression $-y^2 + 4y$. Thus

$$\begin{aligned} x &= -y^2 + 4y \\ &= -(y^2 - 4y + 2^2) + 2^2 \quad -2^2 + 2^2 = 0 \\ &= -(y - 2)^2 + 4 \\ &\leq 4 \end{aligned}$$

The largest value for x is 4 when $y = 2$. That is, the farthest point on the right of the graph is $(4, 2)$. Thus, $(4, 2)$ is the vertex and the line $y = 2$ is the axis of symmetry.

(b) Notice in the equation $x = -y^2 + 4y$ that $a = -1 < 0$ and the parabola opens to the left. For $y = 0, 1, 2, 3, 4$, the corresponding values of x are respectively $x = 0, 3, 4, 3, 0$. The graph contains the points $(0, 0)$, $(3, 1)$, $(4, 2)$, $(3, 3)$, and $(0, 4)$. If the graph is folded along the line $y = 2$, the point $(0, 0)$ coincides with $(0, 4)$, and $(3, 1)$ coincides with $(3, 3)$. A sketch of the parabola is given in Figure 8.11.

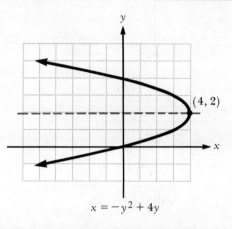

Figure 8.11

$x = -y^2 + 4y$

By completing the square on $x = ay^2 + by + c$, we find that the y-coordinate for the vertex is $-b/2a$.

A summary of the most important features of the parabola $x = ay^2 + by + c$ follows.

The Parabola $x = ay^2 + by + c$

1. Vertex $= (x_0, y_0)$ where
$$\begin{cases} y_0 = -b/2a \\ x_0 = a(y_0)^2 + by_0 + c \end{cases}$$

2. y-intercepts: the solutions of $ay^2 + by + c = 0$ (if any)
3. Axis of symmetry: $y = -b/2a$
4. x-intercept is c
5. Graphs:

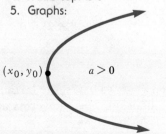

$a > 0$

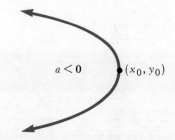

$a < 0$

We can use the preceding information to reexamine Example 4 as follows:
$$x = -y^2 + 4y \quad \text{Here } a = -1, b = 4, \text{ and } c = 0$$
Thus
$$y_0 = \frac{-b}{2a} = \frac{-4}{2(-1)} = \frac{-4}{-2} = 2$$
$$x_0 = -(2)^2 + 4(2) = -4 + 8 = 4$$

The vertex is (4, 2), the axis of symmetry is $y = 2$, and the x-intercept is 0. The y-intercepts are determined as before and the graph appears in Figure 8.11.

Exercises 8.2

Standard Assignment: All odd-numbered **A** exercises

A Graph each of the following quadratic functions by completing the square. Find the vertex and intercepts. See Example 1.

1. $y = x^2 + 2x - 3$
2. $y = x^2 - 2x - 8$
3. $y = x^2 - x + \frac{1}{4}$
4. $y = x^2 + 4x + 4$
5. $y = -x^2 + 4x - 13$
6. $y = -x^2 + x - \frac{1}{2}$
7. $y = 2x^2 - 4x - 6$
8. $y = -2x^2 + 4x - 16$

Graph each parabola by determining the vertex, the axis of symmetry, and the intercepts. See Example 2.

9. $y = -x^2 + 3x - 2$
10. $y = -x^2 + x + 2$
11. $y = x^2 - 2x + 1$
12. $y = x^2 - 4x + 4$
13. $2y = 2x^2 - 2x + \frac{1}{2}$
14. $-2y = 2x^2 - 2x + 1$
15. $-2y = x^2 + 2x + 5$
16. $3y = x^2 - 4x - 8$

Graph each of the following horizontal parabolas. See Examples 3 and 4.

17. $x = y^2$
18. $x = -y^2$
19. $x = -y^2 + 5$
20. $x = y^2 - 6$
21. $x = (y - 2)^2 + 3$
22. $x = -(y + 3)^2 - 4$
23. $x = 3y^2 - 1$
24. $x = \left(\frac{1}{3}\right)y^2$

25. $x = -y^2 - 2y + 3$ 26. $x = y^2 + 6y + 7$

Use the summary chart to find the vertex, axis of symmetry, and x- and y-intercepts and to sketch the graph of each of the following.

27. $y = -2x^2 + 3$ 28. $y = 2x^2 + 4x - 1$

29. $y = x^2 - 7x + 6$ 30. $y = 7x^2 - 42x + 61$

31. $y = 2x^2 - 4x + 3$ 32. $y = 2x^2 + x - 3$

33. $y = 2x^2 - x + 3$ 34. $y = 3x^2 - 5x + 4$

35. $x = 2y^2 + 8y + 3$ 36. $x = 2y^2 - 4y + 8$

37. $x = y^2 - 4y + 3$ 38. $x = -3y^2 + 12$

B 39. If $f(x) = f(-x)$ for all values of x, the graph of $y = f(x)$ is said to be **symmetric with respect to the y axis**. Find the parabolas of Exercises 1–32 that are symmetric with respect to the y-axis.

40. If $g(y) = g(-y)$ for all replacement values of y, the graph of $x = g(y)$ is said to be **symmetric with respect to the x axis**. Find the parabolas of Exercises 1–32 that are symmetric with respect to the x-axis.

8.3 The Distance Formula and Circles

Probably the most widely known fact in geometry is the **Pythagorean theorem**[*] which is attributed to Pythagoras (about 600 B.C.). The theorem holds that if a, b, and c are the lengths of the two legs and hypotenuse, respectively, of a right triangle, then $c^2 = a^2 + b^2$.

In Figure 8.12, P_1P_2 is the hypotenuse of the right triangle P_1P_2Q, with a right angle at $Q(2, -3)$. The lengths of the two legs are

$$a = |2 - (-1)| = 3$$
$$b = |-3 - 1| = 4$$

By the Pythagorean Theorem, the distance $d(P_1, P_2)$ between P_1 and P_2 is

$$d(P_1, P_2) = c = \sqrt{a^2 + b^2} = \sqrt{3^2 + 4^2} = \sqrt{25} = 5$$

Let $P_1(x_1, y_1)$ and $P_2(x_2, y_2)$ be any two points that are not on the same vertical or horizontal line. Then, as in the previous example, they form the

---

[*]Archaeologists have found this same theorem engraved on the tombstone of Chou-Pai-Ching (about 1000 B.C.) in the early Chou Dynasty of ancient China.

Figure 8.12

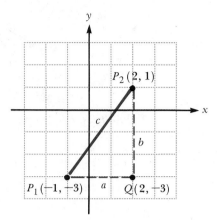

hypotenuse of the right triangle P_1P_2Q (see Figure 8.13) with legs of lengths $|x_2 - x_1|$ and $|y_2 - y_1|$. By the Pythagorean Theorem,

The distance between P_1 and P_2 is

$$d(P_1, P_2) = \sqrt{(x_2 - x_1)^2 + (y_2 - y_1)^2}$$

This formula is called the **distance formula**. You should verify that the formula remains valid even if P_1 and P_2 lie on the same vertical or horizontal line.

Figure 8.13

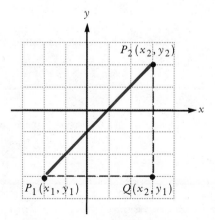

EXAMPLE 1 | Find the distance between the points $(-2, 5)$ and $(3, -4)$.

Solution: Let $P_1 = (-2, 5)$ and $P_2 = (3, -4)$. Then
$$d(P_1, P_2) = \sqrt{[3-(-2)]^2 + (-4-5)^2}$$
$$= \sqrt{5^2 + (-9)^2}$$
$$= \sqrt{106}$$

You should verify that $d(P_2, P_1)$ also equals $\sqrt{106}$.

EXAMPLE 2 | Are $d(P_1, P_2)$ and $d(P_2, P_1)$ always equal?

Solution: Yes, since $(x_2 - x_1)^2 = (x_1 - x_2)^2$ and $(y_2 - y_1)^2 = (y_1 - y_2)^2$ we have
$$d(P_1, P_2) = \sqrt{(x_2 - x_1)^2 + (y_2 - y_1)^2}$$
$$= \sqrt{(x_1 - x_2)^2 + (y_1 - y_2)^2} = d(P_2, P_1)$$

EXAMPLE 3 | Determine whether the three points $P(2, 1)$, $Q(4, 2)$, and $R(6, 3)$ all lie on the same line.

Solution: First, find the lengths of the three segments $\overline{PQ}$, $\overline{QR}$, and $\overline{PR}$.
$$d(P, Q) = \sqrt{(4-2)^2 + (2-1)^2} = \sqrt{5}$$
$$d(Q, R) = \sqrt{(6-4)^2 + (3-2)^2} = \sqrt{5}$$
$$d(P, R) = \sqrt{(6-2)^2 + (3-1)^2} = \sqrt{20} = 2\sqrt{5}$$

The length of the longest segment, $\overline{PR}$, is the sum of the lengths of the two shorter segments, and thus these three points lie on the same line.

In mathematical language, a **circle** is the set of points in a plane that lie at a fixed distance from a fixed point in the plane. The fixed point is called the **center** and the fixed distance is called the **radius** of the circle.

EXAMPLE 4 | Find the equation of the circle with a radius of 5 and center at the origin $(0, 0)$.

Solution: Let (x, y) be any point on the circle. The distance between (x, y) and the center $(0, 0)$ must be 5. By the distance formula we have
$$\sqrt{(x-0)^2 + (y-0)^2} = 5$$
or
$$x^2 + y^2 = 25$$

which is the desired equation. See Figure 8.14.

Figure 8.14

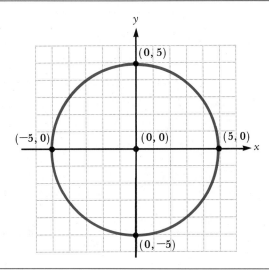

EXAMPLE 5

Find the equation of the circle with a radius of 4 and center at $(-2, 3)$.

Solution: Again, let (x, y) be any point of the circle and use the distance formula.

$$\sqrt{(x + 2)^2 + (y - 3)^2} = 4$$

or

$$(x + 2)^2 + (y - 3)^2 = 16$$

The graph of this circle is shown in Figure 8.15.

Figure 8.15

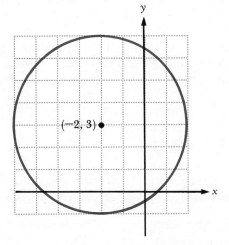

In a manner similar to the solution of Example 5, we can show that the equation

$$(x - a)^2 + (y - b)^2 = r^2$$

represents a circle with a radius r and center (a, b).

EXAMPLE 6 Identify the graph of $x^2 + y^2 - 2x + 4y - 31 = 0$.

Solution: We suspect that the graph is a circle. To be sure, we rewrite the given equation in the form $(x - a)^2 + (y - b)^2 = r^2$. The first step is to put the x-terms and y-terms into two groups by using parentheses, and then move the constant -31 to the right-hand side.

$$(x^2 - 2x +) + (y^2 + 4y +) = 31$$

To complete the squares we fill the blanks by $(-2/2)^2 = 1$ and $(4/2)^2 = 2^2$ resulting in

$$(x^2 - 2x + 1) + (y^2 + 4y + 2^2) = 31 + 1 + 2^2$$

or

$$(x - 1)^2 + (y + 2)^2 = 6^2$$

which represents a circle with a radius of 6 and center $(1, -2)$.

EXAMPLE 7 Identify the graph of $y = \sqrt{25 - x^2}$.

Solution: In order to remove the radical symbol, we square both sides of $y = \sqrt{25 - x^2}$ to get $y^2 = 25 - x^2$ or $x^2 + y^2 = 5^2$, which represents a circle of radius 5 and center at the origin. But, since $\sqrt{25 - x^2} \geq 0$, $y = \sqrt{25 - x^2}$ is non-negative. Therefore, the given equation represents the upper semi-circle of Figure 8.14. Observe that the equation for the lower semicircle is $y = -\sqrt{25 - x^2}$.

Exercises 8.3

Standard Assignment: Exercises 1, 3, 5, 7, 9, 13, 15, 19, 21, 23, 25, 27, 31, 33, 37, 39, 43, 45

A Find the distance between the indicated pairs of points. See Example 1.

1. $(2, 1)$ and $(2, 5)$
2. $(3, 5)$ and $(-2, 5)$

3. (2, 3) and (3, 2)
4. (4, 5) and (5, 4)
5. $(-1, -5)$ and $(2, -3)$
6. $(-4, 1)$ and $(-7, -9)$
7. $(u, u + v)$ and $(u, u - v)$
8. $(v - w, t)$ and $(v + w, t)$
9. (t, k) and (k, t)
10. (m, n) and $(-n, -m)$
11. ▦ $(3.141, 2.067)$ and $(-0.321, -3.367)$
12. ▦ $(-9.88, 7.82)$ and $(-6.45, 5.28)$

Determine whether or not the given points lie on the same line. See Example 3.

13. $(0, 0), (1, 2), (-1, -2)$
14. $(3, 4), (0, 0), (-3, -4)$
15. $(4, -2), (-2, 8), (1, 3)$
16. $(9, 6), (0, -3), (3, 1)$
17. $(-4m, 6n), (10m, -8n), (3m, -n)$
18. $(k, -3k), (-3k, 9k), (2k, -6k)$

Graph each equation. See Examples 4, 5, and 7.

19. $x^2 + y^2 = 4$
20. $x^2 + y^2 = 49$
21. $x^2 + y^2 = 0$
22. $(x - 1)^2 + y^2 = 0$
23. $y = \sqrt{49 - x^2}$
24. $y = \sqrt{4 - x^2}$
25. $y = -\sqrt{36 - x^2}$
26. $y = -\sqrt{16 - x^2}$
27. $(x - 2)^2 + (y - 3)^2 = 5$
28. $(x + 3)^2 + (y - 2)^2 = 7$
29. $(x + 1)^2 + (y + 4)^2 = 8$
30. $(x + 2)^2 + (y + 3)^2 = 6$

Write an equation for each of the following circles. See Examples 4 and 5.

31. center at $(0, 0)$, radius 8
32. center at $(0, 1)$, radius 1
33. center at $(-1, 2)$, radius $\sqrt{2}$
34. center at $(1, -2)$, radius $\sqrt{5}$
35. center at $(-2, -3)$, radius $\sqrt{7}$
36. center at $(-3, -5)$, radius $\sqrt{13}$

Identify the graphs of the following equations. See Examples 6 and 7.

37. $x^2 + y^2 - 2x - 2y - 4 = 0$
38. $x^2 + y^2 - 4x - 2y - 15 = 0$
39. $x^2 + y^2 + 4y = 0$
40. $x^2 + y^2 + 6x = 0$
41. $x^2 + y^2 + 6x - 4y - 20 = 0$
42. $x^2 + y^2 - 4x + 8y - 25 = 0$
43. $y = \sqrt{25 - (x - 1)^2}$
44. $y = \sqrt{16 - (x + 2)^2}$
45. $y - 1 = -\sqrt{16 - x^2}$
46. $y + 2 = -\sqrt{25 - x^2}$
47. $y + 3 = \sqrt{9 - (x + 1)^2}$
48. $y - 3 = -\sqrt{36 - (x - 4)^2}$
49. $y - 2 = -\sqrt{20 + 8x - x^2}$
50. $y + 1 = \sqrt{8 - 2x - x^2}$

Determine whether or not each of the following represents the vertices of a right triangle. (Hint: Use the Pythagorean Theorem.)

51. $A(2, -1), B(3, 2), C(1, 1)$
52. $D(-2, 4), E(-2, -4), F(4, 4)$
53. $P(-3, -10), Q(4, -2), R(-4, 5)$
54. $Q(-6, 2), R(-2, 6), S(3, -3)$
55. $E(-2, 4), F(1, 2), G(1, -3)$
56. $A(a, b), B(a, b + h), C(a + k, b)$

8.4 The Ellipse and the Hyperbola

In Figure 8.14 we see that the upper and lower halves of the graph of the circle with equation $x^2 + y^2 = 25$ coincide when the graph is folded along the x-axis. Similarly, the left and the right halves of this graph coincide when the graph is folded along the y-axis. We will now study another type of graph which has these same features. We examine the graph of the equation

$$\frac{x^2}{25} + \frac{y^2}{16} = 1$$

In order to plot an approximate graph of this equation, we assign $x = 0, \pm 2, \pm 4$, and ± 5 to obtain the corresponding points $(0, \pm 4), (\pm 2, \pm 3.67), (\pm 4, \pm 2.4)$, and $(\pm 5, 0)$. The graph shown in Figure 8.16 looks like a flattened circle and is called an **ellipse**.

Figure 8.16

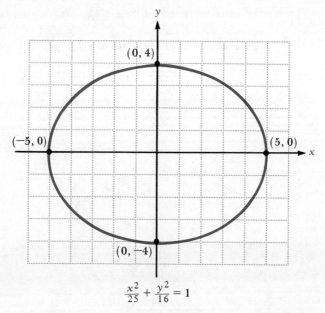

8.4 The Ellipse and the Hyperbola

If a point moves on the plane in such a way that the sum of its distances to two fixed points is a constant, the resulting path of the moving point is an ellipse. The fixed points are called the **foci** (plural of **focus**) of the ellipse.

In general, for distinct positive numbers a and b, the equation

$$\frac{x^2}{a^2} + \frac{y^2}{b^2} = 1$$

represents an ellipse centered at the origin that intersects the coordinate axes at $(\pm a, 0)$ and $(0, \pm b)$. If $a^2 > b^2$, the ellipse is in **horizontal position** and if $a^2 < b^2$, it is in **vertical position.** (See Figure 8.17.) The ellipse in Figure 8.16 is in horizontal position ($a^2 = 25$ and $b^2 = 16$).

Figure 8.17

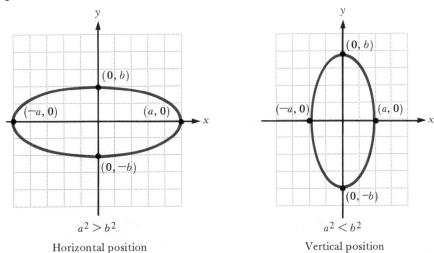

$a^2 > b^2$
Horizontal position

$a^2 < b^2$
Vertical position

EXAMPLE 1

Graph $\dfrac{x^2}{16} + \dfrac{y^2}{25} = 1$.

Solution: This equation is of the form

$$\frac{x^2}{a^2} + \frac{y^2}{b^2} = 1$$

where $a^2 = 16$ and $b^2 = 25$. Since $a^2 < b^2$, the ellipse is in the vertical position. The x-intercepts are $(\pm 4, 0)$ and the y-intercepts are $(0, \pm 5)$. The graph is shown on the following page.

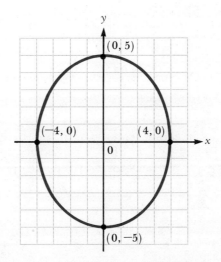

If the plus sign in the equation of an ellipse is changed to a minus sign

$$\frac{x^2}{25} - \frac{y^2}{16} = 1$$

the graph becomes a **hyperbola**. If we set $y = 0$, we find $x = \pm 5$; the graph has x intercepts at 5 and -5. If we set $x = 0$, there are no corresponding real values for y, and so the graph has no y-intercepts. As in the case of the ellipse, the equation remains unchanged when x is replaced by $-x$ and y is replaced by $-y$; this indicates the graph is symmetrical with respect to the coordinate axes.

If we solve for y in terms of x, we will get two functions of x:

$$y = \frac{4}{5}\sqrt{x^2 - 25} \quad \text{and} \quad y = -\frac{4}{5}\sqrt{x^2 - 25}$$

Both functions have the domain $\{x \in \mathbf{R} \mid x^2 \geq 25\} = (-\infty, -5] \cup [5, +\infty)$. Notice that when $|x|$ becomes large, $\sqrt{x^2 - 25}$ and $\sqrt{x^2}$ come closer together. As $|x|$ becomes larger, the graph of the hyperbola gets closer to the lines $y = \frac{4}{5}\sqrt{x^2} = \frac{4}{5}|x|$ or $y = \pm\frac{4}{5}x$. Using all of this information and a few points, such as $x = \pm 6, \pm 7, \pm 8$, we can graph the hyperbola as shown in Figure 8.18. The lines $y = \frac{4}{5}x$ and $y = -\frac{4}{5}x$ are called **asymptotes** of the graph.

The geometric definition of a hyperbola is very similar to that of an ellipse: The path of a point moves in such a way that the *difference* of its distances to two fixed points (called *foci*) is a constant.

The general equation for a hyperbola is

$$\frac{x^2}{a^2} - \frac{y^2}{b^2} = 1 \tag{8.1}$$

Figure 8.18

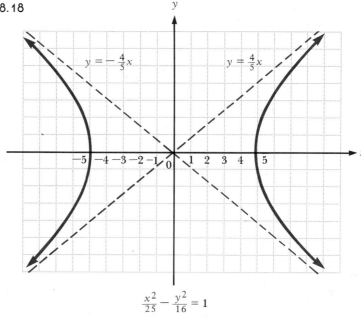

$$\frac{x^2}{25} - \frac{y^2}{16} = 1$$

The two asymptotes are (see Exercise 43):

$$y = \frac{b}{a}x \quad \text{and} \quad y = -\frac{b}{a}x \tag{8.2}$$

A hyperbola that is expressed in the form of Equation (8.1) is said to be written in **standard form**.

EXAMPLE 2   Identify the graph of $9x^2 - 16y^2 - 144 = 0$; if it is a hyperbola, locate the intercepts and find equations of the asymptotes.

Solution: Our first step is to move the constant term to the right side so that

$$9x^2 - 16y^2 = 144$$

Then, divide each term by the constant 144

$$\frac{9x^2}{144} - \frac{16y^2}{144} = 1$$

Simplifying the coefficients, we have

$$\frac{x^2}{16} - \frac{y^2}{9} = 1$$

or
$$\frac{x^2}{4^2} - \frac{y^2}{3^2} = 1$$

which is a hyperbola equation in standard form, where $a = 4$ and $b = 3$. The intercepts are $(-4, 0)$ and $(4, 0)$. Substituting $a = 4$ and $b = 3$ into Equation (8.2), we find the equations for the two asymptotes:

$$y = \frac{3}{4}x$$

and
$$y = -\frac{3}{4}x$$

The graph of this hyperbola is shown in Figure 8.19.

Figure 8.19

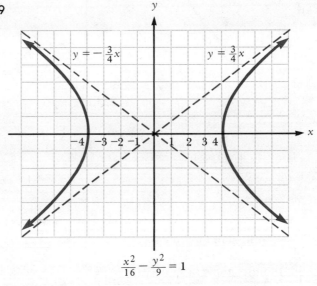

$\frac{x^2}{16} - \frac{y^2}{9} = 1$

If the general equation is in the form

$$\frac{y^2}{b^2} - \frac{x^2}{a^2} = 1$$

the two intercepts would be $(0, -b)$ and $(0, b)$, and the graph should look like the one in Figure 8.20. This equation for the hyperbola is also called standard form.

Circles, ellipses, parabolas, and hyperbolas all come under the general name **conic sections**. This is because they can all be thought of as being the result of cutting a cone with a plane, as is shown in Figure 8.21.

Figure 8.20

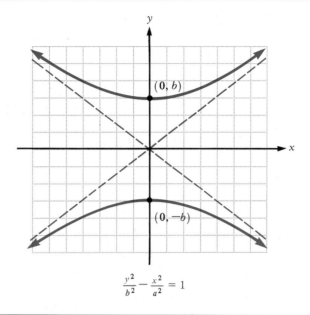

$$\frac{y^2}{b^2} - \frac{x^2}{a^2} = 1$$

Figure 8.21

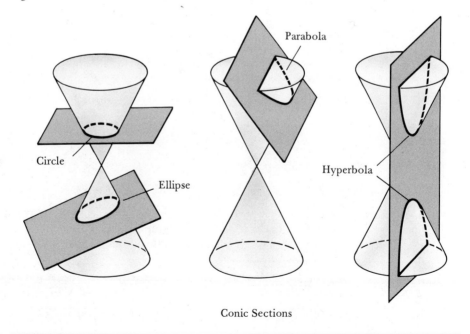

Conic Sections

Exercises 8.4

Standard Assignment: Exercises 1, 3, 5, 7, 9, 11, 13, 23, 27, 31, 33, 35, 37

A Sketch a graph for each of the following equations indicating the *x*- and *y*-intercepts.

1. $\dfrac{x^2}{16} + \dfrac{y^2}{9} = 1$
2. $\dfrac{x^2}{9} + \dfrac{y^2}{4} = 1$
3. $\dfrac{x^2}{4} + \dfrac{y^2}{9} = 1$
4. $\dfrac{x^2}{9} + \dfrac{y^2}{16} = 1$
5. $\dfrac{x^2}{36} + \dfrac{y^2}{49} = 1$
6. $\dfrac{x^2}{64} + \dfrac{y^2}{25} = 1$
7. $y^2 = 16 - 4x^2$
8. $9x^2 = 225 - 25y^2$

An equation of an ellipse is said to be in standard form if it is written in the form of

$$\dfrac{x^2}{a^2} + \dfrac{y^2}{b^2} = 1$$

In Exercises 9–14, express each of the following equations in standard form. State whether the ellipse is horizontal or vertical and give the intercepts.

> EXAMPLE: $9x^2 + 4y^2 - 36 = 0$.
>
> *Solution:* Moving the constant term to the right-hand side, we get
>
> $$9x^2 + 4y^2 = 36$$
>
> Dividing each term by 36, we get
>
> $$\dfrac{x^2}{4} + \dfrac{y^2}{9} = 1 \quad \text{or} \quad \dfrac{x^2}{2^2} + \dfrac{y^2}{3^2} = 1$$
>
> Since $a^2 = 2^2 < 3^2 = b^2$, the ellipse is vertical with intercepts at $(0, -3)$, $(0, 3)$, $(-2, 0)$, and $(2, 0)$.

9. $9y^2 = 36 - 4x^2$
10. $4y^2 = 64 - 16x^2$
11. $2x^2 + y^2 - 32 = 0$
12. $x^2 + 3y^2 - 27 = 0$
13. $98 - 7y^2 - 2x^2 = 0$
14. $100 - 25x^2 - 4y^2 = 0$

B In Exercises 15–22, graph each of the following radical equations and compare it with the graphs of Exercises 1–8; if the graph is a part of any graph that has been previously sketched, indicate the part and the graph it is part of.

15. $y = 2\sqrt{4 - x^2}$
16. $3y = 2\sqrt{9 - x^2}$
17. $3y = -4\sqrt{9 - x^2}$
18. $4y = -3\sqrt{16 - x^2}$
19. $3x = 5\sqrt{9 - y^2}$
20. $3x = 2\sqrt{9 - y^2}$
21. $7x = -6\sqrt{49 - y^2}$
22. $5x = -8\sqrt{25 - y^2}$

A Sketch a graph for each of the following equations.

23. $\dfrac{x^2}{4} - \dfrac{y^2}{9} = 1$

24. $\dfrac{x^2}{9} - \dfrac{y^2}{16} = 1$

25. $\dfrac{x^2}{16} - \dfrac{y^2}{9} = 1$

26. $\dfrac{x^2}{9} - \dfrac{y^2}{4} = 1$

27. $\dfrac{y^2}{9} - \dfrac{x^2}{16} = 1$

28. $\dfrac{y^2}{4} - \dfrac{x^2}{9} = 1$

29. $\dfrac{y^2}{49} - \dfrac{x^2}{25} = 1$

30. $\dfrac{y^2}{36} - x^2 = 1$

31. $y^2 = 16 + x^2$

32. $5y^2 = 25 + x^2$

Rewrite each of the following equations of hyperbolas in standard form. Locate the intercepts and find equations of the asymptotes. See Example 2.

33. $4x^2 = 100 + 25y^2$

34. $25x^2 = 4y^2 + 100$

35. $9y^2 = 25x^2 - 225$

36. $y^2 = 36x^2 - 144$

37. $12 - 3x^2 + 4y^2 = 0$

38. $15 + 5y^2 - 3x^2 = 0$

B 39. Show by a sketch, as in Figure 8.20, how a cone and a plane may intersect at just one point. (Hence, a point may be called a *degenerate circle* or a *degenerate ellipse*.)

40. Show by a sketch how a cone and a plane can intersect in two intersecting lines. (In conic sections, a *degenerate hyperbola* is a pair of intersecting lines.)

41. Sketch the graph of $25x^2 - 16y^2 = 0$.

42. Can a cone and a plane intersect in a line? If so, show this by a sketch.

43. Multiply both sides of the equation

$$\frac{x^2}{a^2} - \frac{y^2}{b^2} = 1$$

by b^2 and divide by x^2 to obtain

$$\frac{b^2}{a^2} - \frac{y^2}{x^2} = \frac{b^2}{x^2}.$$

Argue that for x where $|x|$ is large

$$\frac{y^2}{x^2} \approx \frac{b^2}{a^2},$$

or

$$y^2 \approx \frac{b^2}{a^2} x^2.$$

By taking square roots, deduce that $y \approx \pm (b/a)x$, which are the asymptote lines for the original equation when the $\approx$ sign is replaced by the equal sign.

8.5 The Graphs of Quadratic Inequalities

In Section 7.4 we graphed linear inequalities. We can apply the same ideas to graphing **quadratic inequalities** such as $x^2 + y^2 \leq 25$ or $x^2 + y^2 < 25$. In either case, the boundary $x^2 + y^2 = 25$ is a circle. As with linear inequalities, the boundary divides the coordinate plane into two parts: inside the circle or outside the circle. In order to decide which of the two parts represents the graph, we use a **test point** such as $(0, 0)$. Since $0^2 + 0^2 < 25$ then also $0^2 + 0^2 \leq 25$, so the part containing the origin (inside the circle) represents the graph. The boundary $x^2 + y^2 = 25$ belongs to the graph of $x^2 + y^2 \leq 25$ but *not* of $x^2 + y^2 < 25$, as indicated by the solid and dashed boundaries, respectively (see Figure 8.22).

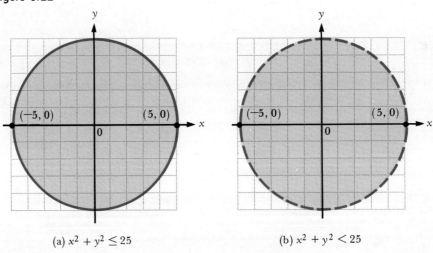

Figure 8.22

(a) $x^2 + y^2 \leq 25$

(b) $x^2 + y^2 < 25$

EXAMPLE 1

Sketch the graph of $y > -x^2 + 4x$.

Solution: The boundary $y = -x^2 + 4x$ is a parabola opening downward (see Figure 8.23). If we use $(0, 2)$ as a test point, we find $2 > -0^2 + 4(0)$ or $2 > 0$, which is *true*. Hence, the part of the plane containing $(0, 2)$ satisfies the inequality. The graph is given in Figure 8.23, where the dashed boundary shows that the boundary is not included in the graph.

8.5 The Graphs of Quadratic Inequalities

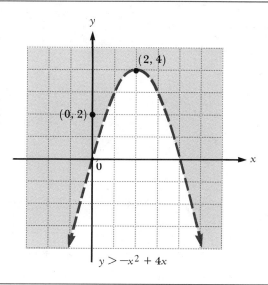

Figure 8.23

$y > -x^2 + 4x$

EXAMPLE 2 Sketch a graph of $\dfrac{y}{6} \leq \sqrt{1 - \dfrac{x^2}{16}}$, with $y \geq 0$.

Solution: Since $y \geq 0$, the graph is in the upper half of the coordinate plane. To find the boundary from the equation

$$\frac{y}{6} = \sqrt{1 - \frac{x^2}{16}} \qquad (y \geq 0)$$

we have

$$\frac{y^2}{36} = 1 - \frac{x^2}{16}$$

$$\frac{x^2}{16} + \frac{y^2}{36} = 1 \qquad (y \geq 0)$$

which represents the upper half of an ellipse with intercepts at $x = \pm 4$, and $y = 6$ as shown in Figure 8.24 on the following page. By using $(0, 0)$ as a test point, we determine that the shaded area under and including the boundary of the upper half ellipse in Figure 8.24 is the graph of the inequality

$$\frac{y}{6} \leq \sqrt{1 - \frac{x^2}{16}} \qquad (y \geq 0)$$

That is,

$$\frac{0}{6} \leq \sqrt{1 - \frac{0^2}{16}}$$

$0 \leq 1$ is *true*.

364   Chap. 8   Functions and Conic Sections

Figure 8.24

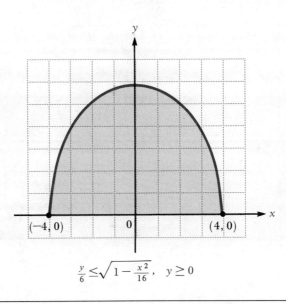

$\frac{y}{6} \leq \sqrt{1 - \frac{x^2}{16}}, \quad y \geq 0$

EXAMPLE 3   Sketch a graph of $16x^2 - 25y^2 \geq 400$.

Solution: The boundary is the hyperbola with equation

$$16x^2 - 25y^2 = 400$$

or

$$\frac{x^2}{25} - \frac{y^2}{16} = 1$$

the graph of which appeared previously in Figure 8.18.

The boundary consists of two curves that divide the plane into three parts as is shown in Figure 8.25(a). To determine which part is in the graph, we use one test point from each of the three parts. We will choose $(-6, 0)$, $(0, 0)$, and $(6, 0)$ from Parts I, II, and III respectively.

Is $16(-6)^2 - 25 \cdot 0^2 \geq 400$ true? Yes.
Is $16 \cdot 0^2 - 25 \cdot 0^2 \geq 400$ true? No, false.
Is $16 \cdot 6^2 - 25 \cdot 0^2 \geq 400$ true? Yes.

The graph is the shaded area of Part I and Part III, including the boundary, as shown is in Figure 8.25(b).

8.5 The Graphs of Quadratic Inequalities 365

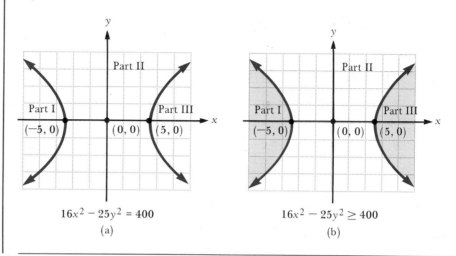

Figure 8.25

$16x^2 - 25y^2 = 400$
(a)

$16x^2 - 25y^2 \geq 400$
(b)

Exercises 8.5

Standard Assignment: Exercises 1, 3, 5, 7, 9, 11, 13, 15, 17, 25, 27, 29, 31, 33, 35, 37

A Sketch a graph for each of the following. See Examples 1–3.

1. $x^2 + y^2 \leq 4$
2. $x^2 + y^2 < 4$
3. $x^2 + y^2 < 9$
4. $x^2 + y^2 \leq 9$
5. $x^2 + y^2 \geq 1$
6. $x^2 + y^2 \geq 2$
7. $x^2 + y^2 > 3$
8. $x^2 + y^2 > 5$
9. $y \leq \sqrt{16 - x^2}, \quad y \geq 0$
10. $y \leq \sqrt{49 - x^2}, \quad y \geq 0$
11. $y + x^2 \leq 0$
12. $y - x^2 \geq 0$
13. $y < -x^2 + 1$
14. $x < -x^2 + 2$
15. $y \geq -x^2 - 1$
16. $y \geq -(x - 1)^2$
17. $y > -(x - 2)^2$
18. $y > -(x + 1)^2$
19. $x \geq -y^2$
20. $x \geq -y^2 - 2$
21. $x < y^2 + 1$
22. $x < (y - 1)^2$
23. $y \leq \sqrt{x}$
24. $y \geq \sqrt{x - 5}$

25. $x > \sqrt{y - 1}$
26. $x < \sqrt{2 - y}$
27. $9x^2 + y^2 \leq 9$
28. $x^2 + 16y^2 \leq 16$
29. $4x^2 + 9y^2 < 36$
30. $3x^2 + 5y^2 < 15$
31. $y^2 \geq 16 - 4x^2$
32. $25y^2 \geq 225 - 9x^2$
33. $y > 2\sqrt{4 - x^2}$
34. $3y > 2\sqrt{9 - x^2}$
35. $3y < 2\sqrt{9 - x^2}$, $\quad x > 0$
36. $y \leq 2\sqrt{4 - x^2}$, $\quad x \leq 0$
37. $9x^2 - 4y^2 \geq 36$
38. $16x^2 - 9y^2 \geq 144$
39. $y^2 < 16 + x^2$
40. $5y^2 < 25 + x^2$
41. $y^2 \geq 36(1 + x^2)$
42. $9y^2 \geq 36 + 4x^2$
43. $\dfrac{y}{7} < \sqrt{1 + \dfrac{x^2}{25}}$, $\quad y > 0$
44. $2x \geq \sqrt{100 + 25y^2}$, $\quad y \leq 0$
45. $5x \geq 2\sqrt{y^2 + 25}$, $\quad y \leq 0$
46. $3y \leq 5\sqrt{x^2 - 5}$, $\quad x < 0$

8.6 Applications

Business Problems

The cost of manufacturing a product is the sum of fixed costs such as rent for the plant, depreciation of machinery, insurance, utilities, and variable costs, such as labor and materials. While the variable costs depend on the number of units produced, the fixed costs remain the same regardless of the production. The company *breaks even* when the total cost (fixed plus variable) equals the total revenue (= income).

EXAMPLE 1 After years of experience and analysis, the ABC Alarm Co. established the cost equation for manufacturing smoke alarms as $C = x^2 - 12x + 41$, with $0 \leq x \leq 9$, where C is the cost per unit in dollars of producing x (thousand) units per day. How many thousand units must be produced per day to minimize the cost per unit produced?

Solution: We use the method of completing the square to rewrite

$$C = x^2 - 12x + 41$$
$$= (x^2 - 12x + 6^2) + 5$$
$$= (x - 6)^2 + 5 \geq 5$$

Since the smallest value $(x - 6)^2$ can take is 0 (when $x = 6$), the smallest value $(x - 6)^2 + 5$ can take is $0 + 5 = 5$, when x is 6. Hence, 6000 units must be produced per day to be cost effective; the minimum cost per unit would then be only $5. Note that the vertex (6, 5) is the low point for the parabola $C = x^2 - 12x + 41$.

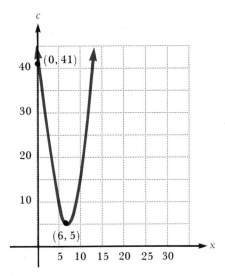

EXAMPLE 2

The Florida Instrument Co. manufactures and sells digital pens. The company spends $8000/week in fixed costs and $2x$ thousand dollars in variable costs for the weekly production of x thousand pens. If the total weekly revenue (the money received) for selling x thousand pens is $R = 10\sqrt{x}$ (thousand dollars/week), what quantity would the company have to sell in order to cover just production costs? (This quantity is called the break-even quantity.) What range of quantity would have to be produced and sold in order for the company to be profitable?

Solution: Since the total weekly cost = fixed costs + variable costs, we have the cost equation

$$C = 8 + 2x \text{ (thousand dollars/week)}$$

The manufacturer breaks even when the total cost = total revenue:

$$C = R$$
$$8 + 2x = 10\sqrt{x}$$
$$x + 4 = 5\sqrt{x} \quad \text{Divide by 2}$$
$$x^2 + 8x + 16 = 25x \quad \text{Square}$$
$$x^2 - 17x + 16 = 0 \quad \text{Standard form}$$
$$(x - 1)(x - 16) = 0 \quad \text{Factor}$$
$$x = 1 \text{ or } 16$$

The company breaks even if the weekly production is either 1000 pens or 16,000 pens.

The graphs of the cost equation and revenue equation on the same coordinate plane are shown in Figure 8.26.

The manufacturer makes a profit if the total revenue is more than the total cost, when $R > C$. From Figure 8.26 we see immediately that Florida Instrument makes a profit if the weekly production is over 1000 pens but less than 16,000 pens.

The results show that there is a *ceiling* level, $x = 16,000$, of profitability due to higher overtime pay and machinery depreciation resulting from overtime use, lack of maintenance, and so forth.

Figure 8.26

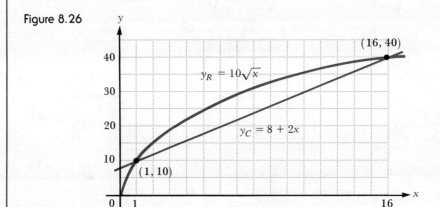

EXAMPLE 3

A photographer at position $(2, 0)$ as indicated in Figure 8.27 is watching a helicopter descending along the path $y = \sqrt{2x}$. The x-axis is at ground

Figure 8.27

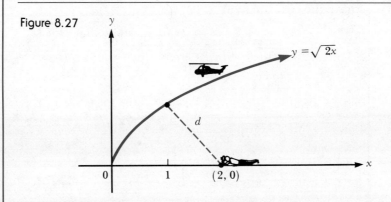

level. At what point of the path is the helicopter closest to the photographer?

Solution: Let (x, y) be the position of the helicopter at a given point of its path $y = \sqrt{2x}$. By the distance formula, the distance d between the helicopter at position (x, y) and the photographer at $(2, 0)$ is given by

$$d = \sqrt{(x - 2)^2 + (y - 0)^2}$$
$$= \sqrt{x^2 - 4x + 4 + y^2}$$

As we prefer to work without a square root sign, we square both sides to get
$$d^2 = x^2 - 4x + 4 + y^2$$

Thus, to minimize the distance d, we minimize d^2 instead. Since the helicopter's position (x, y) is also on its path, $y = \sqrt{2x}$, we have $y^2 = 2x$. To minimize d^2 we first replace y^2 by $2x$ in the previous expression for d^2. Hence

$$\begin{aligned} d^2 &= x^2 - 4x + 4 + y^2 \\ &= x^2 - 4x + 4 + 2x && \text{Since } y^2 = 2x \\ &= (x^2 - 2x + 1) + 3 && \text{Complete the square} \\ &= (x - 1)^2 + 3 \end{aligned}$$

The quantity is minimal when $(x - 1)^2$ is 0, at $x = 1$, when $y = \sqrt{2}$.

Hence, the helicopter is closest to the photographer when it reaches position $(1, \sqrt{2})$. See Figure 8.27.

Application to Agriculture

EXAMPLE 4

After years of study, Southern Agrinomics estimates that if 50 orange trees are planted per acre, each mature tree will yield 60 bushels of oranges per year. For each additional orange tree planted per acre, the number of oranges produced by each tree decreases by 1 bushel per year. How many trees should be planted per acre in order to produce the most oranges?

Solution: If x additional orange trees are planted per acre (for a total of $50 + x$ trees per acre), the yield from each tree decreases to $60 - x$ bushels. Since

$$\text{Total Yield} = \text{Yield Per Tree} \times \text{Number of Trees}$$

we have the yield equation

$$\begin{aligned} y &= (60 - x)(50 + x) \\ &= 3000 + 10x - x^2 && \text{Use FOIL method} \\ &= (3000 + 25) - 25 + 10x - x^2 && 25 - 25 = 0 \\ &= 3025 - (5^2 - 2 \cdot 5x + x^2) && \text{Complete the square} \\ &= 3025 - (5 - x)^2 \end{aligned}$$

The yield is largest when $-(5 - x)^2 = 0$ or $x = 5$. Hence, 5 additional trees, making a total of 55 trees, should be planted per acre.

Application to Physics

EXAMPLE 5

If a ball is thrown straight up from the ground with an initial speed of 64 ft/sec, then its height h (in ft) after t seconds is given by $h = 64t - 16t^2$. What is the maximum height attained by the ball? How many seconds does it take for the ball to reach its maximum height?

Solution:
$$\begin{align} h &= 64t - 16t^2 \\ &= -(4t)^2 + 2 \cdot 8 \cdot (4t) - 8^2 + 8^2 \quad \text{Complete the square} \\ &= 8^2 - [(4t)^2 - 2 \cdot 8 \cdot (4t) + 8^2] \\ &= 64 - (4t - 8)^2 \end{align}$$

The height is highest when $-(4t - 8)^2 = 0$, or $t = 2$. Thus, the ball reaches its maximum height of 64 ft in 2 seconds. Note that the vertex (2, 64) is the high point of the parabola $h = 64t - 16t^2$.

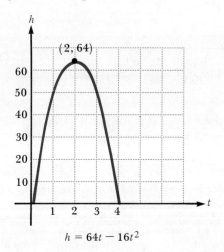

$h = 64t - 16t^2$

Exercises 8.6

Standard Assignment: Exercises 1, 3, 7, 9, 11, 13, 15, 17

A

1. In Example 1, if ABC Alarm Co. decided to make the following, how much would each unit cost?
    (a) 5 thousand units a day
    (b) 7 thousand units a day
    (c) 1 thousand units a day

In Exercises 2–4, follow Example 2.

2. The total revenue less the total cost is called the *profit*. How much profit would Florida Instrument Co. make if 9000 pens were produced and sold in a week?

3. A negative profit is called a *deficit*. What deficit would Florida Instrument Co. incur if only 500 pens were produced and sold in a week?

4. (a) What is the maximum possible weekly profit for Florida Instrument Co.? (Hint: Use the substitution $u = \sqrt{x}$.)
   (b) How many pens must be produced and sold per week to attain the maximum profit?

5. A small neighborhood savings and loan association wishes to encourage small investors by announcing that for x dollars, $1 \leq x \leq 2200$, deposited in a savings account, the association will pay $y\%$ annual interest subject to the following equation

$$100{,}000y = -x^2 + 2000x + 500{,}000$$

What interest rate will an investor receive if he deposits the following?
   (a) $500
   (b) $900
   (c) $2000

6. In Exercise 5, determine the amount of deposit that will yield the maximum (annual) interest rate. What is the maximum interest rate that the association pays to an investor?

7. In Example 4, what is the maximum number of bushels of oranges that an acre of land can produce? If 100 orange trees are planted in an acre, how many bushels of oranges will be produced?

8. An apple orchard now has 80 trees per acre, and the average yield is 900 apples per tree. For each additional tree planted per acre, the average yield per tree is decreased by approximately 10 apples. How many apple trees per acre should be planted to obtain the maximum crop of apples? (See Example 4.)

9. An arrow is shot straight up from the ground with an initial speed of 180 ft/sec. Its height h (in ft) after t seconds is given by $h = 180t - 16t^2$. What is the maximum height that is attained by the arrow? How many seconds does it take for the arrow to reach its maximum height? (See Example 5.)

10. In Exercise 9, how many seconds does it take for the arrow to return to the ground?

11. Mr. Smith gave his son 100 yards of fencing. He asked the son to fence three sides of a rectangular plot that is adjacent to the river, with the side along the river requiring no fencing. Find the largest area that can be fenced.

12. If in Exercise 11, Mr. Smith gave his son 120 yards instead of 100 yards, what are the dimensions of the largest area that can be fenced?

13. Of all pairs of numbers whose sum is 98, which pair has the largest product?

14. Of all rectangles of perimeter 84 cm, which one has the largest area?

15. A photographer at (1, 0) is watching a helicopter descending the path $y = \sqrt{x}$. The x-axis is at ground level. At what point of the path is the helicopter closest to the photographer? (See Example 3.)

16. At what point of the path is the helicopter closest to the photographer if the path of descent in Exercise 15 is $y = 4\sqrt{x}$?

17. A patrolman driving along the road described by the equation $y^2 = 4x$ spots a suspect at (2, 0). At what position of the road is the suspect closest to the patrol? (See Example 3.)

18. What if the road in Exercise 17 was hyperbolical, $y^2 - 2x^2 = 26$, and the suspect was at (1, 0)?

8.7 Inverse Functions

Consider the functions $y = x + 1$ and $y = x^2$. In both cases each x-value determines exactly one y-value. Is it true that each y-value is also determined by exactly one x-value? While this is true for the function $y = x + 1$, it is not true for $y = x^2$. For the function $y = x + 1$, if $y = 5$, x must equal 4; if $y = 4$, x must equal 3, and so on. For $y = x^2$, $y = 9$ results from both $x = 3$ and $x = -3$. Thus, two different values of x yield the same value of y.

A function that has the property that only one x-value is associated with each y-value is called a **one-to-one** function.

EXAMPLE 1

Determine which functions are one-to-one.
(a) $y = 2x - 3$ (b) $y = |x|$ (c) $y = \sqrt{x}$

Solution:
(a) If $y = a$, we have $a = 2x - 3$, which leads to a unique x-value, $x = (a + 3)/2$. The function is one-to-one. For example, if $y = 0$, we have $x = (0 + 3)/2 = 3/2$; if $y = -1$, $x = (-1 + 3)/2 = 1$.
(b) Suppose that $y = 2$; for what values of x does $|x| = 2$? The answer is x may be 2 or -2. The function is not one-to-one; two different values of x correspond to the same value of y.
(c) Since $\sqrt{x} \geq 0$ for all non-negative real values of x, the y-values are non-negative real numbers. Thus, for any $y = a \geq 0$, $a = \sqrt{x}$ leads to only one x-value: $x = a^2$. The function is one-to-one. From $y = 2$, we get $x = 4$; from $y = 3$, we get $x = 9$, and so on.

The requirement for a one-to-one function is that only one x-value is associated with each y-value, $y = a$, and the graph of $y = a$ is a horizontal line. Therefore,

Horizontal Line Test

If each horizontal line intersects the graph of a function in at most one point, the function is one-to-one.

In other words, if a horizontal line intersects the graph of a function in two or more points, then the function is not one-to-one.

EXAMPLE 2 Use the horizontal line test to determine which functions are one-to-one.
(a) $y = ax + b$, $(a \neq 0)$ (b) $y = x^2 - 1$ (c) $y = 2\sqrt{x}$

Solution: From the graphs in Figure 8.28, we see that the functions in (a) and (c) are one-to-one. The function in (b) is not one-to-one since the horizontal line in Figure 8.28(b) intersects the graph at more than one point.

Figure 8.28

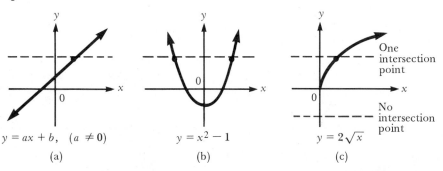

$y = ax + b$, $(a \neq 0)$ $y = x^2 - 1$ $y = 2\sqrt{x}$
(a) (b) (c)

Every one-to-one function $y = f(x)$ defines another one-to-one function, $y = f^{-1}(x)$, which is read **f-inverse of x**, in the following manner. The graph of the inverse function $y = f^{-1}(x)$ consists of those points (b, a) for which the points (a, b) are on the graph of $y = f(x)$. Thus,

$$a = f^{-1}(b) \quad \text{means} \quad b = f(a). \tag{8.3}$$

Each point (b, a) is symmetric with respect to the line $y = x$ to the point (a, b) as is shown in Figure 8.29(a); the graph of the inverse function $y = f^{-1}(x)$ is symmetric to the graph $y = f(x)$ with respect to the line $y = x$. See Figure 8.29(b).

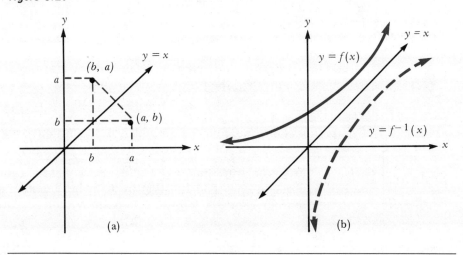

Figure 8.29

To find an equation for the inverse function $y = f^{-1}(x)$ from a given one-to-one function $y = f(x)$, we first interchange x and y in $y = f(x)$. That is, we replace x by y and y by x in the equation $y = f(x)$. Then we solve for y in terms of x to get $y = f^{-1}(x)$. This procedure is demonstrated in the following example.

EXAMPLE 3 Find the inverse functions, if they exist, of
(a) $f(x) = 3x - 5$ (b) $g(x) = x^2 - 1$ (c) $h(x) = 2\sqrt{x}, \quad x \geq 0$

Solution:
(a) Let $y = f(x)$ so that $y = 3x - 5$. We have seen in Example 2(a) that this function is one-to-one. Interchanging x and y we have

$$x = 3y - 5$$

or

$$3y = x + 5 \qquad \text{Add 5 to both sides}$$

$$y = \frac{x + 5}{3} \qquad \text{Solve for } y$$

Thus $$f^{-1}(x) = \frac{x + 5}{3}$$

(b) We write $y = g(x)$, or $y = x^2 - 1$, which is not a one-to-one function as we have seen in Example 2(6). Hence, it does not have an inverse function.

(c) We write $y = h(x)$, or $y = 2\sqrt{x}$; $x \geq 0$, so $y \geq 0$. By examining the graph of h in a manner similar to that of Example 2(c) we see that h is one-to-one. Interchanging x and y,

$$x = 2\sqrt{y} \qquad x \geq 0$$
$$x^2 = 4y \qquad \text{Square both sides; } x \geq 0$$
$$y = \frac{x^2}{4} \qquad \text{Solve for } y; x \geq 0$$

Hence, $h^{-1}(x) = \dfrac{x^2}{4}$, $x \geq 0$

The graphs of $y = h(x) = 2\sqrt{x}$ and $y = h^{-1}(x) = x^2/4$, with $x \geq 0$, are shown in Figure 8.30.

Figure 8.30

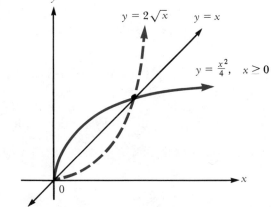

If we consider a function as a set of ordered pairs, as in Section 7.5, then f^{-1}, for a one-to-one function f, is the set of ordered pairs obtained by interchanging the coordinates of each ordered pair in f. For example, if

$$f = \{(1, 2), (2, 3), (3, 4)\}$$

then
$$f^{-1} = \{(2, 1), (3, 2), (4, 3)\}$$

How about $(f^{-1})^{-1}$? We interchange the coordinates of each ordered pair in f to get f^{-1}, so we must interchange the coordinates once again to have

$(f^{-1})^{-1}$. Using the previous example, since
$$f^{-1} = \{(2, 1), (3, 2), (4, 3)\}$$
we have $(f^{-1})^{-1} = \{(1, 2), (2, 3), (3, 4)\} = f$

In general, the inverse of f is f^{-1} and the inverse of f^{-1} is f so f and f^{-1} are inverses of each other.

EXAMPLE 4 Let $f(x) = 1 - x^3$. Find (a) $f(2)$ (b) $f(1)$ (c) $f^{-1}(0)$ (d) $f^{-1}(-7)$

Solution:
(a) $f(2) = 1 - 2^3 = 1 - 8 = -7$
(b) $f(1) = 1 - 1^3 = 0$
(c) From (b) the ordered pair $(1,0)$ belongs to f, so that $(0, 1)$ belongs to f^{-1}. Hence $f^{-1}(0) = 1$.
(d) From (a) the ordered pair $(2, -7)$ belongs to f, hence $(-7, 2)$ belongs to f^{-1}, or $f^{-1}(-7) = 2$.

From Equation (8.3) $f^{-1}(b) = a$ means $b = f(a)$, and so we conclude that

$$f[f^{-1}(b)] = f(a) \qquad f^{-1}(b) = a$$
$$= b \qquad\qquad b = f(a)$$

and
$$f^{-1}[f(a)] = f^{-1}(b) \qquad f(a) = b$$
$$= a \qquad\qquad a = f^{-1}(b)$$

The calculator view of a function such as $f(x) = 3x + 5$ is that when x is keyed in, the calculator "triples x," then "adds 5," and then displays the result. The calculator corresponding to f^{-1} reverses the operations performed by f on x *and* the order in which they are done. Thus, when x is keyed into f^{-1}, the calculator "subtracts 5," then "takes $\frac{1}{3}$ of the resulting number," and then displays the result. The equation for f^{-1} is $f^{-1}(x) = (x - 5)/3$.

Exercises 8.7

Standard Assignment: All odd-numbered exercises

A Determine which of the following are one-to-one functions. See Examples 1 and 2.

1. $y = x + 12$
2. $y = 13 - 2x$
3. $f(x) = x$
4. $f(k) = \sqrt{k} - 2$
5. $g(t) = t^2$
6. $g(t) = 1 - t^4$

7. $y = |x - 1|$
8. $y = |2 - x|$
9. $y = (x + 3)^2$
10. $y = (5 - x)^2$
11. $f(x) = x^3 + 4$
12. $f(x) = (x - 1)^3 + 1$
13. $xy = 1$
14. $(1 - x)y = 3$
15. $y = \sqrt{1 - x^2}$
16. $y = \sqrt{4 + 9x^2}$
17. $y = \sqrt[3]{2x + 1}$
18. $y = \sqrt[3]{2 - 5x}$

Which of the following are one-to-one functions? If a given function is one-to-one, find the equation of the inverse function and sketch the graph of the function and its inverse. See Examples 1–3.

19. $f(x) = 15 - 3x$
20. $g(x) = 2x + 5$
21. $y = \sqrt{4 - x^2}$
22. $y = -\sqrt{9 - x^2}$
23. $y = \sqrt{x} + 3$
24. $y = 4 - \sqrt{x}$
25. $y = \sqrt[3]{x}$
26. $y = \sqrt[3]{1 - x}$
27. $y = \dfrac{1}{x}, \quad x \neq 0$
28. $y = 1 - \dfrac{1}{x}, \quad x \neq 0$

For each of the given one-to-one functions, find its inverse.

29. $f = \{(-3, -2), (-2, -1), (-1, 0), (0, 1)\}$
30. $g = \{(-2, 0), (-1, 1), (0, 2), (1, 3)\}$
31. $h = \{(0, 0), (1, 1), (2, 4), (3, 9), (4, 16)\}$
32. $f = \{(0, 0), (1, -1), (2, -4), (3, -9), (4, -16)\}$
33. $g = \{(-1, 1), (3, -3), (-5, 5), (7, -7)\}$
34. $f = \{(-2, 0), (1, 3), (2, 4), (0, 2)\}$
35. $g = \{(1, -1), (2, -8), (3, -27), (4, 1)\}$
36. $h = \{(a, b), (b, c), (c, d)\}$

Let $f(x) = 1 + x^3$. Find the following values. See Example 4.

37. $f(0)$
38. $f(1)$
39. $f(3)$
40. $f(2)$
41. $f^{-1}(1)$
42. $f^{-1}(2)$
43. $f^{-1}(28)$
44. $f^{-1}(9)$

Chapter 8 Summary

Key Words and Phrases

8.1 Quadratic function
   Parabola
   Opens downward (upward)
   Vertex (vertices)
8.2 Opens to the right (left)
   Symmetric with respect to the x-axis (y-axis)
8.3 Pythagorean Theorem
   Distance formula
   Circle
   Radius
8.4 Ellipse
   Focus (foci)
   Hyperbola
   Asymptotes
   Conic sections
8.5 Quadratic inequality
8.7 One-to-one function
   f-inverse of x
   Horizontal line test
   Inverse function

Key Concepts and Rules

The graph of an equation of the form

$$y = ax^2 + bx + c, \quad (a \neq 0)$$

is a *parabola* that opens upward if $a > 0$, and opens downward if $a < 0$. Similarly, the graph of an equation of the form

$$x = ay^2 + by + c, \quad (a \neq 0)$$

is a parabola that opens to the right if $a > 0$, and opens to the left if $a < 0$.

The Distance Formula: The distance between any two points $P_1(x_1, y_1)$ and $P_2(x_2, y_2)$ is given by

$$d(P_1, P_2) = \sqrt{(x_2 - x_1)^2 + (y_2 - y_1)^2}$$

The Equation of a Circle: An equation of the circle with radius r and center at (a, b) is given by

$$(x - a)^2 + (y - b)^2 = r^2$$

The equation of an *ellipse* is

$$\frac{x^2}{a^2} + \frac{y^2}{b^2} = 1$$

An equation of the *hyperbola* is

$$\frac{x^2}{a^2} - \frac{y^2}{b^2} = 1$$

Review Exercises

If you have difficulty with any of the exercises, look at the section indicated by the number in brackets.

[8.1] Sketch each of the following parabolas and indicate the vertex of each.

1. $y = x^2$
2. $y = x^2 + 4$
3. $y = (x - 1)^2$
4. $y = (x + 2)^2 - 1$
5. $y = x^2 + 5$
6. $y = 4x^2 - 1$
7. $x = y^2 - 2y + 5$
8. $x = y^2 + 2y - 4$
9. $y = -x^2 + 4x - 1$
10. $x = -y^2 - 2y + 1$

11. Which of the parabolas in Exercises 1–10 are symmetric with respect to the x-axis? the y-axis?

[8.2] Find the distance between the following pairs of points.

12. $(2, 5)$ and $(-1, 2)$
13. $(-2, 1)$ and $(3, -2)$
14. (a, b) and (b, a)
15. $(u, u + v)$ and $(u - v, u)$

Graph each of the following equations.

16. $x^2 + y^2 = 36$
17. $(x - 1)^2 + (y - 2)^2 = 0$
18. $y = \sqrt{9 - x^2}$
19. $y = -\sqrt{16 - x^2}$
20. $x = \sqrt{25 - y^2}$

Write an equation for each of the following circles.

21. Center at $(-1, -1)$ and radius 8
22. Center at $(-5, -4)$ and radius $\sqrt{7}$

Identify the graph of each of the following equations.

23. $x^2 + y^2 - 4x - 6y - 1 = 0$
24. $x^2 + y^2 + 6x - 3 = 0$
25. $x^2 + y^2 - 8y - 5 = 0$
26. $x^2 + y^2 - 2x - 2y + 2 = 0$

[8.3] 27. Sketch the graph of $\dfrac{(x - 1)^2}{4} + \dfrac{(y + 2)^2}{9} = 1$.

28. Sketch the graph of $\dfrac{x^2}{16} + \dfrac{(y - 2)^2}{4} = 1$.

Rewrite each of the following equations in standard form and locate the x- and y-intercepts.

29. $9x^2 = 36 - 4y^2$
30. $25y^2 = 100 - 4x^2$
31. $5x^2 + 6y^2 = 150$
32. $y^2 = 36 - 9x^2$

[8.4] Sketch a graph for each of the following equations.

33. $x^2 = 9 + y^2$
34. $4y^2 = 100 + 25x^2$

Rewrite each of the following equations of hyperbolas in standard form, locate the intercepts, and find equations of the asymptotes.

35. $x^2 = 36y^2 + 36$
36. $12 - 3y^2 + 4x^2 = 0$
37. $16(x^2 - 1) = y^2$
38. $9 - x^2 + y^2 = 0$

[8.5] Sketch a graph for each of the following inequalities.

39. $(x - 1)^2 + (y - 2)^2 \le 9$
40. $4 < x^2 + y^2 \le 16$
41. $x + y^2 \le 0$
42. $y < (x - 1)^2$
43. $4x^2 + 9y^2 < 36$
44. $5x^2 < 25 + y^2$

[8.6] 45. Of all rectangles of perimeter 100 cm, which one has the largest area?

46. A spectator at (1, 0) is watching a parachute descending along the path $y = 5\sqrt{x}$. The x-axis is at ground level. At what point of the path is the parachute closest to the spectator? [See Example 3 of Section 8.6.]

47. An arrow is shot straight up from the ground with an initial speed of 200 ft/sec; its height h (in ft) after t seconds is given by $h = 200t - 16t^2$. What is the maximum height attained by the arrow? How many seconds does it take for the arrow to reach the maximum height? (See Example 5 of Section 8.6.)

[8.7] Determine which of the following are one-to-one functions.

48. $y = -2x + 5$
49. $y = 6x - 1$
50. $y = |x - 2|$
51. $y = -|2x + 1|$
52. $y = (2x - 3)^3$
53. $y = -(3 - 2x)^3$
54. $y = 2x^2$
55. $y = 101 - 3x^2$
56. $f(x) = \sqrt[3]{x + 5}$
57. $f(x) = \sqrt[3]{5 - 2x}$
58. $xy = 1$

Which of the following are one-to-one functions? If a given function is one-to-one, find its inverse function and sketch the graph of the function and its inverse.

59. $f(x) = 13 - 4x$
60. $g(x) = 5x - 10$
61. $h(x) = 5 + \sqrt{x}$
62. $f(x) = 10 - \sqrt{x}$
63. $y = \sqrt{9 - x^2}$
64. $y = -\sqrt{4 - x^2}$
65. $y = \sqrt[3]{x - 1}$
66. $y = \sqrt[3]{1 - x}$
67. $y = \dfrac{1}{2x}, \quad x \ne 0$
68. $y = 5 - \dfrac{2}{x}, \quad x \ne 0$

For each of the given one-to-one functions, find its inverse.

69. $f = \{(-3, -1), (-2, 0), (-1, 1), (0, 2)\}$

70. $g = \{(-1, -3), (0, 2), (1, -1), (2, 0)\}$

71. $h = \{(a, 2b), (b, 2c), (c, 2d)\}$

72. $f = \{(2b, a), (2c, b), (2d, c)\}$

Let $f(x) = 2^x$. Find the following values.

73. $f(0)$ 74. $f(1)$ 75. $f(3)$ 76. $f(2)$

77. $f^{-1}(1)$ 78. $f^{-1}(2)$ 79. $f^{-1}(8)$ 80. $f^{-1}(4)$

Practice Test (60 minutes)

1. Sketch a graph of $y = 2x^2 - 6x + 1$ and name the graph. If the graph has a vertex, indicate its location.

2. Show that the points (a, b), $(a + 3, b)$ and $(a + 3, b + 4)$ form the vertices of a right triangle.

3. Write an equation of the circle with a radius of 3 and center at $(0, -3)$.

4. Identify the graph of $x^2 + y^2 - 2x + 4y - 1 = 0$. If it is a circle, find its center and the radius.

5. Rewrite the equation $x^2 + 9y^2 = 144$ in standard form and name the graph. If it is an ellipse, locate the x- and y-intercepts.

6. Rewrite the equation $x^2 = y^2 + 49$ in standard form; sketch and name the graph.

7. Sketch the graph of $36 \leq 4x^2 + 9y^2 < 144$.

8. An orange grove with 40 orange trees per acre will yield 300 oranges per tree. For each additional orange tree planted per acre, the number of oranges produced by each tree decreases by 5 oranges per tree. How many trees should be planted per acre to produce the most oranges?

Which of the following are one-to-one functions? If a given function is one-to-one, find the equation of its inverse function and sketch the graph of the function and its inverse.

9. $y = (2x - 3)^3$ 10. $f(x) = \sqrt[3]{(x - 1)^2}$

11. $g(x) = \sqrt{x - 3} + 1$

Extended Applications

Planetary Orbits

Johannes Kepler (1571–1630) discovered by the analysis of astronomical observations that the planets move in elliptical orbits with the sun as one focus. Isaac Newton (1642–1727) proved this fact mathematically. The orbit of the earth is elliptical with the sun at one focus $(c, 0)$ and a major diameter of $2a = 185.8$ million miles (see diagram). The ratio c/a is approximately 0.0167. How close to the sun does the earth come? How far away? Since $2a = 185.8$, $a = 92.9$ and $c = 0.0167a = 0.0167 \times 92.9 = 1.55143$.

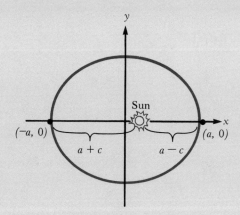

An energetic student will be able to prove that the nearest point in the earth's orbit to the sun at $(c, 0)$ is the endpoint $(a, 0)$ of the major diameter, and that the farthest point is the other end of the major diameter $(-a, 0)$. Thus, the nearest distance between the earth and the sun

$$= a - c$$
$$= 92.9 - 1.55143 \text{ (million miles)}$$
$$= 91{,}348{,}570 \text{ (miles)}$$

The farthest distance between the earth and the sun

$$= a + c$$
$$= 92.9 + 1.55143 \text{ (million miles)}$$
$$= 94{,}451{,}430 \text{ (miles)}$$

The average of these distances is

$$\frac{1}{2}[(a - c) + (a + c)] = a$$

$$= 92.9 \text{ (million miles)}$$

Exercises

1. Find the minor radius b of the earth's orbit and the relative difference,

$$\frac{a - b}{(a + b)/2},$$

of the major and minor radii.

2. From the information you obtained in Exercise 1, is the earth's elliptical orbit around the sun almost like a circle?

3. The planet Mars travels around the sun in an elliptical orbit, with the sun at one focus, whose equation is approximately

$$\frac{x^2}{(142)^2} + \frac{y^2}{(141)^2} = 1$$

where x and y are measured in million mile intervals. How close to the sun does Mars come? How far away from the sun can Mars be? Find the average of these two distances.

9

Systems of Equations

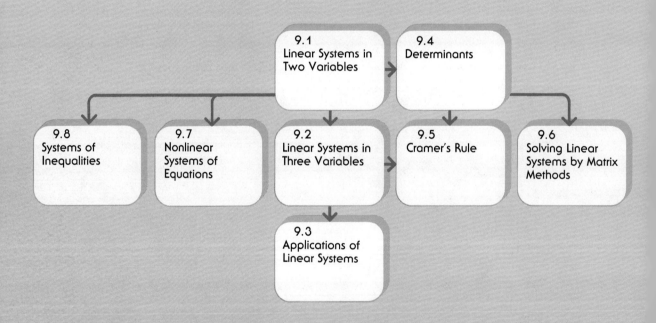

9.1 Linear Systems in Two Variables

In Section 7.1 we saw that a linear equation in two variables has an infinite number of solutions. For example, solutions for $2x + y = 7$ can be found by substituting values for x into the equation and solving for y. If $x = 0$, then $y = 7$ and $(0, 7)$ is one solution; if $x = 1$, then $y = 5$ and $(1, 5)$ is a second solution, and so on.

Problems in business and engineering frequently require the solution of a system of linear equations. The following two equations constitute a system of two linear equations in two variables.

$$2x - y = 5$$
$$x + 2y = 5$$

The ordered pair $(3, 1)$ is a solution of both of these equations as can be verified by replacing x by 3 and y by 1. Thus,

$$2(3) - 1 = 5$$
$$3 + 2(1) = 5$$

are both true. Any ordered pair that is a solution of all the equations in a system of equations is called a **solution of the system**.

The lines that are the graphs of $2x - y = 5$ and $x + 2y = 5$ are shown in Figure 9.1. Notice that $(3, 1)$ is the only point that is common to both lines, and thus it is the only solution of the given system.

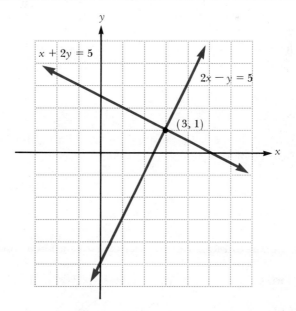

Figure 9.1

Since the graph of any linear equation is a line, solving a system of equations in the (standard) form

$$ax + by = m$$
$$cx + dy = n$$

means finding the common point or points of the lines that are the graphs of these equations. As is indicated in Figure 9.2, there are three possibilities.

1. The graphs intersect at exactly one point; this point is the only solution of the system.
2. The graphs do not intersect; the lines are parallel and there is no solution of the system. The equations are called **inconsistent equations**.
3. The graphs are the same line. Any solution of either equation is a solution of the system. Thus, there are infinitely many solutions of the system. The equations are called **dependent equations**.

Figure 9.2

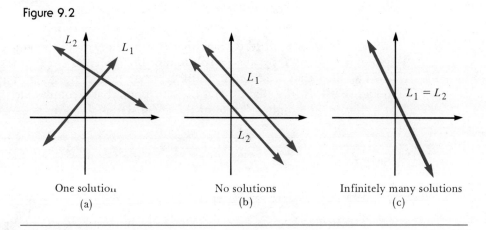

One solution
(a)

No solutions
(b)

Infinitely many solutions
(c)

Because graphing is sometimes inaccurate, particularly if the point of intersection is far away from the origin, mathematicians have developed algebraic methods of solving systems of equations.

Addition Method of Solution

The **addition method** replaces the original system of equations by a simpler, **equivalent system**, which has the same solution set as the original system. Two operations are used in obtaining an equivalent system.

1. Equations may be multiplied by a nonzero constant.
2. An equation may be replaced by the sum of itself and another equation.

Multiplication by a nonzero constant is used to obtain an equivalent system in which the coefficients of one of the variables differ only in sign.

EXAMPLE 1

Solve the system

$$x + y = 7$$
$$3x - 2y = 6$$

Solution:

First: Multiply both sides of $x + y = 7$ by 2 to obtain the equivalent system.

$$2x + 2y = 14$$
$$3x - 2y = 6$$

Second: Find the *sum* of these two equations.

$$\begin{array}{r} 2x + 2y = 14 \\ 3x - 2y = 6 \\ \hline 5x = 20 \end{array}$$

Third: Since any solution must satisfy $5x = 20$, we solve this equation for x, and obtain $x = 4$.

Fourth: Substitute $x = 4$ into either of the original equations and solve for y.

$$x + y = 7$$
$$4 + y = 7$$
$$y = 3$$

The ordered pair (4, 3) is the solution of the system. The solution set is {(4, 3)}.

Check: Substitute $x = 4$ and $y = 3$ into both original equations.

$$\begin{array}{ll} x + y = 7 & 3x - 2y = 6 \\ 4 + 3 = 7 \ \checkmark & 3(4) - 2(3) = 6 \ ? \\ & 12 - 6 = 6 \ \checkmark \end{array}$$

Figure 9.3 displays the geometric solution of the linear system.

Figure 9.3

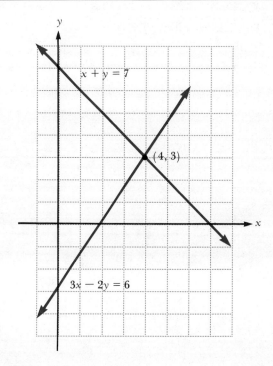

EXAMPLE 2   Solve the system $2x + 3y = 3$
$3x - 4y = 13$

Solution:
  First: Multiply $2x + 3y = 3$ by 3, and $3x - 4y = 13$ by -2 so that the resulting coefficients of x differ only in sign.

$$6x + 9y = 9$$
$$-6x + 8y = -26$$

  Second: Add these equations.

$$\begin{array}{r} 6x + 9y = 9 \\ -6x + 8y = -26 \\ \hline 17y = -17 \end{array}$$

  Third: Solve $17y = -17$ for y, obtaining $y = -1$.

  Fourth: Substitute $y = -1$ into either of the original equations and solve for x.

$$2x + 3(-1) = 3$$
$$2x = 6$$
$$x = 3$$

The ordered pair $(3, -1)$ is a solution of the system. The solution set is $\{(3, -1)\}$.

Check: Substitute $x = 3$ and $y = -1$ into both original equations.

$$2x + 3y = 3 \qquad\qquad 3x - 4y = 13$$
$$2(3) + 3(-1) = 3 \quad ? \qquad 3(3) - 4(-1) = 13 \quad ?$$
$$6 - 3 = 3 \quad \checkmark \qquad\qquad 9 + 4 = 13 \quad \checkmark$$

Each of the systems in Examples 1 and 2 had exactly one algebraic solution, and the geometric solution corresponding to (a) in Figure 9.2 shows intersecting lines. Let's see what happens if the equations of the system describe parallel lines or the same line.

EXAMPLE 3 Solve the system

$$2x + 4y = 12$$
$$-x - 2y = -6$$

Solution:

First: Multiply $-x - 2y = -6$ by 2 so that the coefficients of x differ only in sign, which produces the equivalent system.

$$2x + 4y = 12$$
$$-2x - 4y = -12$$

Second: Add these equations.

$$2x + 4y = 12$$
$$\underline{-2x - 4y = -12}$$
$$0 = 0$$

Third: This results in the equivalent system

$$2x + 4y = 12$$
$$0 = 0$$

Any solution of the equation $2x + 4y = 12$ is a solution of this system and consequently of the original system. *The original equations determine the same line.*

We can now describe the ordered pairs in the solution set in terms of the variable x alone by solving either original equation for y in terms of x.

$$2x + 4y = 12$$
$$4y = 12 - 2x$$
$$y = 3 - \frac{1}{2}x$$

The ordered pair $(x, 3 - \frac{1}{2}x)$ is a solution for any value of x. For example, if $x = 0$,

$$3 - \frac{1}{2}x = 3 - \frac{1}{2}(0) = 3$$

so the ordered pair $(0, 3)$ is a solution. If $x = 2$, $3 - \frac{1}{2}x = 3 - \frac{1}{2}(2) = 2$, and the ordered pair $(2, 2)$ is a solution. In this way we can produce as many solutions as desired. The solution set is $\{(x, 3 - \frac{1}{2}x) \mid x \text{ is a real number}\}$.

> In general, if the addition method results in the equation $0 = 0$, any solution of either original equation is a solution of the system and the geometric solution will show $L_1 = L_2$. See Figure 9.2(c).

EXAMPLE 4 Solve the system

$$5x - 2y = 7$$
$$-10x + 4y = 11$$

Solution:
  First: Multiply $5x - 2y = 7$ by 2 so that the coefficients of x differ only in sign and yield the equivalent system

$$10x - 4y = 14$$
$$-10x + 4y = 11$$

Second: Add these equations.

$$\begin{array}{r} 10x - 4y = 14 \\ -10x + 4y = 11 \\ \hline 0 = 25 \end{array}$$

Third: This results in the equivalent system

$$10x - 4y = 14$$
$$0 = 25$$

As no value of x and y can make both of these equations true, the system has no solution. The original system has no solution; the solution set is $\emptyset$. The lines determined by the original equations are parallel.

> In general, if the addition method results in an equation $0 = c$, where c is not zero, the system has no solution and the geometric solution will show parallel lines. See Figure 9.2(b).

Substitution Method of Solution

A linear system can be solved by the **substitution method** also. We solve either of the equations for one of the variables, and then substitute the solution into the other equation and solve. This method is demonstrated in Example 5.

EXAMPLE 5 Solve the system

$$x - y = -3$$
$$2x + y = 0$$

Solution: In order to use the substitution method, we solve either equation for one of the variables. We will solve $x - y = -3$ for x.

$$x - y = -3$$
$$x = y - 3$$

Now substitute $y - 3$ for x in the other equation, $2x + y = 0$, and solve for y.

$$2x + y = 0$$
$$2(y - 3) + y = 0$$
$$2y - 6 + y = 0$$
$$3y = 6$$
$$y = 2$$

When we solved the first equation for x we learned that $x = y - 3$, and consequently

$$x = y - 3$$
$$x = 2 - 3 \quad \text{Substitute } y = 2$$
$$x = -1$$

The solution of the system is the ordered pair $(-1, 2)$. The solution set is $\{(-1, 2)\}$.

Either the addition method or the substitution method may be used to solve a system of equations. The substitution method may be easier to use if the coefficient of one of the variables is 1.

Exercises 9.1

Standard Assignment: All odd-numbered exercises 1–29

A Solve the following systems by the addition method. See Examples 1–4.

1. $x + y = 0$
   $2x + 3y = 3$

2. $x + y = 3$
   $3x + y = 1$

3. $2x - y = 1$
   $3x - 2y = 0$

4. $x + 3y = 13$
   $x - 2y = 3$

5. $2x - 3y = 1$
   $3x - 4y = 3$

6. $6x + y = 12$
   $-7x + 2y = 5$

7. $5x - y = 5$
   $3x + 2y = -10$

8. $3x + 8y = 0$
   $5x + 2y = 0$

9. $3x - 2y = 1$
   $-7x + 3y = 1$

10. $3x + y = 3$
    $6x + 2y = 11$

11. $x - y = 2$
    $-2x + 2y = 5$

12. $4x + 7y = -3$
    $-8x - 14y = 6$

13. $4x + 6y = 12$
    $2x + 3y = 6$

14. $2x - 5y = 4$
    $5x - 10y = 15$

15. $3x + y = 12$
    $x + 3y = 12$

16. $9x - 3y = 3$
    $-2x + y = -1$

17. $2x - 18y = 6$
    $x - 9y = 3$

18. $6x + 5y = -2$
    $-12x - 10y = 4$

19. $x + y = 5$
    $2x + 2y = 10$

20. $2x + y = 3$
    $4x - 2y = -2$

21. $3x + y = 4$
    $6x - y = 2$

22. $\dfrac{x}{2} - y = 2$
    $2x + y = 3$

23. $\dfrac{x}{5} - \dfrac{y}{2} = 1$
    $x + \dfrac{y}{3} = 5$

24. $\dfrac{x}{3} + \dfrac{y}{2} = 1$
    $\dfrac{x}{5} - y = -2$

Solve the following systems by the substitution method. See Example 5.

25. $x + y = -2$
    $4x + 2y = 2$

26. $6x + y = 1$
    $7x + 2y = 7$

27. $5x - 2y = -3$
    $x + 8y = -9$

28. $x + 6y = 3$
    $-2x - 12y = -6$

29. $3x - y = 7$
    $-9x + 3y = -21$

30. $x + 2y = 5$
    $2x + 4y = 7$

B In Exercises 31–36, let $x = 1/u$ and $y = 1/v$, solve for x and y, and then solve for u and v.

EXAMPLE: $\dfrac{2}{u} + \dfrac{5}{v} = -5$

$\dfrac{3}{u} - \dfrac{2}{v} = -17$

Solution: Replacing $1/u$ by x and $1/v$ by y (reorganizing $2/u = 2(1/u)$, etc.) yields

$$2x + 5y = -5$$
$$3x - 2y = -17$$

Multiplying $2x + 5y = -5$ by 3, and $3x - 2y = -17$ by -2, we get

$$6x + 15y = -15$$
$$-6x + 4y = 34$$

Adding gives

$$6x + 15y = -15$$
$$\underline{-6x + 4y = 34}$$
$$19y = 19$$

Solving $19y = 19$ we have $y = 1$. From

$$2x + 5y = -5$$

we get
$$2x + 5(1) = -5$$
$$2x = -10$$
$$x = -5$$

Now
$$x = \frac{1}{u} \quad \text{and} \quad y = \frac{1}{v}$$

so
$$-5 = \frac{1}{u} \quad \text{and} \quad 1 = \frac{1}{v}$$

$$u = -\frac{1}{5} \quad \text{and} \quad v = 1$$

Thus $\left\{\left(-\frac{1}{5}, 1\right)\right\}$ is the solution set.

31. $\dfrac{2}{u} + \dfrac{1}{v} = 3$
$\dfrac{4}{u} - \dfrac{2}{v} = 0$

32. $\dfrac{3}{u} + \dfrac{1}{v} = 4$
$\dfrac{6}{u} - \dfrac{1}{v} = 2$

33. $\dfrac{6}{u} + \dfrac{3}{v} = 0$
$\dfrac{4}{u} + \dfrac{9}{v} = -1$

34. $\dfrac{5}{u} + \dfrac{10}{v} = 3$
$\dfrac{2}{u} - \dfrac{12}{v} = -2$

35. $\dfrac{3}{u} + \dfrac{4}{v} = 1$
$\dfrac{6}{u} + \dfrac{4}{v} = 3$

36. $\dfrac{7}{u} + \dfrac{2}{v} = -1$
$\dfrac{7}{u} + \dfrac{4}{v} = 0$

Solve the following systems by any method. Here a and b represent nonzero constants.

37. $x + y = a$
$2x + 3y = b$

38. $x + ay = 2b$
$x + y = b$

39. $ax + by = 1$
$ax - by = 1$

40. $ax + by = 1 \quad (a \neq b)$
$bx + ay = -1$

9.2 Linear Systems in Three Variables

A *solution of an equation* in three variables, such as $2x - y + 3z = 11$, is an ordered triple of numbers (a, b, c) such that when x, y, and z are replaced by a, b, and c respectively in the equation, the resulting statement is true. For example, $(3, 1, 2)$ is a solution of $2x - y + 3z = 11$ since $2(3) - (1) + 3(2) = 11$ is *true*. On the other hand, $(1, 2, 1)$ is not a solution of $2x - y + 3z = 11$ because $2(1) - (2) + 3(1) = 11$ is *false*.

A *solution of a system* of three linear equations in three variables, such as

$$x + y - 2z = -3$$
$$3x + 5y + z = 5$$
$$4x - 7y - 2z = 0$$

is an ordered triple that is a solution of all three equations of the system. The triple (1, 0, 2) is a solution of this system since

$$1 + 0 - 2(2) = -3$$
$$3(1) + 5(0) + (2) = 5$$
$$4(1) - 7(0) - 2(2) = 0$$

are all true statements.

The graph of a linear equation in three variables is a plane. The solution (1, 0, 2) of the previously discussed system is the intersection of the three planes determined by the three equations. However, a geometric treatment of a linear system of equations is too involved for our purposes. We will use an extension of the addition method that was discussed in the previous section to solve such systems.

EXAMPLE 1 Solve the system

$$x + 3y + z = 0 \quad \text{Equation (1)}$$
$$2x - y + z = 5 \quad \text{Equation (2)}$$
$$3x - 3y + 2z = 10 \quad \text{Equation (3)}$$

Solution:

First: Use the addition method to eliminate y from Equations (1) and (2). We multiply Equation (2) by 3, and then add

$$\begin{array}{ll} x + 3y + z = 0 & \text{Equation (1)} \\ 6x - 3y + 3z = 15 & 3 \cdot \text{Equation (2)} \\ \hline 7x \qquad + 4z = 15 & \text{Equation (4)} \end{array}$$

Second: Use the addition method to eliminate y from Equations (1) and (3). We need only add

$$\begin{array}{ll} x + 3y + z = 0 & \text{Equation (1)} \\ 3x - 3y + 2z = 10 & \text{Equation (3)} \\ \hline 4x \qquad + 3z = 10 & \text{Equation (5)} \end{array}$$

Third: Solve the new system containing Equations (4) and (5) by the addition method.

$$7x + 4z = 15 \quad \text{Equation (4)}$$
$$4x + 3z = 10 \quad \text{Equation (5)}$$

9.2 Linear Systems in Three Variables

Multiply Equation (4) by 3 and Equation (5) by -4, and then add

$$
\begin{array}{rl}
21x + 12z = & 45 \qquad 3 \cdot \text{Equation (4)} \\
-16x - 12z = & -40 \qquad (-4) \cdot \text{Equation (5)} \\
\hline
5x = & 5
\end{array}
$$

We solve $5x = 5$ to get $x = 1$. Substituting $x = 1$ into Equation (5) yields

$$
\begin{aligned}
4(1) + 3z &= 10 \\
3z &= 6 \\
z &= 2
\end{aligned}
$$

Fourth: Substitute $x = 1$ and $z = 2$ into any of the original three equations and solve for y.

$$
\begin{aligned}
(1) + 3y + 2 &= 0 \qquad \text{Equation (1)} \\
3y &= -3 \\
y &= -1
\end{aligned}
$$

A solution of the system requires that $x = 1$, $y = -1$, and $z = 2$. The solution set is $\{(1, -1, 2)\}$.

Check: We substitute $x = 1$, $y = -1$, and $z = 2$ into all three original equations.

$$
\begin{array}{ll}
x + 3y + z = 0 & 2x - y + z = 5 \\
1 + 3(-1) + 2 = 0 \ ? & 2(1) - (-1) + (2) = 5 \ ? \\
1 - 3 + 2 = 0 \ \checkmark & 2 + 1 + 2 = 5 \ \checkmark
\end{array}
$$

$$
\begin{aligned}
3x - 3y + 2z &= 10 \\
3(1) - 3(-1) + 2(2) &= 10 \ ? \\
3 + 3 + 4 &= 10 \ \checkmark
\end{aligned}
$$

EXAMPLE 2 Solve the system

$$
\begin{array}{rl}
2x + y = 11 & \text{Equation (1)} \\
3y - z = 5 & \text{Equation (2)} \\
x + 2z = 1 & \text{Equation (3)}
\end{array}
$$

Solution:

First: We multiply Equation (1) by -3 so that the coefficients of y in Equations (1) and (2) differ only sign, and then add

$$
\begin{array}{rl}
-6x - 3y = -33 & (-3) \cdot \text{Equation (1)} \\
3y - z = 5 & \text{Equation (2)} \\
\hline
-6x - z = -28 & \text{Equation (4)}
\end{array}
$$

Second: Multiply Equation (3) by 6 and add the resulting equation to Equation (4).

$$-6x - z = -28 \quad \text{Equation (4)}$$
$$\underline{6x + 12z = 6} \quad 6 \cdot \text{Equation (3)}$$
$$11z = -22$$

Third: Solve $11z = -22$ to get $z = -2$.

Fourth: Substitute $z = -2$ into Equation (4) and solve for x.

$$-6x - (-2) = -28$$
$$-6x = -30$$
$$x = 5$$

Fifth: Substitute $x = 5$ into Equation (1).

$$2(5) + y = 11$$
$$y = 1$$

Thus, $x = 5$, $y = 1$, and $z = -2$. The solution set is $\{(5, 1, -2)\}$.

Check: We substitute $x = 5$, $y = 1$, and $z = -2$ into the original equations.

$$2(5) + 1 = 11 \ ? \qquad 3(1) - (-2) = 5 \ ? \qquad 5 + 2(-2) = 1 \ ?$$
$$10 + 1 = 11 \ \checkmark \qquad 3 + 2 = 5 \ \checkmark \qquad 5 + (-4) = 1 \ \checkmark$$

The method for solving a system of three equations in three variables is summarized next.

To Solve Three Equations in Three Variables

1. Use the addition method to eliminate one variable from two of the three given equations.
2. Eliminate the *same* variable from a different pair of equations by the addition method.
3. Solve the system of two equations in two variables that is produced in Steps 1 and 2.
4. Substitute the values obtained in Step 3 into any of the original three equations to obtain the value for the third variable.

If a system of three equations in three variables has no solution (is inconsistent), an equation such as $0 = 5$, which is obviously false, will occur. However, if the equation $0 = 0$ appears, the system may either have infinitely many solutions or no solutions. More investigation is required to determine which possibility is correct. In this book we will only consider systems that have a single solution.

Exercises 9.2

Standard Assignment: Exercises 1, 3, 5, 7, 9, 11, 13, 15, 17, 19, 21

A Solve the following systems of equations. See Examples 1 and 2.

1. $3x + 2y - z = 5$
   $-x + 5y - 2z = 5$
   $2x - y + 3z = 6$

2. $3x - y + z = 1$
   $2x + y + z = 5$
   $4x - y + 2z = 4$

3. $4x + 2y + 3z = 6$
   $x + 2y + 2z = 1$
   $2x - y + z = -1$

4. $2x + y - z = -1$
   $2x - y + 3z = 1$
   $3x + 2y + z = 5$

5. $3x + y + z = 6$
   $x - y + 4z = -3$
   $2x + y + 2z = 3$

6. $4x - 8y - 2z = 12$
   $2x + 5y + 3z = 13$
   $4x - 6y - 3z = 6$

7. $2x + 3y + 2z = 7$
   $x + 3y - z = -2$
   $x - y + 2z = 8$

8. $3x + 3y + 2z = 9$
   $2x - y + 2z = 2$
   $x + 5y - 6z = -9$

9. $x - y + 2z = 3$
   $2x + 2y + z = 3$
   $x + y + 3z = 4$

10. $2x + 3y + 3z = 3$
    $2x - y + z = 5$
    $3x + 2y + 3z = 3$

11. $2x - 3y + z = 2$
    $3x - 3y + 2z = 10$
    $2x + 3y + 2z = 3$

12. $4x - 2y + z = 5$
    $2x + y - 2z = 4$
    $x + 3y - 2z = 6$

13. $2x + y + z = 6$
    $x + y - z = 1$
    $x + y + 2z = 4$

14. $2x + y - 3z = 7$
    $x - y - 2z = 4$
    $3x + 3y + 2z = 4$

15. $x - 3y + 2z = 9$
    $2x + 4y - 3z = -9$
    $3x - 2y + 5z = 12$

16. $x + y = 0$
    $y + 2z = -4$
    $y + z = -3$

17. $2x + 4y + 3z = 6$
    $x + 2z = -1$
    $x - 2y + z = -5$

18. $2y - z = -4$
    $x + z = 3$
    $2x + 3y = -1$

19. $x + y = 9$
    $2y + 3z = 7$
    $x - 2z = 4$

20. $3x - 2z = 11$
    $2x + y = 8$
    $2y + 3z = 1$

21. $2x + y = 4$
    $x + 2z = 3$
    $3y - z = 5$

22. $2.714x - 3.62y + 4.777z = 10.733$
    $4.103x + 4.005y - 6.214z = -4.62$
    $-3.411x + 7.24y - 8.063z = -13.795$

23. $3.412x + 4.715y - 0.924z = -2.865$
    $4.11x + 5.625y + 6.859z = 13.72$
    $1.427x - 1.943y - 2.019z = 8.774$

24. $5.743x + 9.021y + 1.407z = 20.447$
    $3.721x + 0.874y + 2.715z = 8.478$
    $1.936x + 1.859y + 4.555z = 4.023$

25. $3.157x + 4.911y + 1.074z = 3.777$
    $1.433x + 1.433y - 2.743z = -1.87$
    $19.883y + 39.274z = 80.75$

9.3 Applications of Linear Systems

Practical problems often require that more than one unknown be determined. Until now, our only option has been to represent all unknowns in terms of one variable. In order to solve a problem involving two unknowns, it is frequently more convenient to use two variables to represent the unknowns. Then two equations can be written that relate the unknown quantities, and this system of equations can be solved by the methods described in Section 9.1. A similar approach works for problems involving three unknowns.

EXAMPLE 1 Last year, Mrs. Rogers invested $5000. Part was invested in a real estate venture that paid $7\frac{1}{2}\%$ interest per year, and the rest was invested in a small business venture that returned 12% per year. The combined interest for the year totaled $519; how much did she invest at each rate?

Solution: Two amounts are to be determined, so we will represent them by different variables. Let

$$x = \text{amount invested at } 7\frac{1}{2}\%$$
$$y = \text{amount invested at } 12\%$$

Since $7\frac{1}{2}\%$ of x is $0.075x$, and 12% of y is $0.12y$ we have

$$0.075x = \text{interest earned from the } 7\frac{1}{2}\% \text{ investment}$$
$$0.12y = \text{interest earned from the } 12\% \text{ investment}$$

We now write two equations—one relating the two amounts and the other relating the interest earned by these amounts.

$$x + y = 5000 \quad (1)$$
$$0.075x + 0.12y = 519 \quad (2)$$

Multiplying Equation (1) by -75 and Equation (2) by 1000 yields

$$\begin{aligned} -75x - 75y &= -375{,}000 \\ 75x + 120y &= 519{,}000 \\ \hline 45y &= 144{,}000 \quad \text{Add} \end{aligned}$$

$$y = \frac{144{,}000}{45}$$

$$y = \$3200$$

Substituting $y = 3200$ into Equation (1),

$$x + 3200 = 5000$$
$$x = \$1800$$

$3200 is invested at 12% and $1800 is invested at $7\frac{1}{2}\%$.

9.3 Applications of Linear Systems

Check: We check the solution against the original statement of the problem.

$7\frac{1}{2}\%$ of $1800 is $0.075 \times 1800 = 135$

12% of $3200 is $0.12 \times 3200 = 384$

Finally, $135 + $384 = $519, as given.

EXAMPLE 2 At one of the Ivy League colleges, the tuition is $75 more than twice the cost of room and board. The combined cost is $12,225. How much is room and board and what is the tuition?

Solution: We represent the two amounts to be determined by different variables. Let

$$x = \text{cost of tuition}$$
$$y = \text{cost of room and board}$$

Now

$$x + y = 12{,}225 \quad (1)$$

and

$$x = 2y + 75 \quad (2)$$

Rewrite as

$$x + y = 12{,}225 \quad (1)$$
$$x - 2y = 75 \quad (2)$$

Multiply Equation (2) by -1 and add

$$\begin{aligned} x + y &= 12{,}225 \\ -x + 2y &= -75 \\ \hline 3y &= 12{,}150 \quad \text{Add} \\ y &= \$4050 \end{aligned}$$

Substituting $y = 4050$ into our original Equation (2),

$$x = 2(4050) + 75$$
$$x = \$8175$$

Tuition is $8175 and room and board is $4050.

Check: We now check these figures against the original statement of the problem.

Does $8175 + $4050 = $12,225? ✓

Does the tuition exceed twice the room and board, 2×4050, by $75?

$2 \times 4050 = \$8100$ and $8100 + 75 = \$8175$ ✓

EXAMPLE 3 John rowed a distance of 24 miles downstream in 3 hours. The return trip took 6 hours. What is the rate of the boat in still water and what is the rate of the current?

Solution: We represent the two rates to be determined by different variables. Let

$$x = \text{rate of the boat in still water}$$
$$y = \text{rate of the current}$$

When rowing downstream, the rate of the current is added to the rate of the boat in still water so that the rate that the boat is traveling downstream is $x + y$. When rowing upstream, the rate of the current is subtracted from the rate of the boat in still water so that the rate that the boat travels upstream is $x - y$.

Recalling the relation, Distance = Rate × Time for objects traveling at a constant rate, we have

	Rate (mph)	×	Time (hrs)	=	Distance (mi)
Downstream:	$x + y$		3		$3(x + y)$
Upstream:	$x - y$		6		$6(x - y)$

The distance traveled upstream = distance downstream = 24 miles. Thus

$$3(x + y) = 24$$
$$6(x - y) = 24$$

or equivalently

$$\begin{aligned} x + y &= 8 \\ x - y &= 4 \\ \hline 2x &= 12 \quad \text{Add} \\ x &= 6 \end{aligned}$$

Substituting $x = 6$ into $x + y = 8$ gives

$$6 + y = 8$$
$$y = 2$$

$x = 6$ mph is the rate of the boat in still water.

$y = 2$ mph is the rate of the current.

Check: We now check these results against the original statement of the problem.

Downstream rate is $6 + 2 = 8$ mph. Does (8 mph)(3 hrs) = 24 miles? ✔

Upstream rate is $6 - 2 = 4$ mph. Does (4 mph)(6 hrs) = 24 miles? ✔

EXAMPLE 4 A two-digit number equals eight times the sum of its digits. When three times the tens' digit is added to twice the units' digit, the result is 25. Find the number.

Solution: The essential idea in solving digit problems is that the value of a two-digit number is found by multiplying its tens' digit by ten and then adding its units' digit. For example, $63 = 6(10) + 3$. Let

$$t = \text{tens' digit}$$
$$u = \text{units' digit}$$

The two-digit number we seek is $10t + u$. Thus,

$$10t + u = 8(t + u) \quad (1)$$
$$3t + 2u = 25 \quad (2)$$

Equivalently

$$2t - 7u = 0 \quad (1)$$
$$3t + 2u = 25 \quad (2)$$

Multiplying Equation (1) by -3 and Equation (2) by 2 yields

$$-6t + 21u = 0$$
$$\underline{6t + 4u = 50}$$
$$25u = 50 \quad \text{Add}$$
$$u = 2$$

Substituting $u = 2$ into $3t + 2u = 25$ gives

$$3t + 2(2) = 25$$
$$3t = 21$$
$$t = 7$$

The two-digit number is $7(10) + 2 = 72$.

Check: Test 72 in the original statement of the problem.

Does $72 = 8(7 + 2)$? ✔

Does $3(7) + 2(2) = 25$? ✔

If three quantities are to be determined, then we assign three variables and set up three equations.

EXAMPLE 5

A vending machine coin box contains $21.90 in nickels, dimes, and quarters. There are 149 coins in all and there are twice as many quarters as there are nickels. How many of each type of coin does the box contain?

Solution: We use three variables to represent the three quantities to be determined. Let

$$n = \text{the number of nickels in the box}$$
$$d = \text{the number of dimes in the box}$$
$$q = \text{the number of quarters in the box}$$

Our first equation converts the coin values to cents.
$$5n + 10d + 25q = 2190 \quad (1)$$
The next equation totals the number of coins in the box.
$$n + d + q = 149 \quad (2)$$
The final equation represents the relation between the quarters and the nickels.
$$q = 2n \quad (3)$$

We write these equations as a system, and rewrite Equation (3).
$$5n + 10d + 25q = 2190 \quad (1)$$
$$n + d + q = 149 \quad (2)$$
$$-2n + q = 0 \quad (3)$$

Multiply Equation (2) by -10 and add the result to Equation (1).
$$5n + 10d + 25q = 2190$$
$$-10n - 10d - 10q = -1490$$
$$\overline{-5n + 15q = 700} \quad (4)$$

Now form a system using Equations (3) and (4) and solve.
$$-2n + q = 0 \quad (3)$$
$$-5n + 15q = 700 \quad (4)$$

Multiply Equation (3) by -15 and add the result to Equation (4).
$$30n - 15q = 0$$
$$-5n + 15q = 700$$
$$\overline{25n = 700}$$
$$n = \frac{700}{25}$$
$$n = 28$$

Substituting $n = 28$ into Equation (3), we have
$$-2(28) + q = 0$$
$$q = 56$$

Substituting $n = 28$ and $q = 56$ into Equation (2) yields
$$28 + d + 56 = 149$$
$$d = 65$$

There are 28 nickels, 65 dimes, and 56 quarters in the box.

Check: Does $28 + 65 + 56 = 149$? ✓

$$\begin{aligned}
&28 \text{ nickels is } 5 \times 28 = 140 \text{ cents} &&= \$\ 1.40 \\
&65 \text{ dimes is } 10 \times 65 = 650 \text{ cents} &&= 6.50 \\
&56 \text{ quarters is } 25 \times 56 = 1400 \text{ cents} &&= \underline{14.00} \\
&&&\$21.90 \quad ✓
\end{aligned}$$

Finally, the number of quarters is $56 = 2(28) =$ twice the number of nickels.

Exercises 9.3

Standard Assignment: All odd-numbered A exercises

A In each of the following exercises, use two variables to represent the two quantities to be determined. Write two equations in the two variables and solve the resulting system. Check your solution against the stated problem. See Examples 1–4.

1. Mr. Hawkins invested a total of $30,000 in two ventures. The annual interest from one was 8% and the annual interest from the other was $10\frac{1}{2}\%$. He received a total of $2550 in all. How much was invested at 8% and how much was invested at $10\frac{1}{2}\%$?

2. Bill has $73.60 in a collection of silver dollars and silver dimes. Altogether, there are 358 coins. How many of each type does he have?

3. While gambling in Las Vegas, Sally lost twice as much money at the dice table as she did at blackjack. Altogether, she lost $360. How much did she lose at each game?

4. A speculator invested part of $15,000 in a high-risk venture and received a return of 12% at the end of the year. The rest of the $15,000 was invested at 8% annual interest. The combined annual interest from the two ventures was 10% of the total investment. How much was invested at each rate?

5. Admission to the state fair is $2.00 for adults and $1.50 for children. A church group of 57 members paid a total of $98 in admission fees. How many adults and how many children were in the group?

6. A farmer earns a profit of $525 per acre of tomatoes and $475 per acre of soybeans. His soybean acreage is 5 acres more than twice his tomato acreage. If his total profit is $24,500, how many acres of tomatoes and how many acres of soybeans does he have?

7. An herb that sells for 95¢ per pound is mixed with tea that sells for $1.20 per pound to produce an 8-pound mix that is worth $1.15 per pound. How many pounds of herb and how many pounds of tea does the mix contain?

8. A student earns twice as much per hour for tutoring as he does for working at the library. If his average wage is $5.25 per hour, how much does he earn per hour at each job?

9. Two pounds and seven francs are worth $6.20, and three pounds and ten francs are worth $9.20. Find the value of a pound and the value of a franc.

10. Five oranges and 10 grapefruit cost $2.40, and 20 oranges and 22 grapefruit cost $6.00. What is the price of each fruit?

11. With the help of a tail wind, a plane travels 3000 kilometers in 5 hours. The return trip against the wind requires 6 hours. Find both the speed of the plane in still air and the wind speed.

12. A cruiser completes a round trip from its home port to an island and back in 5 hours. Its average rate going was 20 kph and its average rate returning was 30 kph. How long did the trip take in each direction and what was the distance from the port to the island?

13. Two trains leave a terminal at the same time and travel in opposite directions. After 6 hours they are 270 miles apart. One train is traveling 3 mph faster than the other train. Find the speed of each train.

14. Two people who are 12 miles apart will meet in 2 hours if they walk toward each other, and will meet in 4 hours if they walk in the same direction (if the faster person walks toward the slower person). Find the rate of each.

15. A two-digit number is 3 times the sum of its digits. If the units' digit is 3 more than twice the tens' digit, find the number.

16. The tens' digit of a two-digit number is one more than twice its units' digit. Twice the number obtained by reversing the digits of the given number is one more than the given number. What is the number?

17. The tens' digit of a two-digit number is 2 less than the units' digit. If one is subtracted from the number, the result is 8 times the units' digit. Find the number.

18. If 9 is added to a certain two-digit number, the digits are reversed. The number is 4 times the sum of its digits. Find the number.

19. A plumber earns $10 per hour more than his apprentice. For a 40-hour week, their combined earnings were $540. Find the hourly wage of each.

20. A business takes in $26,500 in sales, including tax. If the tax is 6%, how much of the $26,500 is tax?

21. The length of a rectangular field is 30 m longer than its width. The length of the fence around the field is 260 m. Find the length and width of the field.

22. A realtor invested a total of $50,000 in two types of acreage: commercial acreage and residential acreage. The present value of the combined acreage is $50,600, due to a 10% increase in the value of the commercial acreage and a 10% decrease in the value of the residential acreage. What is the *present* value of each type of acreage?

23. The difference of two numbers equals 7 less than the smaller number. Three times the smaller number equals twice the larger number. Find the numbers.

24. The sum of two numbers is 13. The sum of 5 times one of the numbers and 8 times the other number is 50. Find the numbers.

25. The perimeter of an isosceles triangle is 40 cm. The base is twice as long as either equal side. Find the base and side lengths.

26. John's age is one year less than three times his sister's age. Six years ago, John was exactly twice as old as his sister is now. Find John's age and his sister's age.

27. The sum of two numbers equals twice the difference of the numbers. Five times the smaller number equals 10 more than the larger number. Find the numbers.

28. Jack and his father can paint a barn together in 5 hours. Jack leaves after 4 hours and his father requires 3 more hours to finish the job. Find the time required for each of them to paint the barn alone.

29. If $(1, 2)$ and $(3, -4)$ are on the line $y = mx + b$, find m and b by substituting the coordinates into the equation and solving the resulting system.

30. If $(2, 2)$ and $(-\frac{1}{2}, 3)$ are on the graph of $ax + by^2 = 8$, find a and b by substituting the coordinates into the equation and solving the resulting system.

B In each of the following exercises use three variables to represent the three quantities to be determined. Write three equations in the three variables and solve the resulting system. Check your solution against the stated problem. See Example 5.

31. A pay phone's coin box contains nickels, dimes, and quarters. The total number of coins in the box is 300. The number of dimes is three times the number of nickels and quarters together. If there is $30.65 in the box, find the number of nickels, dimes, and quarters that it contains.

32. The Chargers scored a total of 46 points in a football game. Twice the number of points resulting from the sum of field goals and extra points equals two more than the number of points from touchdowns. Five times the number of points scored by field goals equals twice the number of points from touchdowns. Find the number of points resulting from touchdowns, field goals, and extra points. (Note that a touchdown = 6 points; a field goal = 3 points; and an extra point = 1 point.)

33. A man worked 47 hours one week and was paid at three different rates. He earned $4.20 per hour for normal daytime work, $5.00 per hour for night work, and $5.75 per hour if he worked on a holiday. If his total pay for the week was $210, and the number of regular daytime hours he worked exceeded twice the combined hours of night and holiday work by 11 hours, how many hours of each category of work did he perform?

34. A triangle has a perimeter of 25 m. Four times the length of one of the legs equals the sum of the other two legs. Twice the length of this same leg is 2 m more than the length of one of the other legs. Find the length of each leg.

35. A manufacturer buys three components, A, B, and C, for use in making a toaster. He used as many units of A as he did of B and C combined. The cost of A, B, and C is \$4, \$5, and \$6 per unit, respectively. If the manufacturer purchased 100 units of these components for a total cost of \$480, how many units of each was purchased?

36. The equation $x^2 + y^2 + Cx + Dy + E = 0$ describes a circle in a rectangular coordinate system containing the points $(1, -1)$, $(-1, 1)$, and $(0, 2)$. Find C, D, and E.

9.4 Determinants

A **matrix** is a rectangular array of numbers. The numbers in the array are called the **elements** or **entries** of the matrix. In this book, a matrix is displayed by enclosing its entries in brackets. Thus

$$\begin{bmatrix} 3 & 7 \\ 2 & -1 \end{bmatrix}, \quad \begin{bmatrix} 5 & -3 & 2 \\ 1 & 6 & 5 \end{bmatrix}, \quad \text{and} \quad \begin{bmatrix} 3 & 7 \\ 0 & 2 \\ -1 & 1 \end{bmatrix}$$

are matrices. Matrices are described according to the number of rows and columns they contain; rows are horizontal and columns are vertical. The matrices shown are 2×2 (read "two by two"), 2×3, and 3×2 matrices, respectively. The first number gives the number of rows in the matrix and the second number gives the number of columns in the matrix. A matrix that has the same number of rows and columns is called a **square matrix**.

A computation that is associated with a square matrix results in a number called the **determinant**. The determinant has a variety of uses, one of which will be explained in Section 9.5.

The determinant for a 2×2 matrix is called a **second-order determinant** and is written by replacing the matrix brackets by vertical bars:

$$\begin{vmatrix} a & b \\ c & d \end{vmatrix}$$

The determinant is defined by

$$\begin{vmatrix} a & b \\ c & d \end{vmatrix} = ad - bc$$

EXAMPLE 1 Find the following determinants.

(a) $\begin{vmatrix} 2 & 3 \\ 1 & 4 \end{vmatrix} = (2)(4) - (3)(1) = 8 - 3 = 5$

(b) $\begin{vmatrix} -4 & 6 \\ -3 & 9 \end{vmatrix} = (-4)(9) - (6)(-3) = -36 + 18 = -18$

The determinant for a 3×3 matrix is called a **third-order determinant** and is given by the formula

$$\begin{vmatrix} a_1 & b_1 & c_1 \\ a_2 & b_2 & c_2 \\ a_3 & b_3 & c_3 \end{vmatrix} = a_1 b_2 c_3 - a_1 b_3 c_2 + a_2 b_3 c_1 - a_2 b_1 c_3 + a_3 b_1 c_2 - a_3 b_2 c_1$$

The expression on the right-hand side of this equation can be rearranged by grouping and factoring as follows

$$\begin{vmatrix} a_1 & b_1 & c_1 \\ a_2 & b_2 & c_2 \\ a_3 & b_3 & c_3 \end{vmatrix} = a_1(b_2 c_3 - b_3 c_2) - a_2(b_1 c_3 - b_3 c_1) + a_3(b_1 c_2 - b_2 c_1)$$

Notice that the numbers from the first column, a_1, a_2, and a_3, are multiplied by the second-order determinant that is obtained by eliminating the first *column* and the *row* that contains the multiplier.

$$a_1(b_2 c_3 - b_3 c_2) = \begin{vmatrix} \boxed{a_1} & b_1 & c_1 \\ a_2 & b_2 & c_2 \\ a_3 & b_3 & c_3 \end{vmatrix}$$

$$a_2(b_1 c_3 - b_3 c_1) = \begin{vmatrix} a_1 & b_1 & c_1 \\ \boxed{a_2} & b_2 & c_2 \\ a_3 & b_3 & c_3 \end{vmatrix}$$

$$a_3(b_1 c_2 - b_2 c_1) = \begin{vmatrix} a_1 & b_1 & c_1 \\ a_2 & b_2 & c_2 \\ \boxed{a_3} & b_3 & c_3 \end{vmatrix}$$

The second-order determinant that results from deleting the row and column containing any given number in the original 3×3 matrix is called the **minor**, of the number. The minors of a_1, a_2, and a_3 in the previously shown array are, respectively

$$\begin{vmatrix} b_2 & c_2 \\ b_3 & c_3 \end{vmatrix}, \quad \begin{vmatrix} b_1 & c_1 \\ b_3 & c_3 \end{vmatrix}, \quad \text{and} \quad \begin{vmatrix} b_1 & c_1 \\ b_2 & c_2 \end{vmatrix}$$

Finding the determinant by using

$$\begin{vmatrix} a_1 & b_1 & c_1 \\ a_2 & b_2 & c_2 \\ a_3 & b_3 & c_3 \end{vmatrix} = a_1 \begin{vmatrix} b_2 & c_2 \\ b_3 & c_3 \end{vmatrix} - a_2 \begin{vmatrix} b_1 & c_1 \\ b_3 & c_3 \end{vmatrix} + a_3 \begin{vmatrix} b_1 & c_1 \\ b_2 & c_2 \end{vmatrix}$$

is called **expanding the determinant by minors** of the elements in the first column.

EXAMPLE 2   Find the given determinants by expanding by minors of the elements in the first column.

(a) $\begin{vmatrix} 4 & 3 & 1 \\ 4 & 1 & 3 \\ 3 & 4 & 1 \end{vmatrix} = 4 \begin{vmatrix} 1 & 3 \\ 4 & 1 \end{vmatrix} - 4 \begin{vmatrix} 3 & 1 \\ 4 & 1 \end{vmatrix} + 3 \begin{vmatrix} 3 & 1 \\ 1 & 3 \end{vmatrix}$

$= 4[(1)(1) - (3)(4)] - 4[(3)(1) - (1)(4)] + 3[(3)(3) - (1)(1)]$
$= 4(-11) - 4(-1) + 3(8)$
$= -44 + 4 + 24$
$= -16$

(b) $\begin{vmatrix} 4 & 2 & 3 \\ 0 & 3 & 1 \\ 5 & 2 & 3 \end{vmatrix} = 4 \begin{vmatrix} 3 & 1 \\ 2 & 3 \end{vmatrix} - 0 \begin{vmatrix} 2 & 3 \\ 2 & 3 \end{vmatrix} + 5 \begin{vmatrix} 2 & 3 \\ 3 & 1 \end{vmatrix}$

$= 4[(3)(3) - (1)(2)] - 0[(2)(3) - (3)(2)] + 5[(2)(1) - (3)(3)]$
$= 4(7) - 0(0) + 5(-7)$
$= 28 - 0 - 35$
$= -7$

The expression on the right-hand side of the equation in the definition of a third-order determinant on page 407 has the property that each element of the 3×3 matrix appears in exactly two of the terms. We could have factored out the three elements of the second or third columns (or rows) and expressed the determinants in terms of their minors. Therefore, the determinant can be found by expanding by minors of the elements of any column or row. The signs for the three terms in the expansion can be found in the corresponding row or column in the following sign chart.

$$\begin{vmatrix} + & - & + \\ - & + & - \\ + & - & + \end{vmatrix}$$

The signs $+ - +$ from the first column give the sign pattern we used when expanding by minors of the elements in the first column.

EXAMPLE 3   Find the determinant by expanding by minors of the elements of the second column.

Solution: First notice that the sign pattern for the second column is

$$\begin{vmatrix} 1 & 1 & 0 \\ 0 & 3 & -1 \\ 1 & 2 & 1 \end{vmatrix} = -1\begin{vmatrix} 0 & -1 \\ 1 & 1 \end{vmatrix} + 3\begin{vmatrix} 1 & 0 \\ 1 & 1 \end{vmatrix} - 2\begin{vmatrix} 1 & 0 \\ 0 & -1 \end{vmatrix} \quad \text{From } \begin{vmatrix} + & - & + \\ - & + & - \\ + & - & + \end{vmatrix}$$

$$= -(1)[(0)(1) - (-1)(1)] + (3)[(1)(1) - (0)(1)] - 2[(1)(-1) - (0)(0)]$$
$$= -(1)(1) + (3)(1) - (2)(-1)$$
$$= -1 + 3 + 2$$
$$= 4$$

Exercises 9.4

Standard Assignment: All odd-numbered exercises

A Find the following determinants. See Example 1.

1. $\begin{vmatrix} 1 & 2 \\ 2 & 1 \end{vmatrix}$
2. $\begin{vmatrix} 3 & 0 \\ 1 & 2 \end{vmatrix}$
3. $\begin{vmatrix} 7 & -3 \\ 4 & -1 \end{vmatrix}$

4. $\begin{vmatrix} -2 & 1 \\ 0 & 3 \end{vmatrix}$
5. $\begin{vmatrix} 1 & 0 \\ 0 & 1 \end{vmatrix}$
6. $\begin{vmatrix} 0 & 1 \\ 1 & 0 \end{vmatrix}$

7. $\begin{vmatrix} -5 & 0 \\ -4 & 0 \end{vmatrix}$
8. $\begin{vmatrix} -2 & 3 \\ 4 & -1 \end{vmatrix}$
9. $\begin{vmatrix} 3 & -6 \\ 4 & -5 \end{vmatrix}$

Find the following determinants by expanding the determinants by minors of the elements in the first column. See Example 2.

10. $\begin{vmatrix} 2 & 0 & -1 \\ 1 & -6 & 7 \\ -3 & 2 & 5 \end{vmatrix}$
11. $\begin{vmatrix} 3 & 4 & 0 \\ 2 & 1 & 5 \\ 6 & 8 & 0 \end{vmatrix}$
12. $\begin{vmatrix} 1 & 3 & 2 \\ 4 & -1 & 3 \\ -2 & 2 & 1 \end{vmatrix}$

13. $\begin{vmatrix} 1 & 2 & -1 \\ 0 & -3 & 4 \\ 0 & -5 & 3 \end{vmatrix}$
14. $\begin{vmatrix} 0 & 7 & -3 \\ 0 & -2 & 4 \\ 0 & -1 & -8 \end{vmatrix}$
15. $\begin{vmatrix} -3 & 2 & 3 \\ 0 & 4 & -6 \\ 2 & -2 & 1 \end{vmatrix}$

Find the following determinants by expanding using minors of the elements in any column or row. See Examples 2 and 3.

16. $\begin{vmatrix} 1 & 0 & 0 \\ -2 & 3 & 2 \\ 5 & 1 & -1 \end{vmatrix}$
17. $\begin{vmatrix} 1 & 0 & -4 \\ -2 & 0 & -3 \\ 3 & -1 & -8 \end{vmatrix}$
18. $\begin{vmatrix} 3 & -3 & 4 \\ 1 & 2 & -1 \\ 3 & -3 & 5 \end{vmatrix}$

19. $\begin{vmatrix} 3 & 1 & 4 \\ 1 & -2 & -1 \\ 2 & 1 & 3 \end{vmatrix}$
20. $\begin{vmatrix} 1 & 6 & 5 \\ 2 & 0 & 2 \\ -4 & 3 & 1 \end{vmatrix}$
21. $\begin{vmatrix} 1 & -1 & 2 \\ -1 & 2 & -1 \\ 2 & -1 & -1 \end{vmatrix}$

22. $\begin{vmatrix} 5 & 0 & 5 \\ -2 & 5 & -2 \\ 0 & -2 & 5 \end{vmatrix}$
23. $\begin{vmatrix} 7 & -1 & 0 \\ 3 & 8 & -2 \\ 0 & -3 & 0 \end{vmatrix}$
24. $\begin{vmatrix} 2 & 9 & 5 \\ 0 & 3 & 8 \\ 4 & 0 & -3 \end{vmatrix}$

25. $\begin{vmatrix} 0 & 2 & -3 \\ 2 & -3 & 1 \\ 8 & -7 & -3 \end{vmatrix}$
26. $\begin{vmatrix} 3 & 0 & 0 \\ 0 & 2 & 0 \\ 0 & 0 & 1 \end{vmatrix}$
27. $\begin{vmatrix} -3 & -6 & 1 \\ 2 & 7 & 1 \\ 1 & 4 & 2 \end{vmatrix}$

28. $\begin{vmatrix} a & 0 & 0 \\ 2 & b & 0 \\ 3 & -5 & c \end{vmatrix}$
29. $\begin{vmatrix} a & b & c \\ a & b & c \\ 1 & 1 & 1 \end{vmatrix}$
30. $\begin{vmatrix} a & a & 1 \\ b & b & 1 \\ c & c & 1 \end{vmatrix}$

31. Find all real numbers t for which

$$\begin{vmatrix} t-1 & 1 \\ 5 & t-5 \end{vmatrix} = 0$$

32. Find all real numbers t for which

$$\begin{vmatrix} t-2 & 0 \\ 3 & t-7 \end{vmatrix} = 0$$

33. Show that

$$\begin{vmatrix} 1 & a & a^2 \\ 1 & b & b^2 \\ 1 & c & c^2 \end{vmatrix} = (b-a)(c-a)(c-b)$$

34. Verify that

$$\begin{vmatrix} x & y & 1 \\ 2 & 3 & 1 \\ -4 & 7 & 1 \end{vmatrix} = 0$$

is the equation of a line through $(2, 3)$ and $(-4, 7)$.

9.5 Cramer's Rule

A remarkable procedure, which is known as **Cramer's Rule**, uses determinants to solve certain systems of linear equations. Before justifying the procedure, we will state it and use it for a system of two equations in two variables.

Cramer's Rule

For two equations in two variables the system

$$\begin{cases} a_1 x + b_1 y = k_1 \\ a_2 x + b_2 y = k_2 \end{cases} \text{ with } \begin{vmatrix} a_1 & b_1 \\ a_2 & b_2 \end{vmatrix} \neq 0$$

has the unique solution (x, y) given by

$$x = \frac{\begin{vmatrix} k_1 & b_1 \\ k_2 & b_2 \end{vmatrix}}{\begin{vmatrix} a_1 & b_1 \\ a_2 & b_2 \end{vmatrix}} \qquad y = \frac{\begin{vmatrix} a_1 & k_1 \\ a_2 & k_2 \end{vmatrix}}{\begin{vmatrix} a_1 & b_1 \\ a_2 & b_2 \end{vmatrix}}$$

EXAMPLE 1 Use Cramer's Rule to solve

$$5x + 3y = 1$$
$$6x + 4y = 0$$

Solution: Here, $a_1 = 5$, $b_1 = 3$, $k_1 = 1$, $a_2 = 6$, $b_2 = 4$, and $k_2 = 0$. Since

$$\begin{vmatrix} 5 & 3 \\ 6 & 4 \end{vmatrix} = 2 \neq 0$$

we see that Cramer's Rule does apply and

$$x = \frac{\begin{vmatrix} 1 & 3 \\ 0 & 4 \end{vmatrix}}{\begin{vmatrix} 5 & 3 \\ 6 & 4 \end{vmatrix}} = \frac{(1)(4) - (3)(0)}{(5)(4) - (3)(6)} = \frac{4}{2} = 2$$

$$y = \frac{\begin{vmatrix} 5 & 1 \\ 6 & 0 \end{vmatrix}}{\begin{vmatrix} 5 & 3 \\ 6 & 4 \end{vmatrix}} = \frac{(5)(0) - (1)(6)}{(5)(4) - (3)(6)} = \frac{-6}{2} = -3$$

The solution set is $\{(2, -3)\}$.

Check: We replace x by 2 and y by -3 in the original system.

$$5(2) + 3(-3) = 1 \quad \checkmark$$
$$6(2) + 4(-3) = 0 \quad \checkmark$$

In order to justify the use of Cramer's Rule for a system of two equations in two variables, we solve the system by elimination.

$$a_1 x + b_1 y = k_1 \qquad (1)$$
$$a_2 x + b_2 y = k_2 \qquad (2)$$

Multiply Equation (1) by b_2 and Equation (2) by $-b_1$ to obtain

$$
\begin{array}{ll}
a_1 b_2 x + b_1 b_2 y = b_2 k_1 & (1) \\
-a_2 b_1 x - b_1 b_2 y = -b_1 k_2 & (2) \\
\hline
(a_1 b_2 - a_2 b_1)x = b_2 k_1 - b_1 k_2 & \text{Add}
\end{array}
$$

$$x = \frac{b_2 k_1 - b_1 k_2}{a_1 b_2 - a_2 b_1} \qquad \text{If } a_1 b_2 - a_2 b_1 \neq 0$$

Thus

$$x = \frac{\begin{vmatrix} k_1 & b_1 \\ k_2 & b_2 \end{vmatrix}}{\begin{vmatrix} a_1 & b_1 \\ a_2 & b_2 \end{vmatrix}}$$

as given by Cramer's Rule.

Eliminating x instead of y yields the desired result for y. We ask the student to verify this in Exercise 28.

Although we have stated Cramer's Rule for a system of two equations in two variables, the method is quite general. Next we will state the rule for a system of three equations in three variables.

Cramer's Rule

For three equatons in three variables the system

$$a_1 x + b_1 y + c_1 z = k_1$$
$$a_2 x + b_2 y + c_2 z = k_2$$
$$a_3 x + b_3 y + c_3 z = k_3$$

has the unique solution (x, y, z) given by

$$x = \frac{D_x}{D}, \qquad y = \frac{D_y}{D}, \qquad z = \frac{D_z}{D} \qquad \text{if } D \neq 0$$

where

$$D = \begin{vmatrix} a_1 & b_1 & c_1 \\ a_2 & b_2 & c_2 \\ a_3 & b_3 & c_3 \end{vmatrix}, \qquad D_x = \begin{vmatrix} k_1 & b_1 & c_1 \\ k_2 & b_2 & c_2 \\ k_3 & b_3 & c_3 \end{vmatrix}$$

$$D_y = \begin{vmatrix} a_1 & k_1 & c_1 \\ a_2 & k_2 & c_2 \\ a_3 & k_3 & c_3 \end{vmatrix}, \qquad D_z = \begin{vmatrix} a_1 & b_1 & k_1 \\ a_2 & b_2 & k_2 \\ a_3 & b_3 & k_3 \end{vmatrix}$$

EXAMPLE 2 Use Cramer's Rule to solve

$$7x + y + z = 1$$
$$5x - 2y + 3z = 4$$
$$4x + 3y - z = 0$$

Solution: Expanding using column 1 we find

$$D = \begin{vmatrix} 7 & 1 & 1 \\ 5 & -2 & 3 \\ 4 & 3 & -1 \end{vmatrix} = 7\begin{vmatrix} -2 & 3 \\ 3 & -1 \end{vmatrix} - 5\begin{vmatrix} 1 & 1 \\ 3 & -1 \end{vmatrix} + 4\begin{vmatrix} 1 & 1 \\ -2 & 3 \end{vmatrix}$$

$$= 7(-7) - 5(-4) + 4(5) = -9$$

Replacing column 1 in D by the column of constants yields

$$D_x = \begin{vmatrix} 1 & 1 & 1 \\ 4 & -2 & 3 \\ 0 & 3 & -1 \end{vmatrix} = (1)\begin{vmatrix} -2 & 3 \\ 3 & -1 \end{vmatrix} - 4\begin{vmatrix} 1 & 1 \\ 3 & -1 \end{vmatrix} + 0\begin{vmatrix} 1 & 1 \\ -2 & 3 \end{vmatrix}$$

$$= (1)(-7) - 4(-4) + 0(5) = 9$$

Replacing column 2 in D by the column of constants yields

$$D_y = \begin{vmatrix} 7 & 1 & 1 \\ 5 & 4 & 3 \\ 4 & 0 & -1 \end{vmatrix} = 7\begin{vmatrix} 4 & 3 \\ 0 & -1 \end{vmatrix} - 5\begin{vmatrix} 1 & 1 \\ 0 & -1 \end{vmatrix} + 4\begin{vmatrix} 1 & 1 \\ 4 & 3 \end{vmatrix}$$

$$= 7(-4) - 5(-1) + 4(-1) = -27$$

Replacing column 3 in D by the column of constants yields

$$D_z = \begin{vmatrix} 7 & 1 & 1 \\ 5 & -2 & 4 \\ 4 & 3 & 0 \end{vmatrix} = 7\begin{vmatrix} -2 & 4 \\ 3 & 0 \end{vmatrix} - 5\begin{vmatrix} 1 & 1 \\ 3 & 0 \end{vmatrix} + 4\begin{vmatrix} 1 & 1 \\ -2 & 4 \end{vmatrix}$$

$$= 7(-12) - 5(-3) + 4(6) = -45$$

Thus $x = \dfrac{D_x}{D} = \dfrac{9}{-9} = -1$, $y = \dfrac{D_y}{D} = \dfrac{-27}{-9} = 3$, and $z = \dfrac{D_z}{D} = \dfrac{-45}{-9} = 5$.
The solution set is $\{(-1, 3, 5)\}$.

Check: We replace x by -1, y by 3, and z by 5 in the original system.

$$7(-1) + (3) + (5) = 1 \checkmark$$
$$5(-1) - 2(3) + 3(5) = 4 \checkmark$$
$$4(-1) + 3(3) - (5) = 0 \checkmark$$

It is important to realize that Cramer's Rule does not apply if $D = 0$. In this case the system is either dependent or inconsistent.

Exercises 9.5

Standard Assignment: Exercises 1, 3, 5, 9, 11, 19, 21, 23, 25, 27

A Use Cramer's Rule to solve each of the following systems. See Examples 1 and 2.

1. $7x + y = 7$
   $10x + 2y = 6$

2. $4x - y = 0$
   $x + 2y = 9$

3. $3x - 2y = 3$
   $x + y = -4$

4. $5x - 3y = 10$
   $3x - 5y = -10$

5. $x - 2y = -3$
   $-2x + 3y = 1$

6. $2x + y = 15$
   $4x - 7y = 3$

7. $4x + 3y = 3$
   $7x + 4y = 1$

8. $x - 5y = -5$
   $2x - 7y = -1$

9. $x + 2y = 2$
   $2x + 3y = 1$

10. $x + 4y + 3z = 1$
    $2x + 5y + 4z = 4$
    $x - 3y - 2z = 5$

11. $2x + 3y - z = 15$
    $4x - 2y + 3z = 9$
    $3x - y + 2z = 8$

12. $2x - y + z = 1$
    $3x - 3y + 4z = 5$
    $5x - 4y + 5z = 6$

13. $3x + 2y + z = 10$
    $x - 3y + 4z = 7$
    $x + 2y - 3z = -4$

14. $4x + 5z = 6$
    $y - 6z = -2$
    $3x + 4z = 3$

15. $3x + 3y + 3z = 6$
    $4x + 2y + 2z = 6$
    $2x - y + z = -1$

16. $4x + y - z = 13$
    $3x - y + 2z = 9$
    $x - 2y + 3z = 2$

17. $3x + 3y - 2z = 10$
    $4x + y - z = 9$
    $2x + y - 3z = 3$

18. $7x - y + z = 8$
    $3x - 4y - 5z = -4$
    $5x + y - z = 4$

19. $2x - 2y + z = 8$
    $3x - y + 3z = 8$
    $5x + 2z = 10$

20. $x + 2y + z = 9$
    $2x + 3y = 14$
    $y + 2z = 4$

21. $3x - z = 2$
    $2x + y - 3z = 1$
    $4y - 3z = 5$

22. $x + y + z = -1$
    $5x + 3y + 2z = 1$
    $4x - 4z = 12$

23. $5x + y + z = -3$
    $2x - y + 2z = 5$
    $x + y = -2$

24. $6x + y = 10$
    $-3y - z = -11$
    $4x - 2y - z = -1$

25. $3x + y - z = 6$
    $2x + 2y + z = 5$
    $x - z = 2$

26. $x + z = 6$
    $x + y + z = 6$
    $x + y + 2z = 11$

27. $2x + 2y - 2z = 2$
    $3x + y - z = 5$
    $-x + y + z = 7$

28. Use the addition method to eliminate x from the system

$$a_1 x + b_1 y = k_1$$
$$a_2 x + b_2 y = k_2$$

and solve for y. Show that

$$y = \frac{\begin{vmatrix} a_1 & k_1 \\ a_2 & k_2 \end{vmatrix}}{\begin{vmatrix} a_1 & b_1 \\ a_2 & b_2 \end{vmatrix}}$$

29. A linear system of the form

$$a_1x + b_1y + c_1 = 0$$
$$a_2x + b_2y + c_2 = 0$$
$$a_3x + b_3y + c_3 = 0$$

is called a **homogeneous system**. Certainly $x = 0$, $y = 0$, and $z = 0$ solves any such system. Use Cramer's Rule to determine whether $(0, 0, 0)$ is the only solution to each of the following systems. (Note that any homogeneous system is consistent.)

(a) $\quad x + 2z = 0$
$\quad\quad 4y - 3z = 0$
$\quad\quad x + 2y = 0$

(b) $\quad 3x + y = 0$
$\quad\quad x + 2y - z = 0$
$\quad\quad 5x + 5y - 2z = 0$

30. Find all values of k for which the following system does *not* have a unique solution. (See the concluding statements of this section.)

$$3x + 6y = -4$$
$$2x + ky = 17$$

9.6 Solving Linear Systems by Matrix Methods

The variables that are used to represent unknowns in a system of linear equations play only a minor role in solving the system. In this section we will see how to solve a system of linear equations by using a matrix whose elements are the numerical coefficients of the given system of equations. This matrix method of solution is readily implemented on a computer or programmable calculator.

By eliminating the variables and equal signs from the system

$$2x + 4y = 2$$
$$x + 3y = 3$$

we produce the **augmented matrix** for this system

$$\begin{bmatrix} 2 & 4 & | & 2 \\ 1 & 3 & | & 3 \end{bmatrix} \quad (x = 1 \cdot x)$$

The numerical coefficients of the system appear to the left of the dashed line and the constants are on the right. The dashed line itself is merely a convenience that visually separates the coefficients from the constants and can be omitted.

The procedures introduced in Section 9.1 for producing an equivalent system of equations from a given system are applicable to the matrix method of solution. The following **row operations** produce new matrices that correspond to systems that have the same solutions as the original system.

> **Row Operations**
> 1. Multiply the entries of any row by a *nonzero* real number.
> 2. Multiply the entries of any row by a real number and add the results to the corresponding entries of a *different* row.
> 3. Interchange any two rows.

The matrix method of solution progresses as follows

1. The given system

$$ax + by = m$$
$$cx + dy = n$$

2. The augmented matrix

$$\begin{bmatrix} a & b & | & m \\ c & d & | & n \end{bmatrix}$$

3. Row operations are applied to the augmented matrix to obtain the form

$$\begin{bmatrix} 1 & e & | & p \\ 0 & 1 & | & q \end{bmatrix}$$

4. This is the augmented matrix of the equivalent system

$$x + ey = p$$
$$y = q$$

5. The solution of this system is readily obtained by substituting q for y in the equation $x + ey = p$ and solving for x. This yields $x = p - eq$.
6. The solution set is $\{(p - eq, q)\}$.

EXAMPLE 1 Solve the system by the matrix method.

$$2x + 4y = 2$$
$$x + 3y = 3$$

Solution: The augmented matrix is

$$\begin{bmatrix} 2 & 4 & | & 2 \\ 1 & 3 & | & 3 \end{bmatrix}$$

We obtain a 1 in the upper left corner

$$\begin{bmatrix} 1 & 2 & | & 1 \\ 1 & 3 & | & 3 \end{bmatrix} \quad \text{Multiply row 1 by } \frac{1}{2}$$

We obtain a 0 in the lower left corner

$$\begin{bmatrix} 1 & 2 & | & 1 \\ 0 & 1 & | & 2 \end{bmatrix} \quad \text{Add } (-1) \times (\text{row 1}) \text{ to row 2}$$

This results in the equivalent system

$$x + 2y = 1$$
$$y = 2$$

Substituting $y = 2$ into $x + 2y = 1$ we get

$$x + 2(2) = 1$$
$$x = -3$$

The solution set is $\{(-3, 2)\}$.

Check: We replace x by -3 and y by 2 in the original system.

Does $2(-3) + 4(2) = 2$? $\quad -6 + 8 = 2$ ✔

Does $\quad -3 + 3(2) = 3$? $\quad -3 + 6 = 3$ ✔

When the matrix method is applied to three equations in three variables, we seek to use row operations on the augmented matrix to produce a matrix of the form

$$\begin{bmatrix} 1 & a & b & | & p \\ 0 & 1 & d & | & q \\ 0 & 0 & 1 & | & r \end{bmatrix}$$

The resulting system

$$x + ay + bz = p$$
$$y + dz = q$$
$$z = r$$

is then solved by substitution.

EXAMPLE 2 Solve the system by the matrix method.

$$4x + 2y + 4z = 4$$
$$4x + 6y + z = 15$$
$$2x + 2y + 7z = -1$$

Solution: The augmented matrix is

$$\begin{bmatrix} 4 & 2 & 4 & | & 4 \\ 4 & 6 & 1 & | & 15 \\ 2 & 2 & 7 & | & -1 \end{bmatrix}$$

We obtain a 1 in the upper left corner

$$\begin{bmatrix} 1 & \frac{1}{2} & 1 & | & 1 \\ 4 & 6 & 1 & | & 15 \\ 2 & 2 & 7 & | & -1 \end{bmatrix} \quad \text{Multiply row 1 by } \frac{1}{4}$$

We obtain a 0 for the first entry of row 2

$$\begin{bmatrix} 1 & \frac{1}{2} & 1 & | & 1 \\ 0 & 4 & -3 & | & 11 \\ 2 & 2 & 7 & | & -1 \end{bmatrix} \quad \text{Add } (-4) \times \text{(row 1) to row 2}$$

We obtain a 0 for the first entry of row 3

$$\begin{bmatrix} 1 & \frac{1}{2} & 1 & | & 1 \\ 0 & 4 & -3 & | & 11 \\ 0 & 1 & 5 & | & -3 \end{bmatrix} \quad \text{Add } (-2) \times \text{(row 1) to row 3}$$

We obtain a 1 for the second entry of row 2

$$\begin{bmatrix} 1 & \frac{1}{2} & 1 & | & 1 \\ 0 & 1 & 5 & | & -3 \\ 0 & 4 & -3 & | & 11 \end{bmatrix} \quad \text{Interchange row 2 with row 3}$$

We obtain a 0 for the second entry of row 3

$$\begin{bmatrix} 1 & \frac{1}{2} & 1 & | & 1 \\ 0 & 1 & 5 & | & -3 \\ 0 & 0 & -23 & | & 23 \end{bmatrix} \quad \text{Add } (-4) \times \text{(row 2) to row 3}$$

We obtain 1 for the third entry of row 3

$$\begin{bmatrix} 1 & \frac{1}{2} & 1 & | & 1 \\ 0 & 1 & 5 & | & -3 \\ 0 & 0 & 1 & | & -1 \end{bmatrix} \quad \text{Multiply row 3 by } \frac{-1}{23}$$

The system of equations corresponding to this augmented matrix is

$$x + \tfrac{1}{2}y + z = 1$$
$$y + 5z = -3$$
$$z = -1$$

Substituting -1 for z in $y + 5z = -3$

$$y + 5(-1) = -3$$
$$y = 2$$

Substituting -1 for z and 2 for y in $x + \frac{1}{2}y + z = 1$

$$x + \frac{1}{2}(2) + (-1) = 1$$
$$x = 1$$

Thus, $x = 1$, $y = 2$, and $z = -1$ will solve the original system. The solution set for this system is then $\{(1, 2, -1)\}$.

Check:
Does $4(1) + 2(2) + 4(-1) = 4?$ $\quad 4 + 4 - 4 = 4$ ✔
Does $4(1) + 6(2) + (-1) = 15?$ $\quad 4 + 12 - 1 = 15$ ✔
Does $2(1) + 2(2) + 7(-1) = -1?$ $\quad 2 + 4 - 7 = -1$ ✔

Exercises 9.6

Standard Assignment: All odd-numbered exercises

A Complete the steps in the following matrix solutions. See Example 1.

1. $\begin{bmatrix} 4 & 5 & | & -7 \\ 5 & 4 & | & -2 \end{bmatrix}$
2. $\begin{bmatrix} 2 & 6 & | & 8 \\ 3 & -1 & | & 2 \end{bmatrix}$

$\begin{bmatrix} 1 & ? & | & -\frac{7}{4} \\ 5 & 4 & | & -2 \end{bmatrix}$
$\begin{bmatrix} 1 & 3 & | & ? \\ 3 & -1 & | & 2 \end{bmatrix}$

$\begin{bmatrix} 1 & ? & | & -\frac{7}{4} \\ 0 & -\frac{9}{4} & | & ? \end{bmatrix}$
$\begin{bmatrix} 1 & 3 & | & ? \\ 0 & ? & | & -10 \end{bmatrix}$

$\begin{bmatrix} 1 & ? & | & -\frac{7}{4} \\ 0 & 1 & | & ? \end{bmatrix}$
$\begin{bmatrix} 1 & 3 & | & ? \\ 0 & 1 & | & ? \end{bmatrix}$

Solve the following systems by the matrix method. See Examples 1 and 2.

3. $4x - 5y = 50$
   $5x - 2y = 54$

4. $2x - 4y = 10$
   $3y = 18$

5. $6x + 9y = 3$
   $3x - 4y = 10$

6. $3x - 4y = 13$
   $2x + 3y = 3$

7. $3x + 2y = 1$
   $4x + 5y = -1$

8. $2x - y = 1$
   $-x + y = 2$

Complete the steps in the following matrix solutions. See Example 2.

9. $\begin{bmatrix} 1 & 4 & 3 & | & 1 \\ 0 & -3 & -2 & | & 0 \\ 0 & 7 & 5 & | & -3 \end{bmatrix}$
10. $\begin{bmatrix} 1 & 1 & 1 & | & 3 \\ 0 & 1 & 1 & | & 2 \\ 1 & -3 & -2 & | & 5 \end{bmatrix}$

$\begin{bmatrix} 1 & 4 & 3 & | & 1 \\ 0 & 1 & ? & | & 0 \\ 0 & 7 & 5 & | & -3 \end{bmatrix}$
$\begin{bmatrix} 1 & 1 & 1 & | & 3 \\ 0 & 1 & 1 & | & 2 \\ 0 & ? & -3 & | & 2 \end{bmatrix}$

$\begin{bmatrix} 1 & 4 & 3 & | & 1 \\ 0 & 1 & ? & | & 0 \\ 0 & 0 & ? & | & -3 \end{bmatrix}$
$\begin{bmatrix} 1 & 1 & 1 & | & 3 \\ 0 & 1 & 1 & | & 2 \\ 0 & 0 & ? & | & 10 \end{bmatrix}$

$\begin{bmatrix} 1 & 4 & 3 & | & 1 \\ 0 & 1 & ? & | & 0 \\ 0 & 0 & 1 & | & ? \end{bmatrix}$

Solve the following systems by the matrix method. See Example 2.

11. $x + y - z = 4$
$x + 3y + 5z = 10$
$2y + 3z = 3$

12. $4x + 5z = 6$
$y - 6z = -2$
$3x + 4z = 3$

13. $5x - 7y + 4z = 2$
$2x - y + 3z = 4$
$3x + 2y - 2z = 3$

14. $x - y = 1$
$x - z = 7$
$2x - y + z = 0$

15. $2x - y + z = 8$
$x + z = 6$
$x + 2y = 5$

16. $2x - y - z = -4$
$2x + 4y - 3z = 7$
$x + y + z = -5$

9.7 Nonlinear Systems of Equations

A system of equations that includes at least one nonlinear equation is called a **nonlinear system of equations**. The addition and the substitution methods that were used to solve linear systems can also be useful in solving nonlinear systems. The substitution technique is particularly helpful when one of the equations is linear.

EXAMPLE 1 | Solve the system

$$x^2 + y^2 = 25 \quad \text{(1)}$$
$$x - 2y = -5 \quad \text{(2)}$$

Solution: We solve Equation (2) for x and then substitute the resulting expression into Equation (1).

$$x - 2y = -5 \quad (2)$$
$$x = 2y - 5$$

Substituting $2y - 5$ for x in Equation (1), we have

$$(2y - 5)^2 + y^2 = 25$$

$4y^2 - 20y + 25 + y^2 = 25$	Expand $(2y - 5)^2$
$5y^2 - 20y = 0$	Simplify
$y^2 - 4y = 0$	Divide by 5
$y(y - 4) = 0$	Factor

Thus, $y = 0$ or $y = 4$.

Since $x = 2y - 5$, when $y = 0$, $x = -5$ and when $y = 4$, $x = 3$. The solution set for this system is $\{(-5, 0), (3, 4)\}$. The graph of this system is shown in Figure 9.4. The graph of a system is often a valuable aid in checking the number of solutions that a system should have.

Figure 9.4

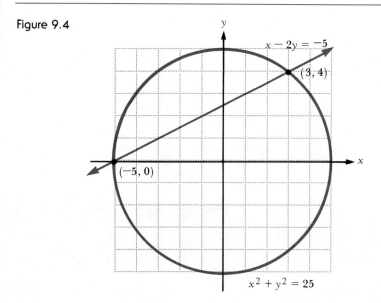

The addition method is usually simpler to use if both of the equations in a system of two equations in two variables are of second degree in both variables.

EXAMPLE 2

Solve the system

$$4x^2 + 9y^2 = 36 \quad (1)$$
$$x^2 + y^2 = 4 \quad (2)$$

Solution: Multiply Equation (2) by -4 so that the coefficients of x^2 differ only in sign.

$$\begin{aligned} 4x^2 + 9y^2 &= 36 \\ -4x^2 - 4y^2 &= -16 \\ \hline 5y^2 &= 20 \quad \text{Add} \\ y^2 &= 4 \\ y &= \pm 2 \end{aligned}$$

Substituting 2 and -2 for y in Equation (2), we have

$$\begin{array}{ccc} y = 2 & & y = -2 \\ x^2 + (\ )^2 = 4 & \text{and} & x^2 + (-2)^2 = 4 \\ x^2 + 4 = 4 & & x^2 + 4 = 4 \\ x^2 = 0 & & x^2 = 0 \\ x = 0 & & x = 0 \end{array}$$

Thus, $(0, 2)$ and $(0, -2)$ are solutions. The solution set is $\{(0, 2), (0, -2)\}$. The graph of the system is shown in Figure 9.5.

Figure 9.5

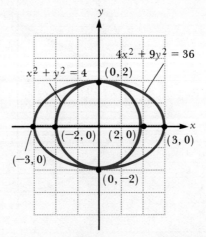

EXAMPLE 3

Solve the system

$$4x^2 + 5y^2 = 180 \quad (1)$$
$$x^2 - y^2 = 9 \quad (2)$$

Solution: Multiply Equation (2) by 5 so that the coefficients of y^2 differ only in sign.

$$
\begin{array}{rl}
4x^2 + 5y^2 =& 180 \\
5x^2 - 5y^2 =& 45 \\
\hline
9x^2 =& 225 \quad \text{Add} \\
x^2 =& 25 \\
x =& \pm 5
\end{array}
$$

Thus, $x = 5$ or $x = -5$. Now substitute 5 and -5 for x in Equation (2).

$$
\begin{aligned}
(5)^2 - y^2 &= 9 & (-5)^2 - y^2 &= 9 \\
-y^2 &= 9 - 25 & \text{and} \quad -y^2 &= 9 - 25 \\
y^2 &= 16 & y^2 &= 16 \\
y = 4 \text{ or } y &= -4 & y = 4 \text{ or } y &= -4
\end{aligned}
$$

If $x = 5$, we have $y = \pm 4$ and if $x = -5$, we have $y = \pm 4$. The solution set is $\{(5, 4), (5, -4), (-5, 4), (-5, -4)\}$. The graph of this system is shown in Figure 9.6.

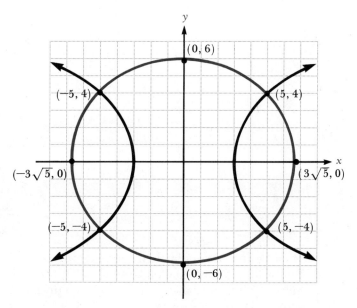

Figure 9.6

Sketching the graph of a system is actually unnecessary for solving the system. The system is solved algebraically; however, the graph sometimes serves as a convenient aid for checking solutions (x, y) when x and y are real numbers. Quadratic equations may have complex numbers as solutions, and so systems containing quadratic expressions may have complex solutions.

Sometimes both the substitution method and the addition method may be used in solving a system of equations.

EXAMPLE 4 Solve the system

$$x^2 - y^2 = 3 \quad (1)$$
$$x^2 - 2xy - y^2 = -1 \quad (2)$$

Solution: Multiplying Equation (2) by -1 yields the system

$$\begin{array}{r} x^2 - y^2 = 3 \\ -x^2 + 2xy + y^2 = 1 \\ \hline 2xy = 4 \end{array} \quad \text{Add}$$
$$xy = 2$$

Solving for y yields

$$y = \frac{2}{x} \quad (3)$$

Substituting $2/x$ for y in Equation (1), we have

$$x^2 - \left(\frac{2}{x}\right)^2 = 3$$

$$x^2 - \frac{4}{x^2} = 3$$

$$x^4 - 4 = 3x^2 \quad (x \neq 0)$$

$$x^4 - 3x^2 - 4 = 0 \quad (x \neq 0)$$

This is a quadratic equation in x^2 that we factor as

$$(x^2 - 4)(x^2 + 1) = 0 \quad (x \neq 0)$$

Thus $\quad x^2 - 4 = 0 \quad$ or $\quad x^2 + 1 = 0$
$(x - 2)(x + 2) = 0 \quad\quad\quad\quad\quad x^2 = -1$
$x = 2 \text{ or } x = -2 \quad\quad\quad\quad x = i \text{ or } x = -i$

None of the values obtained for x is zero, so we substitute the four values found for x into Equation (3) and solve for y. (You will want to review Section 5.7 if you have forgotten how to divide by complex numbers.)

$x = 2$	$x = -2$	$x = i$	$x = -i$
$\frac{2}{x} = \frac{2}{2}$	$\frac{2}{x} = \frac{2}{-2}$	$\frac{2}{x} = \frac{2}{i}$	$\frac{2}{x} = \frac{2}{-i}$
$y = 1$	$y = -1$	$y = -2i$	$y = 2i$

The solution set for the system is $\{(2, 1), (-2, -1), (i, -2i), (-i, 2i)\}$. Each of these solutions may be checked by substitution into the *original* system.

Exercises 9.7

Standard Assignment: All odd-numbered A exercises

A Solve each of the following systems by the substitution method. See Examples 1 and 2.

1. $x^2 - y = 6$
   $x - y = 0$

2. $x^2 - y = 6$
   $5x - y = 0$

3. $x^2 + y^2 = 9$
   $x = 3$

4. $x^2 + y^2 = 9$
   $y = 3$

5. $x^2 + y^2 = 5$
   $x - y = -3$

6. $x^2 + y^2 = 13$
   $2x - 3y = 0$

7. $x^2 - 4x + y^2 = -2$
   $x - y = 2$

8. $x^2 - 8y + y^2 = -6$
   $2x - y = 1$

9. $x - y = -2$
   $xy = 3$

10. $x - 2y = 4$
    $xy = 6$

11. $4x^2 + y^2 = 25$
    $x + y = 5$

12. $2x^2 + 8y^2 = 32$
    $x + 2y = 4$

13. $x^2 - y^2 = 24$
    $5x - 7y = 0$

14. $x^2 - 7y^2 = 9$
    $x - y = 3$

15. $x^2 - 3y = 0$
    $5x^2 - 9y^2 = -36$

Solve each of the following systems by any method or combination of methods. See Examples 3 and 4.

16. $x^2 + 2y^2 = 12$
    $7y^2 - 5x^2 = 8$

17. $x^2 + 8y^2 = 9$
    $3x^2 + y^2 = 4$

18. $x^2 + y^2 = 20$
    $x^2 - y^2 = 12$

19. $3x^2 + 2y^2 = 77$
    $x^2 - 6y^2 = 19$

20. $x^2 + y^2 = 29$
    $xy = 10$

21. $x^2 + y^2 = 25$
    $xy = 12$

22. $x^2 + 3y^2 = 7$
    $xy = 2$

23. $x^2 - xy + y^2 = 9$
    $xy = 9$

24. $3x^2 + xy + y^2 = 9$
    $xy = 2$

25. $y = x^2 - 2$
    $2x = y - 1$

26. $y = x^2 - x - 3$
    $y - 3x = 18$

27. $x^2 + 3y = 0$
    $x - y = -12$

28. $x - 4xy + y = 1$
    $x + y = 1$

29. $2x^2 + y^2 = 3$
    $x - y = 3$

30. $x^2 - xy = -3$
    $4xy - y^2 = 12$

31. $2x^2 - 7xy - 2y^2 = -20$
    $-x^2 + 4xy + y^2 = 11$

32. $x^2 - 2xy - y^2 = 2$
    $-5x^2 + 6xy + 5y^2 = -22$

B 33. $x^2 + y^2 = 10a^2$
    $x - y = 2a$

34. $x^2 - 3xy + 4y^2 = a$
    $xy = a$

35. $x + y = a \quad (a \neq 0)$
    $x^2 - y^2 = a^2$

Write a system of equations for each of the following problems and solve.

36. The sum of two numbers is 17 and their product is 70. Find the two numbers.

37. The perimeter of a rectangle is 54 meters. If the area is 180 square meters, find the length and width of the rectangle.

38. The radius of a circular skating rink is increased by 5%, resulting in an increase of 164π square feet of skating area. What is the area of the enlarged rink?

39. When the radius of a can 20 cm high is increased by 10%, the volume is increased 420π cm³. What is the volume of the original can?

9.8 Systems of Inequalities

In Sections 7.4 and 8.5 we learned how to find and graph the solution sets of linear inequalities and quadratic inequalities. The solution set of a **system of inequalities** is the intersection of the solution sets of the inequalities of the system. The graph of the system of inequalities is the intersection of the graphs of the individual inequalities of the system.

EXAMPLE 1

Graph the system

$$y \geq x^2$$
$$y < x + 2$$

Solution: Graph each inequality separately in the same coordinate plane. (You may need to review Sections 7.4 and 8.5 at this time.)

$y \geq x^2$ $\qquad$ $y < x + 2$
$y = x^2$ $\quad$ Solid graph $\qquad$ $y = x + 2$ $\quad$ Broken graph
Test $(0, 1)$ in $y \geq x^2$. $\qquad$ Test $(0, 1)$ in $y < x + 2$.
$1 \geq 0^2$ $\quad$ True. $\qquad$ $1 < 0 + 2$ $\quad$ True.
Shade as in Figure 9.7. $\qquad$ Shade as in Figure 9.7.

The doubly shaded region in Figure 9.7 is the graph of the system.

9.8 Systems of Inequalities   427

Figure 9.7

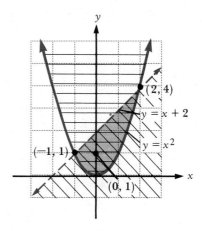

EXAMPLE 2   Graph the system

$$y - x \geq 0$$
$$9x^2 + 16y^2 < 144$$
$$x^2 - y + 2x \leq 2$$

Solution: Graph each inequality separately in the same coordinate plane.

$y - x \geq 0$	$9x^2 + 16y^2 < 144$
$y - x = 0$ Solid graph	$9x^2 + 16y^2 = 144$ Broken graph
Test $(0, 1)$ in $y - x \geq 0$.	Test $(0, 1)$ in $9x^2 + 16y^2 < 144$.
$1 - 0 \geq 0$ True.	$9(0)^2 + 16(1)^2 < 144$
Shade as in Figure 9.8.	$16 < 144$ True.
	Shade as in Figure 9.8.

$$x^2 - y + 2x \leq 3$$
$$x^2 - y + 2x = 3 \quad \text{Solid graph}$$

Test $(0, 1)$ in $x^2 - y + 2x \leq 3$.

$$(0)^2 - (1) + 2(0) \leq 3$$
$$-1 \leq 3 \quad \text{True.}$$

Shade as in Figure 9.8.

The graph of the system is shown in Figure 9.8.

Figure 9.8

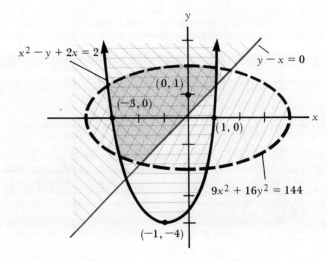

Exercises 9.8

Standard Assignment: All odd-numbered exercises

A Graph each of the following systems of inequalities. See Examples 1 and 2.

1. $2x - y \leq 3$
   $x - 1 \geq 0$

2. $2x + 3y > 6$
   $y - 5x \geq 2$

3. $x + 3y < 9$
   $6y > -2x$

4. $y \leq 1$
   $x + y < 0$

5. $x^2 + y^2 < 25$
   $y > 2x$

6. $x^2 + y^2 > 16$
   $y \leq 3$

7. $x^2 + y^2 \geq 1$
   $x^2 + y^2 \leq 9$

8. $y < x^2 + 2$
   $y \geq 3x + 4$

9. $x^2 - y^2 \leq 9$
   $x^2 + y^2 > 9$

10. $y \geq x^2 + 2x$
    $y^2 - x^2 \leq 9$

11. $x^2 + 4y^2 > 16$
    $y - x^2 \leq 0$

12. $x^2 - y^2 \geq 1$
    $x^2 + y^2 \leq 9$

13. $y \leq 3 - x^2$
    $y \geq x^2 - 3$

14. $y \geq 1$
    $2x + 3y \leq 6$
    $x^2 + y^2 < 9$

15. $4x + 3y \leq 12$
    $y \geq x^2 - 1$
    $x \leq 2$

16. $x \leq 3$
    $y - 3x \leq 1$
    $x^2 + y^2 \geq 4$

17. $4x^2 + 9y^2 > 36$
    $y > x^2 + 6x + 5$
    $y < 3x + 15$

18. $4x^2 + y^2 \leq 4$
    $y \leq 1$
    $y > x^2 - 1$

19. $x^2 - y^2 \leq 1$
    $x^2 + 9y^2 \leq 9$
    $y < 2x$
    $y > -2x$

20. $y^2 - x^2 \geq 9$
    $y - x > 0$
    $y \geq x^2$
    $x^2 + y^2 \leq 25$

21. $y \leq 3$
    $y < 9 - x^2$
    $y > x + 3$
    $x \geq 2$

Chapter 9 Summary

Key Words and Phrases

9.1 System of equations
Solution of a system
Inconsistent equations
Dependent equations
Addition method
Equivalent system
Substitution method
9.4 Matrix
Determinant

Second-order determinant
Third-order determinant
Minor
9.5 Cramer's Rule
9.6 Augmented matrix
Row operations
9.7 Nonlinear system of equations
9.8 System of inequalities

Key Concepts and Rules

Cramer's Rule: The system

$$a_1x + b_1y = k_1$$
$$a_2x + b_2y = k_2$$

with $\begin{vmatrix} a_1 & b_1 \\ a_2 & b_2 \end{vmatrix} \neq 0$

has the unique solution (x, y) given by

$$x = \frac{\begin{vmatrix} k_1 & b_1 \\ k_2 & b_2 \end{vmatrix}}{\begin{vmatrix} a_1 & b_1 \\ a_2 & b_2 \end{vmatrix}} \qquad y = \frac{\begin{vmatrix} a_1 & k_1 \\ a_2 & k_2 \end{vmatrix}}{\begin{vmatrix} a_1 & b_1 \\ a_2 & b_2 \end{vmatrix}}$$

The system

$$a_1x + b_1y + c_1z = k_1$$
$$a_2x + b_2y + c_2z = k_2$$
$$a_3x + b_3y + c_3z = k_3$$

has the unique solution (x, y, z) given by

$$x = \frac{D_x}{D}, \quad y = \frac{D_y}{D}, \quad z = \frac{D_z}{D} \quad \text{if } D \neq 0$$

where

$$D = \begin{vmatrix} a_1 & b_1 & c_1 \\ a_2 & b_2 & c_2 \\ a_3 & b_3 & c_3 \end{vmatrix} \quad D_x = \begin{vmatrix} k_1 & b_1 & c_1 \\ k_2 & b_2 & c_2 \\ k_3 & b_3 & c_3 \end{vmatrix}$$

$$D_y = \begin{vmatrix} a_1 & k_1 & c_1 \\ a_2 & k_2 & c_2 \\ a_3 & k_3 & c_3 \end{vmatrix} \quad D_z = \begin{vmatrix} a_1 & b_1 & k_1 \\ a_2 & b_2 & k_2 \\ a_3 & b_3 & k_3 \end{vmatrix}$$

Review Exercises

If you have difficulty with any of the exercises, look in the section indicated by the number in square brackets.

[9.1] Solve the following systems by either the addition method or the substitution method.

1. $x - y = 3$
   $x - 2y = -1$

2. $3x + 4y = 4$
   $7x - 2y = -2$

3. $2x + y = 17$
   $8x - 2y = 2$

4. $4x - y = 0$
   $6x - 2y = -1$

5. $3x - 2y = 5$
   $-6x + 4y = -10$

6. $\dfrac{x}{3} + \dfrac{y}{2} = 1$
   $2x + 3y = 6$

[9.2] 7. $3x + 3y + z = 8$
   $2x - y + z = 7$
   $-4x + 2y + 2z = 6$

8. $4x + 2y + 3z = 5$
   $x + y + 2z = 1$
   $3x + 2y - z = -1$

9. $x + 6z = 1$
   $y + 5z = -1$
   $x - 2y + z = -2$

[9.3] 10. Jack can weed his garden in 50 minutes working alone. His son takes 1 hour and 15 minutes to do the same job. Jack's son begins helping him weed the garden 25 minutes after Jack started weeding. How much time did the entire job take?

11. Jane averaged 20 miles per hour on her drive to the dentist's office. On her return trip home she averaged 45 miles per hour. The return trip took a half hour less than the trip to the dentist. How far is the dentist's office from Jane's home?

12. Determine the number of gallons of a 40% alcohol solution and the number of gallons of a 25% alcohol solution that must be mixed in order to obtain 30 gallons of a 30% alcohol solution.

13. The sum of Al's age and Debbie's age is 21. Debbie's age three years ago is twice Al's present age. Find their present ages.

14. The sum of the digits of a two-digit number is 13. If the digits are interchanged, the resulting number is 9 less than the original number. Find the original number.

15. The sum of three numbers is 18. The third number is twice the first number and 4 more than twice the second number. Find the numbers.

16. The perimeter of a triangle is 140 centimeters. The sum of the lengths of two of the sides equals the length of the third side. Twice the length of the shortest leg is 10 centimeters less than the length of the longest leg. Find the lengths of the sides of the triangle.

[9.4] Find the following determinants.

17. $\begin{vmatrix} 3 & 2 \\ -1 & 0 \end{vmatrix}$
18. $\begin{vmatrix} -1 & -3 \\ 4 & -4 \end{vmatrix}$
19. $\begin{vmatrix} 13 & 27 \\ 0 & 1 \end{vmatrix}$

20. $\begin{vmatrix} 4 & 1 & 0 \\ 3 & -2 & 1 \\ 1 & 3 & 2 \end{vmatrix}$
21. $\begin{vmatrix} -2 & 1 & 3 \\ 5 & 0 & 4 \\ 4 & -2 & -6 \end{vmatrix}$
22. $\begin{vmatrix} 1 & 2 & 0 \\ -3 & 1 & 0 \\ 4 & 5 & 2 \end{vmatrix}$

[9.5] Use Cramer's Rule to solve the following systems of equations.

23. $3x - 4y = -5$
    $5x - y = 3$

24. $4x + 3y = 1$
    $3x + 2y = 0$

25. $7x + y = -12$
    $5x - 2y = 5$

26. $x + 3y = 7$
    $5x - 12y = 8$

27. $2x - y = 0$
    $6x + 4y = 7$

28. $4x + 2y + 3z = 11$
    $x - y - z = 4$
    $2x - y = 8$

29. $2x + 2y + 7z = -1$
    $4x + 2y + 4z = 4$
    $3x + 3y - z = 10$

30. $4x - 4y + 3z = 7$
    $x + y - z = 4$
    $3x - 2y - 6z = -1$

31. $2x + 3y + 6z = 5$
    $6x + 2y - 3z = -6$
    $2x - 6y + 3z = -6$

[9.6] Solve the following systems by the matrix method.

32. $2x + 2y = 2$
    $x - y = 1$

33. $2x + 11y = 17$
    $7x - 5y = 16$

34. $4x + 5y = 3$
    $3x + 4y = 2$

35. $4x - 6y = -16$
    $3x + 2y = 32$

36. $3x - 2y + z = -2$
    $2x + y + z = 1$
    $x + 3y - z = 3$

37. $2x - y + 3z = 4$
    $5x - y + z = 7$
    $5x - 7y + 4z = 2$

[9.7] Solve each of the following systems.

38. $y - x^2 = 1$
    $y - x = 3$

39. $y = x^2 + 1$
    $y + x^2 = 1$

40. $y = -2x^2 + 16x - 34$
    $y = -4x + 8$

41. $16x^2 - 9y^2 = 144$
    $x^2 - y = 9$

42. $16x^2 - 9y^2 = 144$
    $y = x$

43. $25x^2 + 9y^2 = 225$
    $5x - 3y = 15$

44. $xy = -6$
    $x - y = 5$

45. $y = x^2 - 4$
    $x^2 + y^2 = 16$

46. $x^2 + 3y^2 = 12$
    $x^2 - 8x + y^2 = 14$

[9.8] Graph the following systems of inequalities.

47. $x^2 + y^2 \leq 4$
    $x \geq 3$

48. $y < x^2 - 2x - 8$
    $x^2 + y^2 \leq 25$

49. $3y \geq x^2$
    $y < 2x + 3$

50. $x^2 - 4y^2 < 4$
    $4x^2 + y^2 < 4$

51. $x^2 + y^2 \leq 9$
    $x^2 - y^2 \geq 1$
    $y \geq 0$

52. $y < 3 - x^2$
    $y \geq x$
    $x < 0$

Practice Test (60 minutes)

1. Solve the system
$$3x + y = 2$$
$$5x + 2y = 1$$

2. Solve the system
$$2x + 3y + 4z = 0$$
$$x + y + z = 1$$
$$3x - 2y - 2z = 3$$

3. The perimeter of a parking lot is 200 meters. When the width was doubled and the length extended by 10 meters, the perimeter equaled 300 meters. Find the original width and length of the parking lot.

4. A coin purse contains $2.35 in nickels, dimes, and quarters. There are 18 coins in all and the sum of the number of dimes and nickels equals twice the number of quarters. How many coins of each kind are there?

5. Evaluate
$$\begin{vmatrix} 2 & 4 \\ -1 & 6 \end{vmatrix}$$

6. Evaluate
$$\begin{vmatrix} 1 & 4 & 7 \\ 0 & -2 & 3 \\ -1 & 3 & 2 \end{vmatrix}$$

7. Use Cramer's Rule to solve
$$4x + y + 4z = 3$$
$$2x - 3y + z = 1$$
$$3x - y + 3z = 4$$

8. Solve the system
$$3x^2 - 4y^2 = -1$$
$$2x^2 + 3y^2 = 5$$

9. Graph the system
$$y \leq x^2 - 2$$
$$x < 3y + 6$$

10. Graph the system
$$x^2 + y^2 \geq 9$$
$$y \leq x + 5$$
$$y \leq 5 - x$$
$$y \geq 0$$

Extended Applications

Market Equilibrium

In a given time period, there is usually a relationship between the selling price of a product and the quantity of the product that consumers will buy. Normally, the higher the selling price, the lower the quantity sold. In a similar way, if a manufacturer learns that consumers will buy the product at a higher price, a larger quantity of the product will be produced. When manufacturers are producing the exact quantity of a product that consumers will buy at the going price we have *market equilibrium* for the product. An equation that relates the price per unit of a product to the quantity that can be sold at that price is called a *demand equation*. An equation that relates the price per unit of a product to the quantity that a manufacturer will produce at that price is called a *supply equation*.

Supreme Tapes manufactures a video-cassette tape. The demand equation for the tapes is $p = 4000/q$, and the supply equation for the tapes is $p = (q/20) + 10$. In both equations, p represents the price in dollars and q the number of thousands of tapes. For example, 250,000 tapes could be sold at $16 each ($p = 16$ in the demand equation), but Supreme Tapes would be willing to place only 120,000 tapes on the market at a price of $16 each ($p = 16$ in the supply equation). The graphs of these equations are shown on the following page.

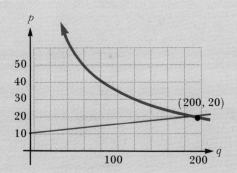

The point (200, 20) is on both graphs. We have market equilibrium at a selling price of $20 ($p = 20$). At this price, Supreme Tapes will supply 200,000 tapes and consumers will buy 200,000 tapes.

The point (200, 20) can be determined algebraically by solving the system

$$p = \frac{4000}{q}$$

$$p = \frac{q}{20} + 10$$

as follows. First set

$$\frac{q}{20} + 10 = \frac{4000}{q}$$

$$q + 200 = \frac{(20)(4000)}{q} \quad \text{Multiply by 20}$$

$$q^2 + 200q = (20)(4000) \quad \text{Multiply by } q$$

$$q^2 + 200q - 80{,}000 = 0 \quad \text{Simplify}$$

$$(q + 400)(q - 200) = 0 \quad \text{Factor}$$

$$q = -400 \quad \text{or} \quad q = 200$$

We discard $q = -400$ as nonsense in this problem, so $q = 200$. Finally,

$$p = \frac{4000}{200}$$

$$p = 20$$

Exercises

Suppose that the first equation is a demand equation for a product and the second is a supply equation for the product, p and q as previously shown.

$$p = 925 - 3q^2$$
$$p = 60q + 25$$

1. What is the demand for the product when the price is set at $p = \$325$? (Remember that q is in thousands of items.)

2. Graph the equations.

3. Find the price and demand at market equilibrium.

10

Exponential and Logarithmic Functions

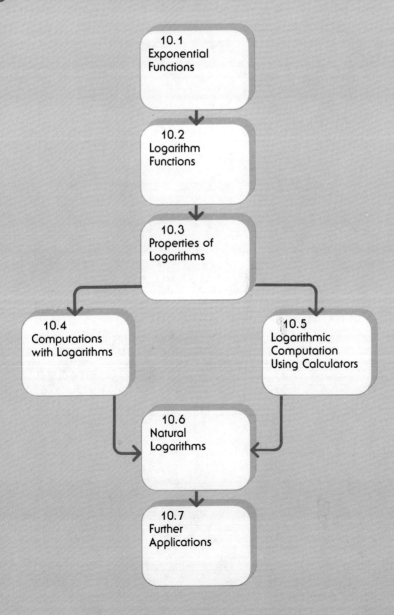

10.1 Exponential Functions

In Chapter 3 we introduced a^x for any real number $a \neq 0$ and any integer x. This concept was generalized in Chapter 5 to cover all rational values of x, where the base number a must be positive to ensure that $a^{1/n}$ is always a real number.

In general, for any $a > 0$, $a \neq 1$, and for any real numbers x, and y, a^x is defined so as to satisfy the following properties.

1. a^x is a unique real number.
2. If $a^x = a^y$, then $x = y$.   (10.1)
3. If $a > 1$ and $x < y$, then $a^x < a^y$.
4. If $0 < a < 1$ and $x < y$, then $a^x > a^y$.

EXAMPLE 1

If a technician knows that initially there are 2000 bacteria in a culture and that the number doubles every hour, then the bacteria count B can be written as a function of t (in hours)

$$B = 2000 \cdot 2^t$$

Solve for t when B is 32,000.

Solution: We have

$$32{,}000 = 2000 \cdot 2^t$$

Dividing both sides by 2000, we get

$$2^t = 16 = 2^4$$

Hence, by Property 2 of Properties (10.1), $t = 4$. That is, 4 hours later the bacteria will have increased to 32,000.

Properties 3 and 4 are used to approximate a^x when x is an irrational number. For example, according to Property 1, 2^π represents a unique real number whose approximate decimal value can be found by Property 3. From a table or a scientific calculator we find $\pi = 3.1415927\ldots$, so we may write $3.141 < \pi < 3.142$; hence by Property 3

$$2^{3.141} < 2^\pi < 2^{3.142}$$

Since both exponents 3.141 and 3.142 are rational numbers, both $2^{3.141}$ and $2^{3.142}$ have already been defined in Chapter 5 and could be evaluated. Using a scientific calculator, we find

$$2^{3.141} \approx 8.82135 \quad \text{and} \quad 2^{3.142} \approx 8.82747$$

so that

$$8.82135 < 2^\pi < 8.82747$$

438   Chap. 10   Exponential and Logarithmic Functions

or, by rounding to two decimal places, $2^\pi = 8.82$. It is actually given by $2^\pi = 8.8249778\ldots$.

By using Properties (10.1), we can define a unique real number a^x for all real numbers x, where $a > 0$ and $a \neq 1$. Thus an **exponential function** is a function of the form

$$y = a^x, \text{ where } a > 0 \text{ and } a \neq 1$$

EXAMPLE 2   Sketch a graph for the exponential function $y = 3^x$.

Solution: First make a table of a few selected values of x and the corresponding values of y.

x	-4	-3	-2	-1	0	1	2	3
$y(=3^x)$	$\frac{1}{81}$	$\frac{1}{27}$	$\frac{1}{9}$	$\frac{1}{3}$	1	3	9	24

Next, plot the points on graph paper, as is shown in Figure 10.1(a), and then draw a smooth curve through them to get the graph in Figure 10.1(b).

Figure 10.1

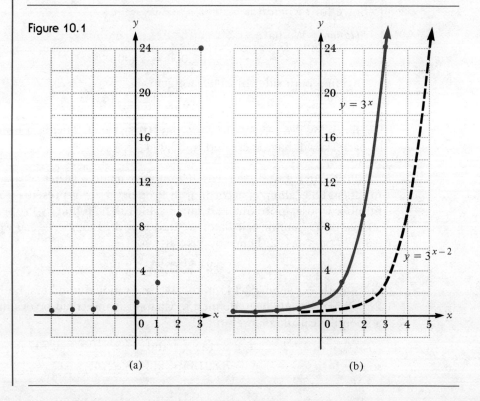

(a)   (b)

As x increases in the positive direction, y ($= 3^x$) increases upward indefinitely. This is typical of the graphs of exponential functions of the form $y = a^x$, where $a > 1$.

EXAMPLE 3 Sketch a graph of $y = \left(\dfrac{1}{2}\right)^x$.

Solution: Make a table similar to that in Example 2.

x	-4	-3	-2	-1	0	1	2	3
$y = \left(\dfrac{1}{2}\right)^x$	16	8	4	2	1	$\dfrac{1}{2}$	$\dfrac{1}{4}$	$\dfrac{1}{8}$

Plotting these points and drawing a smooth curve through them, we get a graph of $y = (\tfrac{1}{2})^x$, as is shown in Figure 10.2. In Figure 10.2, as x increases in the positive direction, $y = (\tfrac{1}{2})^x$ decreases toward 0. This is typical of the graph of an exponential function $y = a^x$ with $0 < a < 1$.

Figure 10.2

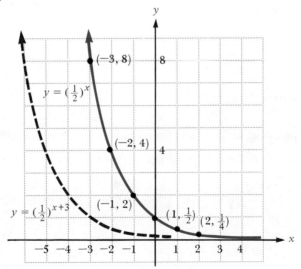

EXAMPLE 4 Discuss the graphs of (a) $y = 3^{x-2}$ (b) $y = \left(\dfrac{1}{2}\right)^{x+3}$

Solution:
(a) Since the exponent $x - 2$ is 0 when $x = 2$, the graph of $y = 3^{x-2}$ can be obtained by (horizontally) shifting the graph of $y = 3^x$ two units to the right, as is shown in Figure 10.1(b).

(b) Since the exponent $x + 3$ is 0 when $x = -3$, the graph of $y = \left(\frac{1}{2}\right)^{x+3}$ is identical to that of $y = \left(\frac{1}{2}\right)^x$ in Figure 10.2, except that it is the result of (horizontally) shifting the graph of $y = \left(\frac{1}{2}\right)^x$ three units to the left as shown in Figure 10.2.

As Figure 10.1(b) and Figure 10.2 indicate, for $a > 0$ and $a \neq 1$, the exponential function $y = a^x$ has domain

$$\{x \mid x \text{ is any real number}\}$$

and range

$$\{y \mid y \text{ is any positive real number}\}$$

Furthermore, each exponential function

$$y = f(x) = a^x \quad \text{with } a > 0 \text{ and } a \neq 1$$

is a one-to-one function, as can be seen from the horizontal line test.

According to Section 8.7, each one-to-one function has an inverse function. The inverse of an exponential function is a very special function that we will explore in the next section.

Exercises 10.1

Standard Assignment: Exercises 1, 3, 5, 7, 11, 19, 21, 23, 25, 27, 29, 31, 33

A Graph each of the following exponential functions. See Examples 2, 3, and 4.

1. $y = 2^x$
2. $y = 4^x$
3. $y = 10^x$
4. $y = \left(\frac{1}{3}\right)^x$
5. $y = \left(\frac{1}{4}\right)^x$
6. $y = \left(\frac{1}{5}\right)^x$
7. $y = 2^{x-3}$
8. $y = 3^{x+1}$
9. $y = 2^{x+(1/2)}$
10. $y = 2^x + \frac{1}{2}$
11. $y = 2^x + 3$
12. $y = 3^x - 2$
13. ▦ $y = 1.87^x$
14. ▦ $y = 2.16^{x+1}$
15. ▦ $y = 2.07^x + 1.84$
16. ▦ $y = 2^{3-x}$
17. ▦ $y = 3^{2-x}$
18. ▦ $y = 4^{(1-x)/2}$

EXAMPLE: Solve the equation $\left(\frac{1}{4}\right)^{1-2x} = 32$.

Solution: Since $\frac{1}{4} = 2^{-2}$ and $32 = 2^5$, we have

$$(2^{-2})^{1-2x} = 2^5$$
$$2^{-2(1-2x)} = 2^5$$

or
$$2^{-2+4x} = 2^5 \quad \text{By the Power Rule}$$

Hence, $\quad -2 + 4x = 5 \quad$ Since $-2(1 - 2x) = -2 + 4x$

We have $\quad x = \dfrac{7}{4} \quad$ By Property 2 of Properties (10.1)

Solve each of the following equations, using $a^x = a^y$ implies $x = y$.

19. $3^x = 81$
20. $5^x = 125$
21. $9^x = 243$
22. $4^x = 128$
23. $3^{-x} = 27$
24. $5^{-x} = 25$
25. $\left(\dfrac{1}{2}\right)^{1-3x} = 16$
26. $\left(\dfrac{1}{3}\right)^{2-x} = 27$
27. $\left(\dfrac{2}{3}\right)^x = \dfrac{9}{4}$
28. $\left(\dfrac{3}{5}\right)^x = \dfrac{125}{27}$
29. $(\sqrt{2})^x = \dfrac{1}{16}$
30. $(\sqrt[3]{3})^x = \dfrac{1}{243}$

31. Rank the following numbers from the largest to the smallest: $2^{7/5}, 2^{2/3}, 2^{\sqrt{2}}$.

32. Rank the following numbers from the smallest to the largest: $7^{5/3}, 7^{7/4}, 7^{\sqrt{3}}$.

33. Suppose a metal block that is cooling under certain conditions has a temperature (in °C) given by $T = 200 \cdot 4^{-0.1t}$ where t is in hours. How long has the cooling been taking place if the block now has a temperature of 100°C?

34. In Example 1 of this section, find t if $B = 128,000$.

10.2 Logarithm Functions

In the last section we noted that for $a > 0$ and $a \neq 1$, the exponential function

$$y = a^x \tag{10.2}$$

is a one-to-one function and, hence, has an inverse function. We let $a = 3$ and try to find an expression for the inverse of $y = 3^x$ by following the procedure introduced in Section 8.7, interchanging x and y in $y = 3^x$ to get $x = 3^y$. We now want to solve the equation $x = 3^y$ for y. For certain values of x, such as $x = 3$ or $x = 9$, we can recognize the corresponding value of y. That is, replacing x first by 3 and then by 9 in $x = 3^y$:

$3 = 3^y$ if $y = 1$, so when $x = 3$, we have $y = 1$

$9 = 3^y$ if $y = 2$, so when $x = 9$, we have $y = 2$

What value of y satisfies $5 = 3^y$? There is a real value of y that solves this equation, but we have not introduced the name for this value yet. (We were in a similar position concerning the solution of $y^3 = 7$ before we introduced the name for the solution, $\sqrt[3]{7}$.)

The notation $\log_3 5$ provides us with a name for the solution of the equation $5 = 3^y$. Thus,

$$y = \log_3 5 \text{ is equivalent to } 5 = 3^y$$

Notice that a "log" is an exponent; $\log_3 5$ is the exponent that is used with the base 3 to obtain the result 5. It is important to recognize that the log notation is a new way to write exponent information. Some further illustrations may be useful. Using $y = \log_3 x$ to mean $x = 3^y$, we have

$4 = \log_3 81$ since $81 = 3^4$ (4 is the exponent)

$0 = \log_3 1$ since $1 = 3^0$ (0 is the exponent)

$-3 = \log_3(\frac{1}{27})$ since $\frac{1}{27} = 3^{-3}$ (-3 is the exponent)

We have now introduced a notation for the inverse function of the exponential function $y = 3^x$, namely

$$y = \log_3 x$$

The previous discussion could be repeated for any real number a instead of 3, provided $a > 0$ and $a \neq 1$. We can define an inverse function for the exponential function of $y = a^x$, which is called a **logarithmic function**, as follows.

Definition of the Logarithmic Function

For $a > 0$ and $a \neq 1$, $y = \log_a x$ means the same as $x = a^y$.   (10.3)

The symbol "log" is the abbreviation for *logarithm*, and $\log_a x$ is read "logarithm of x with base a." The function $y = \log_a x$ is called the **logarithmic function of x with base a.**

EXAMPLE 1   Graph the function $y = \log_3 x$.

Solution: We saw in Section 8.7 that the graphs of a function and its inverse are symmetric with regard to the line $y = x$. That is, given the graph of a function, if we fold the graph paper along the line $y = x$, the graph of the function will coincide with the graph of the inverse function.

10.2 Logarithm Functions

Since $y = \log_3 x$ is the inverse of $y = 3^x$, we first graph the exponential function $y = 3^x$ (see Figure 10.3). Next we sketch a copy of this graph symmetrically about the line $y = x$. This is the graph of $y = \log_3 x$. See Figure 10.3.

Figure 10.3

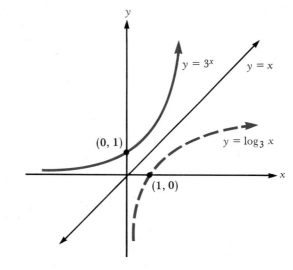

EXAMPLE 2

Rewrite each exponential form in its equivalent logarithmic form:

(a) $2^5 = 32$

(b) $\left(\dfrac{1}{3}\right)^4 = \dfrac{1}{81}$

(c) $\left(\dfrac{1}{4}\right)^{-2} = 16$

(d) $10^3 = 1000$

(e) $10^{-2} = \dfrac{1}{100}$

(f) $a^0 = 1;\ (a > 0,\ a \neq 1)$

Solution: Using Definition (10.3), we have

(a) $\log_2 32 = 5$ (b) $\log_{1/3} \dfrac{1}{81} = 4$ (c) $\log_{1/4} 16 = -2$

(d) $\log_{10} 1000 = 3$ (e) $\log_{10} \dfrac{1}{100} = -2$ (f) $\log_a 1 = 0,\ (a > 0,\ a \neq 1)$

444   Chap. 10   Exponential and Logarithmic Functions

EXAMPLE 3   Rewrite each logarithmic form in its equivalent exponential form.

(a) $\log_{1/5} 25 = -2$
(b) $\log_3 x = 27$
(c) $\log_a a = 1$
(d) $\log_a b = c$

Solution: Again by Definition (10.3), we have

(a) $\left(\dfrac{1}{5}\right)^{-2} = 25$ (b) $3^{27} = x$

(c) $a^1 = a$ (d) $a^c = b$

For any positive a, $a^1 = a$ and $a^0 = 1$. Expressing these two facts in logarithmic notation gives:

$$\log_a a = 1 \quad \text{and} \quad \log_a 1 = 0 \tag{10.4}$$

EXAMPLE 4   Solve the equations:

(a) $\log_5 x = -3$

(b) $\log_3 \dfrac{1}{27} = y$

(c) $\log_x 1000 = 3$

Solution: Rewriting each equation in exponential form, we have

(a) $x = 5^{-3} = \dfrac{1}{5^3} = \dfrac{1}{125}$

(b) $3^y = \dfrac{1}{27}$. Since $\dfrac{1}{27} = \dfrac{1}{3^3} = 3^{-3}$, we have $3^y = 3^{-3}$, which implies $y = -3$.

(c) $x^3 = 1000$. Since $1000 = 10^3$, we have $x^3 = 10^3$ and hence, $x = 10$.

EXAMPLE 5   Evaluate each of the following.

(a) $\log_2 64$
(b) $\log_{10} 10^5$
(c) $\log_{25} \sqrt{5^3}$

Solution:
(a) Let $y = \log_2 64$. Then $2^y = 64 = 2^6$. Hence, $y = 6$.
(b) Let $y = \log_{10} 10^5$. Then $10^y = 10^5$. Hence, $y = 5$.
(c) Let $y = \log_{25} \sqrt{5^3}$. Then

$25^y = \sqrt{5^3} = (5^3)^{1/2} = 5^{3/2}$, or $(5^2)^y = 5^{3/2}$ (since $25 = 5^2$)

That is, $5^{2y} = 5^{3/2}$, which implies that $2y = \dfrac{3}{2}$, or $y = \dfrac{3}{4}$.

A striking and useful formula is found by replacing the second y in the definition

$$y = \log_a x \quad \text{means} \quad x = a^y$$

with $\log_a x$ to obtain the following (for $x > 0$).

$$x = a^{\log_a x} \tag{10.5}$$

Exercises 10.2

Standard Assignment: Exercises 1, 3, 5, 11, 13, 15, 17, 19, 21, 23, 27, 29, 31, 33, 35, 37, 43, 47, 49, 53

A Write each expression in logarithmic form. See Example 2.

1. $5^2 = 25$
2. $81^{1/2} = 9$
3. $49^{-1/2} = \dfrac{1}{7}$
4. $\left(\dfrac{1}{16}\right)^{1/2} = \dfrac{1}{4}$
5. $\left(\dfrac{1}{16}\right)^{-1/2} = 4$
6. $(a^2)^2 = a^4$
7. $10^0 = 1$
8. $10^4 = 10000$
9. $10^{-1} = 0.1$
10. $10^{-3} = 0.001$
11. $a^{-n} = \dfrac{1}{a^n}$
12. $a^{1/n} = \sqrt[n]{a}$

Write each logarithmic expression in exponential form. See Example 3.

13. $\log_2 32 = 5$
14. $\log_7 49 = 2$
15. $\log_{10} 100 = 2$
16. $\log_{10} 10 = 1$
17. $\log_{10} 1 = 0$
18. $\log_{10} 0.1 = -1$
19. $\log_{10} 0.01 = -2$
20. $\log_{1/5} 5 = -1$
21. $\log_{1/5} 25 = -2$
22. $\log_{25} 5 = \dfrac{1}{2}$
23. $\log_8 2 = \dfrac{1}{3}$
24. $\log_a \sqrt[n]{a} = \dfrac{1}{n}$

Graph each of the following logarithmic functions. See Example 1.

25. $y = \log_{1/2} x$
26. $y = \log_2 x$
27. $y = \log_4 x$
28. $y = \log_{1/3} x$

Solve each equation for the given unknown (or variable). See Example 4.

29. $\log_{10} x = 2$
30. $\log_{10} x = -3$
31. $\log_5(x - 2) = 3$
32. $\log_6(y + 1) = -2$
33. $\log_2 \frac{1}{16} = y$
34. $\log_3 \frac{1}{81} = y$
35. $\log_x \frac{1}{5} = -1$
36. $\log_y 625 = 4$
37. ▦ $\log_{16}\sqrt{x - 1} = 0.495$
38. ▦ $\log_{27}\sqrt[3]{1 - x} = 0.342$
39. $\log_a(x - 1) = 0$
40. $\log_a(x^2 - 5x + 7) = 0$
41. $\log_2(x^2 - 7x + 14) = 1$
42. $\log_{10}(x^2 + 5x + 16) = 1$

Evaluate each of the following. See Example 5.

43. $\log_5 125$
44. $\log_9 81$
45. $\log_{10} 10000$
46. $\log_3 \frac{1}{3}$
47. $\log_2 \frac{1}{8}$
48. $\log_4 \frac{1}{64}$
49. $\log_{49} 7$
50. $\log_{27} 3$
51. $\log_{16} 2$
52. $\log_3 \sqrt{27}$
53. $\log_5 \sqrt{125}$
54. $\log_2 \sqrt{32}$

B 55. $\log_5 25^{3/4}$
56. $\log_{10} 1000^{1/2}$
57. $\log_7 49^{-1/4}$
58. $\log_3(\log_5 5)$
59. $\log_4(\log_3 81)$
60. $\log_4[\log_3(\log_2 8)]$
61. $5^{\log_5 7}$
62. $10^{\log_{10} 13}$

10.3 Properties of Logarithms

There are three important properties of logarithms that are very helpful to know when performing numerical calculations.

Properties of Logarithms

For any positive a, $a \neq 1$, we have

1. $\log_a MN = \log_a M + \log_a N$

2. $\log_a \frac{M}{N} = \log_a M - \log_a N$ (10.6)

3. $\log_a M^p = p \log_a M$

EXAMPLE 1

Use Property 1 in Properties (10.6) to evaluate each of the following, given $\log_{10} x = 2.035$ and $\log_{10} 5 = 0.699$.

(a) $\log_{10} 50$
(b) $\log_{10} 25$
(c) $\log_{10} 5x$
(d) $\log_{10} 5x^2$
(e) $\log_{10} 2$

Solution:

(a) $\log_{10} 50 = \log_{10} 10 \cdot 5$ 50 = 10 · 5
$= \log_{10} 10 + \log_{10} 5$ By Property 1
$= 1 + 0.699$ By Properties (10.4), given
$= 1.699$

(b) $\log_{10} 25 = \log_{10} 5 \cdot 5$ 25 = 5 · 5
$= \log_{10} 5 + \log_{10} 5$ By Property 1
$= 0.699 + 0.699$ Given
$= 1.398$

(c) $\log_{10} 5x = \log_{10} 5 + \log_{10} x$ By Property 1
$= 0.699 + 2.035$ Given
$= 2.734$

(d) $\log_{10} 5x^2 = \log_{10} 5 + \log_{10} x^2$ By Property 1
$= 0.699 + \log_{10} x \cdot x$ Given, $x^2 = x \cdot x$
$= 0.699 + \log_{10} x + \log_{10} x$ By Property 1
$= 0.699 + 2.035 + 2.035$ Given
$= 4.769$

(e) Since

$$\log_{10} 10 = \log_{10} 2 \cdot 5 \quad\quad 10 = 2 \cdot 5$$
$$\log_{10} 10 = \log_{10} 2 + \log_{10} 5 \quad\quad \text{By Property 1}$$

By subtracting $\log_{10} 5$ from both sides, we have

$$\log_{10} 2 = \log_{10} 10 - \log_{10} 5$$
$$= 1 - 0.699 \quad\quad \text{By Properties (10.4), given}$$
$$= 0.301$$

You may have foreseen that if Property 2 is used,

$$\log_{10} 2 = \log_{10} \frac{10}{5}$$

$$= \log_{10} 10 - \log_{10} 5$$

which leads directly to the same answer.

EXAMPLE 2   Use Properties 1 and 2 to evaluate each of the following, given $\log_{10} 2 = 0.301$ and $\log_{10} 3 = 0.477$.

(a) $\log_{10} 1.5$

(b) $\log_{10} \frac{1}{20}$

(c) $\log_{10} \frac{3}{40}$

Solution:

(a) $\log_{10} 1.5 = \log_{10} \frac{3}{2}$ $1.5 = \frac{3}{2}$

$\phantom{\log_{10} 1.5} = \log_{10} 3 - \log_{10} 2$ By Property 2

$\phantom{\log_{10} 1.5} = 0.477 - 0.301$ Given

$\phantom{\log_{10} 1.5} = 0.176$

(b) $\log_{10} \frac{1}{20} = \log_{10} 1 - \log_{10} 20$ By Property 2

$\phantom{\log_{10} \frac{1}{20}} = 0 - \log_{10} 2 \cdot 10$ By Properties (10.4), $20 = 2 \cdot 10$

$\phantom{\log_{10} \frac{1}{20}} = -(\log_{10} 2 + \log_{10} 10)$ By Property 1

$\phantom{\log_{10} \frac{1}{20}} = -(0.301 + 1)$ Given

$\phantom{\log_{10} \frac{1}{20}} = -1.301$

(c) $\log_{10} \frac{3}{40} = \log_{10} 3 - \log_{10} 40$ By Property 2

$\phantom{\log_{10} \frac{3}{40}} = 0.477 - (\log_{10} 4 + \log_{10} 10)$ Given, Property 1 ($40 = 4 \cdot 10$)

$\phantom{\log_{10} \frac{3}{40}} = 0.477 - (\log_{10} 2 + \log_{10} 2 + 1)$ $4 = 2 \cdot 2$, Property 1, Properties (10.4)

$\phantom{\log_{10} \frac{3}{40}} = 0.477 - (0.301 + 0.301 + 1)$ Given

$\phantom{\log_{10} \frac{3}{40}} = -1.125$

EXAMPLE 3 Given that $\log_5 z = 210$ and $\log_5 y = 2$, evaluate each of the following.
(a) $\log_5 125z^7$
(b) $\log_5 \sqrt{\dfrac{z}{y}}$
(c) $\log_5 z^{1/30} y^5$

Solution:
(a) $\log_5 125z^7 = \log_5 5^3 \cdot z^7$

$\qquad\qquad\quad = \log_5 5^3 + \log_5 z^7 \qquad$ Property 1

$\qquad\qquad\quad = 3 \log_5 5 + 7 \log_5 z \qquad$ Property 3

$\qquad\qquad\quad = 3 \cdot 1 + 7 \cdot 210 \qquad$ By Properties (10.4), given

$\qquad\qquad\quad = 1473$

(b) $\log_5 \sqrt{\dfrac{z}{y}} = \log_5 \left(\dfrac{z}{y}\right)^{1/2}$

$\qquad\quad\ \ = \dfrac{1}{2} \log_5 \dfrac{z}{y} \qquad\qquad\qquad$ Property 3

$\qquad\quad\ \ = \dfrac{1}{2} (\log_5 z - \log_5 y) \qquad$ Property 2

$\qquad\quad\ \ = \dfrac{1}{2}(210 - 2) \qquad\qquad\quad$ Given

$\qquad\quad\ \ = 104$

(c) $\log_5 z^{1/30} y^5 = \log_5 z^{1/30} + \log_5 y^5 \qquad$ Property 1

$\qquad\qquad\quad\ = \dfrac{1}{30} \log_5 z + 5 \log_5 y \qquad$ Property 3

$\qquad\qquad\quad\ = \dfrac{1}{30} \cdot 210 + 5 \cdot 2 \qquad\quad$ Given

$\qquad\qquad\quad\ = 17$

EXAMPLE 4 Solve each of the following logarithmic equations.
(a) $\log_a(2x - 3) - \log_a(x + 5) = 0$, with $a > 0$, $a \ne 1$
(b) $\log_{10}(x + 8) + \log_{10}(x - 1) = 1$

Solution:
(a) From $\log_a(2x - 3) - \log_a(x + 5) = 0$, we have

$$\log_a \dfrac{2x - 3}{x + 5} = 0 \qquad \text{By Property 2}$$

Hence,

$$\frac{2x-3}{x+5} = a^0 \qquad \text{In exponential notation}$$

or

$$\frac{2x-3}{x+5} = 1 \qquad a^0 = 1$$

$$2x - 3 = x + 5$$
$$x = 5 + 3$$
$$x = 8$$

Check: For $x = 8$, $\quad 2x - 3 = 2(8) - 3 \quad$ and $\quad x + 5 = 8 + 5$
$$= 16 - 3 \qquad\qquad\qquad\quad = 13$$
$$= 13$$

Thus $\log_a(2x - 3) - \log_a(x + 5) = \log_a 13 - \log_a 13 = 0$ as desired. The solution set is $\{8\}$.

(b) By Property 1, we have

$$\log_{10}(x + 8)(x - 1) = 1$$

Rewriting this equation in exponential form gives

$$(x + 8)(x - 1) = 10^1$$

or
$$x^2 + 7x - 8 = 10$$
$$x^2 + 7x - 18 = 0$$
$$(x + 9)(x - 2) = 0$$
$$x = -9 \quad \text{or} \quad x = 2$$

Check: The domain of the log functions is $\{x \mid x > 0\}$, and so the terms $\log_{10}(x + 8)$ and $\log_{10}(x - 1)$ are meaningful only if $x + 8 > 0$ and $x - 1 > 0$. However, $x = -9$ does not satisfy either of these conditions (since $-9 + 8 < 0$ and $-9 - 1 < 0$) while $x = 2$ does satisfy both conditions ($2 + 8 > 0$ and $2 - 1 > 0$).

Hence, $x = -9$ is extraneous, and $x = 2$ is the only solution to the given equation. The solution set is $\{2\}$.

Example 2(b) shows that when solving logarithmic equations, it is essential to check solutions to rule out any extraneous solutions.

WARNING

1. $\dfrac{\log(x - 2)}{\log 7x}$ cannot be simplified, but $\log\left(\dfrac{x - 2}{7x}\right) = \log(x - 2) - \log 7x$
2. $[\log(x - 2)][\log 7x]$ cannot be simplified, but $\log[(x - 2) \cdot 7x] = \log(x - 2) + \log 7x$.

Exercises 10.3

Standard Assignment: Exercises 1, 5, 7, 9, 11, 13, 17, 19, 23, 27, 29, 31, 37, 39, 41, 43

A Given $\log_{10} x = 2.035$, $\log_5 y = 2$, $\log_5 z = 210$, $\log_{10} 2 = 0.301$, and $\log_{10} 3 = 0.477$, evaluate each of the following. See Examples 1–3.

1. $\log_{10} 6$
2. $\log_{10} 5$
3. $\log_{10} 4$
4. $\log_{10} 9$
5. $\log_{10} 3x$
6. $\log_{10} 2x$
7. $\log_5 yz$
8. $\log_5 y^2$
9. $\log_5 z^2$
10. $\log_{10} x^2$
11. $\log_{10} 15x^2$
12. $\log_{10} 25x^2$
13. $\log_5 125 y^2 z$
14. $\log_5 25 y^2 z^2$
15. $\log_{10} 1000 x^3$
16. $\log_{10} 2 \cdot 5$
17. $\log_{10} \dfrac{2}{3}$
18. $\log_{10} \dfrac{5}{3}$
19. $\log_{10} \dfrac{x}{2}$
20. $\log_{10} \dfrac{2}{x}$
21. $\log_{10} \dfrac{6}{x^2}$
22. $\log_5 \dfrac{y^2}{z}$
23. $\log_5 \dfrac{z^2}{y}$
24. $\log_5 \dfrac{25}{z^2 y}$
25. $\log_{10} \sqrt{2}$
26. $\log_{10} \sqrt{3}$
27. $\log_{10} \sqrt{\dfrac{3}{2}}$
28. $\log_{10} \sqrt{\dfrac{x}{2}}$
29. $\log_{10} \sqrt{\dfrac{15}{2x}}$
30. $\log_{10} \sqrt[3]{\dfrac{5x^2}{16}}$
31. $\log_5 y^{11} z^{1/70}$
32. $\log_5 y^{-1/2} z^{-1/35}$
33. $\log_5 25 y^{1/3} z^{1/5}$
34. 🖩 $\log_{10} 25 x^{3.141}$
35. 🖩 $\log_5 y^{\sqrt{2}} z^{\sqrt{3}}$
36. 🖩 $\log_{10} 125 x^{\pi}$

Solve each of the following logarithmic equations. See Example 4.

37. $\log_{10} x + \log_{10}(x + 9) = 1$
38. $\log_5(3x - 1) - \log_5(2x + 7) = 0$
39. $\log_{10}(5x - 2) - \log_{10}(3x + 4) = 0$
40. $\log_{10}(x - 1) + \log_{10}(x + 2) = 1$
41. $\log_6(x + 2) + \log_6(x - 3) = 1$
42. $\log_2(3x - 5) - \log_2(5x + 1) = 3$
43. $\log_3(2x - 7) - \log_3(4x - 1) = 2$
44. $\log_4 \sqrt{x + 3} - \log_4 \sqrt{2x - 1} = \dfrac{1}{4}$

45. $\log_{10}\sqrt{y^2 - 1} - \log_{10}\sqrt{y^2 + 5} = \sqrt{3}$

46. $\log_3\sqrt{x^2 + 1} - \log_3\sqrt{x^2 - 1} = \sqrt{2}$

B 47. A manufacturer's wholesale price, p dollars per unit, is determined by $p = \log_{10}(200 - 0.02q)$, with $q \leq 9500$, where q is the number of units supplied. At what price per unit will the manufacturer supply 5000 units?

48. In Exercise 47, at what price per unit will the manufacturer supply 9500 units?

The Richter scale for measuring the magnitude of an earthquake of intensity I is given by the equation $R = \log_{10}(I/I_0)$, where I_0 is a certain minimum intensity.

49. What is the magnitude of an earthquake with an intensity of 10 times I_0?

50. The San Francisco earthquake of 1906 measured approximately 8 on the Richter scale. What was the intensity of the San Francisco earthquake of 1906?

10.4 Computations with Logarithms

Although the results of the previous section are valid for any positive base $a \neq 1$, the most commonly used base for numerical calculation is the base 10. The reason for the popularity of base 10 is the fact that the number system we commonly use is the decimal system with base 10. Consequently, the base 10 logarithms are called **common logarithms**. For the sake of simplicity, we will write $\log x$ for $\log_{10} x$. Thus,

$$\log 1000 = \log_{10} 10^3 = 3$$
$$\log 100 = \log_{10} 10^2 = 2$$
$$\log 10 = \log_{10} 10^1 = 1$$
$$\log 1 = \log_{10} 10^0 = 0$$
$$\log 0.1 = \log_{10} 10^{-1} = -1 \qquad 0.1 = 10^{-1}$$
$$\log 0.01 = \log_{10} 10^{-2} = -2 \qquad 0.01 = 10^{-2}$$

In order to find common logarithms of numbers other than powers of 10, you will need a scientific calculator or a table of common logarithms. (See Appendix A, Table 2.)

In this section we will use the logarithm table in Appendix A. To find the common logarithm of a number from the tables, first express the number in scientific notation. (Review Section 3.3 now if necessary.) For example, to find $\log 403$, express 403 in scientific notation, 4.03×10^2, and then use

Property 1 to get

$$\log 403 = \log(4.03 \times 10^2)$$
$$= \log 4.03 + \log 10^2 \qquad \text{Property 1}$$
$$= \log 4.03 + 2 \qquad \log 10^2 = 2$$

In order to clarify this explanation, a portion of the common logarithm table has been reproduced here. To find the value of log 4.03, we look at the intersection of the row containing 4.0 under N and the column containing 3.

x	0	1	2	3	4	5	6	7	8	9
1.0	.0000	.0043	.0086	.0128	.0170	.0212	.0253	.0294	.0334	.0374
1.1	.0414	.0453	.0492	.0531	.0569	.0607	.0645	.0682	.0719	.0755
1.2	.0792	.0828	.0864	.0899	.0934	.0969	.1004	.1038	.1072	.1106
1.3	.1139	.1173	.1206	.1239	.1271	.1303	.1335	.1367	.1399	.1430
1.4		.1492		.1553	.1584			.1673	.1703	.1732
...										
3.9	.5911		.5933		.5966	.5977		.5999		
4.0	.6021	.6031	.6042	.6053	.6064	.6075	.6085	.6096	.6107	.6117
4.1	.6128	.6138	.6149	.6160	.6170	.6180	.6191	.6201	.6212	.6222
4.2	.6232	.6243	.6253	.6263	.6274	.6284	.6294	.6304	.6314	.6325
4.3	.6335	.6345	.6355	.6365	.6375	.6385	.6395	.6405	.6415	.6425
4.4	.6435	.6444		.6464	.6474	.6484	.6493	.6503	.6513	.6522

Thus,

$$\log 4.03 = 0.6053$$

and hence

$$\log 403 = 0.6053 + 2 = 2.6053.$$

The decimal part of the logarithm, 0.6053 in the previous example, is called the **mantissa** and the integer part, 2, is the **characteristic**. In general, the characteristic of log x is n if $x = m \times 10^n$, with $1 < m < 10$, in scientific notation. The integer n may be positive, zero, or negative.

EXAMPLE 1 Find each logarithm by using the table.

(a) $\log 4.15 = \log(4.15 \times 10^0)$ $\qquad 4.15 = 4.15 \times 10^0$
$\qquad\qquad\ \ = \log 4.15 + \log 10^0$
$\qquad\qquad\ \ = 0.6180 + 0 \qquad$ Mantissa = 0.6180
$\qquad\qquad\qquad\qquad\qquad\qquad\quad$ Characteristic = 0
$\qquad\qquad\ \ = 0.6180$

(b) $\log 415 = \log(4.15 \times 10^2) \qquad 415 = 4.15 \times 10^2$
$\qquad\qquad\ = \log 4.15 + \log 10^2$
$\qquad\qquad\ = 0.6180 + 2 \qquad$ Mantissa = 0.6180
$\qquad\qquad\qquad\qquad\qquad\qquad$ Characteristic = 2
$\qquad\qquad\ = 2.6180$

(c) $\log 0.000415 = \log(4.15 \times 10^{-4})$ $0.000415 = 4.15 \times 10^{-4}$
$= \log 4.15 + \log 10^{-4}$
$= 0.6180 + (-4)$ Mantissa $= 0.6180$
                       Characteristic $= -4$
$= -3.382$

Notice that the logarithms of 4.15, 415, and 0.000415 all have the same mantissa, 0.6180.

It is possible to reverse the process of finding a logarithm. Given a number y, we can find the number x such that $\log x = y$. The number x is called the **antilogarithm** (or simply **antilog**) of y. Notationally,

antilog $y = x$ means $y = \log x$ (10.7)

For example, from Example 1 we get

antilog$(0.6180) = 4.15$, since $\log 4.15 = 0.6180$
antilog$(2.6180) = 415$, since $\log 415 = 2.6180$
antilog$(-3.382) = 0.000415$, since $\log 0.000415 = -3.382$

To find antilogarithms from a table, just reverse the process of finding a logarithm. For example, to find antilog 2.6191, we look in the table for a mantissa of 0.6191. The number whose mantissa is 0.6191 is 4.16. Finally, since the characteristic is 2,

$$\text{antilog } 2.6191 = \underset{\underset{\text{value}}{\uparrow}}{4.16} \times 10^{\underset{\underset{\text{characteristic}}{\uparrow}}{2}} = 416$$

(table value; characteristic)

A log table contains only positive numbers that are approximations of the log values for numbers between 1 and 10. In order to find the antilog of a negative number such as -2.3716, we *must* rewrite the number in order to recognize a positive mantissa. This is done for -2.3176 by adding and subtracting 3 as follows.

$-2.3716 = (3 - 2.3716) - 3$
$= \underset{\underset{\text{mantissa}}{\uparrow}}{0.6284} + (\underset{\underset{\text{characteristic}}{\uparrow}}{-3})$

Next we locate the mantissa 0.6284 in the table and find the value 4.25 associated with this mantissa. Since the characteristic is -3, we have

antilog$(-2.3716) = 4.25 \times 10^{-3}$
$= 0.00425$

If we had wanted the antilog of -4.3716, we would have added and subtracted 5 and proceeded as shown previously. In general, for negative values of x, add and subtract the smallest integer that is larger than $|x|$ before using the table to find antilog x.

Because any table must contain a finite number of mantissas, we may not find the exact mantissa in the table. If the mantissa m lies somewhere between two consecutive mantissas u and v in the table, there is a method called *linear interpolation* that uses the proportion $(m - u)/(v - m)$ to arrive at an approximate antilog m between antilog u and antilog v.

A detailed discussion of this method is included in Appendix B. For the examples and exercises in this chapter, we shall simply use the number in the table that is closest to our need.

The availability of inexpensive scientific calculators has greatly reduced the need to perform computations using logarithmic tables. Nevertheless, we have included the techniques of performing computations using tables because they illustrate the properties of the logarithmic function.

EXAMPLE 2

Compute $\sqrt[10]{0.976}$

Solution: Let

$$x = \sqrt[10]{0.976} = (0.976)^{1/10}$$

Then

$$\log x = \log (0.976)^{1/10}$$

$$= \frac{1}{10} \log 0.976 \qquad \text{Property 3}$$

$$= \frac{1}{10} \log (9.76 \times 10^{-1})$$

$$= \frac{1}{10} (\log 9.76 + \log 10^{-1}) \qquad \text{Property 1}$$

$$= \frac{1}{10} (0.9894 - 1) \qquad \text{From Table 2}$$

$$= \frac{1}{10} (9.9894 - 10) \qquad \text{Add 9, subtract 9.}$$

$$= 0.9989 - 1$$

Hence,

$$x = \text{antilog}\,(0.9989 - 1) \approx 9.97 \times 10^{-1} \qquad \text{From the Log Table, antilog } 0.9987 \approx 9.97$$

$$= 0.997$$

EXAMPLE 3 Compute $\dfrac{(7.82)^{1/5}(8.64)^{1/2}}{(6.21)^{1/4}}$

Solution: Let
$$x = \dfrac{(7.82)^{1/5}(8.64)^{1/2}}{(6.21)^{1/4}}$$

Then

$$\log x = \log \dfrac{(7.82)^{1/5}(8.64)^{1/2}}{(6.21)^{1/4}}$$

$= \log (7.82)^{1/5} + \log (8.64)^{1/2} - \log (6.21)^{1/4}$ Properties 1 and 2

$= \dfrac{1}{5} \log 7.82 + \dfrac{1}{2} \log 8.64 - \dfrac{1}{4} \log 6.21$ Property 3

From Table 2.

$\dfrac{1}{5} \log 7.82 = \dfrac{1}{5} (0.8932) = 0.1786$

$\dfrac{1}{2} \log 8.64 = \dfrac{1}{2} (0.9365) = 0.4643$ + Add

$\phantom{\dfrac{1}{2} \log 8.64 = \dfrac{1}{2} (0.9365) =\ } 0.6469$ − Subtract

$\dfrac{1}{4} \log 6.21 = \dfrac{1}{4} (0.7931) = 0.1983$

$\phantom{\dfrac{1}{4} \log 6.21 = \dfrac{1}{4} (0.7931) =\ } \log x = 0.4486$

Hence,

$$x = \text{antilog } 0.4486 = 2.81 \quad \text{From the Log Table}$$

Exercises 10.4

Standard Assignment: Exercises 1, 3, 5, 9, 11, 13, 17, 19, 23, 25, 27, 31, 33, 35, 37, 41, 45, 47

A   Identify the characteristic and mantissa for each of the following.

1. $\log 834$
2. $\log 1983$
3. $\log 9$
4. $\log 0.213$
5. $\log 0.599$
6. $\log 0.0064$
7. $\log 0.00008$
8. $\log 268.35$
9. $\log(1.41 \times 10^{-5})$
10. $\log(23.4 \times 10^{-1})$
11. $\log(89.3 \times 10^5)$
12. $\log(145 \times 10^3)$

Find each logarithm or antilogarithm by using the table of common logarithms.

13. log 3.8
14. log 3.85
15. log 3.97
16. log 399
17. log 4510
18. log 42800
19. log 679000
20. log 92.4
21. antilog 0.8299
22. antilog 0.4639
23. antilog 1.5465
24. antilog 2.6015
25. antilog (-0.1703)
26. antilog (-1.3214)

Compute each of the following by using the properties of logarithms and then the Log Table (read values of the nearest entry). See Examples 2 and 3.

27. $(3.24)(2.64)$
28. $(23.95)(4.137)$
29. $(0.345)(0.0256)$
30. $(0.0234)(1.879)$
31. $\dfrac{0.8794}{1.023}$
32. $\dfrac{1.307}{2.045}$
33. $\sqrt{1.414}$
34. $\sqrt{1.732}$
35. $\sqrt[3]{0.6793}$
36. $\sqrt[5]{5.253}$
37. $\sqrt{2.033}\sqrt[3]{5.738}$
38. $\sqrt[5]{2270}\sqrt{3.141}$
39. $\dfrac{13.07}{\sqrt{4.142}}$
40. $\dfrac{\sqrt{3.154}}{\sqrt[3]{1.345}}$
41. $\dfrac{\sqrt[10]{2305}}{1.438}$
42. $\dfrac{\sqrt[3]{27.6}\sqrt{582}}{8.12}$
43. $\sqrt{9.8(15.9-3.3)1.09}$
44. $\dfrac{(3.67)^{1/3}(2.73)^{1/4}}{(3.25)^{1/2}}$
45. $\dfrac{(6.73)^{1/2}(7.32)^{1/3}}{(5.23)^{1/4}}$
46. $\dfrac{(9.803)^{1/10}}{(2.04)^{1/5}(4.03)^{1/2}}$

Answer the following without using the table of logarithms.

47. What is antilog 0? Antilog 1?
48. What is antilog $(\log\sqrt{2})$?
49. Let x be any positive number. What is antilog $(\log x)$?
50. Let y be any real number. What is log (antilog y)?

10.5 Logarithmic Computation Using Calculators

Scientific calculators that contain $\boxed{\log}$, $\boxed{\ln x}$, $\boxed{y^x}$, and $\boxed{\text{INV}}$ keys are very useful in exponential and logarithmic computations. They provide values of exponential and logarithmic values quickly. For example, to find the value of log 439.68 from a calculator, enter the number 439.68 and then press the $\boxed{\log}$ key. Your calculator screen should show something like 2.6431367 (depending on the number of digits your calculator displays and its built-in accuracy).

You should write
$$\log 439.68 \approx 2.6431367$$
or, rounded to four decimal places,
$$\log 439.68 \approx 2.6431$$
However, we will follow customary usage and write
$$\log 439.68 = 2.6431$$
instead of using the $\approx$ sign. You should keep in mind that in logarithms, an equal sign, $=$, usually means approximately equal, $\approx$.

The calculator instructions shown in this chapter apply to calculators that use algebraic logic. If your calculator uses a different logic, read your calculator manual to make the necessary adjustments.

EXAMPLE 1

Find the values of

(a) log 68.7817
(b) log $\sqrt{2}$
(c) log π (use $\pi = 3.1415927$ if there is no π key)
(d) log 0.0014

Solution: You may round each answer to four decimal places.

(a) Enter 68.7817 → Press $\boxed{\log}$ → Display 1.8374729.
By rounding to four decimal places, we have log 68.7817 = 1.8375.

(b) Enter 2 → Press $\boxed{\sqrt{x}}$ → Press $\boxed{\log}$ → Display 0.150515
Hence, log $\sqrt{2}$ = 0.1505.

(c) Enter π → Press $\boxed{\log}$ → Display 0.49714987
Hence, log π = 0.4971.

(d) Enter 0.0014 → Press $\boxed{\log}$ → Display -2.853872
Hence, log 0.0014 = -2.8539.

For the sake of convenience, we have rounded all answers to four decimal places.

It is very simple to find the antilog of any number y from a scientific calculator. To find antilog 1.09,

Enter 1.09 → Press $\boxed{INV}$ → Press $\boxed{\log}$ → Display 12.302688

10.5 Logarithmic Computation Using Calculators

EXAMPLE 2 Find

(a) antilog 4.031
(b) antilog 1.162
(c) antilog (-1.093)

Solution:

(a) Enter 4.031 → Press (INV) → Press (log) → Display 10739.894
Thus, antilog 4.031 = 10739.894

(b) Enter 1.162 → Press (INV) → Press (log) → Display 14.521116
Thus, antilog 1.162 = 14.521116

(c) No special care is required when finding the antilog of a negative number by using a calculator. Thus,

Enter 1.093 → Press (+/−) → Press (INV) → Press (log)
→ Display 0.0807235

Thus, antilog (-1.093) = 0.0807235

Since $y = \log x$ means $x = 10^y$, Equation (10.7) may be restated simply as

$$\text{antilog } y = 10^y \qquad (10.8)$$

Thus, to find antilog (-1.053), or $10^{-1.053}$,

Enter 10 → Press (y^x) → Enter 1.053 → Press (+/−)
→ Press (=) → Display 0.0885116

or   Enter 10 → Press (y^x) → Enter 1.053 → Press (=)
→ Press (1/x) → Display 0.0885116

Exercises 10.5

Standard Assignment: All odd-numbered exercises

Find each logarithm by using a calculator. See Example 1.

1. log 5.124
2. log 7.275
3. log 2436
4. log 314.7
5. log 0.6218
6. log 0.08237
7. log $\sqrt[4]{2}$
8. log π^2
9. log $\sqrt{\pi}$
10. log $(\sqrt{2} + 31)$
11. log$(\pi + \sqrt{3})$
12. log$(\pi^2 - \pi - \sqrt{2})$

Find each antilogarithm. See Example 2.

13. antilog 3.6453
14. antilog 0.5247

15. ▦ antilog 0.3624
16. ▦ antilog 2.8425
17. ▦ antilog (−2.5328)
18. ▦ antilog (−3.5329)
19. ▦ antilog (−1.3415)
20. ▦ antilog (−4.0132)
21. ▦ antilog (log 2.4309)
22. ▦ antilog (log 0.52617)

Find each x.

> EXAMPLE: $\log x = -0.59$ is equivalent to
> $$x = 10^{-0.59}$$
> $= 10 \boxed{y^x}\ 0.59 \boxed{+/-}\ \boxed{=}$ Enter and press keys in the given order
> $= 0.25703958$

23. ▦ $\log x = 2.13059$
24. ▦ $\log x = 0.17327$
25. ▦ $\log x = -9.53021$
26. ▦ $\log x = -7.3217$
27. ▦ $\log x = \sqrt{4.2058}$
28. ▦ $\log x = \sqrt{\pi}$
29. ▦ $\log x = -\sqrt{2.4507}$
30. ▦ $\log x = -\sqrt{3.14159}$

10.6 Natural Logarithms

In Section 10.4 we studied common logarithms, which are logarithms with base 10. There is a second number that regularly appears as a base for exponential and logarithmic functions. This number is an irrational number denoted by the letter e in honor of the Swiss mathematician Leonhard Euler (1707–1783). For example,

$$e \approx 2.71828$$

When **the number e** is used as the base in logarithms, the logarithms are called **natural logarithms** and are denoted by ln x. That is,

$$\ln x = \log_e x$$

Although e may seem to be a strange number to use as a base in a logarithmic function, natural logarithms arise in many applications. Some of these applications will be discussed in the next section.

Let us now see how the natural logarithmic function $\ln x$ ($= \log_e x$) and the common logarithmic function $\log x$ are related. For all $x > 0$, by Equation (10.5) we have

$$x = e^{\log_e x} = e^{\ln x} \qquad (10.9)$$

so that
$$\log x = \log e^{\ln x}$$

which, according to Property 3 in Properties (10.6), can be written as

$$\log x = (\ln x)(\log e)$$

or
$$\ln x = \frac{\log x}{\log e} = \frac{1}{\log e} \log x \qquad (10.10)$$

Since $e \approx 2.718$, we can use a calculator or a table to find $\log e = 0.4343$. Hence, Equation (10.10) can be expressed as

$$\ln x = \frac{1}{0.4343} \log x$$

or since
$$\frac{1}{0.4343} \approx 2.3026$$

we have

$$\ln x = 2.3026 \log x \qquad (10.11)$$

The common logarithm $\log x$ can be converted into the natural logarithm $\ln x$ by multiplying a constant factor of 2.3026.

Tables of natural logarithms require a good deal more space for presentation than common logarithm tables do. This is because only the common logarithms have a convenient characteristic and mantissa form arising from the use of scientific notation.

EXAMPLE 1 Convert the following common logarithms into natural logarithms.

(a) $\log 1000 = 3$
(b) $\log 0.001 = -3$
(c) $\log \pi = 0.4971$
(d) $\log \sqrt{2} = 0.1505$

Solution:
(a) $\ln 1000 = (2.3026)3 = 6.9078$ Use Formula (10.11)
(b) $\ln 0.001 = (2.3026)(-3) = -6.9078$
(c) $\ln \pi = (2.3026)(0.4971) = 1.1446$
(d) $\ln \sqrt{2} = (2.3026)(0.1505) = 0.3465$

We can solve exponential equations involving the base e by using natural logarithms.

EXAMPLE 2 Solve the equation $5 = e^{x-1}$ for x.

Solution: By Definition (10.3), $5 = e^{x-1}$ means
$$x - 1 = \log_e 5$$
that is
$$x = \ln 5 + 1$$
$$= (2.3026) \log 5 + 1$$
$$= (2.3026)(0.6990) + 1 \quad (\log 5 = 0.6990)$$
$$= 2.6094$$

Of course, these results can be found directly if a table of natural logarithm values is available. Also, many calculators have a "ln" key as well as a "log" key. If, for example, a calculator with a "ln" key is available and you want to find ln 606,

Enter 606 → Press $\boxed{\ln x}$ → Display 6.40688

To find $\ln \dfrac{3\sqrt{2}}{\pi}$

Enter 2 → Press $\boxed{\sqrt{x}}$ → Press $\boxed{\times}$ → Enter 3 → Press $\boxed{\div}$
→ Press $\boxed{\pi}$ → Press $\boxed{=}$ → Press $\boxed{\ln}$ → Display 0.30045599

EXAMPLE 3 Use a calculator to find:

(a) $\ln (3/\sqrt{\pi})$
(b) $\ln 10^{\sqrt{2}}$
(c) e

Solution:
(a) Enter π → Press $\boxed{\sqrt{x}}$ → Press $\boxed{\div}$ → Press $\boxed{3}$ → Press $\boxed{=}$ → Press $\boxed{1/x}$ → Press $\boxed{\ln}$ → Display 0.52624735

Rounding to four decimal places we have
$$\ln (3/\sqrt{\pi}) = 0.5262$$

(b) Enter 10 → Press $\boxed{y^x}$ → Enter 2 → Press $\boxed{\sqrt{x}}$ → Press $\boxed{=}$ → Press $\boxed{\ln}$ → Display 3.2563471

Hence, $\ln 10^{\sqrt{2}} = 3.2563$

(c) Since $\ln e = \log_e e = 1$, we have $e = \text{antilog}_e 1 = \text{INV} \ln 1$. Hence,

Enter 1 → Press $\boxed{\text{INV}}$ → Press $\boxed{\ln}$ → Display 2.7182818

Round to four decimal places, $e = 2.7183$

We will next present an application of natural logarithms. Further applications will be given in the next section.

Population Growth

Under many conditions, the rate of increase of population is considered to satisfy the equation

$$P = P_0 e^{kt} \tag{10.12}$$

where P_0 denotes the initial population, P the population after t years, and k is the annual growth rate.

EXAMPLE 4 When the population growth of the city of Temple Terrace was first noted, the population was just about 14,000. Three years later, the population had grown to 15,900. At this rate, how long will it take for Temple Terrace to double its population?

Solution: First we determine the constant k by substituting 3, 14000, and 15900 for t, P_0, and P respectively in Equation (10.12).

$$15900 = 14000 e^{3k}$$

Taking the natural logarithm of both sides, we get

$$\ln 15900 = \ln 14000 + \ln e^{3k}$$
$$= \ln 14000 + 3k$$

Hence,

$$k = \frac{\ln 15900 - \ln 14000}{3}$$
$$= \frac{9.674 - 9.547}{3}$$
$$= 0.0423$$

Next, substitute $2P_0$ for P and the value of k, 0.0423, just obtained, into Equation (10.12).

$$2P_0 = P_0 e^{0.0423t}$$

or $\qquad 2 = e^{0.0423t} \qquad$ Divide both sides by P_0

By Definition (10.3), we have

$$0.0423t = \ln 2 = 0.6931$$

Hence,
$$t = \frac{0.6931}{.0423} \approx 16.39 \text{ (years)}$$

It will take a little over 16 years for Temple Terrace to double its population.

Exercises 10.6

Standard Assignment: All odd-numbered exercises 1–27

A Convert each common logarithm into a natural logarithm. See Example 1.

1. $\log 100 = 2$
2. $\log 10 = 1$
3. $\log 0.01 = -2$
4. $\log \sqrt{\pi} = 0.25$
5. $\log \sqrt{3} = 0.24$
6. $\log 0.0007 = -3.15$
7. $\log 10^{\sqrt{2}} = 1.41$
8. $\log 10^{\pi} = 3.14$

Use Equation (10.8) and $\log e = 0.434$ to convert each of the following natural logarithms into common logarithms.

9. $\ln e = 1$
10. $\ln \sqrt{e} = 0.5$
11. $\ln e^{\pi} = 3.14$
12. $\ln \frac{1}{e} = -1$
13. $\ln 99 = 4.60$
14. $\ln 57 = 4.04$
15. $\ln \sqrt{5} = 0.80$
16. $\ln \pi = 1.14$

Solve each equation for the given unknown. See Example 2.

17. $10 = e^{2x-1}$
18. $e^{3y-2} = 3.9$
19. $e^{(b/2)-3} = \pi$
20. $e^{-(m+4)/3} = \sqrt{2}$
21. $e^{-(m-5)/2} = \sqrt{3}$
22. $e^{\sqrt{5}-7x} = 13$
23. $e^{\sqrt{t}+6} = e^t$
24. $e^{t^2} + \frac{1}{t^2} - e^4 = 0$
25. $(\sqrt{e})^t - 1.57 = 0$
26. $e^{u^2-2u+1} - 1 = 0$
27. $1 - e^{v^2-v-6} = 0$
28. $1 - e^{(t-\sqrt{t}-12)} = 0$

Use a calculator to evaluate each of the following. See Example 3.

29. $\ln \frac{\sqrt{\pi}}{5}$
30. $\ln \frac{\sqrt{11}}{\pi}$
31. $\ln \frac{\sqrt[4]{120}}{7}$
32. $\ln 10^{\sqrt{3}}$
33. $\ln 65^{\sqrt{7}}$
34. $\ln 12^{\sqrt{\pi}}$
35. $\ln \frac{5}{\sqrt{\pi}-1}$
36. $\ln \frac{\pi}{\sqrt{13}+2}$
37. $\ln 15^{-\sqrt[4]{\pi}}$
38. $\ln \left(\frac{3\pi - \sqrt{2}}{6}\right)^{1/2}$
39. $\ln \left(\frac{5-1}{\sqrt{2}}\right)^{-\pi/2}$
40. $\ln \left(\frac{\sqrt{7}-\sqrt{2}}{\sqrt{7}+\sqrt{2}}\right)^{-\sqrt{2}/3}$

Solve each equation for the given unknown.

EXAMPLE: $\ln x + \ln \dfrac{5x}{3} = 8.$

Solution: Using Property 2 of Properties (10.6), we may rewrite the given equation as

$$\ln \dfrac{5x^2}{3} = 8$$

Hence, $\quad \dfrac{5x^2}{3} = \text{Inv} \ln 8 = 2980.958 \qquad$ By a calculator

so $\quad x^2 = 2980.958 \times \dfrac{3}{5} = 1788.575$

$\quad x = \pm 42.292 \qquad$ By a calculator

Since negative values of x make no sense in the given equation, $x = 42.292$ is the only solution.

41. $\ln 2x + \ln \dfrac{x}{4} = 10$

42. $\ln \dfrac{y}{5} + \ln 10y = 7$

43. $\ln 3k^2 - \ln 6k = 9$

44. $\ln 7m^2 - \ln 2m = 5$

45. $\ln 4n - \ln n^2 = 6$

46. $\ln 5p - \ln 2p^2 = 8$

47. $\ln 3x^2 + \ln \dfrac{7x^2}{8} = 6$

48. $\ln 5x^2 - \ln \dfrac{2x}{3} = 7$

49. Use the approach in Equations (10.9) and (10.10) to show that for any positive a, b, and x, with $a \neq 1$ and $b \neq 1$,

$$\log_a x = \dfrac{\log_b x}{\log_b a}$$

50. Use the formula in Exercise 49 to evaluate
    (a) $\log_2 321$
    (b) $\log_3 534$
    (c) $\log_{\sqrt{2}} \pi$
    (d) $\log_2 \sqrt{\pi} + \log_{\sqrt{2}} 65 + \log_{\sqrt{3}} \sqrt{72}$

51. How long will it take for the population of Temple Terrace to triple? (See Example 4.)

52. After decades of statistical observation, the growth rate k for Temple Terrace has been established to be approximately 0.0423. The present population of Temple Terrace is 650,786. What was the population of Temple Terrace 10 years ago? 20 years ago? [Hint: Use Equation (10.12).]

53. The present population of the city of Saint Petersburg, Florida, is estimated to be about 413,000. Twelve years ago, the population of Saint Petersburg was only

332,000. Find the annual growth rate of Saint Petersburg on the basis of the previous information.

54. Use the annual growth rate obtained in Exercise 53 to predict when the population of Saint Petersburg will reach one-half million people.

10.7 Further Applications

Exponential and logarithmic principles have wide application in the real world.

Compound Interest

Suppose P dollars is invested at an annual interest rate r (in decimal notation), compounded annually, and allowed to draw interest for n years. The amount A that is accumulated after n years is given by

$$A = P(1 + r)^n$$

If interest is compounded k times each year (at equal intervals) instead of annually, the total yield A (principal plus interest), is given by

$$A = P\left(1 + \frac{r}{k}\right)^{kn} \qquad (10.13)$$

EXAMPLE 1 $100 is deposited in a bank that pays 6.5% annual interest compounded quarterly. What is the accumulated amount at the end of one year?

Solution: Since "percent" means "parts per hundred," 6.5% is $6.5/100 = 0.065$ in decimal notation. To change % to decimal notation, we shift the decimal point two positions to the left.

We determine $P = \$100$, $r = 0.065$ (in decimal notation), $k = 4$ (quarterly means 4 times per year), and $n = 1$ year.

$$\begin{aligned} A &= 100\left(1 + \frac{0.065}{4}\right)^4 \\ &= 100(1 + 0.01625)^4 \\ &= 100(1.01625)^4 \\ &= 100(1.06660) \qquad \text{By a calculator} \\ &= 106.66 \text{ (dollars)} \end{aligned}$$

The result shows that at the end of one year the interest alone is $66.66, which amounts to 6.66% at a simple annual interest.

10.7 Further Applications

EXAMPLE 2

What simple annual interest rate would yield the same amount of interest as Bank B, which pays 6.5% annual interest compounded daily? (1 year = 365 days)

Solution: If $100 were deposited in Bank B, the accumulated amount at the end of one year would be

$$A = 100\left(1 + \frac{0.065}{365}\right)^{365} \quad \text{By Equation (10.13)} \\ k = 365, n = 1$$

$$= 100(1 + 0.000178)^{365}$$

We can use logarithms to evaluate

$$\log A = \log 100(1 + 0.000178)^{365}$$
$$= \log 100(1.000178)^{365}$$
$$= \log 100 + \log(1.000178)^{365}$$
$$= \log 100 + 365 \log(1.000178)$$
$$= 2.0282136 \quad \text{By a calculator as shown below}$$

> Enter 100 → Press (log) → Press (+) 1.000178 → Press (log)
> → Press (×) 365 → Press (=) → Display **2.0282136**

Hence, A = antilog 2.0282136

= 106.71   By calculator. Round to 2 decimal places as shown below.

> Enter 2.0282136 → Press (INV) → Press (log)
> → Display **106.71209**

The amount of daily compound interest is $6.71, which can be achieved by a simple annual interest rate of 6.71%.

The conclusions of Examples 1 and 2 show that the difference between daily compound interest and quarterly compound interest may not be as big as our intuition might lead us to believe. That is, when $6.66 is returned on each $100 invested when compounded quarterly and $6.71 is returned on each $100 invested when compounded daily at the same annual rate of 6.5%, there is only a difference of 5 cents for a $100 deposit.

Continuous Decay

Each radioactive substance decays continuously according to the exponential expression

$$A = A_0 e^{-kt} \qquad (10.14)$$

where A_0 is the initial amount of the substance, A is the amount of the substance left after t years, and k is the annual rate of decay.

EXAMPLE 3 A certain radioactive substance decays at an annual rate of 0.05. If there are 150 grams of the radioactive substance, how many years will it take for 30 grams of the substance to decay?

Solution: According to Equation (10.14), the amount of the substance left after t years is

$$A = 150 e^{-0.05t}$$

Letting A equal $120 (= 150 - 30)$ grams in the equation we have

$$120 = 150 e^{-0.05t}$$

or

$$e^{-0.05t} = \frac{120}{150} = 0.8$$

so by Definition (10.3) we have

$$-0.05t = \log_e 0.8$$
$$= -0.2231$$

We have
$$-0.05t = -0.05t = -0.2231$$

or
$$t = \frac{0.2231}{0.05} = 4.462$$

It would take almost 4.5 years for 30 grams of the substance to decay.

The time it takes for half of a radioactive substance to decay is called the *half-life* of the substance.

EXAMPLE 4 8.3% of radium disappears in 200 years. Find the half-life of radium.

Solution: Let A_0 = amount of radium initially present. Then $A_0 - 0.083 A_0$ = amount present after 200 years.

By Equation (10.14),
$$A_0 - 0.083A_0 = A_0 e^{-200k}$$
$$A_0(1 - 0.083) = A_0 e^{-200k}$$
$$(1 - 0.083) = e^{-200k}$$
$$e^{-200k} = 0.917$$

To find the value of k, we use Definition (10.3) to get
$$-200k = \ln 0.917$$
$$= -0.0866$$
$$k = \frac{0.0866}{200} = 4.33 \times 10^{-4}$$

which is an approximate annual rate of decay for radium.

Finally, to find the half-life for radium, by substituting $0.5 (= 1/2)$, 1, and 4.33×10^{-4} for A, A_0, and k, respectively, in Equation (10.14) we have
$$0.5 = e^{-4.33 \times 10^{-4} t}$$

or
$$-4.33 \times 10^{-4} t = \ln 0.5 = -0.6931$$

The half-life is
$$t = \frac{0.6931}{4.33} \times 10^4 \approx 1600 \text{ years}$$

pH Values

Scientists define the hydrogen potential of a solution by
$$\text{pH} = -\log [\text{H}^+]$$

where $[\text{H}^+]$ is the concentration of hydrogen ions in an aqueous solution in moles per liter of the solution. The value pH is a measure of the acidity or alkalinity of solutions. The pH value for water is 7. In general, alkaline solutions have pH values greater than 7, and acids have pH values less than 7.

EXAMPLE 5 Calculate to the nearest tenth the pH of grapefruit juice if the hydrogen ion concentration of grapefruit juice is 6.32×10^{-4}.

Solution:

$$\begin{aligned}
\text{pH} &= -\log [\text{H}^+] \\
&= -\log (6.32 \times 10^{-4}) \\
&= -(\log 6.32 + \log 10^{-4}) \\
&= -\log 6.32 + 4 \\
&= -0.8007 + 4 \qquad \text{log 6.32 = 0.8007} \\
&\qquad\qquad\qquad\qquad \text{Table or calculator} \\
&= 3.1992 \\
&\approx 3.2
\end{aligned}$$

EXAMPLE 6   Find the hydrogen ion concentration in beer if the pH value of beer is 4.82.

$$\begin{aligned}
4.82 &= -\log [\text{H}^+] \\
[\text{H}^+] &= \text{antilog} (-4.82) \\
&= \text{antilog} [(5 - 4.82) - 5] \\
&= \text{antilog} [(0.18) - 5] \\
&= 1.51 \times 10^{-5} \qquad \text{From a table or calculator}
\end{aligned}$$

Exercises 10.7

Standard Assignment: All odd-numbered exercises

1. $10,000 is deposited in a bank that pays 6% annual interest compounded quarterly. What is the accumulated amount at the end of two years? (See Example 1.)

2. $5000 is invested in an investment company that pays 14% annual interest compounded semiannually. What is the accumulated amount at the end of five years?

3. $8000 is deposited in a credit union that pays 8% annual interest compounded monthly. What is the total amount of interest earned at the end of three years?

4. How many years will it take for an investment to double at an annual interest rate of 9% compounded yearly?

5. How long will it take for an investment to triple at an annual interest rate of 12% compounded daily?

6. What annual rate of interest (rounded to the nearest $\frac{1}{2}$%) is required so that $1000 will yield $1425 after two and one half years, if the money is compounded monthly?

7. What simple annual interest rate would yield the same amount of interest as an investment paying 9% annual interest compounded monthly? (See Example 2.)

8. What simple annual interest rate would yield the same amount of interest as an investment paying 18% annual interest compounded daily?

9. A trust fund for a child is being set up by a single investment so that at the end of 16 years there will be $42,000. If the investment earns at an annual rate of 9% compounded quarterly, how much should be invested initially?

10. A certain radioactive substance decays at an annual rate of 0.06. The substance initially weighs 185 grams. How long will it take for the substance to decay to 169 grams? (See Example 3.)

11. The radioactive strontium 90 decays from 314 grams to 266 grams in 7 years. Find the annual rate of decay.

12. A certain amount of the radioactive substance B has decayed to 176 grams in 8 years. If the annual rate of decay for the substance B is known to be 0.07, what was the initial amount of the substance?

13. Find the half-life of the radioactive substance described in Exercise 10.

14. Find the half-life of the strontium 90 described in Exercise 11.

15. Find the half-life of the radioactive substance B described in Exercise 12.

16. How long will it take for the radioactive radium described in Example 4 to decay completely?

17. Suppose that 2.4% of carbon 14 has disappeared in 202 years. Find the half-life of carbon 14.

18. Suppose that 1.5% of the radioactive substance B has disappeared in 25 years. Find the half-life of the substance B.

19. The hydrogen ion concentration [H^+] of human blood is approximately 3.98×10^{-8}. Find the pH value of human blood to the nearest tenth. (See Example 5.)

20. The hydrogen ion concentration [H^+] of a solution is approximately 6.4×10^{-7}. Find the pH value of the solution to the nearest tenth.

Find the hydrogen ion concentration [H^+] of the solutions whose pH values are given below. (See Example 6.)

21. wine grapes, pH = 3.15

22. milk, pH = 6.3

23. pure water, pH = 7

24. eggs, pH = 7.78

Chapter 10 Summary

Key Words and Phrases

10.1 Exponential function
10.2 Logarithmic function
    Logarithms with base a
10.3 Common logarithm
    Mantissa

Characteristic
Antilogarithm (antilog)
10.6 The number e
    Natural logarithm ($\ln x$)

Key Concepts and Rules

For any $a > 0$ and $a \neq 1$, it is always true that

$$\log_a a = 1 \quad \text{and} \quad \log_a 1 = 0$$

$$\log_a MN = \log_a M + \log_a N$$

$$\log_a \frac{M}{N} = \log_a M - \log_a N$$

$$\log_a M^p = p \log_a M$$

Applications

Compound Interest: If P dollars is invested for n years at annual rate r compounded k equally-spaced times during the year, then the accumulated amount A is given by

$$A = P\left(1 + \frac{r}{k}\right)^{kn}$$

Continuous Decay: Each radioactive substance decays continuously according to the exponential expression

$$A = A_0 e^{-kt}$$

where A_0 is the initial amount of the substance, A is the amount of the substance left after t years, and k is the annual rate of decay.

Population Growth: Under normal conditions, the rate of increase of population satisfies the expression

$$P = P_0 e^{kt}$$

where P_0 denotes the initial population, P the population after t years, and k is the annual growth rate.

pH Values: Scientists define the hydrogen potential of a solution by

$$\text{pH} = -\log[\text{H}^+]$$

where $[\text{H}^+]$ is the concentration of hydrogen ions in an aqueous solution in moles per liter of solution.

The pH value is a measure of the acidity or alkalinity of solutions. Alkaline solutions have pH values greater than 7, and acids have pH values less than 7.

Review Exercises

If you have difficulty with any of these problems, look in the section indicated by the numbers in brackets.

[10.1] 1. Graph the exponential function $y = 10^x - 8$.

Solve each of the following equations for the given unknown.

2. $3^{-x} = 81$

3. $125^y = 5$

4. $\left(\dfrac{5}{4}\right)^{z-1} = \dfrac{125}{64}$

5. Rank the following numbers from the largest to the smallest:

$$\left(\dfrac{1}{2}\right)^{3/2}, \quad \left(\dfrac{1}{2}\right)^{7/5}, \quad \text{and} \quad \left(\dfrac{1}{2}\right)^{\sqrt{2}}$$

[10.2] Write each exponential expression in logarithmic form and write each logarithmic expression in exponential form.

6. $5^3 = 125$

7. $\left(\dfrac{1}{2}\right)^{-4} = 16$

8. $10^0 = 1$

9. $\log_2 64 = 6$

10. $\log_{49} 7 = \dfrac{1}{2}$

11. $\log_a a^n = k$

Solve each equation for the given unknown.

12. $\log_{10}(x + 99) = 2$

13. $\log_6(y - 1) = -2$

14. $\log_b(z^2 - 7z + 13) = 0$

15. $\log_5(k^2 + 3k - 13) = 1$

Evaluate each of the following.

16. $\log_5 \dfrac{1}{125}$
17. $\log_{10} \sqrt[3]{100}$
18. $a^{\log_a 99}$

[10.3] Solve each of the following logarithmic equations.

19. $\log_2(x + 3) + \log_2(x + 4) = 1$
20. $\log_3(4k - 1) - \log_3(3k + 6) = 2$
21. $\log_{10}(x^2 + 17) - \log_{10}(x + 2) = \log 2 + \log(x - 2)$
22. $\log_2(\log_2 y^2) = 2$

Given $\log_5 x = 2$, $\log_5 y = 3$, and $\log_5 z = 4$, evaluate each of the following.

23. $\log_5 xy$
24. $\log_5 \dfrac{y}{z}$
25. $\log_5 x^{10}$
26. $\log_5 \dfrac{x^2 y}{z^3}$
27. $\log_5 \dfrac{1}{z^5}$
28. $\log_5 \sqrt[3]{\dfrac{y^4 z^5}{x^6}}$

[10.4] Identify the characteristic and mantissa for each of the following.

29. $\log 1984$
30. $\log 0.321$
31. $\log(8.73 \times 10^n)$

Find each logarithm by first using the table of common logarithms and then by using a calculator.

32. $\log 3.88$
33. $\log 485000$
34. $\log 0.000328$
35. $\log(4.67 \times 10^9)$
36. $\log(1.28 \times 10^{-8})$

Answer each of the following questions without using any table of logarithms or a calculator.

37. What is antilog 0? antilog 1?
38. What is antilog $(\log \sqrt{3})$? $\log(\text{antilog } \sqrt{3})$?

[10.5] Find each logarithm or antilogarithm by using a calculator.

39. ▦ $\log 8.386$
40. ▦ $\log \sqrt{6342}$
41. ▦ antilog 0.4735
42. ▦ antilog (-2.4218)

Find each x.

43. ▦ $\log x = 3.24168$
44. ▦ $\log x = -\sqrt{4.23248}$

[10.6] Convert each common logarithmic expression into a natural logarithmic expression and each natural logarithmic expression into a common logarithmic expression. $\left(\text{Hint: } \log x = \dfrac{\ln x}{\ln 10} \quad \text{and} \quad \ln x = \dfrac{\log x}{\log e}\right)$

45. $\log 1000 = 3$
46. $\log 0.001 = -3$
47. $\log 10^{\sqrt{3}} = 1.732$

48. $\ln \pi = 1.14$ 	49. $\ln \sqrt{e} = 0.5$ 	50. $\ln 57^2 = 8.08$

Solve each equation for the given unknown.

51. $e^{2x-1} = 100$ 	52. $e^{-\sqrt{y}+6} = e^y$ 	53. $1 - e^{t^2-12} = 0$

54. $\ln(x + 7) = 1 - \ln x$ 	55. $\log(\log x^3) = 10$

56. $\log(4k + 5) - \log(2k - 3) = \log 7$

57. $e^x - e^{-x} = 15$

[10.7] 58. How long will it take for an investment to double at the annual interest rate of 10% compounded quarterly?

59. A certain amount of the radioactive substance R has been decayed to 96 grams in 12 years. If the annual rate of decay for the substance R is known to be 0.079, what was the initial amount of the substance?

60. Find the pH value of a solution whose hydrogen ion concentration $[H^+]$ is 7.23×10^{-4}.

Practice Test (60 minutes)

1. Sketch the graph of (a) $y = 5^x$ 	(b) $y = \log_5 x$.

2. Write each exponential expression in logarithmic form.

    (a) $\left(\dfrac{1}{3}\right)^{-4} = 81$ 	(b) $10^{-3} = 0.001$ 	(c) $2^{99} = x$

3. Write each logarithmic expression in exponential form.

    (a) $\log_{10} 1 = 0$ 	(b) $\log_5 25 = 2$ 	(c) $\log_2 \dfrac{1}{16} = -4$

For Problems 4–7, solve each equation for the given unknown.

4. $2^{1-x} = 32$ 	5. $\log_5(19 - y) = 2$

6. $\log_{10}(k^2 + 2k + 7) = 1$

7. $\log_3(2x + 1) = \log_3(x + 8) - 1$

In Problems 8–10, given that $\log 2 = 0.301$, $\log 3 = 0.477$, and $\log 7 = 0.845$, evaluate each logarithm.

8. $\log 49$ 	9. $\log \sqrt[3]{14}$ 	10. $\log \sqrt[4]{3000}$

11. How long will it take for an investment to triple at the annual interest rate of 12% compounded semiannually?

12. A certain radioactive substance decayed from 120 grams to 100 grams in 16 years. How long will it take to decay to half the original amount?

Extended Applications

Noise Insulation

The loudness of a sound of intensity I that is measured in *decibels* is defined to be

$$10 \log(I/I_0) \text{ decibels}$$

where I_0 is the minimal intensity detectable by the human ear. Often a larger unit, the *bel*, is more convenient; 10 decibels is equivalent to 1 bel.

The noise level of street sound outside the new concert hall in downtown Chicago is measured to be approximately 7 bels. Using special insulation material, the noise level inside the concert hall is reduced to 29 decibels. How many times greater is the noise intensity outside the concert hall?

Let
$$x = \text{intensity level outside}$$
$$y = \text{intensity level inside}$$

Then
$$70 = 10 \log(x/I_0) \qquad (7 \text{ bels} = 70 \text{ decibels})$$
$$29 = 10 \log(y/I_0)$$

Subtracting these two equations, we get

$$41 = 10[\log(x/I_0) - \log(y/I_0)]$$
$$= 10(\log x - \log I_0 - \log y + \log I_0)$$
$$= 10(\log x - \log y)$$
$$= 10 \log(x/y)$$

or
$$\frac{x}{y} = \text{antilog } 4.1 \ (= 10^{4.1})$$
$$\approx 12{,}600 \qquad \text{By a table of logarithms (or a calculator)}$$

The noise intensity on the outside is approximately 12,600 times greater than that on the inside of the concert hall.

Exercises

1. If the noise level in the concert hall is to be further reduced to 25 decibels, how much must the noise intensity be reduced from its present level?

2. Let I_c represent the intensity corresponding to the noise level associated with ordinary conversation. If

$$60 = 10 \log \left(\frac{I_c}{I_0}\right)$$

how many times greater than I_o is I_c?

3. Let I_w represent the intensity corresponding to the noise level associated with a whisper. If

$$10 = 10 \log \left(\frac{I_w}{I_0}\right)$$

how many times greater than I_0 is I_w?

11
Sequences, Series, Permutations, and Combinations

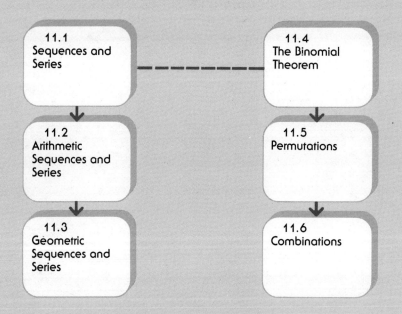

11.1 Sequences and Series

Look at the following orderly listing of numbers.

(a) 1, 5, 9, 13, 17, 21, ☐, ☐, ...
(b) 1, 2, 4, 8, 16, 32, ☐, ☐, ...

What comes next? What comes after 21 in (a)? What comes after 32 in (b)? The student should not have much trouble figuring this out.

We use the word *sequence* freely in ordinary language, as in the phrases "a sequence of car accidents," "a sequence of events," "her attorney gave a sequence of convincing arguments."

In mathematics, a **sequence** is an orderly listing of numbers or expressions that is given by any explicit or implicit rule. For example, the rule defining sequence (a) is that each number, which is called a **term**, is greater than the previous term by 4. Thus, the number that comes after 21 is 25. What is the rule defining the sequence (b)? (See Example 3.)

We shall use the notations

$$a_1, a_2, a_3, a_4, \ldots$$

or

$$b_1, b_2, b_3, b_4, \ldots$$

to represent a sequence. Thus, for the sequence (a)

$$a_1 = 1$$
$$a_2 = 5$$
$$a_3 = 9$$
$$a_4 = 13$$

and so on.

Note that a_1 represents the 1st term and a_4 represents the 4th term. The subscript indicates the place of the term in the sequence. Thus, a_k denotes the kth term, which is called the **general term** of the sequence.

EXAMPLE 1   Find the 8th term a_8 in the sequence (a).

Solution: The nature of the sequence is such that each term is greater than the previous term by 4. We already know that $a_6 = 21$ and $a_7 = 25$; hence $a_8 = 25 + 4 = 29$.

EXAMPLE 2   Write the first four terms of the sequence whose general term is

$$a_k = 3k - 2$$

Solution: The first four terms are

$$a_1 = 3(1) - 2 = 1$$
$$a_2 = 3(2) - 2 = 4$$
$$a_3 = 3(3) - 2 = 7$$
$$a_4 = 3(4) - 2 = 10$$

EXAMPLE 3 Find the general term of the sequence: 2, 4, 8, 16, ...

Solution: Since there are no general methods of finding the general term of a series, we must rely on a careful inspection of the first few terms. In this case,

$$a_1 = 2 = 2^1$$
$$a_2 = 4 = 2^2$$
$$a_3 = 8 = 2^3$$
$$a_4 = 16 = 2^4$$

which indicates that $a_k = 2^k$ is the general term of the given series.

In order to present sequences from a functional point of view, let N_n denote the set consisting of the first n consecutive natural numbers. Thus, $N_4 = \{1, 2, 3, 4\}$, $N_5 = \{1, 2, 3, 4, 5\}$, and so on. Recall that the symbol N always stands for the set $\{1, 2, 3, \ldots\}$ of all natural numbers. In general, there are two kinds of sequences: an **infinite sequence**, which is a function whose domain is the set N, and a **finite sequence**, which is a function whose domain is N_n for some positive integer n. From the functional point of view,

$$2, 4, 6, 8, 10, \ldots$$

is an infinite sequence given by the function $f(k) = 2k$ for each natural number k.

On the other hand,

$$2, 5, 8, 11$$

is a finite sequence given by the function $g(k) = 3k - 1$ with domain $N_4 = \{1, 2, 3, 4\}$. The student should verify that

$$g(1) = 3(1) - 1 = 2$$
$$g(2) = 3(2) - 1 = 5$$
$$g(3) = 3(3) - 1 = 8$$

and

$$g(4) = 3(4) - 1 = 11$$

In the previous examples, the functions $f(k) = 2k$ and $g(k) = 3k - 1$ give the general terms

$$a_k = 2k, \quad k \in \mathbf{N}$$
$$b_k = 3k - 1, \quad k \in \{1, 2, 3, 4\}$$

of each sequence.

EXAMPLE 4 Write the first five terms of the sequence defined by the function $f(k) = \sqrt{k}$, whose domain is $\mathbf{N}$.

Solution: The first five terms are

$$a_1 = f(1) = \sqrt{1} = 1$$
$$a_2 = f(2) = \sqrt{2}$$
$$a_3 = f(3) = \sqrt{3}$$
$$a_4 = f(4) = \sqrt{4} = 2$$
$$a_5 = f(5) = \sqrt{5}$$

As demonstrated in this example, the general term is given by $a_k = f(k)$.

If a plus sign, $+$, is placed between each pair of consecutive terms in a sequence, the resulting expression is called a **series**. If $a_1, a_2, a_3, \ldots, a_n$ are the terms of a sequence, then

$$S_n = a_1 + a_2 + a_3 + \cdots + a_n$$

is a series.

If the sequence is infinite, the corresponding series is said to be an **infinite series**. For example,

$$1 + \frac{1}{2} + \frac{1}{4} + \frac{1}{8} + \frac{1}{16} + \cdots$$

is an infinite series. In this book we will usually be concerned only with finite sequences and series.

We can use a compact **summation notation** Σ to denote a series, as follows.

$$\sum_{k=1}^{n} a_k = a_1 + a_2 + a_3 + \cdots + a_n$$

$$\sum_{k=1}^{5} \frac{1}{k} = 1 + \frac{1}{2} + \frac{1}{3} + \frac{1}{4} + \frac{1}{5}$$

The symbol

$$\sum_{k=1}^{n} a_k$$

is read, "the summation of a_k as k runs from 1 to n."

EXAMPLE 5 Write the following without summation notation.

(a) $\sum_{k=1}^{4} (-2)^k$

(b) $\sum_{k=6}^{10} \frac{(-1)^k}{k}$

Solution:

(a) $\sum_{k=1}^{4} (-2)^k = (-2)^1 + (-2)^2 + (-2)^3 + (-2)^4$

$= -2 + 4 - 8 + 16$

$= 10$

(b) $\sum_{k=6}^{10} \frac{(-1)^k}{k} = \frac{(-1)^6}{6} + \frac{(-1)^7}{7} + \frac{(-1)^8}{8} + \frac{(-1)^9}{9} + \frac{(-1)^{10}}{10}$

$= \frac{1}{6} - \frac{1}{7} + \frac{1}{8} - \frac{1}{9} + \frac{1}{10}$

Exercises 11.1

Standard Assignment: Exercises 1, 3, 5, 7, 9, 13, 15, 17, 19, 21, 23, 27, 29, 31, 35, 37, 41, 45, 49, 51

A Find a pattern in each of the following sequences and fill in the blank spaces.

1. 1, 4, 7, 10, 13, ☐, ☐, . . .

2. 2, 6, 10, 14, 18, ☐, ☐, . . .

3. 2, 7, 12, 17, 22, ☐, ☐, . . .

4. 1, 2, 4, 8, 16, ☐, ☐, . . .

5. 1, 4, 9, 16, 25, ☐, ☐, . . .

6. $1, -\frac{1}{2}, \frac{1}{3}, -\frac{1}{4}$, ☐, ☐, . . .

7. $-1, 8, -27, 64$, ☐, ☐, . . .

8. $3, -6, 12, -24$, ☐, ☐, . . .

Write down the first four terms of each of the following sequences. See Example 2.

9. $a_k = 3k - 1$

10. $a_k = 4 + 3k$

11. $a_k = 4k - 7$

12. $a_k = 5k - 10$

13. $a_k = 5 - 2k$

14. $a_k = 4 - 3k$

15. $a_k = (-1)^k$

16. $a_k = \frac{1}{k}$

17. $a_k = \frac{k-1}{k+1}$

18. $a_k = \frac{2k+1}{3k-1}$

19. $a_k = (-1)^k k(k+1)$
20. $a_k = \dfrac{(-1)^{k+1}}{k(k+1)}$

Find the indicated term for each of the following sequences.

21. $a_k = \dfrac{k}{k+2}$; $\quad a_{20}$
22. $a_k = \dfrac{(-1)^k(2k-1)}{k^2}$; $\quad a_{13}$
23. $a_k = (-1)^{k+1} k(k+1)$; $\quad a_{30}$
24. $a_k = \dfrac{10k-9}{5k+6}$; $\quad a_{25}$
25. $a_k = \dfrac{9k-10}{6k+5}$; $\quad a_{100}$
26. $a_k = (-2)^{2k+1}$; $\quad a_{10}$
27. $a_k = \log 10^{k+1}$; $\quad a_{99}$
28. $a_k = \ln e^{k(k-5)}$; $\quad a_{105}$

Find the general term of each of the following sequences. See Example 3.

29. $1, 5, 9, 13, 17, \ldots$
30. $1, 6, 11, 16, 21, \ldots$
31. $1, -2, 4, -8, 16, \ldots$
32. $1, -4, 7, -10, 13, \ldots$
33. $1 \cdot 2, 2 \cdot 3, 3 \cdot 4, 4 \cdot 5, \ldots$
34. $-1, 4, -9, 16, -25, \ldots$
35. $\dfrac{1}{2}, \dfrac{2}{3}, \dfrac{3}{4}, \dfrac{4}{5}, \ldots$
36. $2, -2, 2, -2, 2, \ldots$

Rewrite each of the following series without summation notation. See Example 5.

37. $\displaystyle\sum_{k=1}^{5} \dfrac{k+3}{k}$
38. $\displaystyle\sum_{k=1}^{6} k(k+2)$
39. $\displaystyle\sum_{k=1}^{6} \dfrac{k}{k+3}$
40. $\displaystyle\sum_{k=1}^{5} \dfrac{k+1}{k^2}$
41. $\displaystyle\sum_{k=1}^{9} \dfrac{(-1)^{k+1}}{2k+1}$
42. $\displaystyle\sum_{k=1}^{7} \dfrac{(-1)^{k-1}}{k(2k-1)}$
43. $\displaystyle\sum_{k=5}^{9} \dfrac{2k+3}{3k-2}$
44. $\displaystyle\sum_{k=4}^{8} \dfrac{2+(-1)^k}{k}$
45. $\displaystyle\sum_{k=4}^{7} (-1)^k k^3$
46. $\displaystyle\sum_{k=5}^{8} (-1)^k \dfrac{1}{10^k}$

Rewrite each series by using summation notation.

47. $1 + 2 + 3 + 4 + 5 + \cdots + 100$
48. $1 + 3 + 5 + 7 + 9 + \cdots + 101$
49. $2 + 4 + 6 + 8 + 10 + 12 + \cdots + 102$
50. $1 - \dfrac{1}{2} + \dfrac{1}{3} - \dfrac{1}{4} + \dfrac{1}{5} - \dfrac{1}{6}$

51. $\dfrac{1}{2} + \dfrac{2}{3} + \dfrac{3}{4} + \dfrac{4}{5} + \cdots + \dfrac{n}{n+1}$

52. $-2 + \dfrac{3}{2} - \dfrac{4}{3} + \dfrac{5}{4} - \cdots + \dfrac{(-1)^n(n+1)}{n}$

11.2 Arithmetic Sequences and Series

Let us look again at the sequence (a) that we examined in the beginning of Section 11.1.

$$1, 5, 9, 13, 17, 21, \ldots$$

Notice that the difference between any two consecutive terms is the constant 4; more specifically,

$$a_2 - a_1 = 5 - 1 = 4$$
$$a_3 - a_2 = 9 - 5 = 4$$
$$a_4 - a_3 = 13 - 9 = 4$$
$$a_5 - a_4 = 17 - 13 = 4$$

This sequence is an example of an **arithmetic sequence**, and the corresponding series

$$1 + 5 + 9 + 13 + 17 + 21 + \cdots$$

is an **arithmetic series**.

An arithmetic sequence is a sequence

$$a_1, a_2, a_3, a_4, \ldots, a_n \quad (\text{or } a_1, a_2, a_3, \ldots, a_n, \ldots)$$

that has the property that

$$a_{k+1} - a_k = d \quad \text{or} \quad a_{k+1} = a_k + d$$

for some constant number d and for each positive integer $k = 1, 2, 3, \ldots, n$ (or $k = 1, 2, 3, \ldots$). The constant number d is called the **common difference** of the arithmetic sequence and of the corresponding arithmetic series.

EXAMPLE 1   Which sequences or series are arithmetic sequences or series? What is the common difference in each of those sequences or series?

(a) $1, 2, 3, 4, \ldots, 99$ (b) $2, 4, 6, 8, 10, \ldots$
(c) $1, 4, 9, 16, \ldots$ (d) $1 + 5 + 9 + 13 + 17 + \cdots$
(e) $1 + 4 + 8 + 16 + 32 + 64$ (f) $1, 1, 1, 1, \ldots$

Solution: (a), (b), and (f) are arithmetic sequences with common differences 1, 2, and 0, respectively; (d) is an arithmetic series with common difference 4; (c) is not an arithmetic sequence, because $4 - 1 = 3$, whereas $9 - 4 = 5$; and (e) is not an arithmetic series, because $4 - 1 = 3$, whereas $8 - 4 = 4$.

In order to write the terms of an arithmetic sequence (or series), it is essential to know the first term, a_1, and the common difference d. For example, if $a_1 = 1$ and $d = 4$, then

$$a_2 = a_1 + d = 1 + 4 = 5$$
$$a_3 = a_2 + d = 5 + 4 = 9$$
$$a_4 = a_3 + d = 9 + 4 = 13$$

and so on.

EXAMPLE 2 Find the first 5 terms of each of the following arithmetic sequences.

(a) $a_1 = 10,\ d = -5$ (b) $b_1 = 2,\ b_2 = 6$

Solution:
(a) For the sequence $a_1 = 10,\ d = 5$

$$a_2 = a_1 + d = 10 + (-5) = 5$$
$$a_3 = a_2 + d = 5 + (-5) = 0$$
$$a_4 = a_3 + d = 0 + (-5) = -5$$
$$a_5 = a_4 + d = -5 + (-5) = -10$$

(b) For the arithmetic sequence $b_1 = 2,\ b_2 = 6$, we first find the common difference.

$$d = b_2 - b_1 = 6 - 2 = 4$$

then

$$b_3 = b_2 + d = 6 + 4 = 10$$
$$b_4 = b_3 + d = 10 + 4 = 14$$

and

$$b_5 = b_4 + d = 14 + 4 = 18$$

Can you find the 126th term, b_{126}, of the previous sequence? If you were to continue the process $b_6, b_7, \ldots$, you would eventually get there, but you would be bored to death. We can use a short cut. Observe that

$$b_1 = b_1 + 0d$$
$$b_2 = b_1 + d = b_1 = 1d$$
$$b_3 = b_2 + d = (b_1 + d) + d = b_1 + 2d$$
$$b_4 = b_3 + d = (b_1 + 2d) + d = b_1 + 3d$$
$$b_5 = b_4 + d = (b_1 + 3d) + d = b_1 + 4d$$

By now, it should be apparent that $b_{100} = b_1 + 99d$, $b_{126} = b_1 + 125d$, and, in general, $b_k = b_1 + (k - 1)d$.

Thus,

$$b_{126} = b_1 + 125d = 6 + 125 \cdot 4 = 6 + 500 = 506$$

> **The General Term of an Arithmetic Sequence**
>
> $$a_k = a_1 + (k - 1)d \qquad (11.1)$$

It has been said that when the great mathematician Karl Friedrich Gauss (1777–1855) was 6, his tutor tried to keep him quiet for a while by asking, "What is the sum of $1 + 2 + 3 + \cdots + 100$?" Within seconds, the young Gauss gave the answer of 5050. How did he do it?

Our logical guess is that Gauss added the first and last terms to get

$$1 + 100 = 101$$

then added the second and the second to the last terms

$$2 + 99 = 101$$

Continuing this way, there would be 50 pairs of numbers with the sum of each pair being 101. Hence,

$$1 + 2 + 3 + \cdots + 98 + 99 + 100 = 50 \cdot 101 = 5050$$

In general, the sum S_n of the arithmetic series

$$S_n = a_1 + a_2 + a_3 + \cdots + a_n$$

with common difference d, is given by

> $$S_n = \frac{n}{2}(a_1 + a_n) \qquad (11.2)$$

According to Equation (11.1), $a_n = a_1 + (n - 1)d$, and so we also have

> $$S_n = \frac{n}{2}[2a_1 + (n - 1)d]$$

11.2 Arithmetic Sequences and Series

EXAMPLE 3

Jim's mother worked as a seamstress. Her wages were gauged according to her productivity and she earned $10 on the first day of work. Her skill (and productivity) improved every day for the first 100 days; the result was that each day she earned 50 cents more than the previous day.

(a) How much was Jim's mother making on her 50th day of work? Her 100th day of work?
(b) If she saved all she earned, how much would she have saved after the first 100 days of work?

Solution: The earning pattern, in this case, is an arithmetic sequence with $a_1 = 10$, $d = 0.5$. By Equation (11.1) we find

$$a_{50} = a_1 + (50 - 1)d$$
$$= 10 + 49 \times 0.5 = 34.50$$
$$a_{100} = a_1 + (100 - 1)d$$
$$= 10 + 99 \times 0.5 = 59.5$$

The answer for (a) is that Jim's mother was making $34.50 on her 50th day of work, and $59.50 on her 100th day of work.

To answer (b) we use Equation (11.2) to find

$$S_{100} = \frac{100}{2}(a_1 + a_{100})$$
$$= 50(10 + 59.5) \quad a_{100} = 59.5, \text{ from (a)}$$
$$= 3475 \text{ (dollars)}$$

EXAMPLE 4

Ben took a 12-month temporary job that had a monthly salary that increased a fixed amount every month. Ben does not remember the starting salary, but remembers that he was paid $820 at the end of the 3rd month and $910 for his last month of work. How much was Ben's starting salary? How much was Ben's total pay for the entire 12 months?

Solution: Ben's earning pattern is an arithmetic sequence with $n = 12$, $a_3 = 820$, and $a_{12} = 910$. According to Equation (11.1),

$$a_3 = a_1 + (3 - 1)d$$
$$a_{12} = a_1 + (12 - 1)d$$

That is

$$a_1 + 2d = 820 \quad (1), (a_3 = 820)$$
$$a_1 + 11d = 910 \quad (2), (a_{12} = 910)$$

Subtracting Equation (1) from Equation (2), we have $9d = 90$ or $d = 10$.

Now, substituting $d = 10$ into Equation (1) we get
$$a_1 + 2 \times 10 = 820$$
or
$$a_1 = 800$$

Ben's starting salary was \$800 per month. In order to find Ben's total pay, we compute the sum S_{12} from Equation (11.2).

$$S_{12} = \frac{12}{2}(a_1 + a_{12})$$

$$= 6(800 + 910)$$

$$= 10{,}260 \text{ dollars}$$

Exercises 11.2

Standard Assignment: Exercises 3, 5, 9, 11, 15, 19, 23, 25, 27, 31, 33, 35, 41, 43

Indicate which of the following sequences and series are arithmetic and state the common differences. See Example 1.

1. 1, 3, 5, 7, 9.
2. 2, 5, 8, 11, 14.
3. 10, 7, 4, 1, −2.
4. 1, −3, 5, −7, 9.
5. 2, 4, 8, 16, 32, . . .
6. $c, c^2, c^3, c^4, \ldots$
7. $x + x^2 + x^3 + \cdots$
8. 7, 5, 3, 1, −1, . . .
9. $1 + \frac{1}{2} + \frac{1}{3} + \cdots$
10. $1 + \frac{1}{2^2} + \frac{1}{3^2} + \cdots$

In Exercises 11–20 find the first 5 terms of each arithmetic sequence. See Example 2.

11. $a_1 = 5, d = 3$
12. $a_1 = 8, d = -4$
13. $a_2 = 7, d = 2$
14. $a_5 = 15, d = 5$
15. $b_5 = 14, b_6 = 16$
16. $b_5 = 17, b_6 = 15$
17. $a_5 = 13, a_7 = 19$
18. $a_5 = 9, a_7 = 5$
19. $a_4 = 5, a_5 = -1$
20. $a_4 = 0, a_5 = -3$

In Exercises 21–30 find the sum of each arithmetic series.

21. $1 + 2 + 3 + \cdots + 1000$
22. $2 + 4 + 6 + \cdots + 100$
23. $1 + 3 + 5 + \cdots + 99$
24. $5 + 10 + 15 + 20 + \cdots + 200$
25. $3 + 6 + 9 + 12 + \cdots + 300$
26. $4 + 7 + 10 + 13 + \cdots + 301$
27. $11 + 8 + 5 + 2 - 1 - 4 - \cdots - 34$
28. $12 + 7 + 2 - 3 - 8 - 13 - \cdots - 48$

29. $50 + 48 + 46 + \cdots - 44$ 30. $75 + 70 + 65 + \cdots - 80$

EXAMPLE: Find the general term of an arithmetic series that is known to have $a_3 = 65$ and $a_{13} = 15$.

Solution: First, by substituting 3 and then 13 for k in Equation (11.1) we get

$$a_3 = a_1 + (3 - 1)d \text{ and } a_{13} = a_1 + (13 - 1)d$$

$$a_1 + 2d = 65 \quad (1)$$
or
$$a_1 + 12d = 15 \quad (2)$$

In order to find a_1 and the common difference d, subtract Equation (1) from Equation (2).

$$10d = -50$$
$$d = -5$$

Substituting $d = -5$ into Equation (1) we get

$$a_1 + 2(-5) = 65$$
$$a_1 - 10 = 65$$
$$a_1 = 75$$

Use Equation (11.1) to write the general term

$$a_k = 75 + (k - 1)(-5)$$
or
$$a_k = 80 - 5k$$

In Exercises 31–40 find the general term for each arithmetic sequence or series.

31. $a_1 = 4, d = 3$ 32. $a_1 = 9, d = -3$
33. $a_2 = 6, d = 2$ 34. $a_3 = 10, d = 5$
35. $a_{10} = 1, a_{15} = -14$ 36. $a_9 = 27, a_{21} = 63$
37. the series of Exercise 26 38. the series of Exercise 27
39. the series of Exercise 28 40. the series of Exercise 29

41. When John started to work as an orange picker, he picked only 10 oranges in the 1st minute, 12 in the second minute, 14 in the third minute, and so on. How many oranges will John pick in half an hour?

42. Walking uphill on a hilly road, Paul walks 60 feet in the 1st minute, 57 feet in the second minute, 54 feet in the third minute, and so on.
    (a) How far will Paul walk in the kth minute?
    (b) How far will Paul walk in 15 minutes?

43. In selecting a job, Mary finds that Company X pays $12,500 the first year and guarantees a raise of $600 each year, while Company Y starts at $14,000 per year with a guaranteed raise of only $200 each year.
    (a) Over the first 5-year period, which company will pay more? How much more?

(b) Over the first 10-year period, which company will pay more? How much more?

44. A pile of logs has 69 logs in the bottom layer, 68 in the second layer, 67 in the next layer, and so on. There are 51 layers in all. How many logs are there in the pile?

11.3 Geometric Sequences and Series

Take a careful look at each of the following sequences.

(a) 1, 2, 4, 8, ...
(b) 2, 10, 50, 250, ...
(c) 4, 2, 1, $\frac{1}{2}$, ...

Each of these sequences shares the common characteristic that each term after the first term is the result of multiplying the preceding term by a fixed number. In sequences (a), (b), and (c), the fixed numbers are 2, 5, and $\frac{1}{2}$ respectively. In sequence (a),

$$a_1 = 1$$
$$a_2 = 2 = 1 \cdot 2 = a_1 \cdot 2$$
$$a_3 = 4 = 2 \cdot 2 = a_2 \cdot 2$$
$$a_4 = 8 = 4 \cdot 2 = a_3 \cdot 2$$

and so on. This type of sequence is called a **geometric sequence**. The fixed number is called the **common ratio** of the sequence.

The sequence $a_1, a_2, a_3, \ldots$ is a **geometric sequence** with **common ratio** r, if

$$a_{k+1} = a_k \cdot r \quad \text{or} \quad \frac{a_{k+1}}{a_k} = r \quad \text{for all } k \geq 1$$

EXAMPLE 1 Find the common ratio for each of the following geometric sequences and write the next three terms of each sequence.

(a) 5, −10, 20, −40, ...
The common ratio

$$r = \frac{a_2}{a_1} = \frac{-10}{5} = -2$$

Multiplying the last given term successively by $r = -2$, we get the next three terms

$$a_5 = (-40) \cdot (-2) = 80$$
$$a_6 = 80 \cdot (-2) = -160$$
$$a_7 = (-160) \cdot (-2) = 320$$

(b) $1, \dfrac{x}{2}, \dfrac{x^2}{4}, \dfrac{x^3}{8}, \ldots$

The common ratio

$$r = \dfrac{a_2}{a_1} = \dfrac{x}{2}$$

The next three terms are

$$a_5 = a_4 r = \dfrac{x^3}{8} \cdot \dfrac{x}{2} = \dfrac{x^4}{16}$$

$$a_6 = a_5 r = \dfrac{x^4}{16} \cdot \dfrac{x}{2} = \dfrac{x^5}{32}$$

$$a_7 = a_6 r = \dfrac{x^5}{32} \cdot \dfrac{x}{2} = \dfrac{x^6}{64}$$

To find the general term of a geometric sequence, let us observe that, if the geometric sequence

$$a_1, a_2, a_3, \ldots, a_k, \ldots$$

has a common ratio r, then

$$a_2 = a_1 r = a_1 r^1$$
$$a_3 = a_2 r = (a_1 r) r = a_1 r^2$$
$$a_4 = a_3 r = (a_1 r^2) r = a_1 r^3$$
$$a_5 = a_4 r = (a_1 r^3) r = a_1 r^4$$

and so on.

You may have noticed that the power of r in the far right column is always 1 less than the subscript number of a in the far left column. In order to find the general term of a geometric sequence, Equation (11.3) is used.

The General Term of a Geometric Sequence

$$a_k = a_1 r^{k-1} \qquad (11.3)$$

To write the terms of a geometric sequence we only need to know the first term and the common ratio.

EXAMPLE 2 Write the first four terms of each indicated geometric sequence.

(a) $a_1 = 32$, $r = \dfrac{1}{2}$

(b) $a_5 = -81$ and $a_6 = 243$

Solution:
(a) We have
$$a_1 = 32$$
$$a_2 = 32 \cdot \left(\dfrac{1}{2}\right) = 16$$
$$a_3 = 32 \cdot \left(\dfrac{1}{2}\right)^2 = 8$$
$$a_4 = 32 \cdot \left(\dfrac{1}{2}\right)^3 = 4$$

(b) The common ratio
$$r = \dfrac{a_6}{a_5} = \dfrac{243}{-81} = -3$$

Then, by substituting the values of a_5 and r into $a_1 r^4 = a_5$, we get
$$a_1 \cdot (-3)^4 = -81$$
$$a_1 \cdot 81 = -81$$
$$a_1 = -1$$

Consequently,
$$a_2 = a_1 r = (-1) \cdot (-3) = 3$$
$$a_3 = a_1 r^2 = (-1) \cdot (-3)^2 = -9$$
$$a_4 = a_1 r^3 = (-1) \cdot (-3)^3 = 27$$

Geometric Series

For each geometric sequence
$$a_1, a_1 r, a_1 r^2, \ldots, a_1 r^{n-1}$$
there is a corresponding **geometric series**
$$S_n = a_1 + a_1 r + a_1 r^2 + \cdots + a_1 r^{n-1} \qquad (11.4)$$

11.3 Geometric Sequences and Series

There is a trick to help in deriving a formula for finding the sum of the geometric series in Equation (11.4). Write S_n and $-S_n r$ as follows

$$S_n = a_1 + a_1 r + a_1 r^2 + \cdots + a_1 r^{n-1}$$
$$-S_n r = \quad - a_1 r - a_1 r^2 - \cdots - a_1 r^{n-1} - ar^n$$

When we add the terms columnwise, we get

$$S_n = a_1 + a_1 r + a_1 r^2 + \cdots + a_1 r^{n-1}$$
$$-S_n r = \quad - a_1 r - a_1 r^2 + \cdots - a_1 r^{n-1} - a_1 r^n \qquad \text{Add}$$
$$\overline{S_n - S_n r = a_1 + 0 \quad + 0 \quad + \cdots + 0 \quad\quad - a_1 r^n}$$

or $S_n(1 - r) = a_1(1 - r^n)$.

If $r \neq 1$, we may divide both sides by $1 - r$ to get

$$S_n = \frac{a_1(1 - r^n)}{1 - r} \qquad (11.5)$$

If $r = 1$, then the geometric series in Equation (11.4) becomes

$$S_n = a_1 + a_1 + \cdots + a_1 \quad (n \text{ terms})$$
$$= na_1$$

EXAMPLE 3 A colony of a certain type of bacteria doubles every day. If there are 1000 bacteria now, how many will there be after 7 days?

Solution: Substituting 1000 for a_1, 2 for r, and 7 for n in Equation (11.5), we have

$$S_7 = \frac{1000(1 - 2^7)}{1 - 2} = \frac{1000(1 - 128)}{-1} = 127{,}000$$

There will be 127,000 bacteria after 7 days.

Can we add an infinite amount of numbers? For example, can we find the sum of the following geometric series?

$$1 + \frac{1}{2} + \frac{1}{4} + \frac{1}{8} + \cdots \qquad (11.6)$$

While we know it would not be physically possible to add an infinite amount of numbers, we can add just the first n terms of this series.

$$S_n = 1 + \frac{1}{2} + \frac{1}{4} + \cdots + \frac{1}{2^{n-1}}$$

Since $a_1 = 1$ and $r = \frac{1}{2}$, by Equation (11.5) we have

$$S_n = \frac{1 - \left(\frac{1}{2}\right)^n}{1 - \frac{1}{2}} = \frac{1}{1 - \frac{1}{2}} - \frac{\left(\frac{1}{2}\right)^n}{1 - \frac{1}{2}}$$

$$= 2 - 2\left(\frac{1}{2}\right)^n$$

If we allow n to become larger and larger, $\left(\frac{1}{2}\right)^n$ will become closer to 0, and consequently S_n becomes closer to 2. These phenomena are denoted, respectively, as

$$\lim_{n \to \infty} \left(\frac{1}{2}\right)^n = 0 \qquad \text{(This is read, ``The limit of } \left(\frac{1}{2}\right)^n \text{ equals 0 as } n \text{ tends to infinity.'')}$$

and
$$\lim_{n \to \infty} S_n = 2$$

Therefore, it would be reasonable to say that the sum of the infinite geometric series in Equation (11.6) is 2. This concept can be generalized as follows.

Let S_n denote the sum of the first n terms of the geometric series.

$$S = a_1 + a_1 r + a_2 r^2 + a_1 r^3 + \cdots, \qquad \text{with } |r| < 1$$

Since

$$\lim_{n \to \infty} r^n = 0$$

we have
$$\lim_{n \to \infty} S_n = \lim_{n \to \infty} \frac{a_1(1 - r^n)}{1 - r} = \frac{a_1}{1 - r}$$

The sum is

$$S = \frac{a_1}{1 - r}$$

See Equation (11.7).

The Sum of the Infinite Geometric Series

$$S = a_1 + a_1 r + a_1 r^2 + \cdots \qquad \text{with } |r| < 1 \qquad (11.7)$$

is

$$S = \frac{a_1}{1 - r}$$

What if $|r| \geq 1$? In this case we have a series such as

$$1 + 2 + 4 + 8 + \cdots + 2^n \qquad (r = 2)$$
$$1 - 3 + 9 - 27 + \cdots + (-3)^n + \cdots \qquad (r = -3)$$

or
$$2 + 2 + 2 + 2 + \cdots \qquad (r = 1)$$

For this kind of series, $\lim_{n \to \infty} S_n$ does not give a definite value. (Usually, S_n increases as n increases, and $\lim_{n \to \infty} S_n$ does not exist.)

EXAMPLE 4 Express the repeating decimal $0.\overline{25} = 0.252525\ldots$ as a fraction.

Solution: We can rewrite the repeating decimal as

$$0.252525\ldots = \frac{25}{100} + \frac{25}{10{,}000} + \frac{25}{1{,}000{,}000} + \cdots$$

$$= \frac{25}{100} + \frac{25}{100^2} + \frac{25}{100^3} + \cdots$$

which is a geometric series with the common ratio

$$r = \frac{1}{100} < 1$$

According to Equation (11.7)

$$0.\overline{25} = \frac{25}{100} + \frac{25}{100^2} + \frac{25}{100^3} + \cdots$$

$$= \frac{\frac{25}{100}}{1 - \frac{1}{100}} \qquad \left(a_1 = \frac{25}{100}, \ r = \frac{1}{100}\right)$$

$$= \frac{25}{99}$$

You may check the answer by dividing 25 by 99.

Exercises 11.3

Standard Assignment: Exercises 1, 3, 5, 7, 9, 11, 13, 15, 17, 21, 23, 25, 27, 29, 33, 35, 37, 41

A Find the common ratio for each of the following geometric sequences and write the next three terms of each sequence. See Example 1.

1. 3, 6, 12, 24, ...
2. 4, 12, 36, 108, ...

3. $1, -3, 9, -27, \ldots$

4. $-1, 2, -4, 8, \ldots$

5. $a, -a, a, -a, \ldots$

6. $1, -b, b^2, -b^3, \ldots$

7. $c^2, c, 1, \dfrac{1}{c}, \ldots$

8. $x, 2, \dfrac{4}{x}, \dfrac{8}{x^2}, \ldots$

9. $\dfrac{x}{a}, -1, \dfrac{a}{x}, -\dfrac{a^2}{x^2}, \ldots$

10. $\dfrac{a}{x}, -\dfrac{a}{xy}, \dfrac{a}{xy^2}, -\dfrac{a}{xy^3}, \ldots$

Write the first four terms of each of the following geometric sequences. See Example 2.

11. $a_1 = \dfrac{1}{16}, r = 2$

12. $a_1 = \dfrac{1}{81}, r = 3$

13. $a_1 = 81, r = -\dfrac{1}{3}$

14. $a_1 = 16, r = -\dfrac{1}{2}$

15. $a_5 = 27, a_6 = 81$

16. $a_5 = 81, a_6 = 27$

17. $a_5 = -1, a_6 = \dfrac{1}{x}$

18. $a_5 = y, a_6 = -y^2$

19. $a_6 = \dfrac{x}{a}, a_7 = \dfrac{x^2}{a^2}$

20. $a_6 = \dfrac{xy^2}{b}, a_7 = -\dfrac{xy^3}{b}$

In Exercises 21–26 write the general term (kth term) of each of the following geometric sequences.

21. $6, 2, \dfrac{2}{3}, \ldots$

22. $\dfrac{1}{10}, \dfrac{1}{2}, \dfrac{5}{2}, \ldots$

23. $10, -\dfrac{1}{10}, \dfrac{1}{1000}, \ldots$

24. $-\dfrac{1}{25}, \dfrac{1}{5}, -1, \ldots$

25. $a, ax, ax^2, \ldots$

26. $a, -x, \dfrac{x^2}{a}, \ldots$

In Exercises 27–32 find the sum of the first 10 terms of each of the following geometric series.

27. $\dfrac{1}{10} + \dfrac{1}{2} + \dfrac{5}{2} + \cdots$

28. $6 + 2 + \dfrac{2}{3} + \cdots$

29. $\dfrac{1}{25} - \dfrac{1}{5} + 1 - \cdots$

30. $-10 + \dfrac{1}{10} - \dfrac{1}{1000} + \cdots$

31. $a + ax + ax^2 + \cdots$

32. $a - x + \dfrac{x^2}{a} + \cdots$

B 33. A colony of a certain type of bacteria triples every day. If there are 500 bacteria now, how many will there be after 5 days?

34. If $1.00 is deposited in a savings bank that pays 12% interest compounded annually, how much will it be worth after 5 years? After n years?

35. If $1.00 is deposited at the beginning of each year in a bank that pays 12% interest compounded annually, what will be the total worth at the end of 8 years?

36. Solve Exercise 35 for the total worth at the end of 10 years.

37. Johnny went to buy a house. The seller offered two choices of payment plans for the house:
    (a) Pay $4000 per month for the next 25 months.
    (b) Pay 1¢ the 1st month, 2¢ the 2nd month, 4¢ the 3rd month, 8¢ the 4th month, and so on for 25 months.
    Without hesitation Johnny bought the house with the payment plan (b). Did Johnny make the correct choice? Why? (A calculator or a log table may be used.)

38. Mike was awarded a one-year scholarship that pays $1 in January, $2 in February, $4 in March, and so on. How much will Mike receive in the year?

39. A particular ball always rebounds $\frac{3}{5}$ the distance it falls. If the ball is dropped from a height of 5 meters, how far will it travel before coming to rest? (Hint: Consider the total downward distance and the total upward distance separately.)

40. Solve Exercise 39 if the ball is dropped from a height of 9 meters.

Express each repeating decimal as a fraction. See Example 4.

41. $3.\overline{14}$
42. $1.4\overline{9}$
43. $3.\overline{33}$
44. $0.\overline{101}$
45. $2.\overline{66}$
46. $9.\overline{99}$

11.4 The Binomial Theorem

The binomial power $(a + b)^n$, where n is a natural number, appears frequently in many branches of mathematics including calculus, probability, statistics, and algebra. In this section we will perform an informal study of the coefficients in the expansion of $(a + b)^n$.

In order to understand the general formula for the expansion of $(a + b)^n$, first observe the following.

$$(a + b)^1 = a + b$$
$$(a + b)^2 = a^2 + 2ab + b^2$$
$$(a + b)^3 = a^3 + 3a^2b + 3ab^2 + b^3$$
$$(a + b)^4 = a^4 + 4a^3b + 6a^2b^2 + 4ab^3 + b^4$$
$$(a + b)^5 = a^5 + 5a^4b + 10a^3b^2 + 10a^2b^3 + 5ab^4 + b^5$$

The first two expansions should be familiar to you; the last three are left as exercises for you to verify.

The expansions for $(a + b)^n$ (with $n = 1, 2, 3, 4, 5$) suggest the following

The Binomial Expansion

1. The expansion of $(a + b)^n$ has $n + 1$ terms:

$$a^n,\ a^{n-1}b,\ a^{n-2}b^2,\ \ldots,\ ab^{n-1},\ b^n$$

2. The exponents of a and b in each term always add up to n.
3. The exponent of a starts at n and decreases by one for each term until it is 0 in the last term, while the exponent of b starts at 0 in the 1st term and increases by one for each term until it is n in the last term.
4. From left to right, the coefficients of the $n + 1$ terms that are given in Step 1, are

$$1,\ \frac{n}{1},\ \frac{n(n-1)}{1\cdot 2},\ \frac{n(n-1)(n-2)}{1\cdot 2\cdot 3},\ \ldots,\ \frac{n(n-1)(n-2)\ldots 2}{1\cdot 2\cdot 3\cdot\ldots\cdot(n-1)},\ 1$$

You may have noticed that the coefficients of the 1st and the last terms are both 1 and the coefficients of the 2nd and the 2nd to the last terms are equal. In general, the coefficients of

$$a^{n-k}b^k \quad \text{and} \quad a^k b^{n-k}$$

are equal for $k = 0, 1, 2, \ldots, n$.

We will now summarize these observations in a formula called the **binomial theorem**, the proof of which is usually given by a method of proof called "mathematical induction," which is a subject discussed in the next higher level of algebra.

The Binomial Theorem

If n is a natural number, then the binomial expansion of $(a + b)^n$ is given by

$$(a + b)^n = a^n + \frac{n}{1}a^{n-1}b + \frac{n(n-1)}{1\cdot 2}a^{n-2}b^2$$

$$+ \frac{n(n-1)(n-3)}{1\cdot 2\cdot 3}a^{n-3}b^3 + \cdots + b^n$$

EXAMPLE 1 Find the expansion of $(x + 2y)^4$ by using the binomial theorem.

Solution: Since $n = 4$, we shall expand $(a + b)^4$ and then replace a by x and b by $2y$.

$$(a + b)^4 = a^4 + \frac{4}{1}a^3b + \frac{4\cdot 3}{1\cdot 2}a^2b^2 + \frac{4\cdot 3\cdot 2}{1\cdot 2\cdot 3}ab^3 + b^4$$

$$= a^4 + 4a^3b + 6a^2b^2 + 4ab^3 + b^4$$

Replacing a by x and b by $2y$ we have

$$(x + 2y)^4 = x^4 + 4x^3(2y) + 6x^2(2y)^2 + 4x(2y)^3 + (2y)^4$$
$$= x^4 + 8x^3y + 24x^2y^2 + 32xy^3 + 16y^4$$

EXAMPLE 2 Find the expansion of $(1 - 2x)^5$.

Solution: Since $n = 5$, we shall expand $(a + b)^5$ and then replace a by 1 and b by $-2x$.

$$(a + b)^5 = a^5 + \frac{5}{1}a^4b + \frac{5 \cdot 4}{1 \cdot 2}a^3b^2 + \frac{5 \cdot 4 \cdot 3}{1 \cdot 2 \cdot 3}a^2b^3 + \frac{5 \cdot 4 \cdot 3 \cdot 2}{1 \cdot 2 \cdot 3 \cdot 4}ab^4 + b^5$$

$$= a^5 + 5a^4b + 10a^3b^2 + 10a^2b^3 + 5ab^4 + b^5$$

Replacing a by 1 and b by $-2x$ we get

$$(1 - 2x)^5 = 1 + 5 \cdot 1 \cdot (-2x) + 10 \cdot 1 \cdot (-2x)^2 + 10 \cdot 1 \cdot (-2x)^3$$
$$+ 5 \cdot 1 \cdot (-2x)^4 + (-2x)^5$$
$$= 1 - 10x + 40x^2 - 80x^3 + 80x^4 - 32x^5$$

EXAMPLE 3 Show that the sum of the coefficients in the binomial expansion of $(a + b)^n$ equals 2^n, that is

$$1 + n + \frac{n(n - 1)}{1 \cdot 2} + \frac{n(n - 1)(n - 2)}{1 \cdot 2 \cdot 3} + \cdots + n + 1 = 2^n$$

Solution: In the binomial expansion

$$(a + b)^n = a^n + \frac{n}{1}a^{n-1}b + \frac{n(n - 1)}{1 \cdot 2}a^{n-2}b^2$$
$$+ \frac{n(n - 1)(n - 2)}{1 \cdot 2 \cdot 3}a^{n-3}b^3 + \cdots + \frac{n}{1}ab^{n-1} + b^n$$

we let both a and b be 1, then $a^{n-k}b^k = (1)^{n-k}(1)^k = 1$.

$$(1 + 1)^n = 1 + n + \frac{n(n - 1)}{1 \cdot 2} + \frac{n(n - 1)(n - 2)}{1 \cdot 2 \cdot 3} + \cdots + n + 1$$

or $\quad 1 + n + \dfrac{n(n - 1)}{1 \cdot 2} + \dfrac{n(n - 1)(n - 2)}{1 \cdot 2 \cdot 3} + \cdots + n + 1 = 2^n$

The coefficients of a binomial expansion are called **binomial coefficients**.

Pascal's Triangle

As early as A.D. 1100, the Chinese scholar Chia Hsien discovered the secret of the binomial coefficients that was later rediscovered by the French philosopher and mathematician Blaise Pascal (1622–1662). In order to understand

Chia Hsien's and Pascal's construction of the binomial coefficients, let us first look at the coefficients of the binomial expansion.

$$(a + b)^0 = 1$$
$$(a + b)^1 = 1a + 1b$$
$$(a + b)^2 = 1a^2 + 2ab + 1b^2$$
$$(a + b)^3 = 1a^3 + 3a^2b + 3ab^2 + 1b^3$$
$$(a + b)^4 = 1a^4 + 4a^3b + 6a^2b^2 + 4ab^3 + 1b^4$$
$$(a + b)^5 = 1a^5 + 5a^4b + 10a^3b^2 + 10a^2b^3 + 5ab^4 + 1b^5$$

If we remove the variables and plus signs and list only the coefficients, we get a triangle of numbers composed of the binomial coefficients that is known as **Pascal's triangle**.

Figure 11.1

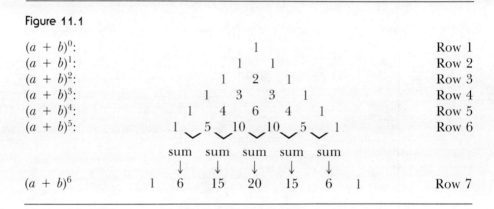

Note the symmetry in Pascal's triangle. If the triangle were folded across the vertical line through the top number 1, the numbers would match. As shown in Figure 11.1, in order to create a new bottom row in the triangle, we must add two neighboring entries in the previous row and always place the number 1 in the first and the last places of the new row.

EXAMPLE 4

Use Pascal's triangle to find the expansion of $(a - 1)^7$.

Solution: We first find the binomial expansion of $(a + b)^7$ by extending the Pascal's triangle in Figure 11.1 from Row 7 to Row 8.

```
(a + b)^6:          1   6   15   20   15   6   1              Row 7
                      sum sum  sum  sum  sum sum
(a + b)^7:       1   7   21   35   35   21   7   1           Row 8
```

Thus,
$$(a+b)^7 = a^7 + 7a^6b + 21a^5b^2 + 35a^4b^3$$
$$+ 35a^3b^4 + 21a^2b^5 + 7ab^6 + b^7$$

Now replace b by -1.
$$(a-1)^7 = a^7 + 7a^6(-1) + 21a^5(-1)^2 + 35a^4(-1)^3 + 35a^3(-1)^4$$
$$+ 21a^2(-1)^5 + 7a(-1)^6 + (-1)^7$$
$$= a^7 - 7a^6 + 21a^5 - 35a^4 + 35a^3 - 21a^2 + 7a - 1$$

Exercises 11.4

Standard Assignment: All odd-numbered exercises

A In Exercises 1–12 use the binomial theorem to expand the given binomial powers. See Examples 1 and 2.

1. $(x + 1)^4$
2. $(x + 2)^4$
3. $(x - 1)^5$
4. $(1 - x)^5$
5. $(x - 2y)^3$
6. $(2x - y)^3$
7. $(2x + 1)^4$
8. $(3x - 2)^4$
9. $\left(u + \dfrac{v}{u}\right)^4$
10. $\left(v - \dfrac{u}{v}\right)^4$
11. $(1 + 0.01)^6$
12. $(1 - 0.01)^7$

13. Use the result of Exercise 11 to evaluate $(1.01)^6$ to three decimal places.

14. Use the result of Exercise 12 to evaluate $(0.99)^7$ to three decimal places.

15. Construct a Pascal's triangle with nine rows.

16. Construct a Pascal's triangle with ten rows.

In Exercises 17–26, use Pascal's triangle to find the expansion. See Example 4.

17. $(a - b)^5$
18. $(x - y)^6$
19. $(1 - y)^6$
20. $(1 - x)^7$

21. $\left(x + \dfrac{1}{x}\right)^7$; simplify your result

22. $\left(x - \dfrac{1}{x}\right)^6$; simplify your result

23. $(2u - 1)^5$
24. $(2u + 3v)^4$
25. $(x - y)^9$
26. $(x + y)^{10}$

Use Pascal's triangle to answer Exercises 27–33.

27. What is the coefficient of a^3b^2 in the expansion of $(a + b)^5$?

28. What is the coefficient of x^3y^4 in the expansion of $(x + y)^7$?

29. What is the coefficient of x^6 in the expansion of $(x + 1)^8$?

30. What is the coefficient of y^5 in the expansion of $(1 - y)^9$?

31. What is the coefficient of x^{99} in the expansion of $(x - 9)^{99}$?

32. What is the coefficient of y^{101} in the expansion of $(8 - y)^{101}$?

33. What is the coefficient of u^9v in the expansion of $(u - 5v)^{10}$?

34. What is the sum of the coefficients in the expansion of $(a - b)^n$? See Example 3.

35. Express the binomial theorem by using the summation notation, Σ.

11.5 Permutations

In scientific analysis, we often need to ask, "What are the possible outcomes?" "How many possible outcomes are there?" We will first consider the number of ways in which a set of k objects can be arranged. An *arrangement* of k different objects in a certain order is called a **permutation** of the k objects. For example, all possible permutations of three objects denoted by P, Q, and R are shown in the following *tree diagram* (Figure 11.2).

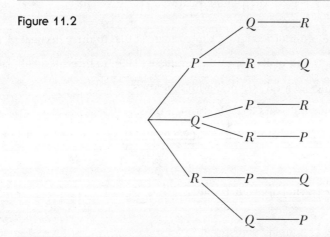

Figure 11.2

Each path in the diagram represents a possible permutation, and there are six paths, or permutations:

$$\text{PQR}, \quad \text{PRQ}, \quad \text{QPR},$$
$$\text{QRP}, \quad \text{RPQ}, \quad \text{RQP}$$

In order to derive a general formula for counting the number of possible permutations of n objects and solving other similar counting problems, we need a general counting principle.

11.5 Permutations

Suppose there are two roads, p and q, connecting Tampa and Orlando, and three roads, x, y, and z, connecting Orlando and Miami. How many routes can we choose from Tampa to Miami that pass through Orlando?

The diagram in Figure 11.3 leads to the following analysis. For each of the 2 routes from Tampa to Orlando, there are 3 choices from Orlando to Miami. Hence, we have $2 \cdot 3$ possible routes from Tampa to Miami. The idea exhibited here is generalized in the *Fundamental Counting Principle*: For a sequence of k events in which the first event can occur in n_1 ways, the second can occur in n_2 ways, and so on, with the kth event occurring in n_k ways, the number of ways in which the sequence of events can occur is

$$n_1 \cdot n_2 \cdot n_3 \cdot \cdots \cdot n_k$$

Figure 11.3

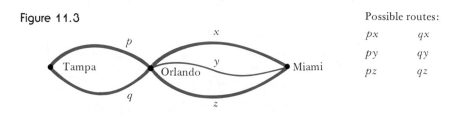

Possible routes:

px	qx
py	qy
pz	qz

EXAMPLE 1

(a) How many four-digit numbers can be formed by digits 1, 2, 3, and 4, using each digit only once?
(b) How many four-digit numbers can be formed if each digit may be repeatedly used?

Solution:
(a) Consider writing a four-digit number from left to right. There are 4 choices for the 1st digit. Since each digit can be used only once, there are 3 choices for the 2nd digit, 2 choices for the 3rd digit, and 1 choice for the last digit. According to the Fundamental Counting Principle, there are $4 \cdot 3 \cdot 2 \cdot 1$, or 24, such four-digit numbers.
(b) If each digit may be repeatedly used, there are the same four choices for each digit; thus there can be $4 \cdot 4 \cdot 4 \cdot 4$, or 256, four-digit numbers.

The number $4 \cdot 3 \cdot 2 \cdot 1$ obtained in Example 1(b) occurs so often that it merits a special symbol

$$4! = 4 \cdot 3 \cdot 2 \cdot 1$$
$$n! = n(n-1)(n-2) \cdots 3 \cdot 2 \cdot 1$$

These are read "four **factorial**" and "n **factorial**," respectively.

Observe that
$$n! = n \cdot (n-1)!$$
However, for $n = 1$, $1! = 1 \cdot 0!$. For this and other reasons, we define $0!$ to be 1.

> $n! = n \cdot (n-1)!$
> $0! = 1$

Our answer to Example 1(a) also indicates that the number of permutations of 4 objects is $4!$. Similarly,

The number of permutations of n objects is $n!$.

EXAMPLE 2

The five semifinalists in a Miss America pageant are to sit in a row for a group picture. How many group pictures can be taken, with each showing a different seating arrangement?

Solution: There are $5!$ ($= 120$) such seating arrangements, and thus 120 pictures can be taken.

Permutations of n Objects Taken r at a Time

Suppose that a class of 20 students is having a meeting to elect 3 officers: a president, a vice president, and a secretary. In how many different ways can the election turn out? The Fundamental Counting Principle comes in handy here. Since there are 20 students, there are 20 possible outcomes in the presidential election. Once the president is elected, there will be one less student available when choosing a vice president. Similarly, after the president and vice president are elected, there will be only 18 choices for the secretary. Thus, the election can have

$$20 \cdot 19 \cdot 18 \; (= 6840)$$

possible outcomes. Each possible result in this example is referred to as a permutation of 20 objects taken 3 at a time.

In general, the number of permutations of n objects taken r at a time is denoted by $P(n, r)$. Thus,

$$P(20, 3) = 20 \cdot (20-1) \cdot (20-2)$$
$$= 6840$$

In general, for $r \leq n$, we have Equation (11.8).

$$P(n, r) = n(n - 1)(n - 2) \ldots (n - r + 1) \qquad (11.8)$$

Notice that there are exactly r factors in the formula for $P(n, r)$, and that $P(n, n) = n!$. We also have Equation (11.9).

$$P(n, r) = \frac{n!}{(n - r)!} \qquad (11.9)$$

EXAMPLE 3   How many signals can a ship send if the ship carries 9 flags and a signal consists of 5 flags hoisted vertically upward on a line?

Solution: Since

$$P(9, 5) = 9 \cdot 8 \cdot 7 \cdot 6 \cdot 5 \quad \text{(5 consecutive integers)}$$
$$= 15120$$

The ship can send as many as 15120 different signals.

Permutations with Some Indistinguishable Objects

We will next determine how many distinguishable permutations can be formed using all 6 letters in the word FREEZE, three of which are identical. Suppose that the number of all distinguishable permutations is P. We want to compare these permutations, which are referred to as the original permutations with the corresponding permutations of the 6 symbols F, R, E, E', Z, E", all of which are different. Using these modified letters, there are 6! distinguishable permutations, which are referred to as modified permutations. If we take any original permutation, such as FEEZER, and leave F, Z, and R fixed while permuting the 3 identical E's, we get nothing new. But if we treat the corresponding modified permutation, FEE'ZE"R, in a similar manner, we get 3! different permutations. Hence, for each original permutation, there are 3! modified permutations. Therefore, $3!P = 6!$, or

$$P = \frac{6!}{3!}$$

This is readily generalized.

The number of distinguishable permutations of n objects in which n_1 are of one kind, n_2 are of a second kind, ..., n_k are of the kth kind, is

$$\frac{n!}{n_1!\, n_2! \cdots n_k!}$$

EXAMPLE 4

(a) Professor Fair has decided to give one A, two B's, three C's, and four F's to the ten students he teaches. In how many ways can he do this?

(b) A pair of identical twins and three other (distinguishable) students are to sit in a row for a picture. How many different pictures can result?

Solution:

(a) There are $\dfrac{10!}{1!\, 2!\, 3!\, 4!}$ (or 12,600) ways.

(b) There are $\dfrac{5!}{2!\, 1!\, 1!\, 1!}$ (or 60) possible pictures.

Exercises 11.5

Standard Assignment: All odd-numbered exercises 1–21

A In Exercises 1–6, how many four-digit numbers can be formed by using the indicated set of digits? Each digit may be used only one time. See Example 1.

1. $\{2, 4, 6, 8\}$
2. $\{1, 3, 5, 7\}$
3. $\{1, 2, 3, 4, 5\}$
4. $\{1, 2, 3, 4, 5, 6\}$
5. $\{0, 1, 2, 3\}$
6. $\{0, 1, 2, 3, \ldots, 9\}$

7. Solve Exercises 4 and 5 if each digit can be used repeatedly. See Example 1(b).

8. Answer Exercises 3 and 6 if each digit can be used repeatedly.

9. The students at Funland University have elected seven officers for their student government. No officer knows in how many ways they (the officers) can line up for a group picture. Can you help?

10. In Exercise 9, the officers consist of four girls and three boys. No girl wants to stand next to another girl, and the boys do not mind standing beside the girls. In how many ways can they line up?

11. How many signals can a ship send if the ship carries 10 flags and a signal consists of 4 flags hoisted vertically upward on a line?

12. How many five-letter code symbols can be made with the letters of the word EQUALITY if letters cannot be repeated?

In Exercises 13–20, calculate.

13. $8!$
14. $7! - 5!$
15. $10!/7!$

16. $P(8, 3)$ 	 17. $P(8, 5)$ 	 18. $P(12, 2)$

19. $P(5, 5)$ 	 20. $P(20, 3) - P(20, 2)$

21. Frank has 5 shirts, 3 pairs of trousers, and 4 ties. Assuming that Frank wears one of each, how many different outfits can he wear?

22. The university cafeteria offers 2 choices of soup, 3 choices of salad, 5 choices of main dishes, and 6 different desserts. How many different four-course meals can you choose?

23. ▦ Maine license plate numbers have 3 letters followed by 3 digits, such as ORH838. What is the maximum number of license plates that could be issued?

24. ▦ Solve Exercise 23, if 2 letters and 4 digits are used.

In Exercises 25–30, in how many distinguishable ways can the letters of each given word be arranged?

25. ▦ ARRANGEMENT 	 26. ▦ AMERICA 	 27. ▦ TAIWAN

28. ▦ SUCCESS 	 29. ▦ GREECE 	 30. ▦ TALLAHASSEE

31. ▦ What are values of $P(5, 0)$, $P(0, 0)$, and $P(n, 0)$? [Use Equation (11.9).]

11.6 Combinations

Sometimes, when we are selecting a certain number of objects from a set, we are not concerned with the order in which the objects are selected or arranged. Such selections or arrangements are called **combinations**.

The number of combinations that are taken r at a time from a set of n objects is denoted by $\binom{n}{r}$ or $C(n, r)$. For example, the combinations that are taken 2 at a time from the set $\{x, y, z\}$ of 3 elements are xy, xz, and yz. Thus,

$$\binom{3}{2} = 3 \quad \text{or} \quad C(3, 2) = 3$$

The Relation Between $\binom{n}{r}$ and $P(n, r)$

Given a set of n objects, for each combination of r objects selected from this set, there are $r!$ permutations of r objects taken from n objects. Thus,

$$\binom{n}{r} r! = P(n, r), \quad (\text{with } r \leq n)$$

So

$$\binom{n}{r} = \frac{P(n, r)}{r!}$$

$$= \frac{n(n - 1)(n - 2) \ldots (n - r + 1)}{1 \cdot 2 \cdot 3 \cdots r} \qquad (11.10)$$

EXAMPLE 1 Find the values of $\binom{4}{0}$, $\binom{4}{1}$, $\binom{4}{2}$, $\binom{4}{3}$, and $\binom{4}{4}$.

Solution: Using Equations (11.10) and (11.9) we get

$$\binom{4}{0} = \frac{P(4, 0)}{0!} = \frac{1}{1} = 1$$

$$\binom{4}{1} = \frac{4}{1} = 4$$

$$\binom{4}{2} = \frac{4 \cdot 3}{1 \cdot 2} = 6$$

$$\binom{4}{3} = \frac{4 \cdot 3 \cdot 2}{1 \cdot 2 \cdot 3} = 4$$

$$\binom{4}{4} = \frac{4 \cdot 3 \cdot 2 \cdot 1}{1 \cdot 2 \cdot 3 \cdot 4} = 1$$

The numbers,

$$1, 4, 6, 4, 1$$

in that order, are exactly the binomial coefficients of $(a + b)^4$ as were seen in row 5 of the Pascal's triangle in Figure 11.1 of Section 11.4.

In general,

$$\binom{n}{0} = \frac{P(n, 0)}{0!} = \frac{1}{1} = 1$$

$$\binom{n}{1} = \frac{n}{1} = n$$

$$\binom{n}{2} = \frac{n(n - 1)}{1 \cdot 2}$$

and

$$\binom{n}{n} = \frac{P(n, n)}{n!} = \frac{n!}{n!} = 1$$

are exactly the binomial coefficients of $(a + b)^n$.

The binomial theorem can be restated as: If n is a natural number, then the expansion of $(a + b)^n$ is given by

11.6 Combinations

$$(a + b)^n = a^n + \binom{n}{1}a^{n-1}b + \binom{n}{2}a^{n-2}b^2 + \cdots$$
$$+ \binom{n}{r}a^{n-r}b^r + \cdots + b^n$$

EXAMPLE 2 The shareholders of ABC Company have elected 10 members to the board of trustees. The 10 board members will be divided into two groups. A group of 4 members will be in charge of the internal affairs and the other group of 6 will oversee the external matters. In how many ways can the 10 board members be divided?

Solution: There are two methods of dividing. One method is to take four members for internal affairs out of the 10 members and leave the remaining 6 for the external affairs. Thus, there are $C(10, 4)$ ways.

The other method is to select 6 members out of 10 for the external matters, leaving the remaining 4 for the internal affairs. Thus, there are $C(10, 6)$ ways.

We expect that the two methods lead to the same answer. Let us now check this.

$$C(10, 4) = \frac{10 \cdot 9 \cdot 8 \cdot 7}{1 \cdot 2 \cdot 3 \cdot 4} = \frac{70 \cdot 9}{3} = 210$$

and

$$C(10, 6) = \frac{10 \cdot 9 \cdot 8 \cdot 7 \cdot 6 \cdot 5}{1 \cdot 2 \cdot 3 \cdot 4 \cdot 5 \cdot 6} = 210 = C(10, 4)$$

Thus, as expected

$$\binom{10}{4} = \binom{10}{6}$$

In general, we have for any two nonnegative integers n and r such that $r \leq n$,

$$\binom{n}{r} = \binom{n}{n - r} \qquad (11.11)$$

Applying Equation (11.11) to the binomial theorem, we now understand why Pascal's triangle is symmetrical.

EXAMPLE 3

Among Conrad's collection of early American coins, he has a half-dollar, a quarter, a nickel, and a penny. He forgot to buy a Christmas gift for his younger brother so he decided to make a gift of some or all of these four coins instead. How many different sets of coins can Conrad give?

Solution: He can give

$C(4, 1)$ different gifts of one coin,

or

$C(4, 2)$ different gifts of two coins,

or

$C(4, 3)$ different gifts of three coins,

or give

$C(4, 4) = 1$ gift of all four coins.

Any set from the total of $C(4, 1) + C(4, 2) + C(4, 3) + C(4, 4) = 4 + 6 + 4 + 1 = 15$ different gift sets may be given.

Exercises 11.6

Standard Assignment: Exercises 1, 3, 5, 7, 11, 13, 15, 17, 19, 21

A In Exercises 1–8 compute each indicated expression.

1. $C(6, 2)$
2. $C(6, 4)$
3. $C(7, 0)$
4. $C(7, 7)$
5. $C(20, 19)$
6. $C(30, 29)$
7. $\binom{5}{0} + \binom{5}{1} + \cdots + \binom{5}{5}$
8. $\binom{6}{0} + \binom{6}{1} + \cdots + \binom{6}{6}$

In Exercises 9–14 compute each given pair and compare the answers.

9. $\binom{3}{1} + \binom{3}{2}$, $\binom{4}{2}$
10. $\binom{4}{2} + \binom{4}{3}$, $\binom{5}{3}$
11. $\binom{5}{2} + \binom{5}{3}$, $\binom{6}{3}$
12. $\binom{5}{1} + \binom{5}{2}$, $\binom{6}{2}$
13. $\binom{5}{3} + \binom{5}{4}$, $\binom{6}{4}$
14. $\binom{5}{4} + \binom{5}{5}$, $\binom{6}{5}$

15. Prove that for any natural numbers n and r such that $r \leq n$,

$$\binom{n}{r-1} + \binom{n}{r} = \binom{n+1}{r}$$ [Hint: Use Equation (11.10).]

16. Prove that
$$\binom{n}{r} = \frac{n!}{r! \cdot (n-r)!}$$

17. Use the result of Exercise 16 to show that
$$\binom{n}{n-r} = \binom{n}{r}$$

18. Can you use the result of Exercise 15 to explain the validity of Pascal's triangle?

19. How many baseball games should be scheduled for a 13-team league if each team plays all other teams once?

20. Solve Exercise 19 if there are 12 teams and each team plays every other team twice.

21. In Professor Smith's test, students are allowed to choose 9 out of 11 problems. In how many ways can the students make their choices?

B 22. Professor Lynn gave a quiz that contains 12 true-false questions. In order to pass the quiz, a student must give correct answers to at least 8 of the 12 questions. In how many ways can a student pass the quiz?

23. In geometry, any two points determine a line. How many lines can be determined by 30 given points, no three of which are collinear?

24. Solve Exercise 23, if 50 points are given, no three of which are collinear.

25. In geometry, any three noncollinear points determine a triangle. How many triangles can be determined by 21 given points, no three of which are collinear?

26. Solve Exercise 25, if 17 points are given, no three of which are collinear.

In Exercises 27–34, solve for the indicated unknown.

27. $\binom{m+1}{m-1} = 3!$

28. $6\binom{n-1}{2} = \binom{n+1}{4}$

29. $2\binom{n-1}{2} = \binom{n}{3}$

30. $\frac{4}{3}\binom{k}{2} = \binom{k+1}{3}$

31. $\binom{n}{0} + \binom{n}{1} + \binom{n}{2} + \cdots + \binom{n}{n} = 64$

32. $\binom{k}{1} + \binom{k}{2} + \binom{k}{3} + \cdots + \binom{k}{k-1} = 126$

33. $\binom{m}{4} = \binom{m}{5}$

34. $\binom{n}{7} = \binom{n}{3}$

35. Brian wants to give some money to his school; he has only one each of $1.00, $5.00, and $10.00 bills. How many different amounts can Brian give? (See Example 3.)

36. Mary has a penny, a nickel, a quarter, and a half dollar. She wishes to give some of her money to her little brother. How many different amounts can she possibly give?

37. In how many ways can a committee of 3 men and 4 women be nominated from a class with 12 women and 10 men?

38. Solve Exercise 37 if the class has 12 men and 10 women.

Chapter 11 Summary

Key Words and Phrases

11.1 Sequence
Term
General term
Infinite sequence
Series
Infinite series
Summation notation
11.2 Arithmetic sequence
Arithmetic series
Common difference
11.3 Geometric sequence

Geometric series
Common ratio
11.4 Binomial theorem
Binomial expansion
Binomial coefficients
Pascal's triangle
11.5 Permutation
Fundamental Counting Principle
Factorial
11.6 Combination

Key Concepts and Rules

The sum of the arithmetic series $a_1 + a_2 + \cdots + a_n$ is given by

$$a_1 + a_2 + \cdots + a_n = \frac{n}{2}(a_1 + a_n)$$

and the sum of geometric series $a_1 + a_2 + \cdots + a_n$ with common ratio r is given by

$$a_1 + a_2 + \cdots + a_n = \frac{a_1(1 - r^n)}{1 - r} \quad \text{(with } r \neq 1\text{)}$$

The Binomial Theorem: If n is a natural number, the expansion of $(a + b)^n$ is given by

$$(a + b)^n = a^n + \frac{n}{1}a^{n-1}b + \frac{n(n-1)}{1 \cdot 2}a^{n-2}b^2$$
$$+ \frac{n(n-1)(n-2)}{1 \cdot 2 \cdot 3}a^{n-3}b^3 + \cdots + b^n$$

The binomial coefficients for $n = 0, 1, 2, \ldots$, form a triangular shape of numbers known as *Pascal's triangle*.

$(a + b)^0$:   1

$(a + b)^1$:   1   1

$(a + b)^2$:   1   2   1

$(a + b)^3$:   1   3   3   1

$(a + b)^4$:   1   sum 4   sum 6   sum 4   1

In general, the number $P(n, r)$ of *permutations* of n objects taken r at a time is given by

$$P(n, r) = \frac{n!}{r!}$$

Theorem: The number of permutations of n objects in which n_1 are of one kind, n_2 are of a second kind, $\cdots$, n_k are of the kth kind, is

$$\frac{n!}{n_1! \, n_2! \, \ldots \, n_k!}$$

The number of *combinations* taken r at a time from a set of n objects is denoted by $\binom{n}{r}$ or $C(n, r)$ and is given by

$$\binom{n}{r} = \frac{n(n-1)(n-2) \cdots (n-r+1)}{r!}$$

Review Exercises

If you have difficulty with any of the problems, look in the section indicated by the numbers in brackets.

[11.1] Find a pattern in each of the given sequences, fill in the blank spaces, then write the general terms.

1. 1, 5, 9, 13, 17, □, □, ...

2. $1, -4, 9, -16, 25, \square, \square, \ldots$

3. $2, \dfrac{3}{2}, \dfrac{4}{3}, \dfrac{5}{4}, \dfrac{6}{5}, \square, \square, \ldots$

4. $1 \cdot 3, -2 \cdot 4, 3 \cdot 5, -4 \cdot 6, 5 \cdot 7, \square, \square, \ldots$

Rewrite each of the following series without summation notation.

5. $\sum_{k=1}^{6} \dfrac{k}{2k-1}$

6. $\sum_{k=2}^{7} \dfrac{(-1)^k}{3k-1}$

7. $\sum_{k=3}^{10} (-1)^{k+1} \dfrac{k^2}{2k-4}$

8. $\sum_{k=1}^{10} \dfrac{1}{k(k+1)}$

Rewrite each series by using summation notation.

9. $1 - 3 + 5 - 7 + 9 - \cdots + 101$

10. $\dfrac{1}{2} - \dfrac{3}{4} + \dfrac{5}{6} - \dfrac{7}{8} + \cdots - \dfrac{99}{100}$

[11.2] Find the sum of each arithmetic series.

11. $2 + 5 + 8 + 11 + \cdots + 98$

12. $1 + 5 + 9 + 13 + \cdots + 101$

13. Find the general term of the series in Exercise 11 and rewrite the series by using summation notation.

14. Find the general term of the series $6 + 11 + 16 + 21 + \cdots + 101$ and rewrite the series by using summation notation.

15. Find the general term of the arithmetic sequence that has $a_3 = 60$ and $a_{13} = 10$.

16. An arithmetic series is known to have 20 terms, $a_{10} = 55$ and $d = 3$. Express this series by using summation notation.

17. After Albert started to work, he managed to save money in such a way that each week's savings was always $2.00 more than the preceding week's savings. After one year Albert was fired. How much was Albert's total savings for the year if his savings for the 13th week was $27.00?

18. Find the common difference of an arithmetic sequence with $a_7 = 24$ and $a_{12} = 39$.

[11.3] 19. What is a geometric sequence?

20. What is the sum of the geometric series that has n terms, initial term a, and the common ratio r?

21. What is the difference between a geometric sequence and a geometric series?

Write the general term (*k*th term) of each indicated geometric sequence.

22. $\dfrac{1}{50}, \dfrac{1}{10}, \dfrac{1}{2}, \dfrac{5}{2}, \ldots$

23. $100, -10, 1, -\dfrac{1}{10}, \ldots$

24. $-\dfrac{1}{9}, \dfrac{1}{3}, -1, \ldots$

25. $a, b, \dfrac{b^2}{a}, \ldots$

Find the sum of each given geometric series.

26. $\dfrac{1}{3} + 1 + 3 + \cdots + 729$

27. $16 + 8 + 4 + \cdots + \dfrac{1}{64}$

28. If a particular ball always rebounds $\dfrac{2}{3}$ the distance it falls, and if the ball is dropped from a height of 7 meters, how far will it travel before coming to rest?

29. Convert the repeating decimal $4.\overline{25}$ into a fraction.

[11.4]

30. How many terms are there in the binomial expansion of $(a + b)^{99}$?

31. What are the terms in the binomial expansion of $(a + b)^7$?

32. Construct an 8-row Pascal's triangle.

33. Use the Pascal's triangle of Exercise 32 to expand

   (a) $(x - 2y)^5$

   (b) $\left(2x + \dfrac{1}{x}\right)^6$

34. Can you express the expansion of $(x - y)^{20}$ by using the summation notation Σ?

35. Find the sum of the binomial coefficients in the expansion of $(x - y)^{21}$.

[11.5]

36. How many four-digit numbers can be formed using 0, 1, 2, 3, 4, and 5? Each digit may be used only once.

37. Solve Exercise 36 if any digit may be used as many times as you wish.

38. What is $P(n, r)$?

39. Calculate $P(11, 4)$.

40. A boarding house offers 3 choices of soup, 4 kinds of entrées, 3 choices of vegetables, and 2 different desserts. How many different four-course meals can you choose?

41. ▦ Iowa Instrument Company makes, among other things, personal computers. Each computer carries a serial number that has 3 letters followed by 4 digits, such as ABA 2005. The production manager reported that last year the company used all but 80 of all possible serial numbers. How many personal computers did Iowa Instrument Company make last year?

42. ▦ In how many distinguishable ways can you arrange the letters of the word REARRANGEMENT?

[11.6] 43. Compute $C(8, 2)$, $C(8, 6)$, and $C(7, 5) + C(7, 6)$.

44. Show that for $r \leq n$, $\binom{n}{n-r} = \binom{n}{r}$.

45. Solve for n, if $\binom{n}{5} = \binom{n}{6}$.

46. Solve for m, if $5\binom{m}{1} = \binom{m}{2}$.

47. Mark wants to donate some money to the Boy Scout Association; he has only one of each of $1.00, $5.00, $10.00, and $20.00 bills. How many different amounts can Mark possibly give?

48. In how many ways can a committee of 4 men and 3 women be elected from a class of 11 men and 9 women?

Practice Test (60 minutes)

1. A mass lecture hall has 32 rows of seats. The front row has 16 seats and each row has one more seat than the one preceding it. How many seats are there?

2. Hal bought a Camaro for $12,000 this year. He estimates that the value of his car will depreciate 20% from the previous year. How much would Hal's Camaro be worth at the end of 5 years?

3. Tim joined the ACR Record Club which requires each member to buy 1 record the first month, 2 records the second month, 4 records the third month, and so on for 12 months. How many records will he have bought by the end of the year?

4. Convert the repeating decimal $3.\overline{21}$ into a fraction.

5. Use the binomial theorem to expand $(2x - 3y)^4$.

6. Show that $\binom{n}{1} + \binom{n}{2} + \cdots + \binom{n}{n} = 2^n - 1$.

7. In the first round of a beauty contest, 4 women will be selected from 51 contestants. How many different choices can be made?

8. In how many distinguishable ways can the letters of the word MISSISSIPPI be arranged?

9. Solve for n, if $5\binom{n}{2} = 3\binom{n}{3}$.

10. In how many ways can a committee of 5 freshmen and 4 sophomores be elected from a group of 12 freshmen and 8 sophomores?

Extended Applications

The Dose-Response Relationship

When a 5 mg dose of a drug that is being given to an animal is increased to 10 mg, it is quite likely that this increase will induce a reaction in the animal. However, if a 100 mg dose is increased by the same 5 mg amount to 105 mg, the increase may not cause a reaction at all. In both cases there is an increased dosage of 5 mg, but the important difference is that in the first case this represented a 100% increase in dosage, whereas in the second case the dosage increase was merely 5%.

In general, when animals are tested for their response to different amounts of a drug, the doses should form a geometric sequence. If d represents the initial dosage and p the percent increase for the second dose (in decimal notation), the drug should be administered in doses:

$$d, \quad d(1 + p), \quad d(1 + p)^2, \quad d(1 + p)^3 \ldots$$

If, for example, the initial dosage is 5 mg and the second dose is to be increased 100% (1 decimal; or $p = 1$) the dosage sequence in mg is

$$5, \quad 5(1 + 1), \quad 5(1 + 1)^2, \quad 5(1 + 1)^3 \ldots$$

or

$$5, 10, 20, 40, 80 \ldots$$

Exercises

1. If the initial dose is 10 mg and the second dose represents an increase of 50%, what should the fourth dose be?

2. If the initial dose is 10 mg and the fourth dose is 40 mg, what was the percent increase from the first to the second dose?

3. Explain why 5 mg, 10 mg, 15 mg, 20 mg ... is not a suitable dosage sequence.

Appendixes

A. Tables

TABLE 1. Square Roots and Cube Roots

n	$\sqrt{n}$	$\sqrt[3]{n}$	n	$\sqrt{n}$	$\sqrt[3]{n}$	n	$\sqrt{n}$	$\sqrt[3]{n}$
1	1.00000	1.00000	35	5.91608	3.27107	69	8.30662	4.10157
2	1.41421	1.25992	36	6.00000	3.30193	70	8.36660	4.12129
3	1.73205	1.44225	37	6.08276	3.33222	71	8.42615	4.14082
4	2.00000	1.58740	38	6.16441	3.36198	72	8.48528	4.16017
5	2.23607	1.70998	39	6.24500	3.39121	73	8.54400	4.17934
6	2.44949	1.81712	40	6.32456	3.41995	74	8.60233	4.19834
7	2.64575	1.91293	41	6.40312	3.44822	75	8.66025	4.21716
8	2.82843	2.00000	42	6.48074	3.47603	76	8.71780	4.23582
9	3.00000	2.08008	43	6.55744	3.50340	77	8.77496	4.25432
10	3.16228	2.15443	44	6.63325	3.53035	78	8.83176	4.27266
11	3.31662	2.22398	45	6.70820	3.55689	79	8.88819	4.29084
12	3.46410	2.28943	46	6.78233	3.58305	80	8.94427	4.30887
13	3.60555	2.35133	47	6.85565	3.60883	81	9.00000	4.32675
14	3.74166	2.41014	48	6.92820	3.63424	82	9.05539	4.34448
15	3.87298	2.46621	49	7.00000	3.65931	83	9.11043	4.36207
16	4.00000	2.51984	50	7.07107	3.68403	84	9.16515	4.37952
17	4.12311	2.57128	51	7.14143	3.70843	85	9.21954	4.39683
18	4.24264	2.62074	52	7.21110	3.73251	86	9.27362	4.41400
19	4.35890	2.66840	53	7.28011	3.75629	87	9.32738	4.43105
20	4.47214	2.71442	54	7.34847	3.77976	88	9.38083	4.44796
21	4.58258	2.75892	55	7.41620	3.80295	89	9.43398	4.46475
22	4.69042	2.80204	56	7.48331	3.82586	90	9.48683	4.48140
23	4.79583	2.84387	57	7.54983	3.84850	91	9.53939	4.49794
24	4.89898	2.88450	58	7.61577	3.87088	92	9.59166	4.51436
25	5.00000	2.92402	59	7.68115	3.89300	93	9.64365	4.53065
26	5.09902	2.96250	60	7.74597	3.91487	94	9.69536	4.54684
27	5.19615	3.00000	61	7.81025	3.93650	95	9.74679	4.56290
28	5.29150	3.03659	62	7.87401	3.95789	96	9.79796	4.57886
29	5.38516	3.07232	63	7.93725	3.97906	97	9.84886	4.59470
30	5.47723	3.10723	64	8.00000	4.00000	98	9.89949	4.61044
31	5.56776	3.14138	65	8.06226	4.02073	99	9.94987	4.62607
32	5.65685	3.17480	66	8.12404	4.04124	100	10.0000	4.64159
33	5.74456	3.20753	67	8.18535	4.06155	101	10.0499	4.65701
34	5.83095	3.23961	68	8.24621	4.08166	102	10.0995	4.67233

TABLE 1. Square Roots and Cube Roots (*continued*)

n	$\sqrt{n}$	$\sqrt[3]{n}$	n	$\sqrt{n}$	$\sqrt[3]{n}$	n	$\sqrt{n}$	$\sqrt[3]{n}$
103	10.1489	4.68755	136	11.6619	5.14256	169	13.0000	5.52877
104	10.1980	4.70267	137	11.7047	5.15514	170	13.0384	5.53966
105	10.2470	4.71769	138	11.7473	5.16765	171	13.0767	5.55050
106	10.2956	4.73262	139	11.7898	5.18010	172	13.1149	5.56130
107	10.3441	4.74746	140	11.8322	5.19249	173	13.1529	5.57205
108	10.3923	4.76220	141	11.8743	5.20483	174	13.1909	5.58277
109	10.4403	4.77686	142	11.9164	5.21710	175	13.2288	5.59344
110	10.4881	4.79142	143	11.9583	5.22932	176	13.2665	5.60408
111	10.5357	4.80590	144	12.0000	5.24148	177	13.3041	5.61467
112	10.5830	4.82028	145	12.0416	5.25359	178	13.3417	5.62523
113	10.6301	4.83459	146	12.0830	5.26564	179	13.3791	5.63574
114	10.6771	4.84881	147	12.1244	5.27763	180	13.4164	5.64622
115	10.7238	4.86294	148	12.1655	5.28957	181	13.4536	5.65665
116	10.7703	4.87700	149	12.2066	5.30146	182	13.4907	5.66705
117	10.8167	4.89097	150	12.2474	5.31329	183	13.5277	5.67741
118	10.8628	4.90487	151	12.2882	5.32507	184	13.5647	5.68773
119	10.9087	4.91868	152	12.3288	5.33680	185	13.6015	5.69802
120	10.9545	4.93242	153	12.3693	5.34848	186	13.6382	5.70827
121	11.0000	4.94609	154	12.4097	5.36011	187	13.6748	5.71848
122	11.0454	4.95968	155	12.4499	5.37169	188	13.7113	5.72865
123	11.0905	4.97319	156	12.4900	5.38321	189	13.7477	5.73879
124	11.1355	4.98663	157	12.5300	5.39469	190	13.7840	5.74890
125	11.1803	5.00000	158	12.5698	5.40612	191	13.8203	5.75897
126	11.2250	5.01330	159	12.6095	5.41750	192	13.8564	5.76900
127	11.2694	5.02653	160	12.6491	5.42884	193	13.8924	5.77900
128	11.3137	5.03968	161	12.6886	5.44012	194	13.9284	5.78896
129	11.3578	5.05277	162	12.7279	5.45136	195	13.9642	5.79889
130	11.4018	5.06580	163	12.7671	5.46256	196	14.0000	5.80879
131	11.4455	5.07875	164	12.8062	5.47370	197	14.0357	5.81865
132	11.4891	5.09164	165	12.8452	5.48481	198	14.0712	5.82848
133	11.5326	5.10447	166	12.8841	5.49586	199	14.1067	5.83827
134	11.5758	5.11723	167	12.9228	5.50688	200	14.1421	5.84804
135	11.6190	5.12993	168	12.9615	5.51785			

TABLE 2. Values of $\text{Log}_{10} x$ and $\text{Antilog}_{10} x$ or (10^x)

x	0	1	2	3	4	5	6	7	8	9
1.0	.0000	.0043	.0086	.0128	.0170	.0212	.0253	.0294	.0334	.0374
1.1	.0414	.0453	.0492	.0531	.0569	.0607	.0645	.0682	.0719	.0755
1.2	.0792	.0828	.0864	.0899	.0934	.0969	.1004	.1038	.1072	.1106
1.3	.1139	.1173	.1206	.1239	.1271	.1303	.1335	.1367	.1399	.1430
1.4	.1461	.1492	.1523	.1553	.1584	.1614	.1644	.1673	.1703	.1732
1.5	.1761	.1790	.1818	.1847	.1875	.1903	.1931	.1959	.1987	.2014
1.6	.2041	.2068	.2095	.2122	.2148	.2175	.2201	.2227	.2253	.2279
1.7	.2304	.2330	.2355	.2380	.2405	.2430	.2455	.2480	.2504	.2529
1.8	.2553	.2577	.2601	.2625	.2648	.2672	.2695	.2718	.2742	.2765
1.9	.2788	.2810	.2833	.2856	.2878	.2900	.2923	.2945	.2967	.2989
2.0	.3010	.3032	.3054	.3075	.3096	.3118	.3139	.3160	.3181	.3201
2.1	.3222	.3243	.3263	.3284	.3304	.3324	.3345	.3365	.3385	.3404
2.2	.3424	.3444	.3464	.3483	.3502	.3522	.3541	.3560	.3579	.3598
2.3	.3617	.3636	.3655	.3674	.3692	.3711	.3729	.3747	.3766	.3784
2.4	.3802	.3820	.3838	.3856	.3874	.3892	.3909	.3927	.3945	.3962
2.5	.3979	.3997	.4014	.4031	.4048	.4065	.4082	.4099	.4116	.4133
2.6	.4150	.4166	.4183	.4200	.4216	.4232	.4249	.4265	.4281	.4298
2.7	.4314	.4330	.4346	.4362	.4378	.4393	.4409	.4425	.4440	.4456
2.8	.4472	.4487	.4502	.4518	.4533	.4548	.4564	.4579	.4594	.4609
2.9	.4624	.4639	.4654	.4669	.4683	.4698	.4713	.4728	.4742	.4757
3.0	.4771	.4786	.4800	.4814	.4829	.4843	.4857	.4871	.4886	.4900
3.1	.4914	.4928	.4942	.4955	.4969	.4983	.4997	.5011	.5024	.5038
3.2	.5051	.5065	.5079	.5092	.5105	.5119	.5132	.5145	.5159	.5172
3.3	.5185	.5198	.5211	.5224	.5237	.5250	.5263	.5276	.5289	.5302
3.4	.5315	.5328	.5340	.5353	.5366	.5378	.5391	.5403	.5416	.5428
3.5	.5441	.5453	.5465	.5478	.5490	.5502	.5514	.5527	.5539	.5551
3.6	.5563	.5575	.5587	.5599	.5611	.5623	.5635	.5647	.5658	.5670
3.7	.5682	.5694	.5705	.5717	.5729	.5740	.5752	.5763	.5775	.5786
3.8	.5798	.5809	.5821	.5832	.5843	.5855	.5866	.5877	.5888	.5899
3.9	.5911	.5922	.5933	.5944	.5955	.5966	.5977	.5988	.5999	.6010
4.0	.6021	.6031	.6042	.6053	.6064	.6075	.6085	.6096	.6107	.6117
4.1	.6128	.6138	.6149	.6160	.6170	.6180	.6191	.6201	.6212	.6222
4.2	.6232	.6243	.6253	.6263	.6274	.6284	.6294	.6304	.6314	.6325
4.3	.6335	.6345	.6355	.6365	.6375	.6385	.6395	.6405	.6415	.6425
4.4	.6435	.6444	.6454	.6464	.6474	.6484	.6493	.6503	.6513	.6522
4.5	.6532	.6542	.6551	.6561	.6571	.6580	.6590	.6599	.6609	.6618
4.6	.6628	.6637	.6646	.6656	.6665	.6675	.6684	.6693	.6702	.6712
4.7	.6721	.6730	.6739	.6749	.6758	.6767	.6776	.6785	.6794	.6803
4.8	.6812	.6821	.6830	.6839	.6848	.6857	.6866	.6875	.6884	.6893
4.9	.6902	.6911	.6920	.6928	.6937	.6946	.6955	.6964	.6972	.6981
5.0	.6990	.6998	.7007	.7016	.7024	.7033	.7042	.7050	.7059	.7067
5.1	.7076	.7084	.7093	.7101	.7110	.7118	.7126	.7135	.7143	.7152
5.2	.7160	.7168	.7177	.7185	.7193	.7202	.7210	.7218	.7226	.7235
5.3	.7243	.7251	.7259	.7267	.7275	.7284	.7292	.7300	.7308	.7316
5.4	.7324	.7332	.7340	.7348	.7356	.7364	.7372	.7380	.7388	.7396
x	0	1	2	3	4	5	6	7	8	9

TABLE 2. Values of $\text{Log}_{10} x$ and $\text{Antilog}_{10} x$ or (10^x) *(continued)*

x	0	1	2	3	4	5	6	7	8	9
5.5	.7404	.7412	.7419	.7427	.7435	.7443	.7451	.7459	.7466	.7474
5.6	.7482	.7490	.7497	.7505	.7513	.7520	.7528	.7536	.7543	.7551
5.7	.7559	.7566	.7574	.7582	.7589	.7597	.7604	.7612	.7619	.7627
5.8	.7634	.7642	.7649	.7657	.7664	.7672	.7679	.7686	.7694	.7701
5.9	.7709	.7716	.7723	.7731	.7738	.7745	.7752	.7760	.7767	.7774
6.0	.7782	.7789	.7796	.7803	.7810	.7818	.7825	.7832	.7839	.7846
6.1	.7853	.7860	.7868	.7875	.7882	.7889	.7896	.7903	.7910	.7917
6.2	.7924	.7931	.7938	.7945	.7952	.7959	.7966	.7973	.7980	.7987
6.3	.7993	.8000	.8007	.8014	.8021	.8028	.8035	.8041	.8048	.8055
6.4	.8062	.8069	.8075	.8082	.8089	.8096	.8102	.8109	.8116	.8122
6.5	.8129	.8136	.8142	.8149	.8156	.8162	.8169	.8176	.8182	.8189
6.6	.8195	.8202	.8209	.8215	.8222	.8228	.8235	.8241	.8248	.8254
6.7	.8261	.8267	.8274	.8280	.8287	.8293	.8299	.8306	.8312	.8319
6.8	.8325	.8331	.8338	.8344	.8351	.8357	.8363	.8370	.8376	.8382
6.9	.8388	.8395	.8401	.8407	.8414	.8420	.8426	.8432	.8439	.8445
7.0	.8451	.8457	.8463	.8470	.8476	.8482	.8488	.8494	.8500	.8506
7.1	.8513	.8519	.8525	.8531	.8537	.8543	.8549	.8555	.8561	.8567
7.2	.8573	.8579	.8585	.8591	.8597	.8603	.8609	.8615	.8621	.8627
7.3	.8633	.8639	.8645	.8651	.8657	.8663	.8669	.8675	.8681	.8686
7.4	.8692	.8698	.8704	.8710	.8716	.8722	.8727	.8733	.8739	.8745
7.5	.8751	.8756	.8762	.8768	.8774	.8779	.8785	.8791	.8797	.8802
7.6	.8808	.8814	.8820	.8825	.8831	.8837	.8842	.8848	.8854	.8859
7.7	.8865	.8871	.8876	.8882	.8887	.8893	.8899	.8904	.8910	.8915
7.8	.8921	.8927	.8932	.8938	.8943	.8949	.8954	.8960	.8965	.8971
7.9	.8976	.8982	.8987	.8993	.8998	.9004	.9009	.9015	.9020	.9025
8.0	.9031	.9036	.9042	.9047	.9053	.9058	.9063	.9069	.9074	.9079
8.1	.9085	.9090	.9096	.9101	.9106	.9112	.9117	.9122	.9128	.9133
8.2	.9138	.9143	.9149	.9154	.9159	.9165	.9170	.9175	.9180	.9186
8.3	.9191	.9196	.9201	.9206	.9212	.9217	.9222	.9227	.9232	.9238
8.4	.9243	.9248	.9253	.9258	.9263	.9269	.9274	.9279	.9284	.9289
8.5	.9294	.9299	.9304	.9309	.9315	.9320	.9325	.9330	.9335	.9340
8.6	.9345	.9350	.9355	.9360	.9365	.9370	.9375	.9380	.9385	.9390
8.7	.9395	.9400	.9405	.9410	.9415	.9420	.9425	.9430	.9435	.9440
8.8	.9445	.9450	.9455	.9460	.9465	.9469	.9474	.9479	.9484	.9489
8.9	.9494	.9499	.9504	.9509	.9513	.9518	.9523	.9528	.9533	.9538
9.0	.9542	.9547	.9552	.9557	.9562	.9566	.9571	.9576	.9581	.9586
9.1	.9590	.9595	.9600	.9605	.9609	.9614	.9619	.9624	.9628	.9633
9.2	.9638	.9643	.9647	.9652	.9657	.9661	.9666	.9671	.9675	.9680
9.3	.9685	.9689	.9694	.9699	.9703	.9708	.9713	.9717	.9722	.9727
9.4	.9731	.9736	.9741	.9745	.9750	.9754	.9759	.9763	.9768	.9773
9.5	.9777	.9782	.9786	.9791	.9795	.9800	.9805	.9809	.9814	.9818
9.6	.9823	.9827	.9832	.9836	.9841	.9845	.9850	.9854	.9859	.9863
9.7	.9868	.9872	.9877	.9881	.9886	.9890	.9894	.9899	.9903	.9908
9.8	.9912	.9917	.9921	.9926	.9930	.9934	.9939	.9943	.9948	.9952
9.9	.9956	.9961	.9965	.9969	.9974	.9978	.9983	.9987	.9991	.9996
x	0	1	2	3	4	5	6	7	8	9

TABLE 3. Values of ln x

x	ln x	x	ln x	x	ln x
		4.5	1.5041	9.0	2.1972
0.1	−2.3026	4.6	1.5261	9.1	2.2083
0.2	−1.6094	4.7	1.5476	9.2	2.2192
0.3	−1.2040	4.8	1.5686	9.3	2.2300
0.4	−0.9163	4.9	1.5892	9.4	2.2407
0.5	−0.6931	5.0	1.6094	9.5	2.2513
0.6	−0.5108	5.1	1.6292	9.6	2.2618
0.7	−0.3567	5.2	1.6487	9.7	2.2721
0.8	−0.2231	5.3	1.6677	9.8	2.2824
0.9	−0.1054	5.4	1.6864	9.9	2.2925
1.0	0.0000	5.5	1.7047	10	2.3026
1.1	0.0953	5.6	1.7228	11	2.3979
1.2	0.1823	5.7	1.7405	12	2.4849
1.3	0.2624	5.8	1.7579	13	2.5649
1.4	0.3365	5.9	1.7750	14	2.6391
1.5	0.4055	6.0	1.7918	15	2.7081
1.6	0.4700	6.1	1.8083	16	2.7726
1.7	0.5306	6.2	1.8245	17	2.8332
1.8	0.5878	6.3	1.8405	18	2.8904
1.9	0.6419	6.4	1.8563	19	2.9444
2.0	0.6931	6.5	1.8718	20	2.9957
2.1	0.7419	6.6	1.8871	25	3.2189
2.2	0.7885	6.7	1.9021	30	3.4012
2.3	0.8329	6.8	1.9169	35	3.5553
2.4	0.8755	6.9	1.9315	40	3.6889
2.5	0.9163	7.0	1.9459	45	3.8067
2.6	0.9555	7.1	1.9601	50	3.9120
2.7	0.9933	7.2	1.9741	55	4.0073
2.8	1.0296	7.3	1.9879	60	4.0943
2.9	1.0647	7.4	2.0015	65	4.1744
3.0	1.0986	7.5	2.0149	70	4.2485
3.1	1.1314	7.6	2.0281	75	4.3175
3.2	1.1632	7.7	2.0412	80	4.3820
3.3	1.1939	7.8	2.0541	85	4.4427
3.4	1.2238	7.9	2.0669	90	4.4998
3.5	1.2528	8.0	2.0794	100	4.6052
3.6	1.2809	8.1	2.0919	110	4.7005
3.7	1.3083	8.2	2.1041	120	4.7875
3.8	1.3350	8.3	2.1163	130	4.8676
3.9	1.3610	8.4	2.1282	140	4.9416
4.0	1.3863	8.5	2.1401	150	5.0106
4.1	1.4110	8.6	2.1518	160	5.0752
4.2	1.4351	8.7	2.1633	170	5.1358
4.3	1.4586	8.8	2.1748	180	5.1930
4.4	1.4816	8.9	2.1861	190	5.2470

TABLE 4. Values of e^x

x	e^x	e^{-x}
0.00	1.0000	1.0000
0.01	1.0101	0.9901
0.02	1.0202	0.9802
0.03	1.0305	0.9705
0.04	1.0408	0.9608
0.05	1.0513	0.9512
0.06	1.0618	0.9418
0.07	1.0725	0.9324
0.08	1.0833	0.9231
0.09	1.0942	0.9139
0.10	1.1052	0.9048
0.11	1.1163	0.8958
0.12	1.1275	0.8869
0.13	1.1388	0.8781
0.14	1.1503	0.8694
0.15	1.1618	0.8607
0.16	1.1735	0.8521
0.17	1.1853	0.8437
0.18	1.1972	0.8353
0.19	1.2092	0.8270
0.20	1.2214	0.8187
0.21	1.2337	0.8106
0.22	1.2461	0.8025
0.23	1.2586	0.7945
0.24	1.2712	0.7866
0.25	1.2840	0.7788
0.26	1.2969	0.7711
0.27	1.3100	0.7634
0.28	1.3231	0.7558
0.29	1.3364	0.7483
0.30	1.3499	0.7408
0.31	1.3634	0.7334
0.32	1.3771	0.7261
0.33	1.3910	0.7190
0.34	1.4050	0.7118
0.35	1.4191	0.7047
0.36	1.4333	0.6977
0.37	1.4477	0.6907
0.38	1.4623	0.6839
0.39	1.4770	0.6771
0.40	1.4918	0.6703
0.41	1.5068	0.6636
0.42	1.5220	0.6570
0.43	1.5373	0.6505
0.44	1.5527	0.6440
0.45	1.5683	0.6376
0.46	1.5841	0.6313
0.47	1.6000	0.6250
0.48	1.6160	0.6188
0.49	1.6323	0.6126
0.50	1.6487	0.6065
0.51	1.6653	0.6005
0.52	1.6820	0.5945
0.53	1.6990	0.5886
0.54	1.7160	0.5827
0.55	1.7333	0.5769
0.56	1.7507	0.5712
0.57	1.7683	0.5655
0.58	1.7860	0.5599
0.59	1.8040	0.5543
0.60	1.8221	0.5488
0.61	1.8404	0.5434
0.62	1.8590	0.5380
0.63	1.8776	0.5326
0.64	1.8965	0.5273
0.65	1.9155	0.5220
0.66	1.9348	0.5169
0.67	1.9542	0.5117
0.68	1.9739	0.5066
0.69	1.9937	0.5016
0.70	2.0138	0.4966
0.71	2.0340	0.4916
0.72	2.0544	0.4868
0.73	2.0751	0.4819
0.74	2.0959	0.4771
0.75	2.1170	0.4724
0.76	2.1383	0.4677
0.77	2.1598	0.4630
0.78	2.1815	0.4584
0.79	2.2034	0.4538
0.80	2.2255	0.4493
0.81	2.2479	0.4449
0.82	2.2705	0.4404
0.83	2.2933	0.4360
0.84	2.3164	0.4317
0.85	2.3396	0.4274
0.86	2.3632	0.4232
0.87	2.3869	0.4190
0.88	2.4109	0.4148
0.89	2.4351	0.4107
0.90	2.4596	0.4066
0.91	2.4843	0.4025
0.92	2.5093	0.3985
0.93	2.5345	0.3946
0.94	2.5600	0.3906
0.95	2.5857	0.3867
0.96	2.6117	0.3829
0.97	2.6379	0.3791
0.98	2.6645	0.3753
0.99	2.6912	0.3716

TABLE 4. Values of e^x (continued)

x	e^x	e^{-x}	x	e^x	e^{-x}
1.0	2.7183	0.3679	5.5	244.69	0.0041
1.1	3.0042	0.3329	5.6	270.43	0.0037
1.2	3.3201	0.3012	5.7	298.87	0.0034
1.3	3.6693	0.2725	5.8	330.30	0.0030
1.4	4.0552	0.2466	5.9	365.04	0.0027
1.5	4.4817	0.2231	6.0	403.43	0.0025
1.6	4.9530	0.2019	6.1	445.86	0.0022
1.7	5.4739	0.1827	6.2	492.75	0.0020
1.8	6.0496	0.1653	6.3	544.57	0.0018
1.9	6.6859	0.1496	6.4	601.85	0.0017
2.0	7.3891	0.1353	6.5	665.14	0.0015
2.1	8.1662	0.1225	6.6	735.10	0.0014
2.2	9.0250	0.1108	6.7	812.41	0.0012
2.3	9.9742	0.1003	6.8	897.85	0.0011
2.4	11.023	0.0907	6.9	992.27	0.0010
2.5	12.182	0.0821	7.0	1096.6	0.0009
2.6	13.464	0.0743	7.1	1212.0	0.0008
2.7	14.880	0.0672	7.2	1339.5	0.0007
2.8	16.445	0.0608	7.3	1480.3	0.0007
2.9	18.174	0.0550	7.4	1636.0	0.0006
3.0	20.086	0.0498	7.5	1808.0	0.0006
3.1	22.198	0.0450	7.6	1998.2	0.0005
3.2	24.533	0.0408	7.7	2208.4	0.0005
3.3	27.113	0.0369	7.8	2440.6	0.0004
3.4	29.964	0.0334	7.9	2697.3	0.0004
3.5	33.115	0.0302	8.0	2981.0	0.0003
3.6	36.598	0.0273	8.1	3294.5	0.0003
3.7	40.447	0.0247	8.2	3641.0	0.0003
3.8	44.701	0.0224	8.3	4023.9	0.0002
3.9	49.402	0.0202	8.4	4447.1	0.0002
4.0	54.598	0.0183	8.5	4914.8	0.0002
4.1	60.340	0.0166	8.6	5431.7	0.0002
4.2	66.686	0.0150	8.7	6002.9	0.0002
4.3	73.700	0.0136	8.8	6634.2	0.0002
4.4	81.451	0.0123	8.9	7332.0	0.0001
4.5	90.017	0.0111	9.0	8103.1	0.0001
4.6	99.484	0.0101	9.1	8955.3	0.0001
4.7	109.95	0.0091	9.2	9897.1	0.0001
4.8	121.51	0.0082	9.3	10938	0.0001
4.9	134.29	0.0074	9.4	12088	0.0001
5.0	148.41	0.0067	9.5	13360	0.0001
5.1	164.02	0.0061	9.6	14765	0.0001
5.2	181.27	0.0055	9.7	16318	0.0001
5.3	200.34	0.0050	9.8	18034	0.0001
5.4	221.41	0.0045	9.9	19930	0.0001

B. Linear Interpolation

The table of logarithms in this text can be used to find the common logarithms of numbers with three digits. A process called **linear interpolation** is used with this table to find logarithms of four-digit numbers. The basic principle of linear interpolation is: **If c is a certain fraction of the distance between a and b, then log c is approximately the same fraction of the distance between log a and log b.**

EXAMPLE 1 Find log 4.555.

Solution: From the table we find log 4.55 = 0.6580 and log 4.56 = 0.6590. The distance between log 4.55 and log 4.56 is 0.6590 − 0.6580 = 0.0010
The distance between 4.55 and 4.56 is 4.56 − 4.55 = .01
The distance between 4.55 and 4.555 is 4.555 − 4.55 = 0.005.
The calculations for log 4.555 are conveniently arranged below:

$$0.01 \left\{ 0.005 \left\{ \begin{array}{l} \log 4.55 = 0.6580 \\ \log 4.555 = \\ \log 4.56 = 0.6590 \end{array} \right\} y \right\} 0.0010$$

To find the value of y we solve

$$\frac{0.005}{0.01} = \frac{y}{0.0010}$$

$$0.5 = \frac{y}{0.0010}$$

$$0.0005 = y$$

We now add 0.0005 to 0.6580 to approximate log 4.555. Thus

$$\log 4.555 = 0.6580 + 0.0005 = 0.6585$$

Notice that 4.555 is *halfway* between 4.55 and 4.56 and that the approximation of log 4.555 is *halfway* between log 4.55 and log 4.56. With practice, this type of interpolation can be done mentally.

The linear interpolation principle can also be applied to the problem of finding the antilog of a number.

EXAMPLE 2 Find antilog 0.5498.

Solution: The mantissa 0.5498 is not a table value. The two table entries that are closest to 0.5498 are 0.5490 and 0.5502. From the table, antilog 0.5490 = 3.54 and antilog 0.5502 = 3.55. The calculations are arranged as before.

$$12\left\{\begin{array}{l}8\left\{\begin{array}{l}0.5490 = \text{antilog } 3.54 \\ 0.5498 = \end{array}\right\}y \\ 0.5502 = \text{antilog } 3.55\end{array}\right\}0.01$$

We are using 8 and 12 in place of 0.0008 and 0.0012, since only their ratios are involved in the calculations.

$$\frac{8}{12} = \frac{y}{0.01}$$

$$y = \left(\frac{8}{12}\right)(0.01) = 0.0066\ldots \approx 0.007$$

Thus antilog $0.5498 = 3.54 + 0.007 = 3.547$.

Exercises

Standard Assignment: All-odd numbered exercises.

Find the following common logarithms by linear interpolation. See Example 1.

1. log 3.225
2. log 7.472
3. log 5.333
4. log 9.5
5. log 5138
6. log 42.89
7. log 812.4
8. log 26350
9. log 0.001522
10. log 0.07893

Find the antilogarithms of the following numbers. See Example 2.

11. antilog 0.5891
12. antilog 0.9093
13. antilog 3.6907
14. antilog 2.7164
15. antilog 1.3713
16. antilog 1.1275
17. antilog $(0.5877 - 1)$
18. antilog $(0.9792 - 3)$

Answers to Selected Exercises

> If you need further help with algebra, you might want to get a copy of the *Student Study Guide* and the *Student Solutions Guide* designed to go along with this book. Your local college bookstore either has these books or can order them for you.

Chapter 1

Exercises 1.1

1. $\{1, 2, 3, 4, 5, 6, 7, 8, 9\}$ **3.** $\emptyset$ **5.** $\{3, 6, 9, 12, 15, 18, 21, 24, 27\}$ **7.** $\{5, 10, 15, 20, \ldots\}$ **9.** $\left\{\frac{1}{2}\right\}$
11. the set of even numbers less than 10   **13.** the set of integer multiples of 2 that are greater than or equal to -6 **15.** the set of natural number multiples of 4 that are less than 21   **17.** the set consisting of the number 1   **19.** $\{0, 5, 6\}$ **21.** $\left\{-\frac{22}{17}, -1, -\frac{1}{3}, 0, \frac{1}{2}, \frac{9}{2}, 5, 6\right\}$ **23.** $\left\{-\frac{22}{17}, -\sqrt{2}, -1, -\frac{1}{3}, 0, \frac{1}{2}, \sqrt{2}\right\}$
25. $\left\{-\frac{22}{17}, -\sqrt{2}, -1, -\frac{1}{3}\right\}$ **27.** $1 \in \{1, 2, 3\}$ **29.** $1 \in \{1\}$ **31.** $W \subset J$ **33.** $\{-1, 0\} \not\subset N$
35. $\{-1\} \subset J$ **37.** $\left\{7\frac{1}{2}, 13\right\} \subset Q$ **39.** $Q \not\subset H$ **41.** $N \subset W$ **43.** $Q \not\subset J$ **45.** $J \not\subset W$ **47.** $W \not\subset N$
49. Yes; if $A \subset B$, then any element belonging to A is in B.   **51.** Yes; if $A = B$, then $A \subset B$ and $B \subset A$.
53. No; the empty set, $\emptyset$, is a subset of every set.

Exercises 1.2

1. 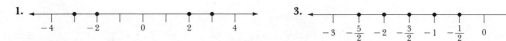   **3.**

5.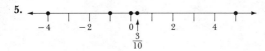

7. $3 > -2$ **9.** $\frac{1}{2} \geq \frac{1}{2}$ **11.** $7 \not< 0$ **13.** $5 \leq 2t$ **15.** $-t > 0$ **17.** $2y + 7 \leq 14$
19. $-6 < z \leq -1$ **21.** $4 = \frac{24}{6}$ **23.** $-\frac{1}{2} > -\frac{3}{2}$ **25.** $-4 < 0$ **27.** true   **29.** true   **31.** true
33. false   **35.** true   **37.** $-1 > -6$ **39.** $-1 > x$ **41.** Reflexive property of equality

43. Substitution property of equality   **45.** Substitution property of equality   **47.** Substitution property of equality   **49.** Transitive property of inequality   **51.** Transitive property of inequality   **53.** Transitive property of inequality

Exercises 1.3

1. Distributive property   **3.** Identity property for multiplication   **5.** Associative property for multiplication   **7.** Commutative property for multiplication   **9.** Zero factor property   **11.** Double negative property   **13.** Inverse property for multiplication   **15.** Distributive property; Inverse property for addition; zero factor property   **17.** Inverse property for addition; Identity property for addition   **19.** Associative property for addition   **21.** Identity property for addition   **23.** Addition property of equality   **25.** Addition property of equality   **27.** Identity property for multiplication   **29.** $1; -1$ **31.** $-1; 1$ **33.** $\frac{2}{9}; -\frac{9}{2}$ **35.** $0.7524; \frac{1}{-0.7524}$ **37.** $-1.35286; \frac{1}{1.35286}$ **39.** -12 **41.** $\frac{1}{2}$ **43.** 10 **45.** 3 **47.** 4 **49.** 2 **51.** 1 **53.** $(5+2)t = 7t$ **55.** 3 **57.** $\frac{1}{3}$ **59.** $-[2] = -2$ **61.** $2 \cdot 2 = 4$ **63.** 12 **65.** 0 **67.** -2 **69.** $\left|\frac{x}{3}\right| = \begin{cases} \frac{x}{3}, & \text{if } \frac{x}{3} \geq 0 \\ -\frac{x}{3}, & \text{if } \frac{x}{3} < 0 \end{cases}$ **71.** $-|x-3| = \begin{cases} -(x-3), & \text{if } x-3 \geq 0 \\ x-3, & \text{if } x-3 < 0 \end{cases}$ **73.** $|x+7| = \begin{cases} x+7, & \text{if } x+7 \geq 0 \\ -(x+7), & \text{if } x+7 < 0 \end{cases}$

75.
$(a+b) \cdot c = c \cdot (a+b)$ Commutative property for multiplication
$c \cdot (a+b) = c \cdot a + c \cdot b$ Distributive property
$c \cdot a = a \cdot c$ Commutative property for multiplication
$c \cdot (a+b) = a \cdot c + c \cdot b$ Substitution property of equality
$c \cdot b = b \cdot c$ Commutative property for multiplication
$c \cdot (a+b) = a \cdot c + b \cdot c$ Substitution property of equality
$(a+b) \cdot c = a \cdot c + b \cdot c$ Commutative property for multiplication

77. closed for multiplication; not closed for addition   **79.** closed for multiplication; not closed for addition   **81.** not closed for multiplication; not closed for addition

Exercises 1.4

1. 3 **3.** -2 **5.** -4 **7.** -12 **9.** $\frac{1}{8}$ **11.** -11 **13.** -29 **15.** -23 **17.** -1 **19.** -9 **21.** -34 **23.** 2 **25.** -6 **27.** -22 **29.** $-\frac{8}{5}$ **31.** 5 **33.** 2 **35.** -6 **37.** 5 **39.** -5 **41.** -17 **43.** 14 **45.** -1 **47.** -134.54352 **49.** 5.79309 **51.** -97.02521 **53.** 22 **55.** 14 **57.** 18 **59.** $\frac{13}{5}$ **61.** -4 **63.** -4 **65.** 0 **67.** -21 **69.** $\frac{3}{2}$ **71.** $-\frac{7}{2}$ **73.** any choice of x **75.** no

Exercises 1.5

1. -6 **3.** 80 **5.** -36 **7.** 72 **9.** -8 **11.** -1 **13.** -640 **15.** -378 **17.** 2 **19.** -2 **21.** 0 **23.** undefined   **25.** $\frac{3}{2}$ **27.** -14 **29.** $\frac{8}{9}$ **31.** $\frac{1}{9}$ **33.** 57.879 **35.** -75.334 **37.** 19 **39.** 3 **41.** 30 **43.** 40 **45.** 19 **47.** -15 **49.** 4 **51.** -88 **53.** -5 **55.** 33 **57.** -4

59. $-\dfrac{40}{13}$ **61.** 0 **63.** $-\dfrac{3}{2}$ **65.** -12 **67.** -89.637 **69.** -26.151 **71.** 0.031 **73.** 98.745
75. $\dfrac{48}{5}$ **77.** $\dfrac{144}{73}$ **79.** 15 **81.** -35 **83.** $\dfrac{-69}{5}$ **85.** $\dfrac{-56}{9}$ **87.** any $x \neq 0$ **89.** any $x \neq 3$
91. no

Review Exercises

1. $\{-2, -1, 0, 1, 2\}$ **3.** $\{1\}$ **5.** $\{-8, -4, -2, 0, 2, 7, 9\}$ **7.** $\{-8, -4, 7, 9\}$ **9.** $\in$
11.

13. $|x| \not< x$ **15.** $-1 < z - 5 < 14$ **17.** Substitution property of equality **19.** Commutative property for addition **21.** Commutative property for multiplication **23.** Multiplication property of equality
25. $-\dfrac{1}{5}; 5$ **27.** $-(x^2 + 1); \dfrac{1}{x^2 + 1}$ **29.** 12 **31.** 5 **33.** -20 **35.** -6 **37.** -9 **39.** 10
41. $\dfrac{51}{35}$ **43.** 28 **45.** -60 **47.** 6 **49.** 2 **51.** -46 **53.** 8 **55.** 1 **57.** 0 **59.** $-\dfrac{22}{9}$
61. -13 **63.** $\dfrac{9}{10}$ **65.** $-\dfrac{5}{8}$

Practice Test

1. (a) $\{3, 4, 5\}$ (b) $\{3, 6, 9, 12, \ldots\}$ (c) $\{-2, 0, 2\}$ **2.** (a) $\in$ (b) $\subset$ (c) $\notin$ (d) $\not\subset$
3. (a)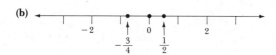

(b)

4. (a) Substitution property (b) Symmetric property (c) Reflexive property (d) Trichotomy property
5. (a) Associative property for addition (b) Distributive property (c) Inverse property for multiplication
(d) Associative property for multiplication (e) Multiplication property of equality
6. (a) -18 (b) 1 (c) 1 (d) -30 (e) undefined (f) 6   **7.** (a) 7 (b) 13 (c) $x^2 + 2$ (d) $x^2 + 1$
8. (a) -1 (b) -18

Extended Applications: Sunscreens

1. 8 hours **3.** 3

Extended Applications: Calories and Dieting

1. 1175 **3.** 3217

Chapter 2

Exercises 2.1

1. $-2x = -3$ 3. $11y + (-3) = 14$ 5. $-3p + 2 = 0$ 7. $-15m + 9 = 4$ 9. $8z + (-8) = 6$
11. $\{3\}$ 13. $\left\{\dfrac{16}{5}\right\}$ 15. $\{7\}$ 17. $\{-1\}$ 19. $\{6\}$ 21. $\{3\}$ 23. $\{12\}$ 25. $\{1\}$ 27. $\{-11\}$
29. $\left\{\dfrac{23}{6}\right\}$ 31. $\left\{\dfrac{7}{8}\right\}$ 33. $\{2\}$ 35. $\{-1.620\}$ 37. $\{-0.530\}$ 39. $\{-21\}$ 41. $\left\{\dfrac{5}{2}\right\}$
43. $\{12\}$ 45. $\{15\}$ 47. $\left\{\dfrac{12}{31}\right\}$ 49. $\left\{\dfrac{7}{4}\right\}$ 51. $\{-38\}$ 53. $\{28\}$ 55. $\left\{\dfrac{4}{5}\right\}$ 57. $\left\{\dfrac{1}{3}\right\}$ 59. $\{7\}$
61. $\left\{\dfrac{15}{7}\right\}$ 63. $\emptyset$ 65. $\left\{\dfrac{17}{4}\right\}$ 67. $k = 6$ 69. $k = 7$

Exercises 2.2

1. $r = \dfrac{d}{t}$ 3. $\pi = \dfrac{C}{2r}$ 5. $R = \dfrac{E}{I}$ 7. $G = \dfrac{gR^2}{m}$ 9. $f = \dfrac{ab}{a+b}$ 11. $h = \dfrac{S - 2\pi r^2}{2\pi r}$
13. 13 feet 15. 57 cm 17. 13 inches 19. 37 cm 21. 3.5 cm 23. $1460
25. 5 years 27. 9% 29. 1200 mm 31. $55,503,200 33. $5238.92 35. 4.875 feet
37. $d = \dfrac{t - a}{m - 1}$ 39. $d = \dfrac{r}{1 + rt}$ 41. $m = \dfrac{Br(n+1)}{2I}$ 43. $R_1 = \dfrac{RR_2}{R_2 - R}$ 45. $y = \dfrac{c - ax}{b}$

Exercises 2.3

1. $x - 3$ 3. $x + 2x$ 5. $\dfrac{x}{x + 2}$ 7. $(x + 7) - \dfrac{1}{2}x = 20$ 9. $7x = \dfrac{1}{2}x + 2$
11. $2 + \dfrac{4}{x + 4} = 123$ 13. $x - 0.20x = 465$ 15. $3x(x + 9) = 214$ 17. $(6x - 3) - 3 = 6(x - 1)$
19. 300 21. $2000 23. 13, 14, 15 25. 25, 27 27. $4000 at 9%; $3000 at 6% 29. $2500
31. 72°, 72°, 36° 33. 40°, 60°, 80° 35. $2\dfrac{1}{2}$ hours 37. 30 mph 39. 30 grams
41. 30 lb of almonds; 10 lb of cashews; 60 lb of pecans 43. 24, 25 45. 20,000
47. 100 mph and 107 mph 49. 2.1 hours 51. 4 m 53. 40 km 55. 231 nickels, 77 dimes, 308 quarters 57. 5%, 10% 59. 86 red, 62 green

Exercises 2.4

1. $\{-1, 1\}$ 3. $\{0\}$ 5. $\{-6, 6\}$ 7. $\emptyset$ 9. $\{-3, 3\}$ 11. $\emptyset$ 13. $\{0\}$ 15. $\{1\}$
17. $\{-5, -3\}$ 19. $\{-3, 7\}$ 21. $\{-3, 4\}$ 23. $\left\{\dfrac{-7}{2}, \dfrac{9}{2}\right\}$ 25. $\left\{\dfrac{-3}{5}, 1\right\}$ 27. $\{3, 5\}$
29. $\left\{-3, \dfrac{-1}{3}\right\}$ 31. $\{-5, 15\}$ 33. $\{4, 8\}$ 35. $\left\{\dfrac{7}{9}\right\}$ 37. $\emptyset$ 39. $\{2\}$ 41. $\{2\}$
43. $\{-2, 36\}$ 45. $\left\{-3, \dfrac{7}{15}\right\}$ 47. $\{0.477, 0.916\}$ 49. $\{0.420\}$ 51. $\{-103, 103\}$
53. $\{-2, 8\}$ 55. $\left\{-3, \dfrac{3}{5}\right\}$

Exercises 2.5

1. $[-1, \infty)$

3. $(-\infty, 3)$

5. $(-1, 3)$
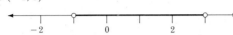

7. $[-2, 0)$

9. $[4, \infty)$

11. $(-6, -3]$

13. $\{x \mid 1 \le x \le 4\}$

15. $\{x \mid -3 < x \le 1\}$

17. $\{x \mid x \ge -3\}$
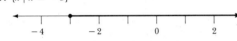

19. $\{x \mid x \le 5\}$

21. $\{x \mid 14 < x < 28\}$

23. $\left\{x \mid -\dfrac{3}{4} < x < \dfrac{9}{4}\right\}$

25. $(0, \infty); (-\infty, 0)$
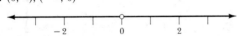

27. $(-2, 3); [4, \infty)$

29. $\{x \mid x < 2\}; \{x \mid x \ge 3\}$
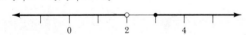

31. $\{x \mid 0 < x \le 1\}; \{x \mid x > 2\}$

Exercises 2.6

1. $\{x \mid x < 7\}$ or $(-\infty, 7)$
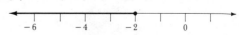

3. $\{k \mid k \le -2\}$ or $(-\infty, -2]$

5. $\left\{x \mid x > -\dfrac{7}{3}\right\}$ or $\left(-\dfrac{7}{3}, \infty\right)$
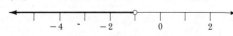

7. $\{y \mid y < -1\}$ or $(-\infty, -1)$

Answers to Selected Exercises

9. The solution set is **R**

11. $\{k \mid k < \frac{5}{2}\}$ or $\left(-\infty, \frac{5}{2}\right)$

13. $\{x \mid x > \frac{11}{4}\}$ or $\left(\frac{11}{4}, \infty\right)$

15. $\{t \mid t \leq \frac{3}{10}\}$ or $\left(-\infty, \frac{3}{10}\right]$

17. $\{z \mid z \leq 1\}$ or $(-\infty, 1]$

19. $\{x \mid x \leq 0\}$ or $(-\infty, 0]$

21. $\{t \mid t > -85\}$ or $(-85, \infty)$

23. $\{k \mid k < 0\}$ or $(-\infty, 0)$

25. $\{x \mid x \geq -21\}$ or $[-21, \infty)$

27. $\{t \mid t > 9\}$ or $(9, \infty)$

29. $\{z \mid z \leq 2\}$ or $(-\infty, 2]$

31. $\{y \mid y \geq -\frac{1}{4}\}$ or $\left[-\frac{1}{4}, \infty\right)$

33. $\{x \mid x \geq -2\}$ or $[-2, \infty)$

35. $\{S \mid S > 8\}$ or $(8, \infty)$

37. $\{t \mid t > 7\}$ or $(7, \infty)$

39. $\{x \mid x > 2\}$ or $(2, \infty)$

41. $\{k \mid k \leq 5\}$ or $(-\infty, 5]$

43. $\{y \mid y \leq -3.844\}$

45. $\{x \mid x < 6.055\}$

47. $\{k \mid k \geq 1.329\}$

49. $\{z \mid z < 0.4166\}$

51. $\{x \mid x > 5\}$
53. $\left\{x \mid x \leq \dfrac{13}{2}\right\}$, assuming "difference" is $2x - 3$
$\left\{x \mid x \geq -\dfrac{7}{2}\right\}$, assuming "difference" is $3 - 2x$
55. $\{x \mid x < 30\}$ **57.** at least 8 dimes   **59.** at least $425

Exercises 2.7

1. ∅   **3.** $\{a, c, d, e, f, g\}$ **5.** $\{a\}$ **7.** $\{a, b, d, e, f, h\}$ **9.** ∅   **11.** $\{a, c, e, g\}$ **13.** ∅
15. $\{a, b, d, e, f, h\}$ **17.** ∅   **19.** $\{a\}$ **21.** $\{b, d, f, h\}$ **23.** $\{a\}$
25. $(1, 3)$ **27.** $(-\infty, 1) \cup [3, \infty)$

29. $(1, \infty)$ **31.** $(-\infty, -1) \cup (2, \infty)$

33. $\{1\}$ **35.** $\left(-\infty, \dfrac{1}{3}\right) \cup \left(\dfrac{1}{3}, \infty\right)$

37. ∅   **39.** $\left(-\infty, -\dfrac{4}{3}\right)$

41. $\left(\dfrac{1}{2}, \dfrac{13}{2}\right)$ **43.** $(-\infty, -1) \cup [3, \infty)$

45. R; the graph is the entire number line.   **47.** $\left(-\infty, \dfrac{3}{2}\right) \cup (4, \infty)$

49. $(-2, 3)$ **51.**

53.   **55.**

57. ∅   **59.** R

61. true   **63.** true   **65.** true   **67.** false   **69.** true

Exercises 2.8

1. $(-3, 3)$ **3.** $(-1, 1)$ **5.** $[-7, 5]$ **7.** $\emptyset$ **9.** $(1, 4)$ **11.** $\left(-\frac{23}{3}, 5\right)$ **13.** $(-4, 8)$ **15.** $(-5, 1)$
17. $(-\infty, -2] \cup [2, \infty)$ **19.** R **21.** $(-\infty, -8] \cup [2, \infty)$ **23.** $(-\infty, 4) \cup (10, \infty)$
25. R **27.** $\left(-\infty, -\frac{1}{4}\right) \cup \left(-\frac{1}{4}, \infty\right)$ **29.** $(-\infty, -2] \cup [3, \infty)$ **31.** $[-2, 4]$ **33.** $(-5, 13)$
35. $\left\{\frac{1}{2}\right\}$ **37.** $\left[-\frac{10}{3}, 10\right]$ **39.** $(-1, 8)$ **41.** $\left(-\frac{8}{9}, \frac{22}{9}\right)$ **43.** $[-1, 5]$ **45.** $[0, 16]$
47. $(-\infty, 0) \cup \left(\frac{1}{2}, \infty\right)$ **49.** $(-0.5878, 1.0971)$ **51.** $(-\infty, -10.299) \cup (-4.653, \infty)$
53. $(-\infty, -1.262) \cup (5.435, \infty)$ **55.** R **57.** $\left(-\frac{7}{2}, \frac{7}{2}\right)$ **59.** $\left(-\infty, -\frac{140}{3}\right) \cup (28, \infty)$

Review Exercises

1. $\{7\}$ **3.** $\{-6\}$ **5.** $\left\{\frac{4}{3}\right\}$ **7.** $\{10\}$ **9.** $\{0\}$ **11.** $\left\{-\frac{11}{5}\right\}$ **13.** $\left\{\frac{2}{5}\right\}$ **15.** $\{-23\}$ **17.** $\left\{\frac{1}{5}\right\}$
19. $x = \frac{y + 7}{3}$ **21.** $B = \frac{T}{T - 2}$ **23.** $r = \frac{11}{\pi}$ cm   **25.** $b = 3$ meters   **27.** $h = 13$ cm
29. $h = 350$ cm   **31.** $x - 2(-x) = -4$ **33.** $x - \frac{1}{x}$ **35.** $5x = \frac{x}{x + 4}$ **37.** $35°, 35°, 110°$
39. 30, 31, 32   **41.** \$400/month for 2 yr lease, \$650/month for $1\frac{1}{2}$ yr lease
43. 5   **45.** 80 km/hr and 85 km/hr   **47.** $\{-7, 7\}$ **49.** $\{-6, 9\}$ **51.** $\left\{-\frac{20}{3}, 0\right\}$ **53.** $\left\{\frac{1}{4}\right\}$
55. $\{-17, 17\}$

57.

59.

61.

63. $(-\infty, 5)$

65. $(2, \infty)$

67. $(-\infty, -1)$

69. $(-\infty, 2]$

71. $(-\infty, -35]$

73. $\left[\frac{9}{10}, \infty\right)$

75. *R*; the graph is the entire number line.   **77.** $\left(\frac{27}{2}, \infty\right)$ **79.** more than 27   **81.** *R*; the graph is the entire number line.   **83.** ∅   **85.** ∅

87.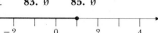

89.

91. (−3, 3)

93. (−∞, 1) ∪ (5, ∞)

95. (1, 15)

97. [−1, 3]

99. ∅   **101.** [−5, 5]   **103.** (−∞, −24) ∪ (24, ∞)

Practice Test

1. {−7}   **2.** $\left\{\frac{8}{5}\right\}$ **3.** $\left\{-\frac{7}{2}\right\}$ **4.** { }   **5.** $n = \frac{Ir}{E - IR}$ **6.** 75 km/hr   **7.** 8 meters

8. $16,500 at 7% and $8500 at $8\frac{1}{2}$%   **9.** 29, 30, 31   **10.** 10 grams   **11.** {3, −3}   **12.** $\left\{1, \frac{-11}{5}\right\}$

13. (−∞, −3]

14. $\left[\frac{1}{4}, \infty\right)$

15. $\left(-\infty, -\frac{5}{2}\right) \cup \left(\frac{7}{2}, \infty\right)$

16. (4, ∞)

17. $\left(-\infty, \frac{7}{4}\right] \cup \left[\frac{7}{2}, \infty\right)$

18. (−∞, −4) ∪ (4, ∞)

19. (−23, −5)

20. (−∞, −6] ∪ [11, ∞)

Extended Applications: Saving Energy

1. $R_1 = \dfrac{SR_0^2}{CFWA - SR_0}$ **3.** $2212.83 **5.** 3.7

Chapter 3

Exercises 3.1

1. base 9; exponent 10 **3.** base 413; exponent 2 **5.** base -2; exponent 4 **7.** base 2; exponent 5 **9.** base x; exponent 1 **11.** base x; exponent 2 **13.** base a; exponent -11 **15.** base $(a + b)$; exponent 4 **17.** base $(3x^2 - 2x + 5)$; exponent -3 **19.** 32 **21.** 16 **23.** 27 **25.** $\dfrac{1}{81}$ **27.** 1 **29.** 0 **31.** undefined **33.** $\dfrac{1}{16}$ **35.** $-\dfrac{1}{32}$ **37.** 16 **39.** 256 **41.** $\dfrac{1}{12}$ **43.** $\dfrac{4}{3}$ **45.** $\dfrac{27}{8}$ **47.** 31415.9 **49.** 5.656843 **51.** 353.556 **53.** $5^7 = 78125$ **55.** 27 **57.** $\dfrac{1}{100}$ **59.** 1 **61.** $9^4 = 6561$ **63.** $\dfrac{1}{13^7}$ **65.** $11^2 = 121$ **67.** $-\dfrac{1}{27}$ **69.** x^3 **71.** $\dfrac{1}{a^4}$ **73.** a^3 **75.** $-6c^5$ **77.** e^m

Exercises 3.2

1. $\dfrac{1}{a^{12}}$ **3.** $\dfrac{1}{a^{12}}$ **5.** 64 **7.** $a^6 b^3$ **9.** $\dfrac{1}{16}$ **11.** $a^{20} b^{30}$ **13.** x^{21} **15.** $\dfrac{a^8 x^4}{b^{12}}$ **17.** $\dfrac{a^{18} b^{24}}{x^{30}}$ **19.** $\dfrac{1}{y - x}$ **21.** ab^2 **23.** $\dfrac{y^2 - x^2}{x^2 y^2}$ **25.** 1 **27.** $\dfrac{8x^{15}}{27y^9}$ **29.** $\dfrac{4b^7}{a^3 c^8}$ **31.** $\dfrac{1}{y^2}$ **33.** $\dfrac{y^2}{t^2 x^{10}}$ **35.** $\dfrac{25}{36}$ **37.** $\dfrac{7s}{p^3}$ **39.** $-\dfrac{1}{x}$ **41.** $\dfrac{1}{x^{12} y^{24} z^6}$ **43.** q **45.** $k^{2n} s^{2n-5}$ **47.** $\dfrac{1}{ab^3}$ **49.** $\dfrac{x^{m^2-1}}{10^{m+1} y^{2m+2}}$ **51.** -5 **53.** -1 **55.** 25

Exercises 3.3

1. 2.43×10^2 **3.** -3.05×10^5 **5.** 3×10^{-3} **7.** 3.78×10^{-6} **9.** -1.09×10^{-4} **11.** 1.03×10^{11}
13.

Earth	12,700	1.27×10^4
Moon	3,480	3.48×10^3
Sun	1,390,000	1.39×10^6
Jupiter	134,000	1.34×10^5
Mercury	4,800	4.80×10^3

15. 270 **17.** 0.015 **19.** $-123,000$ **21.** 0.000000606 **23.** -0.00000000819 **25.** 1.25×10^{-3} or 0.00125 **27.** 2.00×10^5 or 200,000 **29.** 3.00×10^{-6} or 0.000003 **31.** 8.00×10^2 or 800 **33.** 1.25×10^7 or 12,500,000 **35.** 1.00×10^{-2} or 0.01 **37.** 6.30×10^{-3} or 0.0063 **39.** 7.20×10^2 or 720 **41.** 6.45×10^5 or 645,000 to the nearest thousand **43.** 3.73×10^4 or 37,300 to the nearest hundred **45.** 15 **47.** $\dfrac{1}{2}$ **49.** 6 **51.** 9.5×10^{12} km

Answers to Selected Exercises   A21

Exercises 3.4

1. a, c   **3.** a, c   **5.** a, b   **7.** a, c   **9.** b   **11.** degree 2; coefficient 5   **13.** degree 0; coefficient -7
15. degree 3; coefficient $\frac{1}{2}$ **17.** degree 7; coefficient $-\frac{1}{8}$ **19.** degree 3; coefficient $\frac{1}{12}$
21. degree 20; coefficient π
23. $x^3 - x^2 + x - 5$;   **25.** $x^5 + \sqrt{3}\,x^3 - 2x^2 + \frac{1}{2}x$;
$-5 + x - x^2 + x^3$
$\frac{1}{2}x - 2x^2 + \sqrt{3}\,x^3 + x^5$
27. $t^{99} + 5t^{91} - \frac{\pi}{6}t^{49} - 3t^3$;   **29.** $v^{10} - v^8 - \frac{1}{2}v^4 - 5v + \sqrt{2}$;
$-3t^3 - \frac{\pi}{6}t^{49} + 5t^{91} + t^{99}$
$\sqrt{2} - 5v - \frac{1}{2}v^4 - v^8 + v^{10}$
31. $-x^2$; monomial; degree 2   **33.** $5z$; monomial; degree 1   **35.** $6x^2 - 11$; binomial; degree 2
37. $\frac{1}{6}x^3 + \frac{51}{5}$; binomial; degree 3   **39.** $-5x^2 + 10x - 13$; trinomial; degree 2   **41.** $4u^3 + 17u^2 + 15$;
trinomial; degree 3   **43.** $2x^2 - 3x - 1$ **45.** $x^3 - x^2 + 5x - 9$ **47.** $6x - y + zw^2$
49. $4x^3 + x^2 - 2x + 5$ **51.** $5x^3 - 4x^2 - 20x + 8$ **53.** $-2x^3 - 9x^2 - 14x - 5$
55. $15x^2 + xy - 17y^2 + 12x + 29y + 69$ **57.** $6.893x^2 - 5.094x - 0.455$ **59.** -7 **61.** 0
63. -248.05

Exercises 3.5

1. $2x^2 + 10x$ **3.** $4t^2 - 8t$ **5.** $12z^2 + 18z$ **7.** $30m^3 - 40m^2$ **9.** $-16x^4 + 8x^2y^2$
11. $6x^3 - 12x^2 + 50x$ **13.** $-72t^2 - 60t + 78t^4$ **15.** $12.844x^3 - 24.402x^2 + 47.916$
17. $z^2 + 3z + 2$ **19.** $6k^2 + 7k + 2$ **21.** $20x^2 + x - 12$ **23.** $55t^2 - 134t + 63$
25. $x^4 - 4$ **27.** $4r^2 - 1$ **29.** $\frac{1}{4}x^2 - \frac{9}{4}y^2$ **31.** $x^2 + 10x + 25$ **33.** $4v^2 - 4v + 1$
35. $16k^2 + 40k + 25$ **37.** $25x^2 + 110xy + 121y^2$ **39.** $\frac{1}{25}s^2 - \frac{1}{15}st + \frac{1}{36}t^2$
41. $x^4 - 7x^3 + 15x^2 - 32x + 35$ **43.** $8t^3 + 27$ **45.** $64y^3 - 125$ **47.** $6x^4 + 19x^3 + 32x^2 + 26x + 7$
49. $15u^5 - 22u^4 + 42u^3 - 23u^2 + 13u + 5$ **51.** $7x^7 + 21x^6 - 12x^5 - 15x^4 + 3x^3 + 2x^2 + 26x - 8$
53. $130.616x^4 - 646.619x^3 + 2338.786x^2 - 2443.405x + 1020.51$ **55.** $x^3 + 3x^2 + 3x + 1$
57. $8z^3 + 60z^2 + 150z + 125$ **59.** $s^3 + 3s^2t + 3st^2 + t^3$ **61.** $a^2 + 2ab + b^2 - 36$
63. $9x^2 - 6xy + y^2 - 81$ **65.** $16z^4 - 89z^2 + 100$ **67.** $64 - 49x^2 + 42x^3 - 9x^4$
69. $16x^2 + 24xy + 9y^2 - 96x - 72y + 144$ **71.** $x^{3n} - 3x^{2n-1}$ **73.** $z^{2n} - 2z^n + 1$
75. $10x^m - 15x^{-2m}$ **77.** $k^{4r} - 4$ **79.** $8z^{3n} + 12z^{2n} + 6z^n + 1$ **81.** $9p^{2n} - 12 + 4p^{-2n}$
83. $16x^{2m} - 49y^{2n}$ **85.** $12t^{2m} + 10t^{n+m-1} - 18t^{n+m+1} - 15t^{2n}$ **87.** $27r^{3n} + 135r^n + 225r^{-n} + 125r^{-3n}$
89. $(a + b)^{-z} + 1$

Exercises 3.6

1. $5(x + 2)$ **3.** $27(y + 1)$ **5.** $x(x + 2)$ **7.** $5(yz - 1)$ **9.** $2x(2 - x)$ **11.** $11x(3a + 5x)$
13. $2x^4(x^2 - 3x - 7)$ **15.** $5y(2yz + 7y^2 + 1)$ **17.** $3y^2(1 + y - 2x)$ **19.** $2a^2(a - 4 + 3a^3)$
21. $3a^2b^2(a - 6a^3b + 2)$ **23.** $8a^2x^3(4x^2 - 2a + a^3x)$ **25.** $13x^5y^2 + 11xz^4 + 13y^3z^2$
27. $xz(x^2y^3z^2 - 3xy + z)$ **29.** $x^2y^2(5 - 2xy^2 + 3y^3 + x)$ **31.** $(2a + b)(x + y)$
33. $(2x - b)(3a - 5y)$ **35.** $(a + b)(5x - y)$ **37.** $(3x + y)(2a - 1)$ **39.** $3(2x + 1)$
41. -2 **43.** $-3x(-x - (-3y))$ **45.** -1 **47.** $(2x - 3y)(x - 3y)$ **49.** $(x + y)(a - b)$
51. $(2y + z)(x + w)$ **53.** $(x - 2)(x - 3y)$ **55.** $(y + 1)(y^2 + 1)$ **57.** $x^n(x^n + 1)$ **59.** $x^{3n}(1 - x^{3n})$
61. $2p^r(1 - 4p^r)$ **63.** $y^{p+3}(1 + y - y^2)$ **65.** $(1 + y)(x^r y^p + z^p)$

Exercises 3.7

1. $(x + 2)(x - 2)$ 3. $(t + 6)(t - 6)$ 5. $(2k + 1)(2k - 1)$ 7. $(11 + 3x)(11 - 3x)$
9. $(y + 3a)(y - 3a)$ 11. $(2t + 5s - 5)(2t - 5s + 5)$ 13. $kn(k + n)(k - n)$
15. $5x(x + 5)(x - 5)$ 17. $x^2y^2(x^2 + y^2)(x + y)(x - y)$ 19. $st^2(s^2 + t^2)(s + t)(s - t)$
21. $2x(x^2 + 4)(x + 2)(x - 2)$ 23. $(y + 7)^2$ 25. $(x + 5)^2$ 27. $\left(z - \dfrac{1}{2}\right)^2$
29. $(t - 8)(t + 1)$ 31. $(3x - 2y)^2$ 33. $(x + y + 12)^2$ 35. $(2z + 3)^2$ 37. $(3x + 5)(x + 3)$
39. $-(2x + 3)(x - 7)$ 41. $-(3y - 1)(y - 15)$ 43. $(r - 5)(r - 3)$ 45. $(3n + 1)(n - 5)$
47. $(q^2 + 1)(q + 2)(q - 2)$ 49. $(x + y - 3)(x + y + 1)$ 51. $(x^2 + xy + y^2)(x^2 - xy + y^2)$
53. $(r + 1)(r^2 - r + 1)$ 55. $(y - 3)(y^2 + 3y + 9)$ 57. $\left(3k + \dfrac{1}{2}\right)\left(9k^2 - \dfrac{3}{2}k + \dfrac{1}{4}\right)$
59. $\left(\dfrac{1}{3}n - 5\right)\left(\dfrac{1}{9}n^2 + \dfrac{5}{3}n + 25\right)$ 61. $(2a + 3b)(4a^2 - 6ab + 9b^2)$ 63. $27xy(x - 2y)(x^2 + 2xy + 4y^2)$
65. $(s + t)(s - t)(s^2 - st + t^2)(s^2 + st + t^2)$

Exercises 3.8

1. $60(a + 3)$ 3. $7(5x - 10y + 2)$ 5. $27x(xy + 3)$ 7. $(a + b)(x - y)$ 9. $x^2y^2(ay + bx + c)$
11. $(11p + 1)(11p - 1)$ 13. $(10m + 11)(10m - 11)$ 15. $(9p + q + r)(9p - q - r)$
17. $(x^2 + 9)(x + 3)(x - 3)$ 19. $(z + 9)(z^2 - 9z + 81)$ 21. $(6 - p)(36 + 6p + p^2)$
23. $(5x + 7y)(25x^2 - 35xy + 49y^2)$ 25. $(2p - q - r)(4p^2 + 2pq + 2pr + q^2 + 2qr + r^2)$
27. $(2x + 1)^2$ 29. $(y - 6)^2$ 31. $2b(5a + 2)^2$ 33. $z^2(z - 1)^2$ 35. $(4x - 1)(x - 1)$
37. $10y(y + 5)(y + 2)$ 39. $2(2x - y)(x + 2y)$ 41. $(z^2 + 6z + 10)^2$ 43. $(2p + 2q - 3)^2$
45. $2(8x + 8y + 3)(x + y - 3)$ 47. $(x^2 + y^2)(x^4 - x^2y^2 + y^4)$ 49. $(2p + 3)(q - 5)$
51. $(x^3 - 5)(y + 1)(y - 1)$ 53. $x^{m+1}y^n(y - 1)$ 55. $(5x^m + 11y^n)(5x^m - 11y^n)$
57. $(3z^n + 2w^m)(9z^{2n} - 6z^n w^m + 4w^{2m})$ 59. $(4x^m - 3y^n)^2$

Review Exercises

1. base 6; exponent 5   3. base 4; exponent -5 5. 16   7. $\dfrac{1}{125}$ 9. 20,327   11. 17.1734
13. 36   15. $\dfrac{1}{7}$ 17. a^3 19. $\dfrac{1}{a^{12}}$ 21. $5^6 = 15625$ 23. $\dfrac{a^{10}}{b^{15}c^{20}}$ 25. 2.4×10^8 27. 9.3×10^7
29. 5.13×10^{-6} 31. 2,130,000   33. 4.2×10^{-3} 35. degree 2; coefficient 3
37. degree 0; coefficient 125   39. $-3x^3 + 5x^2 + 2x + 6$ 41. $20x^2 + 23x + 6$
43. $x^3 + 4x^2 + 20x - 25$ 45. $25(4x - 1)$ 47. $(a + b)(x + 3)$ 49. $(x + 10)(x - 10)$
51. $(3x + 5y - 15)(3x - 5y + 15)$ 53. $(x + 3)(x^2 - 3x + 9)$ 55. $\left(\dfrac{1}{5}x + \dfrac{1}{2}\right)\left(\dfrac{1}{25}x^2 - \dfrac{1}{10}x + \dfrac{1}{4}\right)$
57. $(2x + 3y)(2x - 3y + 2)$ 59. $(2x - 1)(x + 2)$ 61. $(x^2 + 2xy - y^2)(x^2 - 2xy - y^2)$
63. $(3p + 3q + 1)(p + q - 5)$ 65. $3(5x^n + 11y^m)(5x^n - 11y^m)$

Practice Test

1. -125 2. $\dfrac{1}{81}$ 3. $\dfrac{1}{10}$ 4. $\dfrac{1}{5}$ 5. $\dfrac{81}{64}$ 6. $\dfrac{3^{k-1}y^{2k-2}}{4^{k-1}x^{k^2-1}}$ 7. 6.957×10^{-8}
8. 2.4695×10^{11} 9. 0.214 10. $7xy(3x - y + 4)$ 11. $(5x - 3)(x + \sqrt{3})(x - \sqrt{3})$
12. $(2x + 9y)(2x - 9y)$ 13. $(9x^2 + 4y^2)(3x + 2y)(3x - 2y)$ 14. $(x + 5)(x^2 - 5x + 25)$
15. $(5y - 2)(25y^2 + 10y + 4)$ 16. $(2x - 5)(2x + 3)$ 17. $(3x - 3y + 1)(x - y + 1)$
18. $(a - b)(x + y)$

Extended Applications: Poiseville's Law

1. $V = 9260(1.44 \times 10^{-4} - r^2)$; 1.33

Chapter 4

Exercises 4.1

1. -5 **3.** 2 **5.** ± 1 **7.** none **9.** none **11.** $-2, 1$ **13.** $\dfrac{x}{3y}$ **15.** $\dfrac{7}{10x^2}$ **17.** $\dfrac{1}{5x^2}$
19. $-\dfrac{5x}{3y}$ **21.** $\dfrac{7x^2}{8yz^3}$ **23.** $\dfrac{1}{u}$ **25.** $\dfrac{2x}{3}$ **27.** -1 **29.** $\dfrac{3}{t-1}$ **31.** $-\dfrac{2}{x+3}$ **33.** $\dfrac{10x}{x+1}$
35. $\dfrac{3}{2x-1}$ **37.** $x^2 + xy + y^2$ **39.** $\dfrac{x^2 + x + 1}{x-1}$ **41.** $\dfrac{u-2}{u+2}$ **43.** $\dfrac{x^2 + y^2}{x^2 - y^2}$
45. $\dfrac{x+1}{x-1}$ **47.** $\dfrac{10x}{15x^2}$ **49.** $\dfrac{9x}{6xy}$ **51.** $\dfrac{6xy^2}{15x^2y^2}$ **53.** $\dfrac{3x^2 - 5x + 2}{x^2 - 1}$ **55.** $\dfrac{5x^2 - 7x + 2}{x^3 - 1}$
57. $\dfrac{3x^4 - 3x^3 + 8x^2 - 5x + 5}{x^3 + 1}$ **59.** $\dfrac{6u - 15v}{4u^2 - 25v^2}$ **61.** $\dfrac{125u^2 + 50uv + 20v^2}{125u^3 - 8v^3}$ **63.** $3x^2 - 3xy$
65. $6x^3 + 15x^2$ **67.** $x^2 + y^2$

Exercises 4.2

1. $\dfrac{2x+1}{5}$ **3.** $\dfrac{2r-1}{3}$ **5.** $\dfrac{5t}{3}$ **7.** $\dfrac{2}{x-1}$ **9.** 3 meters **11.** 104° **13.** $6z^2$ **15.** $18a^2b$
17. $70r(r-1)$ **19.** $12xyz$ **21.** $12(r-2)$ **23.** $(2k+1)^2(2k-1)$ **25.** $2(4s-5t)(4s+5t)$
27. $(2k+3)(4k^2 - 6k + 9)$ **29.** $(x+4)(x-4)(x-3)$ **31.** $(4t+3)(16t^2 - 12t + 9)$
33. $(z-1)(z-2)(z-3)$ **35.** $(r-5)(r-4)(r-1)$ **37.** $\dfrac{5x+2}{25x^2}$ **39.** $\dfrac{4s-3r}{42r^2s}$
41. $\dfrac{3a^2 + 2a + 2}{72a(a+1)}$ **43.** $\dfrac{4z - 3x + 6y}{24xyz}$ **45.** $\dfrac{11k}{21(3k+1)}$ **47.** $\dfrac{2(9n^2 - 5n + 1)}{(3n+1)(3n-1)^2}$
49. $\dfrac{1}{3(7a+6b)}$ **51.** $-\dfrac{1}{16x^2 + 12x + 9}$ **53.** $\dfrac{9(r-3)}{(r+5)(r-5)(r+4)}$ **55.** $\dfrac{1}{4x^2 + 6x + 9}$
57. $\dfrac{2}{(p-3)(p-1)}$ **59.** $\dfrac{1}{(y-1)(y-4)}$

Exercises 4.3

1. $\dfrac{1}{6x}$ **3.** $\dfrac{1}{8st}$ **5.** $-\dfrac{8b}{5a}$ **7.** $-\dfrac{3}{4x^2yz}$ **9.** 1 **11.** $-\dfrac{4v^5}{u^3w^2}$ **13.** $-\dfrac{1}{2x^5y^5}$ **15.** $\dfrac{1}{6z}$
17. 1 **19.** $\dfrac{r(r-1)}{5s}$ **21.** $\dfrac{k+4}{3k(k-4)}$ **23.** $\dfrac{1}{x}$ **25.** $\dfrac{1}{y^2 + 5y + 25}$ **27.** $\dfrac{(x+1)^2}{(x+2)(x+3)}$
29. $\dfrac{1}{6}$ **31.** $-\dfrac{3}{t}$ **33.** $\dfrac{2x^2}{5y^5}$ **35.** $\dfrac{v}{u(u+v)}$ **37.** $\dfrac{9}{8k}$ **39.** $\dfrac{y-5}{y^2}$ **41.** $\dfrac{2}{z+1}$ **43.** $\dfrac{1}{x-5}$
45. $\dfrac{(2z+1)(2z-1)}{(z+3)(z+4)}$ **47.** $\dfrac{(y-2)^2}{(y-4)^2}$ **49.** $\dfrac{u+v}{u-v}$ **51.** $\dfrac{6}{a^4b^3}$ **53.** $\dfrac{25a^8b^5}{216}$ **55.** $\dfrac{x-3}{x(x+4)}$
57. 1 **59.** 1 **61.** no

Exercises 4.4

1. proper 3. improper 5. improper 7. proper 9. improper 11. improper
13. $x - 1 + \dfrac{1}{x}$ 15. $3 - 2t$ 17. $n - 1 + \dfrac{1}{n-1}$ 19. $y + 3$ 21. $3 + \dfrac{x+5}{x^2+x}$
23. $4 + \dfrac{5u+2}{u^2+1}$ 25. $3t + 1 + \dfrac{-12}{3t+1}$ 27. $3x - \dfrac{3}{7} + \dfrac{59}{7(7x+1)}$ 29. $2x + 3 + \dfrac{-7}{3x-2}$
31. $4x^2 - 2x - 1 + \dfrac{3x+2}{2x^2+1}$ 33. $t^2 - 5t + 13 + \dfrac{-t+12}{t^2+2t-1}$ 35. $-2u - 7 + \dfrac{10u+34}{-u^2+u+3}$
37. $2x^2 - x + \dfrac{1}{2} + \dfrac{-3(x-1)}{2(2x^2+x-1)}$ 39. $-24z^2 + 72z - 200 + \dfrac{543z+209}{z^2+3z+1}$
41. $u - 1 + \dfrac{-5u+12}{u^2-2u+5}$ 43. $y - 2 + \dfrac{7y^2-2y+6}{y^3+2y^2-y-2}$ 45. $2t - 8 + \dfrac{57t^2-35t+53}{t^3+6t^2-3t+5}$
47. $-y - 2 + \dfrac{6y^2-6y-1}{y^3-2y^2-5}$ 49. $-2x + 1 + \dfrac{6x^2-38x-10}{x^3+3x-10}$

Exercises 4.5

1. $\dfrac{4}{5}$ 3. $\dfrac{1}{5}$ 5. $\dfrac{25}{24}$ 7. $\dfrac{3b}{4a}$ 9. $\dfrac{x-1}{3x+1}$ 11. $\dfrac{xy+1}{xy-2}$ 13. $\dfrac{x^2+y}{4x-3y}$ 15. $\dfrac{2k-30}{2k+25}$
17. $\dfrac{9m+2n}{m-12n}$ 19. $\dfrac{4y-3x^2}{6x+15y^2}$ 21. $\dfrac{p^3+2p}{p^4+4}$ 23. $\dfrac{q}{5q-2}$ 25. $\dfrac{xy}{x+y}$ 27. $\dfrac{xy}{4x+3y}$
29. $\dfrac{12pq}{3p+4q}$ 31. $\dfrac{2x+1}{(x-1)(4x+5)}$ 33. $\dfrac{-x}{x^2+x-2}$ 35. $\dfrac{5(q+1)}{3q-2}$ 37. $\dfrac{26x-21y}{-29x+21y}$
39. $-\dfrac{9x^2+40x+44}{2x^2-4x+8}$ 41. $\dfrac{1}{2p-q}$ 43. $-\dfrac{1}{5}$ 45. $\dfrac{34x}{51x+15}$ 47. $\dfrac{x(3x-2y)(7y-1)}{28x^2y-4x^2-5y^2}$
49. $\dfrac{y^2-5y+3}{y-4}$ 51. $\dfrac{4-3z}{1-z}$ 53. $\dfrac{(L_1-M)(L_2-M)}{L_1+L_2-2M}$

Exercises 4.6

1. quotient $x^3 - x^2 + x - 1$; remainder 2   3. quotient $x^4 + x^3 + x^2 + x + 1$; remainder 0
5. quotient $x - 2$; remainder 3   7. quotient $y + 1$; remainder 4   9. quotient $2z + 4$; remainder -1
11. quotient $5w - 17$; remainder 51   13. quotient $x^2 + 4x + 18$; remainder 62
15. quotient $2y^2 + 5y + 15$; remainder 60   17. quotient $3z^2 - 19z + 103$; remainder -526
19. quotient $-4w^2 - 21w - 131$; remainder -786 21. quotient $x^3 - 3x^2 + 10x + 15$; remainder -105
23. quotient $y^4 - 2y^3 + 3y^2 - 21y + 64$; remainder 97   25. $6x + 9 + \dfrac{25}{x-2}$
27. $t^2 + 3t + 16 + \dfrac{75}{t-5}$ 29. $x^2 - \dfrac{1}{2}x - \dfrac{7}{4} + \dfrac{39}{4(2x+1)}$ 31. $3x^2 + 2x - \dfrac{23}{3} + \dfrac{197}{3(3x-2)}$
33. $2k^2 - k + \dfrac{9}{2} + \dfrac{7}{2(2k-3)}$ 35. -26 37. 2   39. -21 41. 52   43. 76   45. 2348
47. 6   49. 2 is a solution.   51. -2 is a solution.

Review Exercises

1. 4   3. 2, 3   5. $\dfrac{1}{y-1}$ 7. $\dfrac{2x^2+5x+3}{x^2-1}$ 9. $2x^3 + 5x^2$ 11. $z + 3$ 13. $12(2x-1)$
15. $2(4z+5)(4z-5)$ 17. $\dfrac{17}{21(x+3)}$ 19. $\dfrac{1}{(z-1)(z-4)}$ 21. $\dfrac{1}{6y}$ 23. $\dfrac{w-1}{w(w-3)}$

25. $4x - 3 + \dfrac{5}{2x}$ **27.** $3 + \dfrac{-z+1}{z^2+2}$ **29.** $\dfrac{2t+1}{t-1}$ **31.** $\dfrac{z}{2y^2+yz}$
33. quotient $x - 4$; remainder -5 **35.** quotient $z^4 - 3z - 2$; remainder -9 **37.** 1 **39.** 2318

Practice Test

1. $\dfrac{3x^2 + 7x - 6}{x^2 - x - 12}$ **2.** $2x^2 + 9x + 10$ **3.** $\dfrac{6(x^2 + 2x - 1)}{(5x - 3)(3x + 1)}$ **4.** $\dfrac{2y^2 + 35y - 3}{(2y+3)^2(2y-3)}$ **5.** $\dfrac{1}{x-5}$
6. $\dfrac{(y+6)^2}{(y+4)^2}$ **7.** $3 + \dfrac{4x - 12}{x^2 + 1}$ **8.** $\dfrac{-7x - 28}{5x^3 + 17x^2 - 82x - 264}$ **9.** 0 **10.** $p(c) = 2856$

Extended Applications: Mathematics in the Space Program

1. 1.6 m/sec²   **3.** 550 km; 7580 m/sec

Chapter 5

Exercises 5.1

1. 5 **3.** -9 **5.** 6 **7.** -3 **9.** 5 **11.** does not exist **13.** 2 **15.** 1 **17.** -2 **19.** 2
21. $\dfrac{7}{9}$ **23.** $\dfrac{3}{4}$ **25.** 31.432 **27.** -0.567 **29.** 9.327 **31.** 1.331 **33.** $9 \cdot |x|$ **35.** $-12z^2$
37. $0.4 \cdot |s|$ **39.** $6a$ **41.** $-5c$ **43.** $1.1e$ **45.** $-\dfrac{1}{5}g^2|h|$ **47.** $11p^4q^2|r|$ **49.** $-3rs^2$
51. $-4u^4v^6w$ **53.** $.2\,a^4|b|^3c^4$ **55.** $2x^5y^6z^7$ **57.** $2a^2$ **59.** $\dfrac{1}{2}a^2|b|^3$ **61.** $8e^2|f|^3$
63. $\dfrac{11k^4}{13m^2|n|^3}$ **65.** $\left|\dfrac{90u}{v-w}\right|$ **67.** $\dfrac{25(c^2+1)}{4d^2e^4|f|^5}$ **69.** 10% **71.** 21 m/sec **73.** 3.12 units

Exercises 5.2

1. $345^{1/2}$ **3.** $(-606)^{1/3}$ **5.** $a^{2/5}$ **7.** $b^{3/7}c^{5/7}$ **9.** $(-11)^{2/3}$ **11.** $|s|^{3/10}$ **13.** $\sqrt{32}$ **15.** $\sqrt[9]{56}$
17. $\sqrt[9]{14^2}$ **19.** $\sqrt[9]{91^{-3}}$ **21.** $\sqrt[3]{\left(\dfrac{7}{8}\right)^2}$ **23.** $\sqrt[5]{\left(\dfrac{17}{31}\right)^3}$ **25.** 36 **27.** $-\dfrac{1}{3}$ **29.** $-\dfrac{2}{11}$
31. $\dfrac{1}{b^3c^9}$ **33.** $\dfrac{c^6}{9d^2}$ **35.** $\dfrac{9x^4y^3}{z}$ **37.** $100\dfrac{1}{2}$ **39.** $5x^2$ **41.** ab **43.** $x^{127/128}$ **45.** $x^{1/3}(9x+1)$
47. $3z^{-1/5}(7z - 2)$ **49.** $x^{n/2}(7x^n + 8x^{2n})$ **51.** $7z^{-n/5}(2z^n - 3z^{2n})$ **53.** x^{n+1} **55.** z^{n+1} **57.** $\dfrac{b^{2n+2}}{5a^{m+1}}$
59. $u^m + 1$ **61.** 1.74×10^6 meters **63.** 40 cm **65.** 1.64 lb/ft²

Exercises 5.3

1. $2\sqrt{2}$ **3.** $10\sqrt{5} - 3\sqrt{2}$ **5.** $4\sqrt{3} + 7\sqrt{5}$ **7.** $45 - 4\sqrt{2}$ **9.** $-\sqrt[9]{2}$ **11.** $-17x\sqrt{2x}$
13. $16z\sqrt{2z} - 14z\sqrt{2}$ **15.** $22\sqrt[4]{2}$ **17.** $-18\sqrt[9]{2}$ **19.** $-4x\sqrt{2} + 3y\sqrt{5}$ **21.** $8x\sqrt{2y}$
23. $6 - 3\sqrt{2}$ **25.** $5\sqrt{3} - 6$ **27.** 14 **29.** 1 **31.** 4 **33.** 14 **35.** 2 **37.** -43
39. $m - n$ **41.** 17 **43.** $16 + 6\sqrt{7}$ **45.** $52 - 14\sqrt{3}$ **47.** $7 + 2\sqrt{10}$ **49.** $53 - 20\sqrt{7}$
51. $3\sqrt{5} + \sqrt{35} + 6 + 2\sqrt{7}$ **53.** $20 + 4\sqrt{5} - 5\sqrt{3} - \sqrt{15}$ **55.** $3\sqrt{2} + \sqrt{42} + 3 + \sqrt{21}$
57. $5\sqrt{35} - \sqrt{5} - 105 + 3\sqrt{7}$ **59.** $6 - 2\sqrt{77}$ **61.** $12u + 11\sqrt{u} - 15$ **63.** $8v + 2\sqrt{vw} - 15w$

65. 10   **67.** $7 - k$ **69.** $\dfrac{5 + 3\sqrt{5}}{2}$ **71.** $\dfrac{3\sqrt{7} - 4}{5}$ **73.** $\dfrac{\sqrt{3} + \sqrt{5}}{3}$ **75.** $2k - 3$ **77.** -7
79. 1

Exercises 5.4

1. $2\sqrt{3}$ **3.** $\dfrac{\sqrt{5}}{5}$ **5.** $-\dfrac{2\sqrt{6}}{3}$ **7.** $\dfrac{1}{2}$ **9.** $-\dfrac{\sqrt{21}}{7}$ **11.** $-4\sqrt{15}$ **13.** $\dfrac{\sqrt{15}}{6}$ **15.** $\dfrac{4\sqrt{5a}}{a}$
17. $\dfrac{3\sqrt{cd}}{d}$ **19.** $\dfrac{\sqrt{2} - m}{2 - m^2}$ **21.** $3(2 + \sqrt{3})$ **23.** $\sqrt{3} - \sqrt{2}$ **25.** $\dfrac{\sqrt{15p} + 3}{5p - 3}$
27. $\dfrac{42 + 16x\sqrt{7}}{63 - 64x^2}$ **29.** $\dfrac{15\sqrt{7} - 14\sqrt{5}y}{45 - 28y}$ **31.** $\dfrac{a\sqrt{b} - b}{a^2 - b}$ **33.** $\dfrac{u^2 - 2uv\sqrt{w} + v^2 w}{u^2 - v^2 w}$
35. $\dfrac{a\sqrt{x} - b\sqrt{y}}{a^2 x - b^2 y}$ **37.** $\dfrac{\sqrt[3]{4} - \sqrt[3]{2} + 1}{3}$ **39.** $-\dfrac{4\sqrt[3]{9} + 6\sqrt[3]{3} + 9}{3}$ **41.** $\dfrac{5\sqrt[3]{3} - \sqrt[3]{225} + 3\sqrt[3]{5}}{8}$
43. $\dfrac{\sqrt[3]{x^2} - \sqrt[3]{x} + 1}{x + 1}$ **45.** $\dfrac{c(a^2\sqrt[3]{x^2} + ab\sqrt[3]{x} + b^2)}{a^3 x - b^3}$ **47.** $\dfrac{a^2\sqrt[3]{x^2} - a\sqrt[3]{xy} + \sqrt[3]{y^2}}{a^3 x + y}$
49. $\dfrac{c^2 u\sqrt[3]{v} + cd\sqrt[3]{u^2 v^2} + d^2 v\sqrt[3]{u}}{c^3 u - d^3 v}$ **51.** $\dfrac{\sqrt[4]{125} - \sqrt[4]{5} + \sqrt{5} - 1}{4}$ **53.** $(\sqrt{3} + \sqrt[4]{6})(\sqrt{3} + \sqrt{2})$
55. $\dfrac{\sqrt[4]{15}(2\sqrt[4]{3} - \sqrt[4]{5})(4\sqrt{3} + \sqrt{5})}{43}$

Exercises 5.5

1. $\{10\}$ **3.** $\{12\}$ **5.** $\{17\}$ **7.** $\emptyset$ **9.** $\{1\}$ **11.** $\{2\}$ **13.** $\{63\}$ **15.** $\{-4\}$ **17.** $\{172\}$
19. $\{6\}$ **21.** $\{8\}$ **23.** $\emptyset$ **25.** $\{5\}$ **27.** $\left\{\dfrac{24}{11}\right\}$ **29.** $\{14\}$ **31.** $\{7\}$ **33.** $\left\{-\dfrac{1}{5}\right\}$ **35.** $\left\{-\dfrac{1}{5}\right\}$
37. $\{0\}$ **39.** $\left\{\dfrac{1}{16}\right\}$ **41.** $\emptyset$ **43.** $\emptyset$ **45.** $\{6\}$ **47.** $\left\{\dfrac{6}{5}\right\}$

Exercises 5.6

1. real part 3; imaginary part 4   **3.** real part 5; imaginary part -2 **5.** real part -7; imaginary part -1 **7.** real part -11; imaginary part 5   **9.** real part 0; imaginary part 1   **11.** real part -1; imaginary part 0   **13.** real part -1; imaginary part 0   **15.** real part $\sqrt{3}$; imaginary part $\sqrt{3}$ **17.** $7 + 9i$ **19.** $16 - 8i$ **21.** $(a + b)i$ **23.** $(c - d)i$ **25.** $(u + w) + vi$
27. $q + (p - r)i$ **29.** $10 - i$ **31.** $4 + 12i$ **33.** $2a - c$ **35.** yes   **37.** 0   **39.** no   **41.** yes
43. $\{-2i, 2i\}$ **45.** $\left\{-\dfrac{3\sqrt{2}}{2}i, 1 + \dfrac{3\sqrt{2}}{2}i\right\}$ **47.** $\{1 - \sqrt{3}i, 1 + \sqrt{3}i\}$ **49.** $\{2 - \sqrt{7}i, 2 + \sqrt{7}i\}$
51. $\{-3, 3, -3i, 3i\}$ **53.** $x = 3; y = -1$ **55.** $u = 0; v = 1$

Exercises 5.7

1. 1   **3.** i **5.** -1 **7.** i **9.** 10   **11.** $-15i$ **13.** $-21i$ **15.** $30i$ **17.** $-4 + 3i$
19. $35 - 7i$ **21.** 2   **23.** 26   **25.** $173 + 569i$ **27.** $(18 + \sqrt{6}) + (6\sqrt{2} - 3\sqrt{3})i$
29. $(\sqrt{14} - 15) - (5\sqrt{7} + 3\sqrt{2})i$ **31.** $-164 - 26\sqrt{5}i$ **33.** $-54 + 72i$ **35.** $\sqrt{3} - 2i$
37. $-\dfrac{1}{2} + \dfrac{1}{2}i$ **39.** $-\dfrac{3}{5} + \dfrac{6}{5}i$ **41.** $-\dfrac{35}{26} + \dfrac{7}{26}i$ **43.** $\dfrac{5}{2} + \dfrac{1}{2}i$ **45.** $\dfrac{1}{13} - \dfrac{5}{13}i$ **47.** $\dfrac{83}{89} - \dfrac{26}{89}i$
49. $166.068 + 93.826i$ **51.** 15632.125 **53.** $-1.285 + 0.418i$ **55.** yes
57. $\dfrac{a}{a^2 + b^2} - \dfrac{b}{a^2 + b^2}i$ **59.** $x = 1$ **61.** $x = 7; y = 5$ **63.** 0; yes; yes

Review Exercises

1. 6   **3.** -11 **5.** 3   **7.** -8 **9.** $9x^2$ **11.** $3z^2$ **13.** $\dfrac{12k^2}{k^2+4}$ **15.** -4 **17.** $-\dfrac{1}{2}$
19. -6 **21.** $\dfrac{1}{2}m^2$ **23.** $a^{121/243}$ **25.** $4-16\sqrt[3]{2}$ **27.** $16k^2\sqrt{2}-21k^2$ **29.** $10-6\sqrt{51}$
31. 19   **33.** $10-2\sqrt{21}$ **35.** $2ac+3bd+(ad+bc)\sqrt{6}$ **37.** 1   **39.** $\dfrac{-17\sqrt{5}}{3}$ **41.** $10-4\sqrt{6}$
43. $\dfrac{\sqrt[3]{y^2}-c\sqrt[3]{y}+c^2}{y+c^3}$ **45.** $\dfrac{(a\sqrt[4]{x}-1)(a^2\sqrt{x}+1)}{a^4x-1}$ **47.** $\{7\}$ **49.** $\{9\}$ **51.** $\{50\}$ **53.** $\emptyset$
55. real part -3; imaginary part -8 **57.** real part $\sqrt{5}$; imaginary part $-\sqrt{5}$
59. real part -1; imaginary part 0   **61.** real part 0; imaginary part -1 **63.** $28+4i$
65. $6m-(5m-5n)i$ **67.** yes   **69.** $\left\{-\dfrac{1}{2},\dfrac{1}{2},-\dfrac{1}{2}i,\dfrac{1}{2}i\right\}$ **71.** $1-5i$ **73.** $12+18i$
75. $-29-12\sqrt{7}\,i$ **77.** $-\dfrac{\sqrt{5}}{2}-2i$ **79.** $\dfrac{98}{125}+\dfrac{39}{125}i$ **81.** $\dfrac{9}{85}+\dfrac{2}{85}i$ **83.** 0

Practice Test

1. (a) -5 (b) $3a^2$ (c) $\dfrac{11b^4}{c^2+5}$ **2.** (a) $\dfrac{1}{9}$ (b) $\dfrac{s^8 t^4}{4}$ (c) $3k^2$ **3.** $-35+16\sqrt[3]{2}$
4. $156+59\sqrt{6}$ **5.** $\dfrac{10\sqrt{5}-27}{6}$ **6.** $\dfrac{6\sqrt{35}+24\sqrt{10}}{225}$ **7.** $\emptyset$ **8.** $\dfrac{4\sqrt[3]{135}-6\sqrt[3]{275}+45\sqrt[3]{3}}{159}$
9. $3i$ **10.** $u=\pm 2, v=-\dfrac{7}{2}$

Extended Applications: Mathematics in Biology

1. 1.653 sq meters   **3.** 0.053 sq meters

Chapter 6

Exercises 6.1

1. $\{-5,-2\}$ **3.** $\left\{-1,\dfrac{3}{2}\right\}$ **5.** $\left\{-\dfrac{5}{2},\dfrac{1}{3}\right\}$ **7.** $\{-4,0\}$ **9.** $\left\{-2,\dfrac{9}{2}\right\}$ **11.** $\{-1,0,2\}$
13. $\{-1,2\}$ **15.** $\{-5,1\}$ **17.** $\{2,3\}$ **19.** $\{1,10\}$ **21.** $\{-6,-3\}$ **23.** $\left\{\dfrac{1}{2},1\right\}$ **25.** $\left\{-\dfrac{2}{3},1\right\}$
27. $\left\{-4,\dfrac{1}{2}\right\}$ **29.** $\left\{-\dfrac{5}{2},\dfrac{1}{6}\right\}$ **31.** $\left\{-\dfrac{1}{3},\dfrac{5}{2}\right\}$ **33.** $\left\{-\dfrac{7}{3},-1\right\}$ **35.** $\{-\sqrt{3},\sqrt{3}\}$ **37.** $\{0,7\}$
39. $\left\{-1,\dfrac{5}{4}\right\}$ **41.** $\{-5,0\}$ **43.** $\left\{-2,\dfrac{7}{4}\right\}$ **45.** $\{1,2\}$ **47.** $\{-4,3\}$ **49.** $\{-2,2\}$
51. $x^2-5x+6=0$ **53.** $3x^2+7x+2=0$ **55.** $x^2-6x+9=0$ **57.** $x^2+3x=0$
59. $4x^2+25x-21=0$ **61.** 15 cm   **63.** 10 cm $\times$ 24 cm or 12 cm $\times$ 20 cm   **65.** $\left\{-\dfrac{7}{2}b,\dfrac{7}{2}b\right\}$
67. $\{4a\}$ **69.** $\left\{\dfrac{b}{4},\dfrac{2b}{3}\right\}$ **71.** $\{a,b\}$

Exercises 6.2

1. $\{-5, 5\}$ **3.** $\left\{-\frac{1}{4}, \frac{1}{4}\right\}$ **5.** $\{-2i, 2i\}$ **7.** $\{-i, i\}$ **9.** $\{-2, 2\}$ **11.** $\{-3, -1\}$
13. $\{5 - \sqrt{3}, 5 + \sqrt{3}\}$ **15.** $\{7 - 2i, 7 + 2i\}$ **17.** $\{-3, -1\}$ **19.** $\{2 - i, 2 + i\}$ **21.** $\left\{-\frac{3}{4}, \frac{2}{3}\right\}$
23. $\left\{\frac{1}{5}\right\}$ **25.** $\{5\}$ **27.** $\{-3, 2\}$ **29.** $\left\{\frac{5 - \sqrt{13}}{6}, \frac{5 + \sqrt{13}}{6}\right\}$ **31.** $\{1 - 4i, 1 + 4i\}$ **33.** $\left\{-\frac{9}{4}, -\frac{1}{4}\right\}$
35. $\left\{\frac{-3 - \sqrt{3}\,i}{6}, \frac{-3 + \sqrt{3}\,i}{6}\right\}$ **37.** $\left\{\frac{9}{2}, -1\right\}$ **39.** $\left\{\frac{1 - \sqrt{15}}{7}, \frac{1 + \sqrt{15}}{7}\right\}$ **41.** $\left\{\frac{2 - \sqrt{7}}{3}, \frac{2 + \sqrt{7}}{3}\right\}$
43. $\left\{\frac{-1 - \sqrt{22}}{3}, \frac{-1 + \sqrt{22}}{3}\right\}$ **45.** $\{6 - \sqrt{33}, 6 + \sqrt{33}\}$ **49.** $\left\{-\frac{\sqrt{2b}}{2}, \frac{\sqrt{2b}}{2}\right\}$ **51.** $\{b - 5, b + 5\}$
53. $\left\{\frac{-a - i}{3}, \frac{-a + i}{3}\right\}$ **55.** $\{-\sqrt{1 - a^2}, \sqrt{1 - a^2}\}$ **57.** $\{-\sqrt{b^2 - a^2}, \sqrt{b^2 - a^2}\}$
59. $\{-\sqrt{a^2 + b^2}, \sqrt{a^2 + b^2}\}$

Exercises 6.3

1. $\{-1 - \sqrt{5}, -1 + \sqrt{5}\}$ **3.** $\left\{-\frac{1}{2}, \frac{5}{3}\right\}$ **5.** $\left\{\frac{-3 - \sqrt{11}\,i}{2}, \frac{-3 + \sqrt{11}\,i}{2}\right\}$ **7.** $\{2 - \sqrt{3}, 2 + \sqrt{3}\}$
9. $\left\{-\frac{5}{2}, \frac{2}{3}\right\}$ **11.** $\left\{\frac{4 - \sqrt{5}\,i}{7}, \frac{4 + \sqrt{5}\,i}{7}\right\}$ **13.** $\left\{\frac{4 - 2i}{5}, \frac{4 + 2i}{5}\right\}$ **15.** $\left\{-4, \frac{1}{2}\right\}$
17. $\left\{\frac{-3 - \sqrt{7}\,i}{4}, \frac{-3 + \sqrt{7}\,i}{4}\right\}$ **19.** $\left\{\frac{5 - \sqrt{17}}{4}, \frac{5 + \sqrt{17}}{4}\right\}$ **21.** $\left\{\frac{-7\sqrt{2}}{2}, \frac{7\sqrt{2}}{2}\right\}$
23. $\left\{\frac{-3 - i}{2}, \frac{-3 + i}{2}\right\}$ **25.** $\left\{-\frac{5}{3}i, \frac{5}{3}i\right\}$ **27.** $\left\{\frac{-1 - \sqrt{5}\,i}{3}, \frac{-1 + \sqrt{5}\,i}{3}\right\}$
29. $\left\{\frac{1 - \sqrt{7}\,i}{4}, \frac{1 + \sqrt{7}\,i}{4}\right\}$ **31.** $\{-3 - \sqrt{11}, -3 + \sqrt{11}\}$ **33.** $\{(2 - \sqrt{3})i, (2 + \sqrt{3})i\}$
35. $\{1 - \sqrt{7} + i, 1 + \sqrt{7} + i\}$ **37.** $\{-0.591, 2.257\}$ **39.** $\{-1.175, 0.425\}$ **41.** $\{-3.991, 3.991\}$
43. two unequal real roots **45.** two unequal imaginary roots **47.** two unequal real roots
49. two unequal real roots **51.** two unequal real roots **53.** $2 + 2\sqrt{2}$ cm
55. $2\left(x - \frac{11 + \sqrt{217}}{4}\right)\left(x - \frac{11 - \sqrt{217}}{4}\right)$ **57.** $10\left(y - \frac{1}{5}\right)\left(y + \frac{1}{2}\right)$ **59.** $(z + 4 + \sqrt{11})(z + 4 - \sqrt{11})$
61. $r_1 + r_2 = \frac{-b + \sqrt{b^2 - 4ac}}{2a} + \frac{-b - \sqrt{b^2 - 4ac}}{2a} = \frac{-2b}{2a} = -\frac{b}{a}$
63. $a(x - r_1)(x - r_2) = a(x^2 - (r_1 + r_2)x + r_1 r_2) = a\left(x^2 + \frac{b}{a}x + \frac{c}{d}\right) = ax^2 + bx + c$

Exercises 6.4

1. $\{-2, 2\}$ **3.** $\left\{\frac{1}{2}\right\}$ **5.** $\emptyset$ **7.** $\{40 - 10\sqrt{15}, 40 + 10\sqrt{15}\}$ **9.** $\{16\}$ **11.** $\{1, 13\}$ **13.** $\{18\}$
15. $\{6\}$ **17.** $\{0\}$ **19.** $\left\{-1, \frac{5}{3}\right\}$ **21.** $\{-7, 3\}$ **23.** $\{3, 7\}$ **25.** $\{0, 8\}$ **27.** $\emptyset$ **29.** $\{2\}$
31. $\emptyset$ **33.** $\{0, 1\}$ **35.** $\{13, 5\}$ **37.** $\{4\}$ **39.** $\{2\}$

Exercises 6.5

1. $\{-1, 1\}$ **3.** $\{-3, -2, 2, 3\}$ **5.** $\left\{-\frac{\sqrt{2}}{2}, \frac{\sqrt{2}}{2}, -i, i\right\}$ **7.** $\left\{\frac{1}{9}, 1\right\}$ **9.** $\{-4 - \sqrt{7}, -4 + \sqrt{7}\}$
11. $\{4, 9\}$ **13.** $\{16, 121\}$ **15.** $\{-2, 2\}$ **17.** $\{-5, -\sqrt{13}, \sqrt{13}, 5\}$ **19.** $\left\{-\frac{10}{7}, -\frac{2}{7}\right\}$ **21.** $\{-27, 8\}$

23. $\left\{-\frac{8}{27}, 1\right\}$ **25.** $\{1\}$ **27.** $\left\{-\frac{10}{33}, -\frac{5}{18}\right\}$ **29.** $\{3, 13\}$ **31.** $\left\{\frac{1-\sqrt{6}}{5}, \frac{1+\sqrt{6}}{5}, -\frac{\sqrt{5}}{5}, \frac{\sqrt{5}}{5}\right\}$
33. $\{1 - \sqrt{3}, 1 - \sqrt{2}, 1 + \sqrt{2}, 1 + \sqrt{3}\}$ **35.** $\left\{\frac{1}{4}\right\}$ **37.** $\left\{-\frac{3}{10}, 6\right\}$ **39.** $\left\{-2, -\frac{1}{3}, \frac{1}{2}, 3\right\}$

Exercises 6.6

1. numerator 6 and denominator 8 (or numerator -8 and denominator -6)   **3.** 7, 13
5. 10 in, 24 in   **7.** 20   **9.** 9 min   **11.** 55 mph   **13.** 6 hr   **15.** 2%   **17.** 6, 8, 10   **19.** 1 yd
21. 12   **23.** $\sqrt{7}$ min   **25.** $\dfrac{4 + \sqrt{106}}{5}$ hr $\approx$ 2 hr 52 min   **27.** 5 hr   **29.** 45   **31.** $\dfrac{27}{6}$ or $\dfrac{14}{-7}$
33. \$12   **35.** 9 hr 6 min

Exercises 6.7

1. $(-\infty, -4) \cup (1, \infty)$ **3.** $(0, 3)$ **5.** $\left(-3, \frac{1}{2}\right)$ **7.** $(-\infty, 1] \cup [2, \infty)$ **9.** $(-3, 7)$ **11.** $(-2, 3)$
13. $\left[-2, \frac{3}{5}\right]$ **15.** $\left(-\infty, \frac{3}{4}\right) \cup (2, \infty)$ **17.** $(-\infty, -\sqrt{3}] \cup [\sqrt{3}, \infty)$ **19.** $\left(-\frac{3}{2}, \frac{1}{2}\right)$
21. $(-\infty, -3] \cup \left[\frac{2}{5}, \infty\right)$ **23.** R $\left[-\frac{2}{3}, 1\right]$ **27.** $\left(-\infty, -\frac{5}{3}\right] \cup \left[\frac{7}{2}, \infty\right)$ **29.** $\left[\frac{1}{2}, 3\right]$
31. $(5 - \sqrt{22}, 5 + \sqrt{22})$ **33.** $(-\infty, 4 - 2\sqrt{5}) \cup (4 + 2\sqrt{5}, \infty)$ **35.** $(-2 - \sqrt{2}, -2 + \sqrt{2})$
37. at most 4000   **39.** \$100,000   **41.** $(-\infty, -2) \cup (0, 2)$ **43.** $(-5, -3) \cup (2, \infty)$
45. $\left(-\frac{5}{2}, \infty\right)$ **47.** $\left(\frac{2}{5}, \infty\right)$ **49.** $(-\infty, 2) \cup [3, \infty)$ **51.** $\left(-\infty, \frac{4}{3}\right] \cup (2, \infty)$
53. $(-\infty, -2) \cup (1, \infty)$ **55.** $\left(-\frac{1}{2}, \frac{3}{4}\right]$ **57.** $(-\infty, 0) \cup \{1\}$ **59.** If $ax^2 + bx + c = 0$ has no real roots then $b^2 - 4ac < 0$. Then since $4a^2 > 0$, we have $\dfrac{b^2 - 4ac}{4a^2} < 0$. This implies, since $\left(x + \dfrac{b}{2a}\right)^2 \geq 0$ for all x, that the solution set for $\left(x + \dfrac{b}{2a}\right)^2 < \dfrac{b^2 - 4ac}{4a^2}$ is $\emptyset$. The inequality $\left(x + \dfrac{b}{2a}\right)^2 < \dfrac{b^2 - 4ac}{4a^2}$ is equivalent to $ax^2 + bx + c < 0, (a > 0)$ and hence the solution set for $ax^2 + bx + c < 0, (a > 0)$ is $\emptyset$.

Review Exercises

1. $\{-1\}$ **3.** $\{-i, i\}$ **5.** $\{0, 7\}$ **7.** $\left\{-\frac{5}{4}, \frac{5}{4}\right\}$ **9.** $\{-6, 5\}$ **11.** $\left\{-\frac{1}{3}, 15\right\}$ **13.** $\left\{\frac{4k}{3}, 5k\right\}$
15. $x^2 - 3x - 10 = 0$ **17.** $2x^2 - x = 0$ **19.** $\{-9, -1\}$ **21.** $\left\{\dfrac{7 - 2\sqrt{3}}{2}, \dfrac{7 + 2\sqrt{3}}{2}\right\}$
23. $\left\{\dfrac{1 - \sqrt{5}i}{3}, \dfrac{1 + \sqrt{5}i}{3}\right\}$ **25.** $\left\{-5, \frac{2}{3}\right\}$ **27.** $\{3 - \sqrt{3}, 3 + \sqrt{3}\}$ **29.** $\left\{\dfrac{2 - \sqrt{14}}{2}, \dfrac{2 + \sqrt{14}}{2}\right\}$
31. $\{-i, i\}$ **33.** $\left\{\dfrac{-5 - \sqrt{53}}{2}, \dfrac{-5 + \sqrt{53}}{2}\right\}$ **35.** $\left\{\dfrac{3 - \sqrt{7}i}{4}, \dfrac{3 + \sqrt{7}i}{4}\right\}$ **37.** $\{-5\}$
39. $\left\{3 - \frac{1}{2}\sqrt{38}i, 3 + \frac{1}{2}\sqrt{38}i\right\}$ **41.** $\{2 - i, 2 + i\}$ **43.** $\left\{0, \dfrac{3i}{4}\right\}$ **45.** $\left\{-3i, \dfrac{2}{3}i\right\}$ **47.** one real solution   **49.** two distinct imaginary solutions   **51.** one real solution   **53.** $\left\{\frac{9}{4}\right\}$ **55.** $\{16\}$
57. $\left\{-\frac{1}{2}, 2\right\}$ **59.** $\left\{\frac{3}{5}\right\}$ **61.** $\{4, 25\}$ **63.** $\{-1, 4\}$ **65.** $\left(-\infty, -\frac{1}{2}\right] \cup [3, \infty)$
67. $(-\infty, 1 - 2\sqrt{3}) \cup (1 + 2\sqrt{3}, \infty)$ **69.** R **71.** $\left(-2, -\frac{1}{3}\right)$ **73.** 9 ft   **75.** $\frac{1}{2}$ ft   **77.** 10.56 mph

Practice Test

1. $\left\{-2, \dfrac{1}{3}\right\}$ 2. $\left\{\dfrac{i-4}{2}, \dfrac{-i-4}{2}\right\}$ 3. $\{2, 8\}$ 4. $\left\{\dfrac{3 \pm \sqrt{47}\, i}{4}\right\}$ 5. $\{1\}$ 6. $\left\{\dfrac{2}{7}\right\}$
7. $\{-1, 1, -2i, 2i\}$ 8. $\{-6, 2\}$ 9. 60 shares 10. 2, 4, or 6, 8 11. $\left[-2, \dfrac{7}{2}\right]$ 12. $\left(\dfrac{4}{3}, \dfrac{18}{5}\right)$

Extended Applications: Rescue Mission

1. 427 ft 3. 3.75 sec

Chapter 7

Exercises 7.1

1. I 3. on x-axis 5. origin 7. on y-axis

9.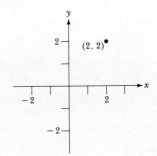

11.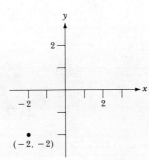

13.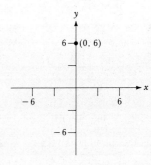

15.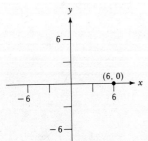

17.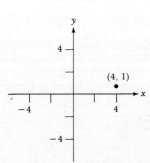

19.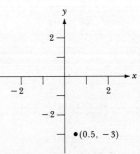

21. Only (1, 1) and (3, 6) are on the line. 23. Only (0, 0) and (2, 4) are on the line. 25. All but (2, 8) are on the line. 27. (0, −4), (4, 0), (4, 0), (5, 1) 29. (0, −4), (8, 0), (2, −3), (4, −2)

31. x-intercept $-\frac{3}{2}$; y-intercept 3

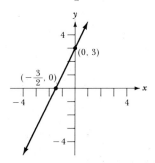

33. x-intercept $\frac{4}{3}$; y-intercept -4

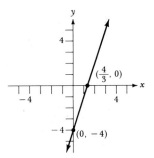

35. x-intercept 2; y-intercept 3

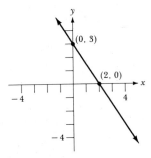

37. x-intercept 4; no y-intercept

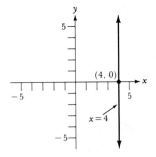

39. y-intercept 6; no x-intercept

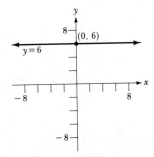

41.

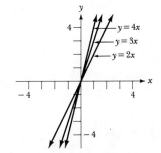

43. -3 **45.** 1   **47.** (a) $11,200 (b) $5600; after 10 years

A32   Answers to Selected Exercises

49. (a) $2200
(b) $2212
(c) $4600
(d) $4612

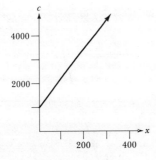

51. $m = 4; b = -8$

53.

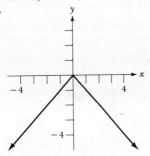

Exercises 7.2

1. $m = -2$

3. $m = 0$

5. $m = 1$

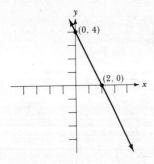

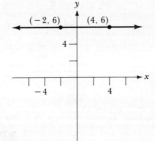

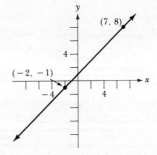

7. $m = -\frac{3}{2}$

9.

11.

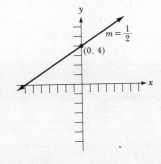

13. 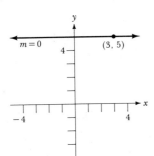 **15.**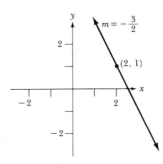

17. $m = 2$; rising   **19.** $m = \frac{1}{2}$; rising   **21.** m is undefined; vertical   **23.** $m = \frac{3}{2}$; rising

25. parallel   **27.** neither   **29.** parallel   **31.** perpendicular   **33.** neither   **35.** parallel   **37.** $\frac{3}{2}$

39. (a) $6400 (b) $5600 (c) $800 (d) -800

Exercises 7.3

1. $y = 3$ **3.** $y - 1 = \frac{1}{7}(x + 1)$ **5.** $y - 4 = -\frac{1}{3}x$ **7.** $y = \frac{5}{2}(x - 4)$ **9.** $x = -\frac{1}{2}$ **11.** $x = 5$

13. $y = 0$ **15.** $y = x$; slope 1; y-intercept 0   **17.** $y = -\frac{1}{2}x + \frac{1}{2}$; slope $-\frac{1}{2}$; y-intercept $\frac{1}{2}$

19. $y = 2x + 6$; slope 2; y-intercept 6   **21.** $y = -\frac{1}{2}x - 2$; slope $-\frac{1}{2}$; y-intercept -2

23. $y = \frac{2}{5}x - \frac{11}{5}$; slope $\frac{2}{5}$; y-intercept $-\frac{11}{5}$ **25.** $x + y = 1$ **27.** $3x - 2y = 1$ **29.** $y = 3$

31. $7y - 5x = 39$ **33.** $9y - x = 37$ **35.** $7x + 2y = 4$ **37.** $y = 14$ **39.** $y = -\frac{1}{2}x$

41. $y = \frac{5}{3}x + 2$ **43.** $y = -\frac{2}{7}x - 1$ **45.** $y = -8$ **47.** $y = 6x - 2$ **49.** $y = -\frac{1}{2}x$

51. $x = 14$ **53.** $y = \frac{2}{3}(x - 1)$ **55.** $x = -2$ **57.** $5600; $4200

Exercises 7.4

1. $x + y > 0$ **3.** $x + 2y \geq 8$ **5.** $2x - 5y > 10$

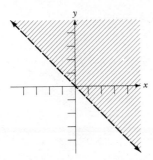 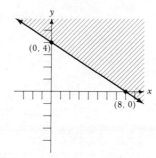 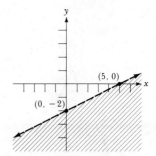

7. $3y - 9 \le 2$

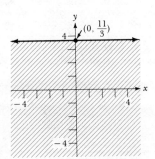

9.

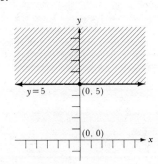

11.

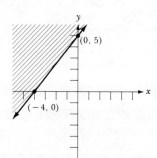

13.

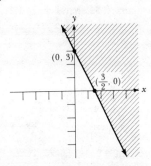

15.

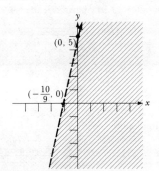

17.

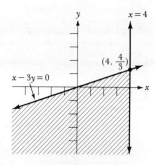

19.

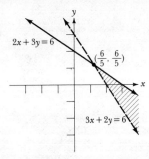

21.

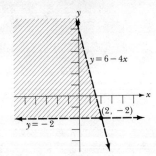

23.

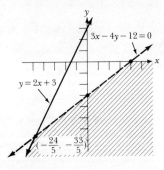

25.

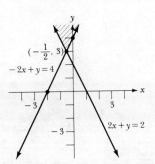

27. Intersection is the empty set.

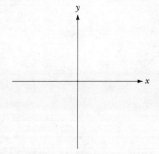

29. 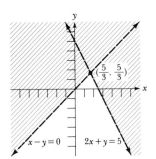 **31.** 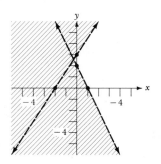 **33.**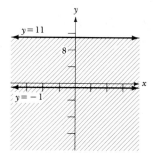

35.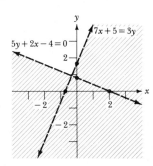

Exercises 7.5

1. domain $\{-2, 0, 2\}$; range $\{0, 2\}$; function **3.** domain $\{-2, 0, 1, 2\}$; range $\{-2, 0, 1, 2\}$; function
5. domain $\{4, 9\}$; range $\{-3, -2, 2, 3\}$; not a function **7.** domain R; function **9.** domain $\{x | x \geq 0\}$; function **11.** domain $\{x | x \neq 0\}$; function **13.** domain $\{x | x > 1 \text{ or } x < -1\}$; function
15. domain $\left\{x | x \geq -\dfrac{1}{2}\right\}$; function **17.** domain R; function **19.** domain R; function
21. domain $[0, \infty)$; function **23.** $f(x) = \dfrac{x}{x-1}; \dfrac{4}{3}$ **25.** $f(x) = \dfrac{3}{x-2}; \dfrac{3}{2}$ **27.** $f(x) = \dfrac{\sqrt{x}}{x^2+2}; \dfrac{1}{9}$
29. 1 **31.** 2 **33.** $\sqrt{2}$ **35.** 2 **37.** 10.6363 **39.** 2.484 **41.** no **43.** no **45.** yes
47. (a) $2x + 2h + 1$ (b) $2h$ (c) $2, h \neq 0$
49. (a) $x^2 + 2xh + h^2 - 3x - 3h$ (b) $2xh + h^2 - 3h$ (c) $2x + h - 3, h \neq 0$
51. $x_2 + x_1$ **53.** $2(a + b) - 3$

Exercises 7.6

1. $\dfrac{1}{2}$; $x = \dfrac{1}{2}y$; 56 **3.** 33; $r = \dfrac{33}{u}$; 99 **5.** 1; $m = \dfrac{q}{p}$; $\dfrac{1}{2}$ **7.** 72; $u = \dfrac{72}{t^3}$; $\dfrac{1}{3}$
9. $\dfrac{1}{17}$; $P = \dfrac{1}{17}TQ^2$; $\dfrac{324}{17}$ **11.** 8.4668 **13.** 85 **15.** 200 **17.** 784 ft **19.** 12 newtons
21. 1280 candlepower **23.** 16 ft **25.** 0.37 meter **27.** 43.1 ohms

Review Exercises

1. $(-4, 17)$ is on the line.

3. x-intercept: -2
y-intercept: -5

5. x-intercept: none
y-intercept: 10

7. x-intercept: $\dfrac{12}{5}$
y-intercept: -4

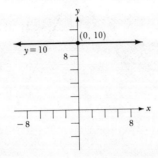
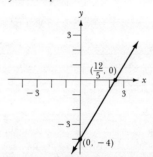

9. $m = -4$

11.
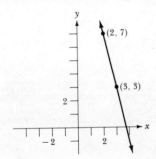

13.
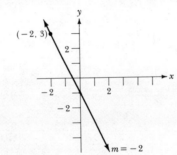
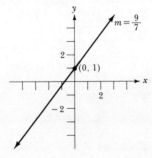

15. $m = -\dfrac{5}{8}$; falling   **17.** perpendicular   **19.** $y = 4x - 1$ **21.** $y = \dfrac{3}{5}x$ **23.** $3y - 5x = 4$

25. $x = 1$ **27.** $4y - x = 2$

29.
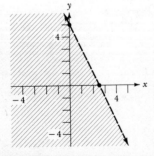

31.

33.
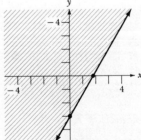

35. domain $\{x | x \geq -1\}$; not a function   **37.** domain R; function   **39.** domain $\left\{x | x \geq -\dfrac{3}{2}\right\}$; function

41. -1 **43.** 0 **45.** -1 **47.** 100 **49.** 60 megacycles/sec

Practice Test

1. $m = \dfrac{3}{5}$

x-intercept: $-\dfrac{1}{3}$

y-intercept: $\dfrac{1}{5}$

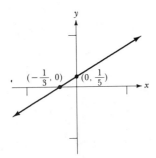

2. $m = -7$

x-intercept: $\dfrac{5}{7}$

y-intercept: 5

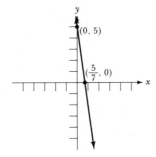

3.

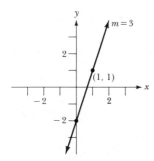

4. $x - 2y = -4$

5. $2x + 7y = 22$ **6.** $x = 9$ **7.** $y = 3(x - 5)$

8.

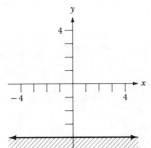

9.

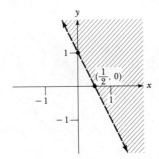

10. domain: all reals; not a function   **11.** domain: all real except 0; function

12. domain: all reals > 1; function   **13.** -1 **14.** $\dfrac{1}{3}$ **15.** 15

A38 Answers to Selected Exercises

Extended Applications: Radar

1. 931.5 miles **3.** 232.9 miles

Chapter 8

Exercises 8.1

1.
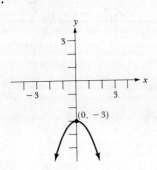

3.
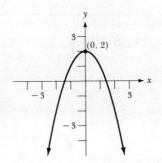

5.
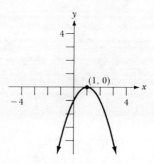

7.
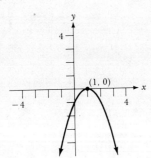

9.
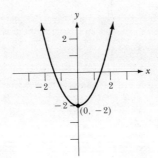

11.
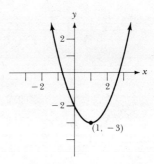

13.
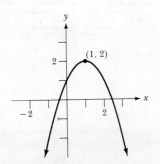

15.
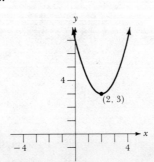

17.
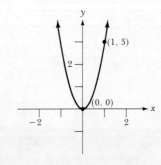

19.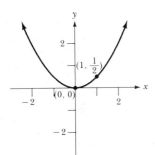

21.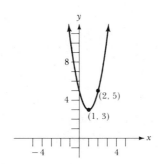

23. 30 kids **25.** 550 watts

27.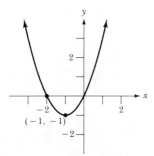

29.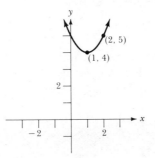

31.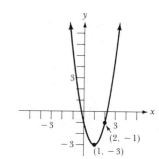

33.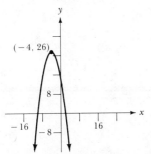

35.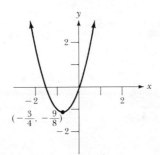

37.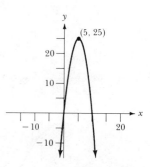

39.

41.

Exercises 8.2

1.
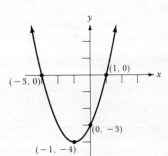

3.
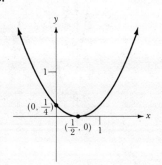

5.
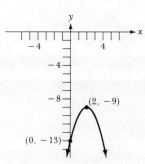

7.
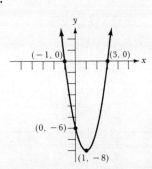

9.
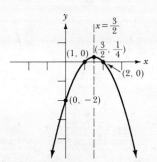

11.
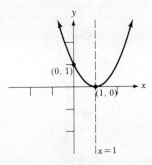

13. See Figure for exercise 3. Axis of symmetry is $x = \frac{1}{2}$.

15.
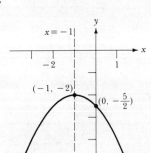

17.
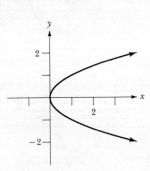

19.
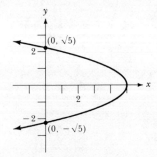

21.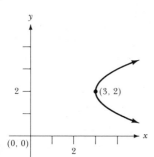

23.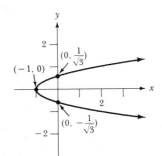

25.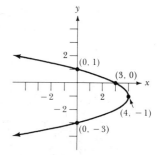

27. vertex: (0, 3)
axis of symmetry: $x = 0$
x-intercept: $\pm \sqrt{\dfrac{3}{2}}$
y-intercept: 3

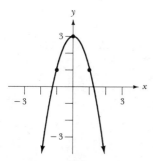

29. vertex: $\left(\dfrac{7}{2}, -\dfrac{25}{4}\right)$
axis of symmetry: $x = \dfrac{7}{2}$
x-intercept: 1, 6
y-intercept: 6

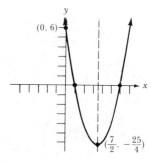

31. vertex: (1, 1)
axis of symmetry: $x = 1$
x-intercept: none
y-intercept: 3

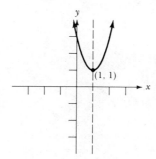

33. vertex: $\left(\dfrac{1}{4}, \dfrac{23}{8}\right)$
axis of symmetry: $x = \dfrac{1}{4}$
x-intercept: none
y-intercept: 3

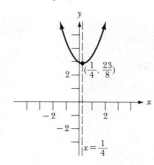

35. vertex: $(-5, -2)$
axis of symmetry: $y = -2$
x-intercept: 3
y-intercept: $\dfrac{-4 \pm \sqrt{10}}{2}$

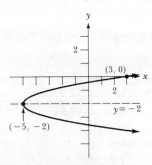

37. vertex: $(-1, 2)$
axis of symmetry: $y = 2$
x-intercept: 3
y-intercept: 1, 3

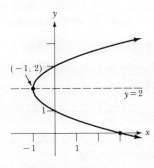

39. 27

Exercises 8.3

1. 4   **3.** $\sqrt{2}$ **5.** $\sqrt{13}$ **7.** $2|v|$ **9.** $\sqrt{2}|k-t|$ **11.** 6.443   **13.** yes   **15.** yes   **17.** yes

19. 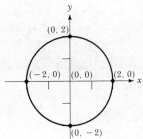   **21.**   **23.**

25. 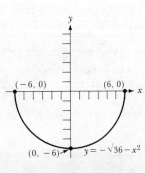   **27.** 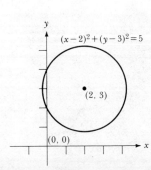   **29.**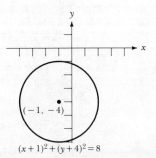

31. $x^2 + y^2 = 64$ **33.** $(x+1)^2 + (y-2)^2 = 2$ **35.** $(x+2)^2 + (y+3)^2 = 7$
37. circle: center at $(1, 1)$, radius $\sqrt{6}$ **39.** circle: center at $(0, -2)$, radius 2   **41.** circle: center at $(-3, 2)$, radius $\sqrt{33}$ **43.** upper semicircle: center at $(1, 0)$, radius 5   **45.** lower semicircle: center at $(0, 1)$,

radius 4 **47.** upper semicircle: center at $(-1, -3)$, radius 3 **49.** lower semicircle: center at $(4, 2)$, radius 6 **51.** yes **53.** yes **55.** no

Exercises 8.4

1. 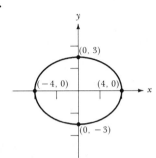 **3.** 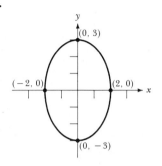 **5.**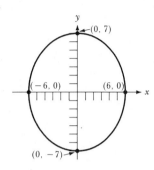

7.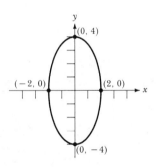

9. $\dfrac{x^2}{9} + \dfrac{y^2}{4} = 1$; horizontal; intercepts $(3, 0)$, $(-3, 0)$ and $(0, 2)$, $(0, -2)$ **11.** $\dfrac{x^2}{16} + \dfrac{y^2}{32} = 1$; vertical; intercepts $(0, 4\sqrt{2})$, $(0, -4\sqrt{2})$ and $(4, 0)$, $(-4, 0)$ **13.** $\dfrac{x^2}{49} + \dfrac{y^2}{14} = 1$; horizontal; intercepts $(7, 0)$, $(-7, 0)$ and $(0, \sqrt{14})$, $(0, -\sqrt{14})$

15. upper half of graph in exercise 7 **17.** lower half of graph in exercise 4

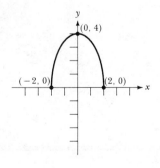 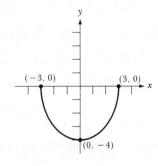

19. right half of graph in exercise 8

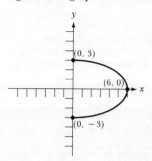

21. left half of graph in exercise 5

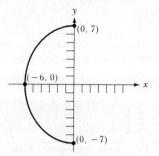

23.

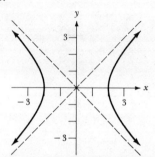

25.

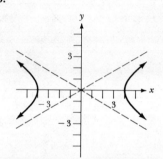

27.

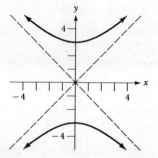

29.

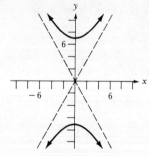

31.

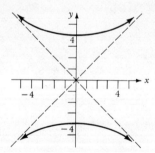

33. $\dfrac{x^2}{25} - \dfrac{y^2}{4} = 1$; intercept: $(\pm 5, 0)$; asymptotes: $y = \pm\dfrac{2}{5}x$ **35.** $\dfrac{x^2}{9} - \dfrac{y^2}{25} = 1$; intercepts: $(\pm 3, 0)$; asymptotes: $y = \pm\dfrac{5}{3}x$ **37.** $\dfrac{x^2}{4} - \dfrac{y^2}{3} = 1$; intercepts: $(\pm 2, 0)$; asymptotes: $y = \pm\dfrac{\sqrt{3}}{2}x$

39. 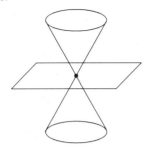   **41.**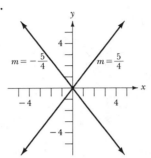

Exercises 8.5

1. $x^2 + y^2 \leq 4$ 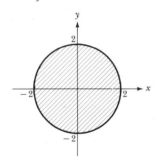   **3.** $x^2 + y^2 < 9$ 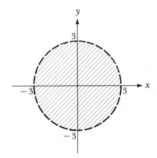   **5.** $x^2 + y^2 \geq 1$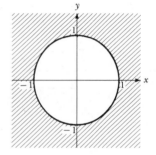

7. $x^2 + y^2 > 3$ 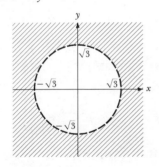   **9.** $y \leq \sqrt{16 - x^2}, y \geq 0$ 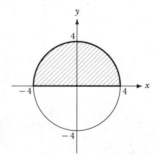   **11.** $y + x^2 \leq 0$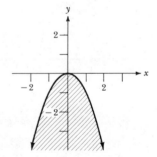

13. $y < -x^2 + 1$ 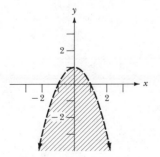   **15.** $y \geq -x^2 - 1$ 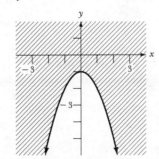   **17.** $y > -(x - 2)^2$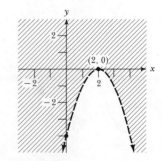

19. $x \geq -y^2$

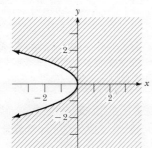

21. $x < y^2 + 1$

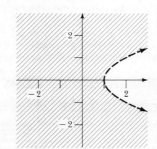

23. $y \leq \sqrt{x}$

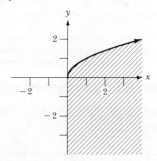

25. $x > \sqrt{y-1}$

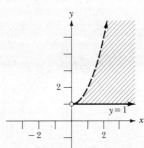

27. $9x^2 + y^2 \leq 9$

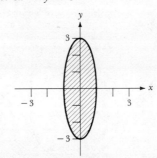

29. $4x^2 + 9y^2 < 36$

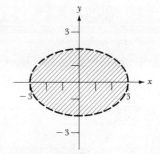

31. $y^2 \geq 16 - 4x^2$

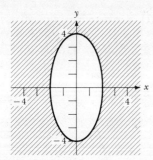

33. $y > 2\sqrt{4 - x^2}$

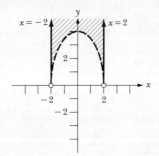

35. $3y < 2\sqrt{9 - x^2}$

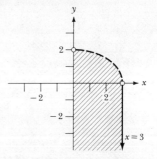

37. $9x^2 - 4y^2 \geq 36$

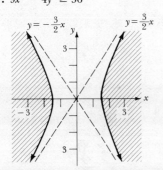

39. $y^2 < 16 + x^2$

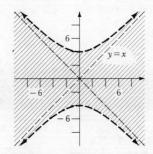

41. $y^2 \geq 36(1 + x^2)$

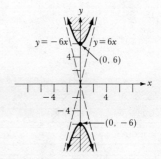

43. $\frac{y}{7} < \sqrt{1 + \frac{x^2}{25}}$

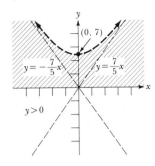

45.

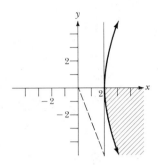

Exercises 8.6

1. (a) $6 (b) $6 (c) $30   **3.** $1929   **5.** (a) 12.5% (b) 14.9% (c) 5%
7. 3025; 1000   **9.** $506\frac{1}{4}$ ft; $5\frac{5}{8}$ sec   **11.** 1250 sq yd   **13.** 49, 49   **15.** $\left(\frac{1}{2}, \frac{\sqrt{2}}{2}\right)$ **17.** at the origin

Exercises 8.7

1. yes   **3.** yes   **5.** no   **7.** no   **9.** no   **11.** yes   **13.** yes   **15.** no   **17.** yes
19. $f^{-1}(x) = -\frac{1}{3}x + 5$ **21.** not one-to-one   **23.** $f^{-1}(x) = (x-3)^2, \; x \geq 3$

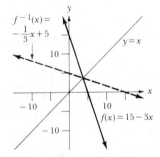

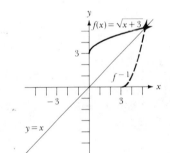

25. $f^{-1}(x) = x^3$ **27.** $f(x) = \frac{1}{x} = f^{-1}(x)$

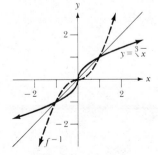

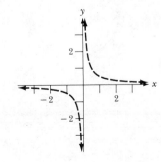

29. $f^{-1} = \{(-2, -3), (-1, -2), (0, -1), (1, 0)\}$ **31.** $h^{-1} = \{(0, 0), (1, 1), (4, 2), (9, 3), (16, 4)\}$
33. $g^{-1} = \{(1, -1), (-3, 3), (5, -5), (-7, 7)\}$ **35.** $g^{-1} = \{(-1, 1), (-8, 2), (-27, 3), (1, 4)\}$ **37.** 1
39. 28   **41.** 0   **43.** 3

Review Exercises

1. $y = x^2$ **3.**   **5.** $y = x^2 + 5$

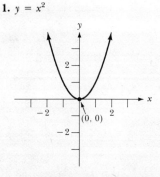 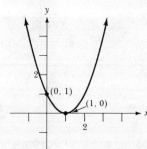 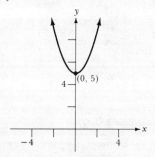

7. $x = y^2 - 2y + 5$ **9.** $y = -x^2 + 4x - 1$

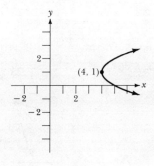 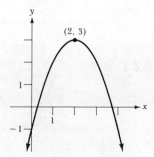

11. None are symmetric to x-axis; 1, 2, 5, 6 are symmetric with respect to y-axis.   **13.** $\sqrt{34}$ **15.** $\sqrt{2}\,|v|$
17. $(x - 1)^2 + (y - 2)^2 = 0$ **19.** $y = -\sqrt{16 - x^2}$

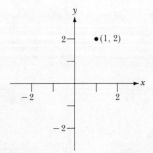 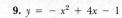

21. $(x + 1)^2 + (y + 1)^2 = 64$ **23.** circle with center at (2, 3) and radius $\sqrt{14}$ **25.** circle with center at (0, 4) and radius $\sqrt{21}$

27.

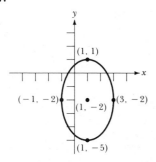

29. $\dfrac{x^2}{4} + \dfrac{y^2}{9} = 1$; x-intercepts $(\pm 2, 0)$; y-intercepts $(0, \pm 3)$

31. $\dfrac{x^2}{30} + \dfrac{y^2}{25} = 1$; x-intercepts $(\pm \sqrt{30}, 0)$; y-intercepts $(0, 5), (0, -5)$

33. $x^2 = 9 + y^2$

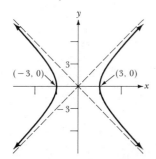

35. $\dfrac{x^2}{6^2} - y^2 = 1$; x-intercepts $(\pm 6, 0)$; y-intercepts: none; asymptotes: $y = \pm \dfrac{1}{6}x$ **37.** $x^2 - \dfrac{y^2}{4^2} = 1$; x-intercepts $(\pm 1, 0)$; y-intercepts: none; asymptotes: $y = \pm 4x$

39. $(x - 1)^2 + (y - 2)^2 \leq 9$ **41.** $x + y^2 \leq 0$ **43.** $4x^2 + 9y^2 < 36$

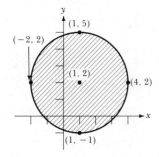 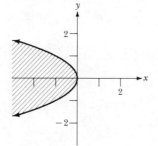 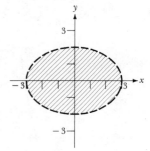

45. 25 cm × 25 cm **47.** 625 ft; $\dfrac{25}{4}$ sec **49.** yes **51.** no **53.** yes **55.** no **57.** yes

59. $f^{-1}(x) = -\frac{1}{4}x + \frac{13}{4}$ **61.** $h^{-1}(x) = (x - 5)^2, x \geq 5$

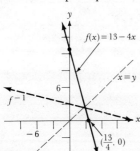
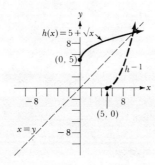

63. not 1 to 1
65. $f^{-1}(x) = x^3 + 1$ **67.** $f^{-1}(x) = \frac{1}{2x}$ $x \neq 0$

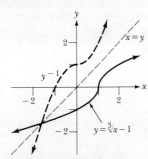
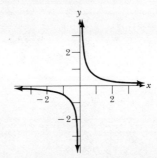

69. $f^{-1} = \{(-1, -3), (0, -2), (1, -1), (2, 0)\}$ **71.** $h^{-1} = \{(2b, a), (2c, b), (2d, c)\}$ **73.** 1 **75.** 8
77. 0 **79.** 3

Practice Test

1. parabola

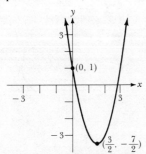

2. $\sqrt{[(a + 3) - a]^2 + (b - b)^2} = \sqrt{3^2} = 3, \sqrt{[(a + 3) - a]^2 + [(b + 4) - b]^2} = \sqrt{3^2 + 4^2} = 5$
$\sqrt{[(a + 3) - (a + 3)]^2 + [(b + 4) - b]^2} = \sqrt{0^2 + 4^2} = 4$ and $3^2 + 4^2 = 5^2$ **3.** $x^2 + (y + 3)^2 = 9$
4. $c = (1, -2), r = \sqrt{6}$ **5.** $\frac{(x - 1)^2}{3} + \frac{(y - 1)^2}{2} = 1$; x-intercepts $1 \pm \frac{1}{2}\sqrt{6}$; y-intercepts $1 \pm \frac{2}{3}\sqrt{3}$; ellipse

6. $\dfrac{x^2}{7^2} - \dfrac{y^2}{7^2} = 1$, hyperbola

7.

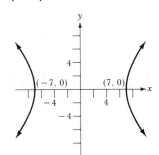

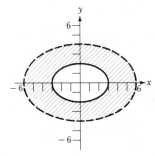

8. 50 trees

9. $y^{-1} = \dfrac{1}{2}(\sqrt[3]{x} + 3)$

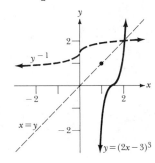

10. not one to one

11. $g^{-1}(x) = (x - 1)^2 + 3,\ x \geq 1$

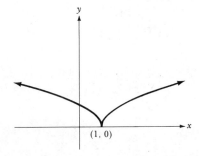

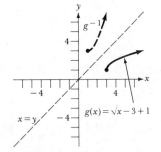

Extended Applications: Planetary Orbits

1. 92.887 million miles; 0.00014   **3.** 125 million miles; 159 million miles; 142 million miles

Chapter 9

Exercises 9.1

1. $\{(-3, 3)\}$ **3.** $\{(2, 3)\}$ **5.** $\{(5, 3)\}$ **7.** $\{(0, -5)\}$ **9.** $\{(-1, -2)\}$ **11.** $\emptyset$
13. $\{(x, -\frac{2}{3}x + 2) | x \in R\}$ **15.** $\{(3, 3)\}$ **17.** $\{(x, \frac{1}{9}(x - 3)) | x \in R\}$ **19.** $\{(x, 5 - x) | x \in R\}$
21. $\{(\frac{2}{3}, 2)\}$ **23.** $\{(5, 0)\}$ **25.** $\{(3, -5)\}$ **27.** $\{(-1, -1)\}$ **29.** $\{(x, 3x - 7) | x \in R\}$ **31.** $\{(\frac{4}{3}, \frac{2}{3})\}$
33. $\{(14, -7)\}$ **35.** $\{(\frac{3}{2}, -4)\}$ **37.** $\{(3a - b), b - 2a)\}$ **39.** $\{(\frac{1}{a}, 0)\}$

Exercises 9.2

1. $\{(1, 2, 2)\}$ **3.** $\{(3, 3, -4)\}$ **5.** $\{(2, 1, -1)\}$ **7.** $\{(3, -1, 2)\}$ **9.** $\{(1, 0, 1)\}$
11. $\{(-19, -\frac{13}{3}, 27)\}$ **13.** $\{(3, -1, 1)\}$ **15.** $\{(1, -2, 1)\}$ **17.** $\{(-1, 2, 0)\}$ **19.** $\{(10, -1, 3)\}$
21. $\{(1, 2, 1)\}$ **23.** $\{(4.484, -3.435, 2.130)\}$ **25.** $\{(-4.543, 3.643, 0.212)\}$

Exercises 9.3

1. $24,000 at 8%; $6,000 at $10\frac{1}{2}$% **3.** $120 at blackjack; $240 at dice table **5.** 25 adults; 32 children
7. 1.6 pounds of the herb; 6.4 pounds of tea **9.** A pound is $2.40; a franc is $0.20
11. 550 km/hr; 50 km/hr **13.** 21 mph; 24 mph **15.** 27 **17.** 57
19. $11.75/hr for plumber; $1.75/hr for apprentice **21.** 80 m by 50 m **23.** 14; 21
25. base 20 cm; sides 10 cm each **27.** 5; 15 **29.** $m = -3; b = 5$
31. 53 nickels, 22 quarters, 225 dimes **33.** 35 hr regular; 8 hr night; 4 hr holiday
35. 50 A's; 20 B's; 30 C's

Exercises 9.4

1. -3 **3.** 5 **5.** 1 **7.** 0 **9.** 9 **11.** 0 **13.** 11 **15.** -24 **17.** 11 **19.** 0
21. -6 **23.** -42 **25.** -2 **27.** -11 **29.** 0 **31.** $\{0. 6\}$

Exercises 9.5

1. $\{(2, -7)\}$ **3.** $\{(-1, -3)\}$ **5.** $\{(7, 5)\}$ **7.** $\{(-\frac{9}{5}, \frac{17}{5})\}$ **9.** $\{(-4, 3)\}$ **11.** $\{(4, 2, -1)\}$
13. $\{(1, 2, 3)\}$ **15.** $\{(1, 2, -1)\}$ **17.** $\{(2, 2, 1)\}$ **19.** $\{(2, -2, 0)\}$ **21.** $\{(1, 2, 1)\}$ **23.** $\{(-1, -1, 3)\}$
25. $\{(1, 2, -1)\}$ **27.** $\{(2, 4, 5)\}$ **29.** (a) $(0, 0, 0)$ is only solution (b) $(0, 0, 0)$ is not only solution

Exercises 9.6

1. $\frac{5}{4}; \frac{5}{4}; \frac{27}{4}; \frac{5}{4}; -3$ **3.** $\{(10, -2)\}$ **5.** $\{(2, -1)\}$ **7.** $\{(1, -1)\}$ **9.** $\frac{2}{3}; \frac{2}{3}; \frac{1}{3}; \frac{2}{3}; -9$ **11.** $\{(5, 0, 1)\}$
13. $\{(1, 1, 1)\}$ **15.** $\{(3, 1, 3)\}$

Exercises 9.7

1. $\{(3, 3), (-2, -2)\}$ **3.** $\{(3, 0)\}$ **5.** $\{(-2, 1), (-1, 2)\}$ **7.** $\{(3, 1), (1, -1)\}$ **9.** $\{(-3, -1), (1, 3)\}$
11. $\{(0, 5), (2, 3)\}$ **13.** $\{(7, 5), (-7, -5)\}$ **15.** $\{(3, 3), (-3, 3)\}$ **17.** $\{(1, 1), (1, -1), (-1, 1), (-1, -1)\}$
19. $\{(5, 1), (5, -1), (-5, 1), (-5, -1)\}$ **21.** $\{(3, 4), (-3, -4), (4, 3), (-4, -3)\}$
23. $\{(3, 3), (-3, -3)\}$ **25.** $\{(3, 7), (-1, -1)\}$ **27.** $\emptyset$ **29.** $\emptyset$ **31.** $\{(1, 2), (-1, -2)\}$
33. $\{(-a, -3a), (3a, a)\}$ **35.** $\{(a, 0)\}$ **37.** 15 m by 12 m **39.** 2000π cm^3

Exercises 9.8

1.

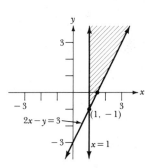

3.

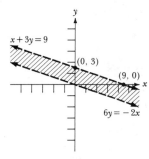

5.

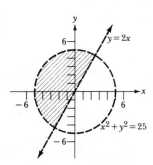

7.

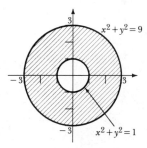

9.

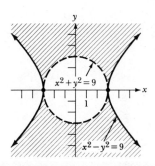

11.

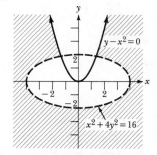

13.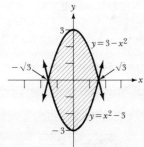

15.

17.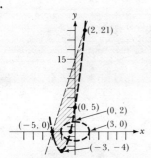

19.

21. The solution set is empty.

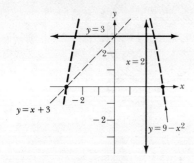

Review Exercises

1. $\{(7, 4)\}$ **3.** $\{(3, 11)\}$ **5.** $\left\{\left(x, \dfrac{3x - 5}{2}\right) \mid x \in \mathbf{R}\right\}$ **7.** $\{(1, 0, 5)\}$ **9.** $\{(7, 4, -1)\}$

11. 18 miles   **13.** Al is 6 and Debbie is 15.   **15.** 3, 5, 10   **17.** 2   **19.** 13   **21.** 0

23. $\{(1, 2)\}$ **25.** $\{(-1, -5)\}$ **27.** $\left\{\left(\dfrac{1}{2}, 1\right)\right\}$ **29.** $\{(1, 2, -1)\}$

31. $\left\{\left(-1, 1, \dfrac{2}{3}\right)\right\}$ **33.** $\{(3, 1)\}$ **35.** $\left\{\left(\dfrac{80}{13}, \dfrac{88}{13}\right)\right\}$ **37.** $\left\{\left(\dfrac{103}{69}, \dfrac{83}{69}, \dfrac{51}{69}\right)\right\}$ **39.** $\{(0, 1)\}$

41. $\left\{(3, 0), (-3, 0), \left(\dfrac{\sqrt{97}}{3}, \dfrac{16}{9}\right), \left(-\dfrac{\sqrt{97}}{3}, \dfrac{16}{9}\right)\right\}$ **43.** $\{(3, 0), (0, -5)\}$ **45.** $\{(0, -4), (\sqrt{7}, 3), (-\sqrt{7}, 3)\}$

47.

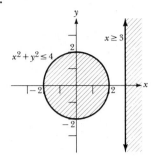

49.

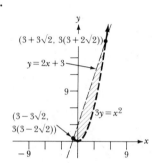

51.

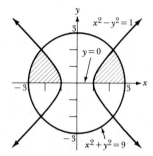

Practice Test

1. $\{(3, -7)\}$ **2.** $\{(1, 2, -2)\}$ **3.** 40 m × 60 m **4.** 7 nickels, 5 dimes, 6 quarters **5.** 16
6. -39 **7.** $\{(-3, -1, 4)\}$ **8.** $\{(1, 1), (-1, 1), (1, -1), (-1, -1)\}$

9.

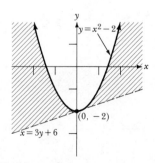

10.

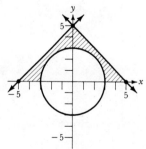

Extended Applications: Market Equilibrium

1. 14142 items   **3.** $625; 10,000 items

Chapter 10

Exercises 10.1

1. $y = 2^x$

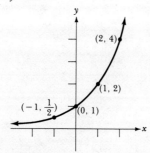

3. $y = 10^x$

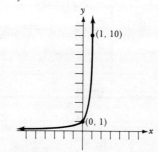

5. $y = \left(\dfrac{1}{4}\right)^x$

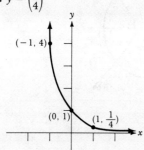

7. $y = 2^{x-3}$

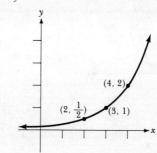

9. $y = 2^x + \dfrac{1}{2}$

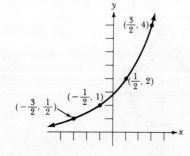

11. $y = 2^x + 3$

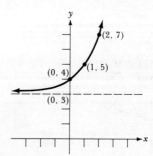

13. $y = 1.87^x$

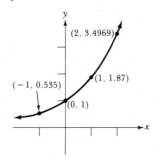

15. $y = 2.07^x + 1.84$

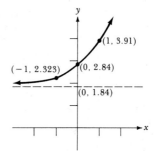

17. $y = 3^{2-x}$

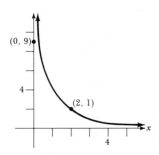

19. $\{4\}$ **21.** $\{2.5\}$ **23.** $\{-3\}$ **25.** $\left\{\dfrac{5}{3}\right\}$ **27.** $\{-2\}$ **29.** $\{-8\}$ **31.** $2^{\sqrt{2}}, 2^{7/5}, 2^{2/3}$
33. 5 hours

Exercises 10.2

1. $\log_5 25 = 2$ **3.** $\log_{49} \dfrac{1}{7} = -\dfrac{1}{2}$ **5.** $\log_{1/16} 4 = -\dfrac{1}{2}$ **7.** $\log_{10} 1 = 0$ **9.** $\log_{10} 0.1 = -1$
11. $\log \dfrac{1}{a^n} = -n$ **13.** $2^5 = 32$ **15.** $10^2 = 100$ **17.** $10^0 = 1$ **19.** $10^{-2} = 0.01$
21. $\left(\dfrac{1}{5}\right)^{-2} = 25$ **23.** $8^{1/3} = 2$
25. $y = \log_{1/2} x$

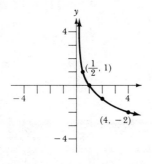

27. $y = \log_4 x$

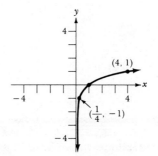

29. {100} **31.** {127} **33.** {−4} **35.** {5} **37.** {16.562} **39.** {2} **41.** {3, 4} **43.** 3
45. 4 **47.** −3 **49.** $\frac{1}{2}$ **51.** $\frac{1}{4}$ **53.** $\frac{3}{2}$ **55.** $\frac{3}{2}$ **57.** $-\frac{1}{2}$ **59.** 1 **61.** 7

Exercises 10.3

1. 0.778 **3.** 0.602 **5.** 2.512 **7.** 212 **9.** 420 **11.** 5.246 **13.** 217 **15.** 9.105
17. −0.176 **19.** 1.734 **21.** −3.292 **23.** 418 **25.** 0.1505 **27.** 0.088 **29.** −0.580
31. 25 **33.** $44\frac{2}{3}$ **35.** 366.56 **37.** {1} **39.** {3} **41.** {4} **43.** ∅ **45.** ∅ **47.** \$2 **49.** 1

Exercises 10.4

1. 2; 0.9212 **3.** 0; 0.9542 **5.** −1; 0.7774 **7.** −5; 0.9031 **9.** −5; 0.1492 **11.** 6; 0.9509
13. 0.5798 **15.** 0.5988 **17.** 3.6542 **19.** 5.8319 **21.** 6.76 **23.** 35.2 **25.** 0.676
27. 8.55 **29.** 0.00883 **31.** 0.862 **33.** 1.19 **35.** 0.879 **37.** 2.55 **39.** 6.44 **41.** 1.51
43. 11.6 **45.** 3.33 **47.** 1; 10 **49.** x

Exercises 10.5

1. 0.7096 **3.** 3.3867 **5.** −0.2063 **7.** 0.0753 **9.** 0.2486 **11.** 0.6879 **13.** 4418.8
15. 2.304 **17.** 0.002932 **19.** 0.04555 **21.** 2.4309 **23.** $x = 135.08$ **25.** $x = 2.9498 \times 10^{-10}$
27. $x = 112.41$ **29.** $x = 0.0272$

Exercises 10.6

1. 4.6052 **3.** −4.6052 **5.** 0.5526 **7.** 3.2467 **9.** 0.4343 **11.** 1.3637 **13.** 1.9978
15. 0.3474 **17.** {1.6513} **19.** {8.2894} **21.** {3.9014} **23.** {9} **25.** {0.9022} **27.** {−2, 3}
29. −1.0371 **31.** −0.7490 **33.** 11.0444 **35.** 1.8676 **37.** −3.6053 **39.** −1.6332
41. {209.89} **43.** {16206} **45.** {0.00992} **47.** {−3.521, 3.521} **49.** Let $y = \log_a x$. Then $a^y = x$.
Taking logarithm of both sides to the base b, we have

$$y \log_b a = \log_b x$$

$$y = \frac{\log_b x}{\log_b a}$$

$$\log_a x = \frac{\log_b x}{\log_b a}$$

51. approx 26 years **53.** 0.0182

Exercises 10.7

1. \$11264.93 **3.** \$2161.90 **5.** 9 years 57 days **7.** 9.38% **9.** \$10111 **11.** 0.0237
13. 11.55 years **15.** 9.9 years **17.** 5764 years **19.** 7.4 **21.** 7.08×10^{-4} **23.** 1.00×10^{-7}

Review Exercises

1.

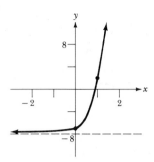

3. $\left\{\dfrac{1}{3}\right\}$ **5.** $\left(\dfrac{1}{2}\right)^{7/5}, \left(\dfrac{1}{2}\right)^{\sqrt{2}}, \left(\dfrac{1}{2}\right)^{3/2}$ **7.** $\log_{1/2} 16 = -4$ **9.** $2^6 = 64$ **11.** $a^k = a^n$
13. $\left\{\dfrac{37}{36}\right\}$ **15.** $\{-6, 3\}$ **17.** $\dfrac{2}{3}$ **19.** $\{-2\}$ **21.** $\{5\}$ **23.** 5 **25.** 20 **27.** -20
29. characteristic 3; mantissa 0.2975 **31.** characteristic 7; mantissa 0.9410 **33.** 5.6857 **35.** 9.6693
37. 1; 10 **39.** 0.9236 **41.** 2.9751 **43.** 1744.54 **45.** 6.9077 **47.** 3.9880 **49.** 0.2171
51. $\{2.8026\}$ **53.** $\{-2\sqrt{3}, 2\sqrt{3}\}$ **55.** $\{\sqrt[9]{10^{10}}\}$ **57.** $\left\{\ln\left(\dfrac{15 + \sqrt{229}}{2}\right)\right\}$ **59.** 247.7 grams

Practice Test

1. (a) $y = 5^x$ (b) $y = \log_5 x$

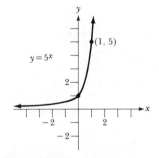 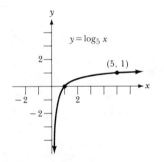

2. (a) $\log_{1/3} 81 = -4$
(b) $\log_{10} .001 = -3$
(c) $\log_2 x = 99$
3. (a) $10^0 = 1$
(b) $5^2 = 25$
(c) $2^{-4} = \dfrac{1}{16}$
4. $\{-4\}$ **5.** $\{-6\}$ **6.** $\{-3, 1\}$ **7.** $\{1\}$ **8.** 1.690 **9.** 0.2292 **10.** .5795
11. 9 years 6 months **12.** 60.83 years

Extended Applications: Noise Insulation

1. $478 I_0$ **3.** 10 (it is 9 times greater)

Chapter 11

Exercises 11.1

1. 16, 19 **3.** 27, 32 **5.** 36, 49 **7.** $-125, 216$ **9.** 2, 5, 8, 11 **11.** $-3, 1, 5, 9$
13. $3, 1, -1, -3$ **15.** $-1, 1, -1, 1$ **17.** $0, \frac{1}{3}, \frac{1}{2}, \frac{3}{5}$ **19.** $-2, 6, -12, 20$ **21.** $\frac{20}{21}$ **23.** -930
25. $\frac{178}{121}$ **27.** 100 **29.** $a_k = 4k - 3$ **31.** $a_k = (-2)^{k-1}$ **33.** $a_k = k(k+1)$ **35.** $a_k = \frac{k}{k+1}$
37. $4 + \frac{5}{2} + 2 + \frac{7}{4} + \frac{8}{5}$ **39.** $\frac{1}{4} + \frac{2}{5} + \frac{1}{2} + \frac{4}{7} + \frac{5}{8} + \frac{2}{3}$
41. $\frac{1}{3} - \frac{1}{5} + \frac{1}{7} - \frac{1}{9} + \frac{1}{11} - \frac{1}{13} + \frac{1}{15} - \frac{1}{17} + \frac{1}{19}$ **43.** $1 + \frac{15}{16} + \frac{17}{19} + \frac{19}{22} + \frac{21}{25}$
45. $4^3 - 5^3 + 6^3 - 7^3$ **47.** $\sum_{k=1}^{100} k$ **49.** $\sum_{k=1}^{51} 2k$ **51.** $\sum_{k=1}^{n} \frac{k}{k+1}$

Exercises 11.2

1. arithmetic; $d = 2$ **3.** arithmetic; $d = -3$ **5.** not arithmetic **7.** not arithmetic **9.** not arithmetic **11.** 5, 8, 11, 14, 17 **13.** 5, 7, 9, 11, 13 **15.** 6, 8, 10, 12, 14 **17.** 1, 4, 7, 10, 13
19. $23, 17, 11, 5, -1$ **21.** 500,500 **23.** 2500 **25.** 15,150 **27.** -184 **29.** 144
31. $a_k = 1 + 3k$ **33.** $a_k = 2 + 2k$ **35.** $a_k = 31 - 3k$ **37.** $a_k = 1 + 3k$ **39.** $a_k = 17 - 5k$
41. 1170 oranges **43.** (a) Y; \$3500
(b) X; \$3000

Exercises 11.3

1. $r = 2$; 48, 96, 192 **3.** $r = -3$; $81, -243, 729$ **5.** $r = -1$; $a, -a, a$ **7.** $r = \frac{1}{c}$; $\frac{1}{c^2}, \frac{1}{c^3}, \frac{1}{c^4}$
9. $r = -\frac{a}{x}$; $\frac{a^3}{x^3}, -\frac{a^4}{x^4}, \frac{a^5}{x^5}$ **11.** $\frac{1}{16}, \frac{1}{8}, \frac{1}{4}, \frac{1}{2}$ **13.** $81, -27, 9, -3$ **15.** $\frac{1}{3}, 1, 3, 9$ **17.** $-x^4, x^3, -x^2, x$
19. $\frac{a^4}{x^4}, \frac{a^3}{x^3}, \frac{a^2}{x^2}, \frac{a}{x}$ **21.** $6\left(\frac{1}{3}\right)^{k-1}$ **23.** $10\left(-\frac{1}{100}\right)^{k-1}$ **25.** ax^{k-1} **27.** 244140.6
29. -65104.16 **31.** $\frac{a(1-x^{10})}{1-x}$ **33.** 121,500 **35.** \$13.78 **37.** no; (b) costs \$235,544.31 more
39. 20 meters **41.** $\frac{311}{99}$ **43.** $\frac{10}{3}$ **45.** $\frac{8}{3}$

Exercises 11.4

1. $x^4 + 4x^3 + 6x^2 + 4x + 1$ **3.** $x^5 - 5x^4 + 10x^3 - 10x^2 + 5x - 1$ **5.** $x^3 - 6x^2y + 12xy^2 - 8y^3$
7. $16x^4 + 32x^3 + 24x^2 + 8x + 1$ **9.** $u^4 + 4u^2v + 6v^2 + \frac{4v^3}{u^2} + \frac{v^4}{u^4}$ **11.** 1.061520150601 **13.** 1.062

15. Row 1   1 $(a + b)^0$:
      Row 2   1   1 $(a + b)^1$:
      Row 3   1   2   1 $(a + b)^2$:
      Row 4   1   3   3   1 $(a + b)^3$:
      Row 5   1   4   6   4   1 $(a + b)^4$:
      Row 6   1   5   10   10   5   1 $(a + b)^5$:
      Row 7   1   6   15   20   15   6   1 $(a + b)^6$:
      Row 8   1   7   21   35   35   21   7   1 $(a + b)^7$:
      Row 9   1   8   28   56   70   56   28   8   1 $(a + b)^8$:

17. $a^5 - 5a^4b + 10a^3b^2 - 10a^2b^3 + 5ab^4 - b^5$ **19.** $1 - 6y + 15y^2 - 20y^3 + 15y^4 - 6y^5 + y^6$
21. $x^7 + 7x^5 + 21x^3 + 35x + \dfrac{35}{x} + \dfrac{21}{x^3} + \dfrac{7}{x^5} + \dfrac{1}{x^7}$ **23.** $32u^5 - 80u^4 + 80u^3 - 40u^2 + 10u - 1$
25. $x^9 - 9x^8y + 36x^7y^2 - 84x^6y^3 + 126x^5y^4 - 126x^4y^5 + 84x^3y^6 - 36x^2y^7 + 9xy^8 - y^9$ **27.** 10
29. 28   **31.** 1   **33.** -50 **35.** $\displaystyle\sum_{k=0}^{n} \dfrac{n(n-1)(n-2)\cdots(n-k+1)}{1\cdot 2\cdot 3\cdots k} a^{n-k}b^k$

Exercises 11.5

1. 24   **3.** 120   **5.** 18   **7.** 1296; 192   **9.** 5040   **11.** 5040   **13.** 40320   **15.** 720
17. 6720   **19.** 120   **21.** 60   **23.** 17,576,000   **25.** 2,494,800   **27.** 360   **29.** 120   **31.** 1; 1; 1

Exercises 11.6

1. 15   **3.** 1   **5.** 20   **7.** 32   **9.** 6; 6; same   **11.** 20; 20; same   **13.** 15; 15; same

15. $\binom{n}{r-1} + \binom{n}{r} = \dfrac{n(n-1)\cdots(n-(r-1)+1)}{1\cdot 2\cdot 3\cdots(r-1)} + \dfrac{n(n-1)\cdots(n-r+1)}{1\cdot 2\cdot 3\cdots r}$

$= \dfrac{n(n-1)\cdots(n-r+2)}{1\cdot 2\cdot 3\cdots(r-1)}\cdot\dfrac{r}{r} + \dfrac{n(n-1)\cdots(n-r+1)}{1\cdot 2\cdot 3\cdots r}$

$= \dfrac{n(n-1)\cdots(n-r+2)\,[r + (n-r+1)]}{r!}$

$= \dfrac{(n+1)n(n-1)\cdots((n+1)-r+1)}{r!}$

$= \binom{n+1}{r}$

17. $\binom{n}{n-r} = \dfrac{n!}{(n-r)!\,(n-(n-r))!} = \dfrac{n!}{(n-r)!\,r!} = \dfrac{n!}{r!\,(n-r)!} = \binom{n}{r}$

19. 78   **21.** 55   **23.** 435   **25.** 1330   **27.** {3}   **29.** {6}   **31.** {6}   **33.** {9}   **35.** 7 different amounts   **37.** 59400

Review Exercises

1. 21, 25; $4k - 3$ **3.** $\dfrac{7}{6}; \dfrac{8}{7}; \dfrac{(k+1)}{k}$ **5.** $1 + \dfrac{2}{3} + \dfrac{3}{5} + \dfrac{4}{7} + \dfrac{5}{9} + \dfrac{6}{11}$

7. $\dfrac{9}{2} - \dfrac{16}{4} + \dfrac{25}{6} - \dfrac{36}{8} + \dfrac{49}{10} - \dfrac{64}{12} + \dfrac{81}{14} - \dfrac{100}{16}$

9. $\displaystyle\sum_{k=1}^{51} (-1)^{k+1}(2k - 1)$

11. 1650   **13.** $3k - 1;\ \displaystyle\sum_{k=1}^{33}(3k-1)$ **15.** $a_k = 75 - 5k$ **17.** $2808

19. A sequence of $a_1, a_2, a_3, \cdots$ is a geometric sequence if $\dfrac{a_{k+1}}{a_k} = r$ for all $k \geq 1$.   **21.** A geometric series is a sum of the form $a_1 + a_1 r + a_1 r^2 + \cdots$.   **23.** $a_k = 100\left(-\dfrac{1}{10}\right)^{k-1}$ **25.** $a_k = a\left(\dfrac{b}{a}\right)^{k-1}$
27. $\dfrac{2047}{64}$ **29.** $\dfrac{421}{99}$ **31.** $a^7 + 7a^6b + 21a^5b^2 + 35a^4b^3 + 35a^3b^4 + 21a^2b^5 + 7ab^6 + b^7$
33. (a) $x^5 - 10x^4y + 40x^3y^2 - 80x^2y^3 + 80xy^4 - 32y^5$
(b) $64x^6 + 192x^4 + 240x^2 + 160 + \dfrac{60}{x^2} + \dfrac{12}{x^4} + \dfrac{1}{x^6}$
35. 0   **37.** 1080 (assuming the number may not start with 0)   **39.** 7920   **41.** 175,759,920 computers
43. 28; 28; 28   **45.** $\{11\}$ **47.** 15

Practice Test

1. 1008   **2.** $3929.60   **3.** 4095   **4.** $\dfrac{318}{99}$ **5.** $16x^4 - 96x^3y + 216x^2y^2 - 216xy^3 + 81y^4$
6. $\binom{n}{0} + \binom{n}{1} + \cdots + \binom{n}{n} = 2^n$ so $\binom{n}{1} + \cdots + \binom{n}{n} = 2^n - 1$ **7.** 249,900   **8.** 34650
9. $n = 7$ **10.** 55440

Extended Applications: The Dose-Response Relationship

1. 33.75 mg   **3.** It is not a geometric sequence.

Index

Absolute value, 14, 64
Addend, 19
Addition
   associative property for, 12
   closure property for, 12
   commutative property for, 12
   identity property for, 12
   inverse property for, 12
Addition property of equality, 14
Additive inverse, 13
Algebraic fraction, 150
Antilogarithm, 454
Arithmetic sequence, 484
   common difference of, 484
   general term of, 486
Augmented matrix, 415

Base
   of a logarithm, 442
   of a power, 98
Binomial, 117
Binomial coefficients, 499
Binomial expansion, 498
Binomial theorem, 498

C, 220
Cartesian Coordinate System, 283
Circle, 350
   center of, 350
   radius of, 350
Closure property
   for addition, 12
   for multiplication, 12
$C(n, r)$, 507
Combination, 507
Common logarithm, 452
   characteristic of, 453
   mantissa, 453
Completing the square, 243
Complex conjugate, 224
Complex fraction, 176

Complex number, 219
   imaginary part of, 219
   real part of, 219
Compound interest, 466
Compound sentence, 79
Conic sections, 358
Constant, 2
Continuous decay, 468
Coordinate, 8
Cramer's Rule, 410, 412
Cube root, 196

Denominator, 27, 150
   rationalizing, 210
Determinant, 406
   minor, 407
   second-order, 406
   third-order, 407
Distance between a and b, 23
Distance formula, 349
Distributive property, 12
Dividend, 27, 150
Division algorithm, 171
Divisor, 27, 150
Double negative property, 13

e, 460
Element, 2
Ellipse, 354
   focus (foci) of, 355
Empty set, 2
Equality, 7
   properties, 7
Equal sets, 2
Equation(s), 41
   dependent, 386
   equivalent, 42
   equivalent system of, 386
   inconsistent, 386
   linear, 41
   nonlinear system of, 420

Equation(s) (*continued*)
    radical, 214
    root(s) of, 41
    solution (set) of, 41
Exponent, 98
Exponential function, 438

Factor, 26, 98, 128
Factorial, 503
Factoring
    of $Ax^2 + Bx + C$, 134
    of difference of squares, 132
    by grouping, 128, 141
    by perfect squares, 133
    by substitution, 141
    of sum/difference of cubes, 137
    by trial and error, 134
    of trinomials, 133
FOIL method, 123
Formula, 48
Fraction
    complex, 176
    improper, 170
    proper, 170
Function, 312
    exponential, 438
    inverse, 372
    linear, 316
    logarithmic, 442
    one-to-one, 372
Fundamental Counting Principle, 503

Geometric sequence, 490
    common ratio of, 490
    general term of, 491
Geometric series, 492
Graph, 8
    of an equation, 283
    of an ordered pair, 283
    symmetric with respect to *x*-axis, 348
    symmetric with respect to *y*-axis, 348

H, 4
Hyperbola, 356
    asymptotes of, 356

i, 218
Identity property
    for addition, 12
    for multiplication, 12
Imaginary number, 220
Inconsistent equations, 386
Inequalities
    equivalent, 72
    linear, 72
    quadratic, 362
    solution set of, 72
    system of, 462
Inequality properties, 9
Integers, 3
Intersection, 79
Interval, 69
Interval notations, 71
Inverse
    additive, 12
    multiplicative, 12
Inverse function, 372
Inverse property
    for addition, 12
    for multiplication, 12
Irrational numbers, 4

J, 3

Least common denominator (LCD), 158
Least common multiple (LCM), 158
Least terms, 201
Like signs, 26
Like terms, 117
Linear equation
    point-slope form, 299
    slope-intercept form, 301
Linear function, 316
Linear inequality, 306
Linear system in three variables, 393
Logarithmic function, 442
Lowest terms, 142

Market equilibrium, 433
Matrix, 406
    augmented, 415
    elements of, 406
    entries of, 406
    row operations of, 416
Member, 2
Monomial, 117
Multiplication
    associative property for, 12

closure property for, 12
commutative property for, 12
identity property for, 12
inverse property for, 12
of rational expressions, 164
sign rule for, 27
Multiplication property of equality, 14
Multiplicative inverse, 13

N, 3
nth root, 196
  principal, 196
Natural logarithm, 460
Natural number, 2
Negative, 12
Null set, 2
Number line, 8
Numerator, 27, 150

Opposite, 13
Ordered pair, 283
Order of operations, 29

Parabola, 333
  axis of symmetry, 340, 344, 345
  horizontal, 344
  opens to the right, 344
  opens upward, 333
  vertex of, 333, 343, 345
Pascal's triangle, 500
Permutation, 502
pH value, 469
Planetary orbits, 382
$P(n, r)$, 505
Poiseville's Law, 148
Polynomial, 115
  ascending order, 116
  coefficients, 116
  degree of, 116
  descending order, 116
  terms of, 116, 117
  value of, 121
Power rule, 106
Principal nth root, 196
Product, 26
  factor of, 26
Product rule, 100
Proof, 12
Properties of logarithms, 446

Pure imaginary number, 219
Pythagorean theorem, 348

Q, 4
Quadrants, 283
Quadratic equation, 233
  discriminant of, 251
  standard form of, 233
Quadratic formula, 247
Quadratic function, 333
Quadratic inequality, 269
Quotient, 27, 166, 172
Quotient rule, 101

R, 4
Radical, 196
Radical equation, 214
Rational expression, 150, 205
  lowest terms, 152
  multiplication of, 164
Rational numbers, 4
Real numbers, 4
  properties of, 12
Reciprocal, 12, 166
Reflexive property, 7
Relation, 311
  domain of, 311
  range of, 311
Remainder, 172
Remainder theorem, 186
Root, 41
  extraneous, 255

Scientific notation, 110
Sequence, 479
  arithmetic, 484
  common difference, 484
  finite, 480
  general term, 486
  geometric, 490
  infinite, 480
  terms of, 479
Series, 481
  arithmetic, 484
  geometric, 492
  infinite, 481
Set(s), 2
  elements of, 2
  empty, 2

Set(s) *(continued)*
  equal, 2
  intersection of, 79
  members of, 2
  null, 2
  union of, 82
Set-builder notation, 2
Sign chart, 269
Slope, 291
Solution of the system, 385, 394
Square matrix, 406
Subset, 3
Substitution method, 391
Substitution property, 8
Subtraction, 21
Sum, 19
Summation notation, 481
Symmetric property, 7
Synthetic division, 182
Synthetic divisor, 184
System of inequalities, 426

Test point, 307, 362
Theorem, 12
Transitive property, 8, 9
Trichotomy property, 9
Trinomial, 117

Union, 82
Unlike signs, 26

Variable, 2
Variation, 319
  constant of, 320
  direct, 320
  inverse, 321
  joint, 322

W, 3
Whole numbers, 3

x-coordinate, 283
x-intercept, 287

y-coordinate, 283
y-intercept, 287

Zero factor property, 14

METRIC UNITS

- **Standard Units of Metric Measure**

  Meter(m): length (approx. 3.28 ft)
  Liter(L): volume (approx. 1.06 qt)
  Gram(g): weight (approx. 0.035 oz)

- **Important Prefixes**

  kilo (× 1000)   deci (× 1/10)
  hecto (× 100)   centi (× 1/100)
  deka (× 10)   milli (× 1/1000)

- **Abbreviations**

Length		**Volume**		**Weight**	
m	meter	L	liter	g	gram
km	kilometer	kL	kiloliter	kg	kilogram
hm	hectometer	hL	hectoliter	hg	hectogram
dam	dekameter	daL	dekaliter	dag	dekagram
dm	decimeter	dL	deciliter	dg	decigram
cm	centimeter	cL	centiliter	cg	centigram
mm	millimeter	mL	milliliter	mg	milligram

- **English-Metric Conversions**

  Length

  1 in = 2.540 cm   1 cm = 0.3937 in
  1 ft = 30.48 cm   1 cm = 0.03281 ft
  1 yd = 0.9144 m   1 m = 1.0936 yd
  1 mi = 1.609 km   1 km = 0.6215 mi

  Volume (based on U.S. measurements)

  1 pt = 0.4732 liter   1 liter = 2.1133 pt
  1 qt = 0.9464 liter   1 liter = 1.0567 qt
  1 gal = 3.785 liters   1 liter = 0.2642 gal

  Weight

  1 oz = 28.35 g   1 g = 0.0353 oz
  1 lb = 453.6 g   1 g = 0.002205 lb
  1 lb = 0.4536 kg   1 kg = 2.205 lb